Kinetics of
Ion-Molecule
Reactions

NATO ADVANCED STUDY INSTITUTES SERIES

A series of edited volumes comprising multifaceted studies of contemporary scientific issues by some of the best scientific minds in the world, assembled in cooperation with NATO Scientific Affairs Division.

Series B: Physics

RECENT VOLUMES IN THIS SERIES

The series is published by an international board of publishers in conjunction with NATO Scientific Affairs Division

A	Life Sciences	Plenum Publishing Corporation
B	Physics	New York and London
C	Mathematical and Physical Sciences	D. Reidel Publishing Company Dordrecht and Boston
D	Behavioral and Social Sciences	Sijthoff International Publishing Company Leiden
E	Applied Sciences	Noordhoff International Publishing Leiden

Kinetics of Ion-Molecule Reactions

Edited by
Pierre Ausloos
United States Department of Commerce
National Bureau of Standards
Washington, D.C.

PLENUM PRESS • **NEW YORK AND LONDON**
Published in cooperation with NATO Scientific Affairs Division

Library of Congress Cataloging in Publication Data

Nato Advanced Study Institute on Kinetics of Ion-Molecule Reactions, La Baule, 1978.
Kinetics of ion-molecule reactions.

(Nato advanced study institutes series: Series B, physics; v. 40)
Includes bibliographical references and index.
1. Chemical reaction, Conditions and laws of—Congresses. 2. Ions—Congresses. 3. Molecules—Congresses. I. Ausloos, Pierre J. II. Title. III. Series.
QD501.N37 1978   541'.39   79-367
ISBN-13: 978-1-4613-2933-6   e-ISBN-13: 978-1-4613-2931-2
DOI: 10.1007/978-1-4613-2931-2

Proceedings of the NATO Advanced Study Institute on Kinetics of Ion—Molecule Reactions held at La Baule, France, September 4—15, 1978

© 1979 Plenum Press, New York
Softcover reprint of the hardcover 1st edition 1979

A Division of Plenum Publishing Corporation
227 West 17th Street, New York, N.Y. 10011

Preface

The investigation of the elementary reactions of reactive intermediate species began about half a centruy ago with the advent of free radical kinetics as an active area of chemical research. In spite of the relatively greater ease of detection of a species carrying an electrical charge, and the fact that organic chemists had for decades postulated mechanisms involving ionic intermediates, the systematic study of the elementary reactions of ions was delayed for more than twenty years after the first beginnings of free radical kinetics. Even at this writing, in 1978, the word "kinetics" is considered by many chemists to be synomomous with "kinetics of neutral species".

Yet in spite of the relatively late start and separation from the mainstream of kinetics, the field of ion physics and chemistry is fluorishing, and growing at an ever faster pace. Instrumentalists devise ever more sophisticated apparatuses with capabilities of delving into nearly every aspect of the interactions between ions and molecules. Even satellites orbiting the earth are now being used effectively to determine rate coefficients of ionospheric ion-neutral reactions, some of which can not as yet be measured in the laboratory.

This book contains discussions of some of these new experimental approaches and the resulting knowledge which has been acquired since the earlier 1974 NATO Advanced Study Institute on Interactions between Ions and Molecules, which was held in Biarritz, France. Besides the more fundamental aspects of ion kinetics and the characterization of ions with regard to structure and energy, ionic processes occurring in interstellar space, the earth's atmosphere, lasers, and combustion, are discussed here. In addition, readers will find chapters describing nucleation phenomena and the relationships between ionic processes occurring in the gas and liquid phases.

I would like to express my appreciation to everyone who contributed to the success of the NATO Advanced Study Institute on Kinetics of Ion-Molecule Reactions held at La Baule, France, in

September, 1978. This includes not only all of the lecturers,
those who contributed to the discussion panels, and those who
chaired the sessions, but, just as important, all of those in
attendance whose interest and enthusiasm contributed to the
stimulating atmosphere of the Institute. I would also like to
give recognition to the efforts of the members of the Organizing
Committee Rose Marx, Sharon Lias, Keith Jennings, Paul Kebarle,
Eldon Ferguson, and especially Tom Govers, who was largely responsible for the smooth functioning of the physical details of the
meeting. On behalf of all participants, I would like to thank the
Scientific Affairs Division of the North Atlantic Treaty Organization
without whose financial assistance the occurrence of such a constructive summer school would not have been possible. I personally
would like particularly to acknowledge the moral support supplied
by the secretary of the Scientific Affairs Division, Dr. Tilo
Kester. We also thank the National Science Foundation of the
U.S.A. for travel grants.

P. Ausloos
Washington, DC
November, 1978

Contents

POTENTIAL ENERGY SURFACES FOR ION-MOLECULE REACTIONS:

SUMMARY OF THE PANEL DISCUSSION

Joyce J. Kaufman

Department of Chemistry
The Johns Hopkins University
Baltimore, Maryland, 21218, U.S.A.

SUMMARY

An introductory background lecture on symmetry and spin restrictions and quantum chemical computational techniques for potential energy surfaces of ion-molecule reactions was presented
by Joyce J. Kaufman. Symmetry restrictions for both L-S and j-j
coupling were outlined and the concept she had previously introduced of carrying extra "ghost" pseudosymmetry information for
lower symmetry intermediates(1-3)was described and examples of its
usefulness in interpreting and predicting ion-neutral and neutral-
neutral reactions were given. The various ab-initio and other
computational procedures for potential energy surfaces were described and their relative merits evaluated. This was followed by
a presentation by D. M. Hirst on ab-initio configuration interaction calculations of potential energy surfaces for triplet
states of NH_2^+(4). J. J. Leventhal discussed some of his recent
results on spin-conservation or apparent non-conservation in ion-
neutral reactions observed by collision produced luminescence in
a crossed beam apparatus(5).

SYMMETRY AND SPIN RESTRICTIONS AND QUANTUM CHEMICAL
COMPUTATIONAL TECHNIQUES FOR POTENTIAL ENERGY SURFACES

A. Introduction

Since the last NATO Advanced Study Institute on Ion-Molecule

1

Reactions held in Biarritz, France in 1974, while there have been
many ab-initio configuration interaction quantum chemical calculations carried out on various systems, there still have been very
few configuration interaction or even MC-SCF calculations carried
out for potential energy surfaces of even the simplest of the ion-
molecule reactions, $A^+ + H_2 \rightarrow AH^+ + H$. This is not due to lack of
the computational techniques or their implementation but rather to
the low priority given to most such projects. Thus, while it is
much more feasible now than it was several years ago to carry out
such ab-initio configuration interaction quantum chemical calculations for ion-molecule reactions of interest, the insight gained
from the application of symmetry and spin restrictions will remain
a valuable tool in the interpretation and prediction of such reactions.

B. Symmetry and Spin Restrictions and Carrying
of Extra "Ghost" Information

The application of symmetry and spin correlation rules allows
one to determine what electronic states of an intermediate complex
are permitted from the electronic states of the separated reactants and into what electronic states of products these intermediates are permitted to dissociate. These restrictions are the
extension to three dimensional polyatomic molecules of the original Wigner-Witmer quantum mechanical derivations for diatomics[6].
A comprehensive and thorough discussion of these is given by
Herzberg[7].

For all collisional phenomena, reactive or non-reactive, the
reactants must approach each other and the products must recede
along the electronically permitted potential energy surfaces of
the intermediate quasimolecule. The mechanisms and kinetics of
reactive collisions are determined by the shapes of these potential energy surfaces and by the types of curve crossings, electronically permitted or forbidden, which take place. There are
two types of curve crossing phenomena: 1) adiabatic, where the
reactants approach each other relatively slowly, and at least for
states of the same symmetry, the curves do not cross; 2) diabiatic, where the reactants approach each other rapidly and may tend
to remain on the potential energy surface of the same electronic
configuration occupation even though it may not be the lowest
energy potential surface of that particular symmetry and spin in
that region[8].

The Landau-Zener probabilities of transition between states
of different symmetry[9,10] had been shown to be inapplicable in
certain instances and has been revised[11]. The Landau-Zener
formula as such is not applicable to polyatomic molecules and

modifications were made to allow it to be used in certain specific cases(12).

1. Symmetry and Spin Correlation Rules

a. <u>Diatomics.</u> The two most useful schemes for building up the electronic states of diatomic molecules are from the electronic states of the separated atoms and from the molecular orbitals of the diatomic molecule(13). The correlation between these two schemes is general and is valid for correlation rules for polyatomic systems. The correlation rules ordinarily are applied for L-S or Russell-Saunders coupling for lighter atoms. However, by use of the theory of double groups these correlation rules can be extended to j-j coupling of heavier atoms also. The correlation rules are valid for diabiatic potential surfaces as well as for the more customarily considered adiabiatic potential surfaces (where the electronic motion adjusts itself to each change in nuclear configuration without undergoing electronic transition to a higher energy state of the same symmetry).

b. <u>Polyatomics.</u> The most useful schemes for building up the electronic states of polyatomic molecules are: 1) from separated atoms or molecular fragments, and 2) from the molecular orbitals of the polyatomic molecule. In addition the correlations from geometrical conformations of different symmetry (both higher and lower) are valuable. For L-S coupling, where spin-orbit coupling is small, there is spin correlation. What must be conserved in these cases of spin correlation of the intermediate quasi-molecule is vector addition of the total spin. For j-j coupling, where spin-orbit coupling is large, spin by itself is no longer a good quantum number. In cases where j-j coupling is dominant, the only good quantum number of the intermediate quasimolecule is the double group product of the double group representation of the spin times the symmetry. In principle to derive the proper correlations for j-j coupling is a little more complicated than for L-S coupling. The double group representations of the spin are given in the Tables for multiplication tables of each group in Herzberg's Polyatomics (7). The major complication arises from the fact that the double group correlations are in some cases so much less restrictive than the single correlations. For example, even for the simplest $A^+ + H_2$ (or the complementary $A + H_2^+$) reactions in unsymmetrical C_s symmetry it is not possible to discern whether an intermediate $E_{\frac{1}{2}}$ state arose from the cognate in plane $^2A'$ or out of plane $^2A''$ state in single group symmetry and in C_{2v} symmetry whether an intermediate $E_{\frac{1}{2}}$ state arose from a $^2B_2$, $^2A_1$ or $^2B_1$ state. We have been able to solve that dilema by deriving the symmetry and spin correlations first in L-S coupling, then by resolving the spin into the double group representation and then performing the double group multiplication term by term retaining the identity between the terms in single and double group notation.

(1) From separated atoms or molecular fragments
     (going through geometrical conformations of different
     symmetries).

For polyatomic molecules, the most useful correlations are those between separated molecular fragments or separated atom plus molecular fragment. For general polyatomic molecules if the symmetry of at least one of the fragments is lower than that of the complete molecule and if during the approach of the fragments the molecule has the geometrical conformation of the fragment with the lower symmetry Herzberg(7) suggests obtaining the resulting states in terms of the lower symmetry, then if in the final molecule the symmetry is now higher, using the correlation rules between deformed conformations to find the relation of the states of the intermediate to the states of the final molecule. An ambiguity may arise however because while the correlations of species of a point group of higher symmetry into those of a point group of lower symmetry are always unambiguous, the reverse correlation is not always unambiguous. We have been able to circumvent this problem by a combination of several ways. One is by tracing the reverse decomposition from higher symmetry intermediate into lower symmetry fragments. For intermediate quasimolecule where the final symmetry is lower than that of a component fragment but higher than that of a less symmetrical geometrical conformation along the path of its formation we have been able to trace the path uniquely by carrying extra "ghost" information. For symmetry labelling purposes it attributes to the intermediate complex an extra pseudo-symmetry of a component fragment which may be of higher order than the true symmetry of the intermediate quasimolecule. For $A^+ + H_2$ reactions this enables us to ascertain the appropriate correlations going from linear symmetry through C_s symmetry to the correct states on C_{2v} symmetry both for L-S(1-3, 14,15) and j-j coupling (2,3,16). We have previously called attention to other modes of behavior which are general characteristics of ion-molecule reactions. For instance, there is no switching of charge in diabatic correlations. This tracing of diabatic paths is particularly pertinent in view of a previously reported tragectory calculation by Chapman and Preston(17) of an ion-molecule system ($Ar^+ + H_2$) in which it was shown that even at lower energies there was a tendency to remain on the diabiatic surface rather than switch to the adiabatic surface.

(a) Examples

1'. L-S Coupling. We have earlier given examples of carrying L-S coupling "ghost" information for $C^+ + H_2$(1,14) and

F^+ + H_2(15). Essentially for the linear intermediate even though it was $C_{\infty v}$ and did not retain the complete $D_{\infty h}$ symmetry of the H_2. we retained for information purposes the extra (g) and (u) subscript information. It is this which enabled us to ascertain the appropriate correlation from linear symmetry going through unsymmetrical C_S symmetry to the correct states in C_{2v} symmetry. (See the cited earlier publications for detailed tables.) The L–S coupling portion of the conditions for the Cl^+ + H_2 described la- in this section are identical to those for F^+ + H_2.

   2'. j-j Coupling. In our laboratory, Professor Walter Koski has studied experimentally the dynamics of the F^+(D_2,D)FD^+(15) and the Cl^+(D_2,D)ClD^+ (16) reactions.   The interpretation of the dynamics of this former reaction is complicated not only by the lack of knowledge of the properties of the intermediate H_2F^+ but also by the fact that the ionization potential of fluorine is higher than that of H_2 so the energy of F^+ + H_2 is higher than that of H_2^+ + F resulting in curve crossings which complicate the interpretation. This complication is eliminated in the Cl^+ + H_2 reaction as a result of the lower ionization potential of chlorine.

   In Professor Koski's recent study of the Cl^+ + D_2 reaction, Cl^+(^3P_g, ^1D_g and ^1S_g) states were present in the beam. The presence of Cl^+(^1S_g) was demonstrated by the threshold for formation of CCl^+ from the reaction of Cl^+ with CO. Since the yield from Cl^+(^1S_g) + CO reaction prevented seeing the onsets of reactions involving Cl^+(^1D_g) or Cl^+(^3P_g) other experiments which consisted of measuring the energy spectra of Cl^+ resulting from Cl^+ + D_2 and Cl^+ + Ar collisions were carried out to demonstrate the presence of these states. Two scattered Cl^+ peaks, were readily observable. Their spacing corresponds to the difference in energy between Cl^+(^1D_g) and Cl^+(^3P) indicating that they arise from the inelastic and superelastic collisions between Cl^+ ions and the target gases. The reactions can be represented by the equations

$$Cl^+(^3P_g) + Ar \rightarrow Ar + Cl^+(^1D_g)$$

$$Cl^+(^1D_g) + Ar \rightarrow Ar + Cl^+(^3P_g)$$

   Conversion of electronic to translational energy and vice versa is expected to be most favored in collisions where there are avoided crossings of states of same symmetry (and spin). The ordinary spin restrictions in L–S coupling make it difficult to rationalize the basis for the observed conversion of electronic to translational energy and vice versa in the Cl^+(^3P_g) + Ar(^1S_g) and Cl^+(^1D_g) + Ar(^1S_g) because the only $ArCl^+$ states allowed from the first set of reactants are $^3\Sigma^-$ and $^3\Pi$ and from the second set of reactants are $^1\Sigma^+$, $^1\Pi$ and $^1\Delta$. However, examination of the re-

strictions arising from the application of double groups (j-j or
spin-orbit coupling) indicates that there are several states arising from each set of reactants which have symmetries in common.

(2) From the molecular orbitals of the polyatomic molecule.
      Molecular orbitals and the techniques for computing them
      are discussed in the following section.

         C. Quantum Chemical Computational Techniques

For molecules which possess symmetry group theoretical considerations place a severe restriction on the possible eigenfunctions of the system since these must form bases for some irredu-
cible representations of the group of symmetry operations and must
transform in a precise manner under the operations of the point
group.

Molecular orbitals (MO's) ϕ_i, are taken to be linear combinations of atomic orbitals (LCAO's) χ_r, usually centered on each
atom.

$$\phi_i = \sum_r \chi_r C_{ri}$$

A single configurational molecular electronic wave function ϕ is
represented by an antisymmetrized product wave function(18),

$$\Phi = (N!)\phi_1{}^{[1}\phi_2 -- \phi_N{}^{N]}$$

since only states can occur whose eigenfunctions are antisymmetric
with respect to exchange of any two electrons. There are non-empirical ab-initio quantum chemical calculational procedures and
less rigorous semi-empirical quantum chemical calculational procedures. Of these only the ab-initio quantum chemical procedures
incorporating configuration interaction are inherently capable of
giving reliable potential energy surfaces for ion-molecule reactions (or for any collisional processes, reactive or non-reactive).

         1. Non-Empirical Ab-Initio SCF, MC-SCF and CI Techniques

The well known Hartree-Fock-Roothaan LCAO-MO-SCF procedure
(linear combination of atomic orbitals to form molecular orbitals
by a self-consistent technique(19)to form the best single determinant electronic wave function) may be supplemented by MC-SCF
[multiconfiguration SCF(20)where both the coefficients of the AO's
in the MO's and the coefficients of a combination of determinants
are varied simultaneously] or by CI [configuration interaction, a
superposition of electronic configurations of a specific symmetry
and spin when a single determinant is insufficient to describe a
molecular system or a potential energy surface satisfactorily(21)].

In CI calculations the wave function is constructed from a linear combination of Φ's

$$\Psi = \sum_i \Phi_i C_i$$

and the coefficients C_i are varied to minimize the total energy.

For reliable potential energy surface calculations on ion-molecule reactions it seems imperative to use ab-initio configuration interaction calculations for several reasons. There are often several crossings or avoided crossings of surfaces of the same symmetry and spin. Thus a single determinant wave function is insufficient and there is also no guarantee that an MC-SCF calculation of the higher states will give the states correctly ordered.

While in the earlier days for calculations on diatomics Slater orbitals with an $e^{-\alpha r}$ dependence were used for the basis atomic functions, it is not possible to solve the three-and-four-center integrals over Slater orbitals analytically and they must be solved by tedious numerical integration. For polyatomic systems Gaussian atomic basis functions, with an $e^{-\alpha r^2}$, dependence, have been used for more than the past decade and a half. While Gaussian atomic orbitals are not as good approximations as Slater orbitals to the physical atomic basis orbitals, the product of two Gaussian functions is still a Gaussian function and the three-and-four-center integrals can be evaluated analytically and hence rapidly. While longer Gaussian expansions are necessary, for comparable calculated total energies for polyatomic molecules it is possible to evaluate the necessary integrals over Gaussian basis functions at least ten times faster than the integrals over the corresponding Slater orbitals. It is customary to use linear combinations of contracted Gaussian orbitals in which the contraction coefficients for each individual Gaussian basis function are optimized for the atom or small molecule fragment but then not varied again for use in a general molecular calculation.

One of the most significant developments in recent years has been the introduction of ab-initio effective core model potentials which allow calculations of integrals involving the valence electrons only, yet explicitly and accurately (22) or various other types of pseudopotentials(23-37). Several years ago Bonifac and Huzinaga gave a new formulation of the model potential (MODPOT) method (ab-initio effective core model potentials) for carrying out non-empirical ab-initio LCAO-MO-SCF calculations considering only the valence electrons. The effect of the core electrons is taken into account through the use of a modified model potential (MODPOT) Hamiltonian.

As an example for molecules composed of first and second row atoms and hydrogen it has the following form:

$$H = \sum_{i}^{n_v} h(i) + \sum_{i>j}^{n_v} 1/r_{ij};$$

$$h(i) = -\frac{1}{2}\Delta_i - \sum_{k}^{n_h} 1/r_{ik}$$

$$- \sum_{k}^{n_c}(Z_k - N_c^k)(1 + A_{1k}e^{-\alpha_{1k}r_{ik}^2}$$

$$+ A_{2k}e^{-\alpha_{2k}r_{ik}^2}$$

$$+ \sum_{k}^{n_c}\{B_{1s}^k\ |1s_k><1s_k| + B_{2s}^k\ |2s_k><2s_k|$$

$$+ B_{2p}^k(\ |2p_{xk}><2p_{xk}| + |2p_{yk}><2p_{yk}|$$

$$+ |2p_{zk}><2p_{zk}|)\}.$$

In equations (1) and (2), n_v is the number of valence electrons, n_h is the number of hydrogen atoms and n_c is the number of atoms with core electrons. N_c^k is the number of core electrons in the k-th core. $|1s_k>$, $|2s_k>$ and $|2p_k>$ are inner shell atomic orbitals for the K-th core.

Bonifacic and Huzinaga(22) have introduced Gaussian terms into the one-electron part of the Hamiltonian to account for the screening of the nuclei by the core electrons and projection operator terms to prevent the buildup of the valence molecular orbitals in the core regions. Bonifacic and Huzinaga derived model potential expressions for inner s, p and d orbitals. We extended the derivation also to inner f orbitals. Since these MODPOT procedures are not yet in general use, we will describe our own findings in some detail as a guide for others who may be interested in exploring the use of such functions. The MODPOT procedure of Bonifacic and Huzinaga was introduced as an option into our MOLASYS(38) and GIPSY(39) computer programs. With the GIPSY computer program we can calculate the usual one-electron overlap, kinetic energy, nuclear attraction and two-electron repulsion integrals between s-, p-, d- and f- contracted Gaussian functions. In addition the new type of one-electron integrals resulting from the use of the Bonifacic-Huzinaga Hamiltonian with s-, p-, d- and f type core atomic orbitals can be calculated.

We verified the accuracy of the MODPOT method for a variety of molecules(40) by carrying out the reference calculations with two-electron integrals accurate to 6 decimal places. Then the MODPOT calculations were carried out with the valence shell two-

electron integrals accurate to 6 decimal places and the MODPOT
modified one-electron integrals. The same reference atomic basis
set was used for both calculations. Comparison of the MODPOT results to the completely ab-initio ones (including inner shell
electrons) showed that even for small or medium size atomic basis
sets the valence orbital energies and gross atomic populations
are accurate[Δ] to more than 2 decimal places (the average error is
0.005 a.u.). The maximum error in the orbital energies for the
largest molecule studied to date is only 0.011 a.u. and in the
gross atomic populations is only 0.010 a.u. It should be emphasized that the accuracy[Δ] obtained with the MOTPOT approximation
gets better and better as the size of the atom basis sets is increased. Moreover, for molecules composed even of two second row
atoms, such as Cl_2, the savings in time between the MODPOT method
compared to the completely ab-initio is a factor of ten. The savings in time also increase dramatically as the size of the molecule or the heaviness of the atoms increase.

We also introduced a charge conserving integral prescreening
approximation especially effective for spatially extended systems.
This is a rigorous option to our completely ab-initio programs; it
can either be used or not used. This approximation, which we
named VRDDO(41), variable retention of diatomic differential overlap was inspired in part by a suggestion from solid state physics
by Wilhite and Euwema(42). The approximation consists of neglect-
ing all one-electron integrals (both energy and overlap) and two-
electron integrals that involve basis function pairs $\Phi_i(1)\Phi_j(1)$
whose pseudo-overlap:

$$S^*_{ij} = \int \Phi^*_i(1)\Phi^*_j(1)dv_1$$

is less than some threshold τ_1. The pseudo-function $\Phi^*_i(1)$ is represented by a single normalized spherically-averaged Gaussian
function whose exponent is equal to that of the most diffuse
primitive in the contracted Gaussian function $\Phi_i(1)$. Our method
for calculating the pseudo-overlap S^*_{ij} for the basis function
pair $\Phi_i(1)\Phi_j(1)$ differs somewhat from that used by Wilhite and
Euwema(43), however our VRDDO procedure is still charge conserving.

The VRDDO approximation was introduced into our MOLASYS(38)
and GIPSY(39) computer programs. A second threshold τ_2 that controls the accurace of the two-electron repulsion integrals (integrals are accurate to n decimal places where $\tau_2 = 10^{-n}$) can be

_____

Δ Where accurate implies accuracy with respect to the completely
  ab-initio calculations with the same atomic basis set retaining all integrals.

specified in the input data. All two-electron integrals whose absolute magnitude is less than τ_2 are neglected. The results(40,
41)show this second approximation to be numerically accurate. The
computer CPU time required to generate the two-electron integral
list for a large molecule is substantially reduced as τ_2 is increased.

We have investigated the effect of varying the thresholds τ_1
and τ_2 in order to arrive at values that yield a good compromise
between accuracy and computer running time. We tested the accuracy of the VRDDO method by computing the completely ab-initio
reference calculations and then the VRDDO calculations with the
same atomic basis set and compared the numerical results. The
introduction of the VRDDO approximation with $\tau_1 = 10^{-2}$ and $\tau_2 =$
10^{-4} hardly affects the accuracyΔ of the computed properties at
all. The valence orbital energies and gross atomic populations
are essentially accurateΔ to 3 decimal places (in a.u.). The error in the total energy with respect to the completely ab-initio
calculation increases slightly on going from the smaller to the
larger molecules. However, the VRDDO relative energy differences
between isomers or along a potential curve agree to about the
fourth decimal place (in a.u.) compared to the same relative energy differences from completely ab-initio calculations. The maximum error found in the total overlap population is 0.008 and in
the dipole moment is 0.009 a.u.

If both the MODPOT and VRDDO approximations are introduced,
the accuracyΔ with respect to the reference calculations is about
the same as that obtained using only the MODPOT approximation.
The maximum error in the orbital energies is 0.012 a.u., in the
gross atomic populations is 0.009, in the total overlap populations is 0.027 and in the dipole moment is 0.009 a.u.

The relative accuracy of the MODPOT/VRDDO method along a potential energy surface (such as for molecular conformational
changes or interactions between two species, 0.0001 - 0.0002 a.u.)
is even greater than its absolute accuracy. The relative increase
in speed goes up even more dramatically as the molecules get larger. Moreover, for processes or phenomena which involve energy
difference (ionization potentials, electron affinities, spectral
transition energies) the MODPOT method meshed in with the CI programs gives an accuracy of 0.005 eV as compared to experiment(40e).

In these CI programs, each configuration is a spin- and sym-
metry-adapted linear combination of Slater determinants in the
terms of the spin-bonded functions of Boys and Reeves(43-45)as
formulated originally by Shavitt, Pipano and co-workers(46-48) and
subsequently modified and extended in our laboratory(49-51). The
determinants can also be obtained from single or MC-SCF wave
functions.

The routine calculations of CI wave functions containing thousands of configurations is possible. Huge lists of configurations can be reduced to a tractable number by an application of an orbital weighting procedure and/or a perturbation procedure which estimate the energy contributions of individual configuration functions in CI calculations (52-54). Energy extropolation procedures can also be used(55). The determination of several of the lowest eigenvalues and corresponding eigenvectors of these large CI matrices can be done efficiently by the method of Shavitt.(56-58). This method is well suited for states which have a mixture of more than one dominant configuration function, such as excited states. [If the energy eigenvalues come close together such as at a curve crossing, it becomes increasingly difficult to obtain satisfactory convergence in solving for the respective eigenvalues of the CI matrix by this method. A new method by Davidson(59) circumvents this problem.]

For basis configurations in the CI program in addition to Hartree-Fock orbitals iterative natural orbitals (INO)(60) or pseudo-natural orbitals (PNO),(61) or MC-SCF (multi-configuration self-consistent field)(20)functions may be used if desired or the Brillouin theorem may be used (62). [Natural orbitals are obtained by diagonalizing the first order density matrix from the CI wave function (usually but not necessarily, allowing both selected single and double excitations). Pseudo-natural orbitals are obtained by diagonalizing the first order density matrices from the CI calculation when usually only a single pair of electrons is excited at once. A "pseudonatural orbital" procedure can then be repeated for other electron pairs and the results then meshed together and Schmidt orthogonalized; care must be taken to insure proper orthonormalization. In the MC-SCF procedure the coefficients of the atomic orbitals in the molecular orbitals and the coefficients of the molecular orbitals in the CI determinants are varied simultaneously. A particular truncated form of MC-SCF is the use of OVC(63)(optimized valence configurations) which attempt to account for the molecular extra correlation energy (MECE) by determining only the correlation energy in the valence shell which varies strongly as the atoms combine from infinite separation.]

For the preliminary SCF calculations attention must be paid to the stability of the Hartree-Fock solutions(64)and to methods for high spin open shell systems(65-66).

The Brillouin theorem states that for a closed shell system, there are no matrix elements connecting the single excitations with the ground state configuration. Brillouin's theorem is applicable in a restricted sense to open shell systems - there are no matrix elements between singly excited configurations, which retain the open shell spin coupling, with the open shell SCF wave functions(67). Use of the Brillouin theorem is also helpful for obtaining wave functions where the ordinary Hartree Fock SCF gives convergence problems. The main purpose of all these techniques (INO, PNO, MC-SCF, OVC and the Brillouin theorem) is to choose a truncated set of fairly optimal or reference Slater determinants to describe the occupied and virtual orbital space of the system. Desirably, a larger CI is then performed on the system using excitations to and from the reference functions.

Improved virtual orbitals can also be calculated for use in CI calculations. One method of calculating improved virtual orbitals (IVO's) is to solve the Hartree-Fock equations for the N-electron occupied space and then solve the virtual orbitals in the field of N-1 electrons (keeping the occupied orbitals frozen) and properly orthogonalize the virtual orbital space to the occupied orbital space. This technique results in virtual orbitals which are more appropriate to describe excited states. We have shown this numerically by CI calculations of energies of the ground and first two excited $^1A_1$ states of $H_3O^+(68)$, using either the virtual orbitals from the N-electron problem or from the N-1 electron problem. The total energy of the ground state was lower using the virtual orbitals of the N-electron system - as anticipated - how, ever, the energy of the excited states were lower using the virtual orbitals of the N-1 electron system as anticipated. [This method is somewhat similar in concept to that originally propossed (69), but slightly different.] This procedure for N-electron excited state virtual orbitals in the framework of the SCF calculation of the N-1 electron problem closely resembles those proposed by Huzinaga(70). Huzinaga's recent method for improved virtual orbitals in the extended basis function space(71)is also a useful procedure where there are convergence problems for the Hartree-Fock calculations for the N-electron occupied space of the excited states. This should also be helpful in optimizing virtual orbitals to use them in perturbation theory expressions.

The most time and space consuming part and the limiting factor of a CI calculation is the transformation of integrals from integrals over atomic basis functions to integrals over molecular orbitals. We have implemented(72) a technique which incorporates into an effective CI Hamiltonian operator the effect of all molecular orbitals from and to which no excitations are made and which makes it more feasible to do CI calculations on these systems requiring large basis function sets.

When appropriate Davidson's formula(73)may be used to estimate the contribution of quadruple excitations to the CI energy.

There has been recent work reported on the self-consistent electron pairs in CI calculations. The use of natural orbitals was introduced by Meyer(74). Self-consistent electron pairs (SCEP) is a subsequent development which uses an operator formalism and iterative variational method to obtain correlated wave functions(75). The coupled electron pair approximation (CEPA)(76) can be also incorporated into SCEP and a semi-empirical extrapolation technique has been suggested and applied for recovering the correlation defect from such wave functions(77). The attraction of SCEP lies in the avoidance of integral transformations as well as avoidance of construction and diagonalization of large CI matrices.

From our previous experience we already know we need triple, quadruple and higher excitations for many ion-molecule systems. The SCEP formulation would have to be extended to include these. We shall also explore the desirability of using perturbation theory(78-82) - however, our initial analysis indicates perturbation theory is more tractable to apply for ground states (especially closed shell systems) than for the excited high multiplicity potential energy curves and surfaces most ion-molecule reactions.

In 1970-1971, we carried out perhaps the first large scale ab-initio configuration interaction calculations for an ion-molecule reaction, that of the $O^+ + N_2 \rightarrow NO^+ + N$ reaction(83). Our results showed that even for the linear attack of O^+ on N_2, the $^4\Pi$ state had several different dominant molecular orbital configurational occupancies in different regions of the surface. Other symmetry and spin states for this system, both reactants and products, had similar behaviors. We solved both for the lowest eigenvalue root of the large CI matrix and for a number of the next higher roots. The results indicated that in each case of curve crossing (or pseudo-crossing or avoided crossing) the higher energy roots were the diabiatic continuation of the molecular orbital occupancies of the lower energy branches. In the intervening years there have been surprisingly few additional large scale CI calculations on potential energy surfaces of ion-molecule reactions. The subsequent calculations by other researchers of the $O^+ + N_2$ system have been mostly modest basis set SCF calculations with a few MC-SCF calculations(84). The following ab-initio CI calculations on other ion-molecule systems deserve mention. There have been several CI calculations of the $C^+ + H_2$ system(85,86). The importance of using polarization functions in the basis set as emphasized by the subsequent reinvestigation of the original LBS $C^+ + H_2$ potential energy surface(85) by Pearson and Roueff(86). Using a larger basis set in which polarization functions were added necessitated a revision of the original LBS surface(85). Near the intersection of the $^2A_1$ and $^2B_2$ surfaces (C_{2v} symmetry) the system can pass to the strongly bound CH_2^+ well with a minimum

of activation energy. The line of interaction located by LBS was
calculated to have an energy 10.3 kcal higher than that of the
isolated reactants. The subsequent calculation by Pearson and
Roueff show the line of intersection close to the LBS prediction
but reaching an energy 15 kcal below isolated reactants. [The
italics (underlined) are those by Pearson and Rcueff.]

A recent CI calculation of the $N^+ + H_2$ surface(4)(the results
of which were presented in this panel discussion by D. M. Hirst)
used a double zeta basis set plus polarization functions. A few
MC-SCF points were calculated on the $F^+ + H_2$ surface to calibrate
a non-rigorous DIM (diatomics in molecules) calculation(87).

2. Less Rigorous Molecular Orbital Methods

Semi-rigorous molecular orbital methods (these methods include electron-electron repulsion but neglect three-and-four-
center integrals and make approximations for the other integrals
and for the form of the Hamiltonian equations to be solved) en-
compass the zero differential overlap (ZDO) methods such as CNDO
(complete neglect of differential overlap)(88), INDO(88)(intermediate neglect of differential overlap), NDDO(89)(neglect of diatomic differential overlap), MINDO(90)(modified INDO), MDNO(91)
(modified NDDO), etc. Previously we had shown by comparison with
accurate ab-initio calculations the non-applicability of CNDO and
INDO supermolecule potential energy surfaces for ion-molecule
interactions even when the infinite separation of the interacting
supermolecule system is correctly expressible by a single determinant wave function(92). The same critique applies to the other
ZDO methods.

Other semi-rigorous methods include PCILO(93)(perturbative
configuration interaction based on localized orbitals using a
CNDO/2 parametrization) developed by Professor Pullman and coworkers, which while it has proven remarkably reliable for angular conformational analyses of large drugs and biomolecules(94) is
not recommended by them for calculation of accurate intermolecular
interactions. There appear no comparisons with accurate ab-initio
calculations for potential energy surfaces using other semi-rigorous techniques such as PRDDO(95)(partial retention of diatomic
differential overlap), NEMO(96)(non-empirical molecular orbital
method using Fock matrix elements taken from ab-initio calculations on smaller molecules) or transfer of Fock matrix elements
(97,98) to enable an appropriate evaluation of their validity for
reliable potential energy surfaces for ion-molecule reactions.
At the very best the single determinant formalism for which all
these were originally derived would have to be supplemented by
some type of CI procedure. Whether even such an ansatz would be

reliable is speculative. The recent introduction of ab-initio effective core model potentials and meshing these into proper CI programs would seem to mitigate the **appropriateness from here** on of using any semi-rigorous techniques for the calculations of the potential energy surfaces of ion-molecule reactions.

Semi-empirical techniques such as the extended Hückel(99) and the iterative extended Hückel procedures(100-103)(which neglect electron-electron repulsion) often do not even give a minimum in the energy for intramolecular bond distance variations of bonded atoms. Thus such techniques are completely unsuitable for potential energy surfaces of ion-molecule reactions.

3. Diatomics in Molecules (DIM)

The diatomics in molecules technique(104) is a valence bond formulation based on experimental or calculated potential energy curves for the diatomic partners. It has been used as originally derived and in an extended formalism(105)to generate potential energy surfaces for some ion-molecule reactions. Among them, $H^+ + H_2$(106), $Ar^+ + H_2$(107), and most recently $F^+ + H_2$(87).

4. Electrostatic Molecular Potential Contour Maps

An intriguing recent technique, the generation of electrostatic molecular potential contour maps has great conceptual appeal in investigating mechanisms of ion-molecule reactions, especially those involving medium or larger size molecules. The electrostatic potential arising from molecule A is completely defined at every point of the space if one knows the charge distribution (electronic and nuclear) of the molecule(108). Customarily a test positive charge is used to probe the potential.

For molecular interactions involving molecules having net charges or permanent dipoles, useful information may be drawn by examination of the electrostatic potential, arising from one of the partners, and by simple electrostatic calculations, involving a potential and a simplified description of the charge distribution of the other molecule involved in the interaction(108).

The correct definition of the electrostatic potential
$$V(\underline{r}) = \sum_\alpha^{\text{nuclei}} Z_\alpha / |\underline{r}_\alpha - \underline{r}| - \int dr_1 \, \rho(r_1) / |\underline{r}_1 - \underline{r}|$$ is given by the electron distribution $\rho(r_1)$ derived from molecular wave functions and by the distribution of the nuclear charges Z_α, specific of a given molecular geometry without further simplification. $V(\underline{r})$ gives more detailed and less ambiguous information than a popula-

tion analysis, being a function computed in the overall molecular surrounding space. From the potential definition, the first order interaction energy between the molecular charge distribution and a point charge distribution is $W(\{r_i\}) = \sum_i q(r_i) V(r_i)$.

Such an expression has previously been used for comparative purposes, for the study of interaction between two molecular species, by computing the electrostatic potential of the first partner and by assuming some point charge model as representative of the charge distribution of the second partner. Thus this method is an intriguing model for ion-molecule reactions. This concept can be also extended in a more subtle way by using an electron density contour map to describe the charge distribution of the second partner as a function of the space surrounding this second partner.

ACKNOWLEDGEMENT

The research of Joyce J. Kaufman was performed in collaboration with Professor Walter S. Koski, The Johns Hopkins University and supported by the U. S. Department of Energy.

REFERENCES

1. Joyce J. Kaufman, "Theoretical Considerations of Potential Energy Surfaces for Ion-Molecule Reactions," Presented at the NATO Advanced Study Institute for Ion-Molecules, Biarritz, France, June 1974. In Interactions Between Ions and and Molecules, Ed. P. Ausloos, Plenum Press, New York, 1975, pp. 185-213.

2a. Joyce J. Kaufman, "Extension of Symmetry and Spin Restrictions to Carry "Ghost" Information," Presented at American Chemical Society, National Meeting, Division of Physical Chemistry, Chicago, August 1974.

 b. Joyce J. Kaufman, "Extension of Symmetry and Spin Restrictions to Carry "Ghost" Information," IX International Conference on the Physics of Electronic and Atom Collisions, Eds. J. Risley and R. Geballe, Seattle, July 1975, p. 586.

 c. Joyce J. Kaufman, "Extension of Spin and Symmetry Restrictions to Carry "Ghost" Information," Presented at the Second Summer Conference on Electronic Transition Lasers, Woods Hole, Mass., Sept. 1975. In Electronic Transition Lasers, Ed. J. I. Steinfeld, The Mass. Inst. of Technology, 1976, pp. 286-292.

3. Joyce J. Kaufman, "L-S and j-j Coupling Considerations in Some Ion-Molecule Reactions," an invited lecture presented at the 9th Int. Hot Atom Symposium, Blacksburg, Va., 1977.

4. D. M. Hirst, Mol. Phys. _35_, 1559 (1978).
5. G. D. Myers, J. G. Ambrose, P. B. James and J. J. Leventhal, Phys. Rev. A. _18_, 85 (1978).
6. E. Wigner and E. E. Witmer, Z. Physik. _51_, 859 (1928).
7. G. Herzberg, Molecular Spectra and Molecular Structure. III. Electronic Spectra and Electronic Structure of Polyatomic Molecules, D. Van Nostrand Co., Inc., New York, 1966. See subject Index for tables for individual symmetry groups. (In the text of the present article this book is often referred to as "Herzberg Polyatomics".)
8. W. Lichten, "Advances in Chemical Physics," _3_, Ed. I. Prigogine, Interscience Publishers, New York, 1968, p.p. 382-423.
9a. L. Landau, Soviet Phys. _1_, 89 (1932).
 b. L. Landau, Phys. Sowj. _2_, 46 (1932).
10. E. G. C. Stueckelberg, Helv. Phys. Acta, _5_, 369 (1932).
11a. D. R. Bates, Proc. Roy. Soc., _A257_, 22 (1960).
  b. D. R. Bates, Proc. Roy. Soc., _84_, 517 (1964).
12. C. A. Coulson and K. Zalewski, Proc. Roy. Soc., _A268_, 437 (1962).
13. G. Herzberg, Molecular Spectra and Molecular Structure. I. Spectra of Diatomic Molecules, D. Van Nostrand Co., Inc., New York, 1950.
14. C. A. Jones, K. L. Wendell, Joyce J. Kaufman and W. S. Koski, J. Chem. Phys. _65_, 2345 (1976).
15. K. Wendell, C. A. Jones, Joyce J. Kaufman, and W. S. Koski, J. Chem. Phys. _63_, 750 (1975).
16. C. A. Jones, I. Sauers, Joyce J. Kaufman, and W. S. Koski, J. Chem. Phys. _67_, 3599 (1977).
17. S. Chapman and R. K. Preston, J. Chem. Phys. _60_, 650 (1974).
18. J. C. Slater, Quantum Theory of Molecules and Solids, Vol. I., McGraw Hill Book Co., New York, 1963.
19. C. C. J. Roothaan, Rev. Mod. Phys. _23_, 69 (1951); _32_, 179 (1960).
20. J. Hinze, J. Chem. Phys. _59_, 6424 (1973).
21. I. Shavitt, "The Method of Configuration Iteraction," In Modern Theoretical Chemistry, Vol II, Electronic Structure Ab-Initio Methods, Ed. H. F. Schaefer, Plenum Press, New York, 1976., pp. 189-275.
22a. V. Bonifacic and S. Huzinaga, J. Chem. Phys. _60_, 2779 (1974).
  b. Ibid, _62_, 1507 (1975).
  c. Ibid, _62_, 1509 (1975).
  d. D. McWilliams, and S. Huzinaga, J. Chem. Phys. _63_, 4678 (1975).
  e. S. Huzinaga, Personal discussion, August, 1975.
  f. S. Huzinaga, "Atomic and Molecular Calculations with Model Potential Method," Presented at Quantum Chemical Symposium, 1st North American Chemical Congress, Mexico City, December, 1975.

g. V. Bonifacic and S. Huzinaga, J. Chem. Phys. $\underline{64}$, 956 (1976).

h. S. Huzinaga, Private communication, July, 1977.

23a. C. F. Melius and W. A. Goddard, III, Phys. Rev. $\underline{A10}$, 1528 (1974).

 b. C. F. Melius, B. D. Olafson and W. A. Goddard, III, Chem. Phys. Letts. $\underline{28}$, (1974).

 c. C. F. Melius, W. A. Goddard, III and L. A. Kahn, J. Chem. Phys. $\underline{56}$, 3342 (1972).

24. L. Kahn, P. Baybutt and D. C. Truhlar, J. Chem. Phys. $\underline{65}$, 3826 (1976).

25a. S. Topiol, J. W. Moskowitz, C. F. Melius, M. D. Newton, and J. Jafri, "Ab-Initio Effective Potentials for Atoms of the First Three Rows of the Periodic Table," Courant Institute Report (00-3077-105), ERDA Research and Development Report, New York University, Chemistry, January 1976.

 b. M. A. Ratner, J. W. Moskowitz and S. Topiol, Chem. Phys. Letts. $\underline{50}$, 233 (1977).

 c. C. F. Melius, A. P. Mortola, M. B. Baille, J. W. Moskowitz and M. A. Ratner, Surface Science 59, 279 (1976).

 d. S. Topiol, M. A. Ratner and J. W. Moskowitz, Chem. Phys. Letts. $\underline{46}$, 256 (1977).

 e. M. A. Ratner, J. W. Moskowitz and S. Topiol, Chem. Phys. Letts. $\underline{46}$, 495 (1977).

 f. S. Topiol, A. Zunger and M. A. Ratner, Chem. Phys. Letts. $\underline{49}$, 367 (1977).

 g. S. Topiol, J. W. Moskowitz and C. F. Melius, J. Chem. Phys. $\underline{68}$, 2364 (1978).

26a. J. C. Barthelat and Ph. Durand, J. Chem. Phys. $\underline{71}$, 505 (1974).

 b. J. C. Barthelat and Ph. Durand, J. Chem. Phys. $\underline{71}$, 1105 (1974).

 c. J. C. Barthelat and Ph. Durand, Chem. Phys. Letts. $\underline{16}$, 63 (1972).

 d. Ph. Durand and J. C. Barthelat, Chem. Phys. Letts. $\underline{27}$, 191 (1974).

 e. Ph. Durand and J. C. Barthelat, Theor. Chim. Acta. $\underline{38}$, 283 (1975).

 f. G. Nicholas, J. C. Barthelat and Ph. Durand, J. Am. Chem. Soc. $\underline{98}$, 1346 (1976).

 g. J. C. Barthelat, Ph. Durand and A. Serafina, Mol. Phys. $\underline{33}$, 159 (1977).

 h. C. Teichteil, J. P. Malrieu and J. C. Barthelat, Mol. Phys. $\underline{33}$, 181 (1977).

 i. C. Teichteil and J. P. Malrieu, Chem. Phys. Letts. $\underline{49}$, 152 (1977).

27a. J. N. Bardsley, "Pseudopotentials in Atomic and Molecular Physics," Case Studies in Atomic Physics $\underline{4}$, 299 (1974).

 b. J. N. Bardsley, Personal discussion at the International Conference on the Physics of Electronic and Atomic Collisions Meeting, Seattle, Washington, July, 1975.

c. J. N. Bardsley, B. N. Junker and D. W. Norcross, Chem. Phys. Letts. $\underline{37}$, 502 (1976).

28a. P. Coffey, C. S. Ewig and J. R. Van Wazer, J. Am. Chem. Soc. $\underline{97}$, 1656 (1975).

b. C. S. Ewig and J. R. Van Wazer, Inorg. Chem. $\underline{14}$, 1848 (1975).

c. C. S. Ewig and J. R. Van Wazer, J. Chem. Phys. $\underline{63}$, 4035 (1975).

d. P. Coffey, C. S. Ewig and J. R. Van Wazer, Chem. Phys. Letts. $\underline{39}$, 27 (1976).

e. C. S. Ewig, R. Osman and J. R. Van Wazer, J. Chem. Phys. $\underline{66}$, 3557 (1977).

29a. T. C. Chang, P. Habitz, B. Pittell and W. H. E. Schwarz, Theor. Chim. Acta $\underline{24}$, 263 (1974).

b. P. Habitz, W. H. E. Schwarz and R. Alrichs, J. Chem. Phys. $\underline{66}$, 5117 (1977).

c. T. C. Chang, P. Habitz and W. H. E. Schwarz, Theor. Chim. Acta $\underline{44}$, 61 (1977).

d. B. Pittel and W. H. E. Schwarz, Chem. Phys. Letts. $\underline{46}$, 121 (1977).

e. P. Hafner and W. H. E. Schwarz, J. Phys. B. $\underline{11}$, 217 (1978).

30a. M. E. Schwartz and J. D. Switzlski, J. Chem. Phys. $\underline{57}$, 4125 (1972).

b. M. E. Schwartz and J. D. Switzlski, J. Chem. Phys. $\underline{57}$, 4132 (1972).

31. R. G. Hyde and J. B. Peel, Mol. Phys. $\underline{33}$, 887 (1977).

32. J. G. Snyders and F. J. Baerends, Mol. Phys. $\underline{33}$, 1651 (1977).

33a. P. J. Hay, W. R. Wadt and L. R. Kahn, "Ab-Initio Effective Core Potentials for Molecular Calculations II," preprint, November 1977.

b. L. R. Kahn, P. J. Hay and R. D. Cowan, "Relativistic Effects in Ab-Initio Core Potentials for Molecular Calculations - Applications to the Uranium Atom," preprint, November 1977.

c. L. R. Kahn, P. J. Hay and R. D. Cowan, "Relativistic and Non-Relativisitic Effective Core Potentials for Xenon - Applications to XeF, Xe_2 and Xe_2^+," Preprint, November 1977.

d. L. R. Kahn, P. J. Hay and R. D. Cowan, J. Chem. Phys. $\underline{68}$, 2386 (1978).

34. G. Das and A. C. Wahl, J. Chem. Phys. $\underline{64}$, 4672 (1976).

35a. D. F. Mayers, I. P. Grant and N. C. Pyper, J. Phys. $\underline{B9}$, 2777 (1976).

b. N. C. Pyper, I. P. Grant and R. B. Gerber, Chem. Phys. Letts. $\underline{49}$, 479 (1977).

36. Y. S. Lee, W. C. Ermler and K. S. Pitzer, J. Chem. Phys. $\underline{67}$, 5861 (1977).

37. J. Demuznk, Chem. Phys. Letts. $\underline{45}$, 74 (1977).

38a. H. E. Popkie, "MOLASYS: Computer Programs for Molecular Orbital Calculations on Large Systems," The Johns Hopkins University, 1974.

b. H. E. Popkie, "MOLASYS - MERGE," The Johns Hopkins University, 1978.

39a. H. E. Popkie, "GIPSY (Gaussian Integral Program System): A System of Computer Programs for the Evaluation of One-and Two-Electron Integrals Involving s-, p-, d-type Contracted Cartesian Gaussian Functions," The Johns Hopkins University, 1975.

b. H. E. Popkie, "GIPSY (Gaussian Integral Program System): A System of Computer Programs for the Evaluation of One-and Two-Electron Integrals Involving s-, p-, d- and f-type Contracted Cartesian Gaussian Functions," The Johns Hopkins University, 1976.

40a. H. E. Popkie and Joyce J. Kaufman, "Molecular Calculations With the VRDDO/MODPOT Procedure: Preliminary Results for Formamide, Pyyrole, Pyrazole, Imidazole and Nitrobenzene," Invited lecture presented at Summer Research Conference on Theoretical Chemistry, Boulder, Colo., June 1975.

b. H. E. Popkie and Joyce J. Kaufman, Int. J. Quantum Chem. $\underline{S10}$, 47 (1976).

c. H. E. Popkie and Joyce J. Kaufman, "Quantum Chemical Calculations Using the MODPOT/VRDDO Method," Presented at Int. Congress on Quantum Chemistry, New Orleans, La., April 1976.

d. H. E. Popkie and Joyce J. Kaufman, J. Chem. Phys. $\underline{66}$, 4827 (1977).

e. H. E. Popkie and Joyce J. Kaufman, Chem. Phys. Letts. $\underline{47}$, 55 (1977).

41. H. E. Popkie and Joyce J. Kaufman, Int. J. Quantum Chem. $\underline{QBS2}$, 279 (1975).

42a. D. L. Wilhite and R. N. Euwema, Chem. Phys. Letts. $\underline{20}$, 610 (1973).

b. D. L. Wilhite and R. N. Euwema, J. Chem. Phys. $\underline{61}$, 375 (1974)

c. R. N. Euwema and R. L. Green, J. Chem. Phys. $\underline{62}$, 4445 (1975).

43. C. M. Reeves, Ph.D. Thesis, Cambridge Univ. (1957).

44. C. M. Reeves, Commun. ACM (Assoc. Comput. Mach.) $\underline{9}$, 276 (1966).

45. I. L. Cooper and R. McWeeny, J. Chem. Phys. $\underline{45}$, 226 (1966).

46. The basic CI programs described below are those written originally by I. Shavitt and A. Pipano and continued by A. Pipano while he was a postdoctoral in our group at The Johns Hopkins University 1970-1971. In addition we have subsequently written a number of additional features to these CI programs 1971-present.

47. Z. Gershgorn and I. Shavitt, Int. J. Quantum Chem. $\underline{1S}$, 403 (1967).

48. A. Pipano and I. Shavitt, Int. J. Quantum Chem. $\underline{2}$, 741 (1968).

49. A. Pipano and Joyce J. Kaufman, The Johns Hopkins University, Unpublished, 1971.

50. H. J. T. Preston and Joyce J. Kaufman, The Johns Hopkins University, 1972-present.

51. R. C. Raffenetti, H. J. T. Preston and Joyce J. Kaufman, The Johns Hopkins University, 1973.

52. Z. Gershgorn and I. Shavitt, Int. J. Quantum Chem. $\underline{2}$, 751 (1968).

53a. I. Shavitt, "Energy Contributions of Configurations in Configuration Interaction Wave Functions," preprint, 1972.

   b. I. Shavitt, K. Hsu, R. C. Raffenetti, L. R. Kahn and P. J. Hay, "Energy Contributions and Selection of Configurations in Large Configuration Interaction Calculations," 1974.

54. I. Shavitt, R. C. Raffenetti and K. Hsu, Theor. Chim. Acta $\underline{45}$, 33 (1977).

55. R. J. Buenker, Sigrid D. Peyerimhoff and W. Butscher, "Applicability of the Multi-Reference Double Excitation CI (MRD-CI) Method to the Calculation of Electron Wavefunctions and Comparison with Related Techniques," preprint, November 1977.

56a. C. F. Bender, I. Shavitt and A. Pipano, private communication, 1971.

   b. I. Shavitt, C. F. Bender, A. Pipano and R. P. Hosteny, J. Comp. Phys. $\underline{11}$, 90 (1973).

57. I. Shavitt, private communication, visit to the Johns Hopkins University, 1971.

58. I. Shavitt, J. Comput. Phys. $\underline{6}$, 124 (1970).

59a. E. R. Davidson, preprint, 1974.

   b. E. R. Davidson, J. Comput. Phys. $\underline{17}$, 87 (1975).

60. C. F. Bender and E. R. Davidson, J. Phys. Chem. $\underline{70}$, 2675 (1966).

61. C. Edmiston and M. Krauss, J. Chem. Phys. $\underline{45}$, 1833 (1966).

62. L. Brillouin, Les Champs Self Consistents de Hartree et de Fock, Herman, Paris, 1934, p. 19.

63. A. C. Wahl and G. Das, Adv. Quantum Chem. $\underline{5}$, 261 (1970).

64. R. Seeger and J. A. Pople, J. Chem. Phys. $\underline{66}$, 3045 (1977).

65. J. S. Binkley, J. A. Pople and P. A. Dobash, Mol. Phys. $\underline{28}$, 1423 (1974).

66. J. W. Caldwell and M. S. Gordon, Chem. Phys. Letts. $\underline{43}$, 493 (1976).

67. G. M. Schwenzer, S. V. O'Neil, H. F. Schaefer, III, C. P. Baskin and C. F. Bender, J. Chem. Phys. $\underline{60}$, 2787 (1974).

68. R. C. Raffenetti, H. J. T. Preston and Joyce J. Kaufman, Chem. Phys. Letts. $\underline{46}$, 573 (1977).

69. W. J. Hunt and W. A. Goddard, III, Chem. Phys. Letts. $\underline{3}$, 413 (1969).

70a. S. Huzinaga and C. Arnau, Phys. Rev. $\underline{A1}$, 1285 (1970).

   b. S. Huzinaga and C. Arnau, J. Chem. Phys. $\underline{54}$, 1948 (1971).

   c. K. Kirao and S. Huzinaga, Chem. Phys. Letts. $\underline{45}$, 55 (1977).

71. S. Huzinaga and K. Kirao, J. Chem. Phys. $\underline{66}$, 2157 (1977).

72a. R. C. Raffenetti, The Johns Hopkins University, 1973.

b.   R. C. Raffenetti, ICASE, 1974-1976

73. S. R. Langhoff and E. R. Davidson, Int. J. Quantum Chem.
     $\underline{8}$, 61 (1974).

74a. W. Meyer, Int. J. Quantum Chem. $\underline{S5}$, 341 (1971).

 b. W. Meyer, Private communication, visit to the Johns Hopkins
     University, 1971.

 c. W. Meyer, J. Chem. Phys. $\underline{58}$, 1017 (1973).

75a. C. E. Dykstra, H. F. Schaefer, III and W. Meyer, "A Theory
     of Self-Consistent Electron Pairs. Computational Methods
     and Preliminary Applications," LBL-5110 preprint, April 1976.

 b. C. E. Dykstra, H. F. Schaefer, III and W. Meyer, J. Chem.
     Phys. $\underline{65}$, 2740 (1976).

 c. C. E. Dykstra, M. Hereld, R. R. Luchese, H. F. Schaeffer,
     III and W. Meyer, J. Chem. Phys. $\underline{67}$, 4071 (1974).

 d. C. E. Dykstra, J. Chem. Phys. $\underline{67}$, 4176 (1977).

76. W. Meyer, J. Chem. Phys. $\underline{64}$, 2901 (1976).

77. E. L. Mehler and W. Meyer, Chem. Phys. Letts. $\underline{38}$, 144 (1976).

78. C. Moeller and M. S. Plesset, Phys. Rev. $\underline{46}$, 618 (1934).

79a. K. A. Brueckner, Phys. Rev. $\underline{97}$, 1353 (1955).

 b. Ibid, $\underline{100}$, 36 (1955).

80. J. Goldstone, Proc. Royal Soc. London Series $\underline{A239}$, 267
     (1957).

81. J. A. Pople, J. S. Binkley and R. Seeger, Int. J. Quantum
     Chem. $\underline{S10}$, 1 (1976).

82a. R. J. Bartlett and D. M. Silver, J. Chem. Phys. $\underline{62}$, 3258
     (1975).

 b. Ibid, $\underline{64}$, 1260 (1976).

 c. Ibid, $\underline{64}$, 4578 (1976).

 d. R. J. Bartlett and D. M. Silver, Private communication,
     April, 1976.

83. A. Pipano and Joyce J. Kaufman, J. Chem. Phys. $\underline{56}$, 5258
     (1972).

84. D. G. Hopper, J. Am. Chem. Soc. $\underline{100}$, 1019 (1978).

85. D. H. Liskow, C. F. Bender and H. F. Schaefer III, J. Chem.
     Phys. $\underline{61}$, 2507 (1974).

86. P. K. Pearson and B. Roueff, J. Chem. Phys. $\underline{64}$, 1240 (1976).

87. J. Kendrick, P. J. Kuntz and I. H. Hillier, J. Chem. Phys.,
     $\underline{68}$, 2373 (1978).

88. J. A. Pople and D. A. Beveridge, Approximate Molecular Orbi-
     tal Theory, McGraw-Hill, New York, 197C.

89a. J. A. Pople, D. P. Santry, and G. A. Segal, J. Chem. Phys.
     $\underline{43}$, S129 (1965).

 b. J. A. Pople and G. A. Segal, J. Chem. Phys. $\underline{43}$, S136 (1965).

90. M. J. S. Dewar and R. C. Dougherty, The PMO Theory of Organ-
     ic Chemistry, Plenum Publishing Corp., New York, 1975.

91a. M. J. S. Dewar and W. Theil, J. Amer. Chem. Soc. $\underline{99}$, 4899
     (1977).

 b. Ibid, 4970 (1977).

92.   Joyce J. Kaufman and R. Predney, Int. J. Quantum Chem. 5, 235 (1971).

93a. S. Diner, J. P. Malrieu, F. Jordan, and M. Gilbert, Theor. Chim. Acta 15, 100 (1968) and references therein.

  b. J. P. Malrieu, B. Huron, and P. Rancurel, J. Chem. Phys. 58, 5745 (1973).

94.   B. Pullman and P. Courriere, Conformation of Biological Molecules and Polymers, Jerusalem Symp. on Quantum Chem. and Biochemistry, Vol. 5, Eds., E. D. Bergmann and B. Pullman, Jerusalem Academic Press, Jerusalem, Israel, 1972, pp. 547-565.

95a. T. A. Halgren and W. N. Lipscomb, Proc. Nat. Acad. Sci. 69, 652 (1972).

  b. T. A. Halgren and W. N. Lipscomb, J. Chem. Phys. 58, 1569 (1973)

  c. D. A. Dixon, D. A. Kleier, T. A. Hall and W. N. Lipscomb, J. Amer. Chem. Soc. 96, 2293 (1974).

96a. M. D. Newton, F. P. Boer, and W. N. Lipscomb, J. Amer. Chem. Soc. 88, 2353 (1966).

  b. F. P. Boer, M. D. Newton, and W. N. Lipscomb, J. Amer. Chem. Soc. 88, 2361 (1966).

  c. M. D. Newton, F. P. Boer, and W. N. Lipscomb, J. Amer. Chem. Soc. 88, 2367 (1966)

  d. W. E. Palke and W. N. Lipscomb, J. Amer. Chem. Soc. 88, 2384 (1966).

  e. W. N. Lipscomb and D. B. Boyd, J. Chem. Phys. 48, 4955 (1968).

97a. Ph. Degan, G. Leroy, and D. Peeters, Theor. Chim. Acta 30, 243 (1973)

  b. G. Leroy, personal discussion at the 1st Int. Cong. of Quantum Chemistry, Meton, France, 1973.

98.   J. E. Eilers and D. R. Whitman, J. Amer. Chem. Soc. 95, 2067 (1973).

99.   R. Hoffman, J. Chem. Phys. 39, 1397 (1963)

100. L. C. Cusachs and D. J. Miller, Adv. in Chem. Ser., 110, 1 (1972).

101a. R. Rein, N. Fukuda, H. Win, G. A. Clarke, and F. E. Harris, J. Chem. Phys. 45, 4743 (1966).

  b. R. Rein, N. Fukuda, G. A. Clarke and F. E. Harris, J. Theor. Biol. 21, 88 (1968).

102. M. Zerner and M. Gouterman, Theor. Chim. Acta 4, 44 (1966).

103. J. H. Corrington, H. S. Aldrich, C. W. McCurdy, and L. C. Cusachs, Int. J. Quantum Chem. 5S, 307 (1971).

104a. F. O. Ellison and J. C. Patel, J. Amer. Chem. Soc. 86, 2115 (1960).

  b. A. A. Wu and F. O. Ellison, J. Chem. Soc. Faraday Trans II. 68, 259 (1972).

105. J. C. Tully, in Modern Theoretical Chemistry, Ed. H. F. Schaefer III, Plenum Press, New York, 1977, pp. 173-200.

106. R. K. Preston and J. C. Tully, J. Chem. Phys. <u>54</u>, 4297
      (1971).
107. P. J. Kuntz and A. C. Roach, J. Chem. Soc. Faraday Trans. II
      <u>68</u>, 259 (1972).
108. E. Scrocco and J. Tomasi, Topics in Current Chemistry <u>21</u>,
      no. 9, 97 (1973) and references therein.

AB-INITIO CONFIGURATION INTERACTION (CI) CALCULATIONS OF
POTENTIAL SURFACES FOR TRIPLET STATES OF NH_2^+ - D. M. HIRST

These triplet NH_2^+ surfaces(H-1) are relevant to the dynamics
of the reaction $N^+ + H_2 \rightarrow NH^+ + H$ studied by Mahan and his co-
workers(H-2)and to the reaction $N + D_2^+ \rightarrow ND^+ + D$ which has been
investigated by Gentry and his co-workers(H-3). Of particular interest is the possibility of complex formation in the first reaction at low energies.

A. Computational Details

The starting point for the configuration interaction calculations is the restricted Hartree-Fock wavefunction Φ_0 for the
symmetry state being considered. These calculations were made
with ATMOL 3 suite of programmes of Saunders and Guest(H-4) using
Dunning's 9s 5p/4s 2p contracted Gaussian basis for nitrogen and
the 4s/3s basis for hydrogen supplemented with polarization functions.

Molecular orbital wavefunctions are inadequate for potential
surface calculations because they do not dissociate correctly.
This deficiency can be remedied by the inclusion of configuration
interaction. One set of functions Φ_i which have the same spin
and symmetry as Φ_0 was generated by exciting electrons from occupied orbitals to unoccupied or virtual orbitals. Usually one
considers only single and double excitations although higher excitations may be of some importance. The total number, n, of configurations can be restricted, without loss of accuracy, by keeping core orbitals (i.e. 1s for N) doubly occupied and excluding
excitations to the corresponding virtual orbitals of high energy.

In potential surface calculations Φ_0 may not be the dominant
configuration at all points on the surface. In the case of NH_2^+,
the $^3B_1$ state of the ion is reasonably well described by the function $\Phi_0 = (1a_1)^2(2a_1)^2(1b_2)^2 3a_1{}^1 1b_1{}^1$ but dissociation to $N^+(^3P)$ +
H_2 ($^1\Sigma_g^+$) is described by the function $\Phi_1 = (1a_1)^2(2a_1)^2(3a_1)^2 1b_1{}^1$
$4a_1{}^1$. In such cases it is necessary to include configurations resulting from additional root configurations such as Φ_1 as well as

from Φ_0. Ideally one should include all root configurations which have a coefficient 0.1 in the final wavefunction. If states other than the ground state are to be calculated further root functions must be included.

However this often results in a prohibitively large number of configurations and some selection procedure has to be employed. This usually involves the use of perturbation theory to estimate the contribution of each configuration to the final energy. Those configurations contributing less than a certain threshold energy are then eliminated.

The CI calculations were made using the SPLICE suite of programs(H-5)implemented on the IBM 370/195 computer at the Rutherford Laboratory.

B. Potential Surfaces for NH_2^+

Calculations were made for the collinear $^3\Sigma$ state, the C_{2v} $^3A_2$ and $^2B_1$ states and for $^3A''$ states of C_s symmetry. Surfaces correlating with $N^+(^3P) + H_2 (^1\Sigma_g^+)$ and $N(^4S) + H_2^+(^2\Sigma_g^+)$ were considered. Details of the sets of root configurations used are

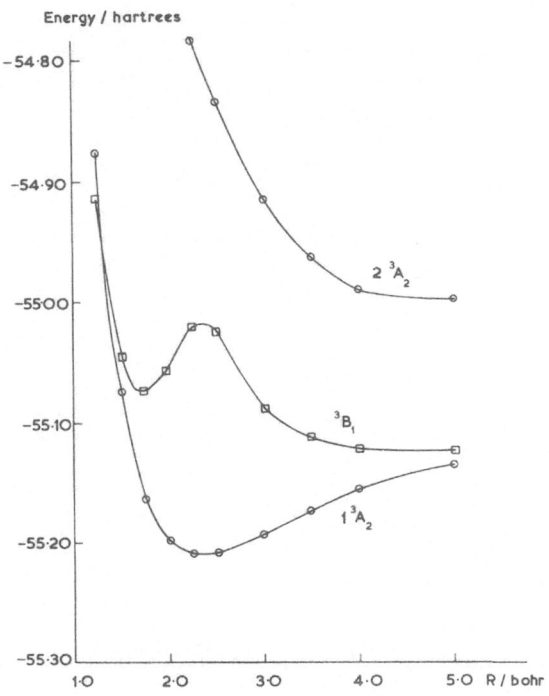

Figure 1. Energy in hartrees of C_{2v} (HNH)$^+$ for $r_{HH}=1.4$ bohr.

given elsewhere. The numbers of configurations generated (before selection) were 5312 ($^3\Sigma-$), 4923 ($^3A_2$), 9763 ($^3\mathring{B}_1$) and 12104 ($^3A''$) and these were reduced to 1200–1800 for the final CI. Figure 1 shows a cut through the entrance channel for the perpendicular approach of N^+ to H_2.

The lowest $^3A_2$ surface has a shallow well (of depth 2.6 eV) whereas the $^3B_1$ surface has a local maximum of height of 2.9 eV. The existence of this barrier, which arises from an avoided intersection between the configurations in Φ_0 and Φ_1 gives us some understanding of why complex formation is relatively unimportant in the first reaction despite the fact that the $^3B_1$ state of NH_2^+ is 6 eV stable with respect to $N^+(^3P) + H_2$.

The second surface of $^3A_2$ symmetry (which correlates with $N(^4S) + H_2^+$ ($^2\Sigma_g^+$) is repulsive and unlikely to be important in the second dynamics reaction at low energies.

As the geometry of approach is changed from C_{2v} through C_s symmetry to the collinear $C_{\infty v}$ approach, the 1^3A_2 surface correlates through the $1^3A''$ surface to the $1^3\Sigma^-$ surface shown in Figure 2.

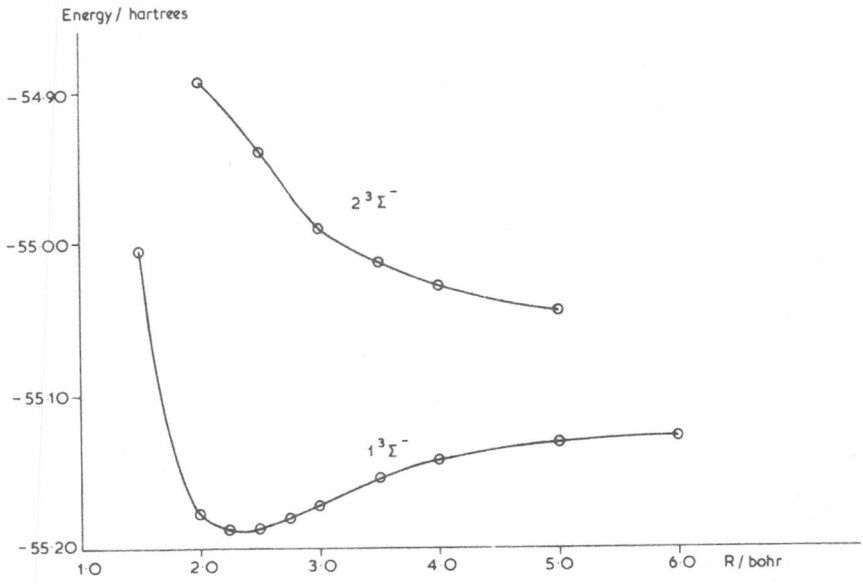

Figure 2. Energy in hartrees of linear $^3\Sigma^-(NHH)^+$ for r_{HH}=1.4 bohr.

The $^3B_1$ surface correlates with the repulsive $^3\Pi$ surface (H-6) and the 2^3A_2 surface correlates with the $2^3\Sigma^-$ surface which is also repulsive. The question of whether motion on the $^3A_2$ – $^3A''$ – $^3\Sigma^-$ surface could give rise to complex formation can only be answered by dynamical calculations. However the occurrence of complex formation can be understood if we consider a distortion from C_{2v} symmetry to C_s symmetry. Surfaces of $^3A_2$ and $^3B_1$ symmetry are now both $^3A''$ symmetry and an avoided intersection results. In the entrance channel (r_{HH} = 1.4 bohr) this occurs on the repulsive part of the potential. However, if r_{HH} is increased to about 3 bohr, the intersection moves to the attractive part of the potential and a slight distortion from C_{2v} geometry results in a low energy path to the deep potential well of $^3B_1$ NH_2^+. This is illustrated in Figure 3 by a cut through the surface with r_{HH} = r bohr for an approach at 10 degrees to the perpendicular.

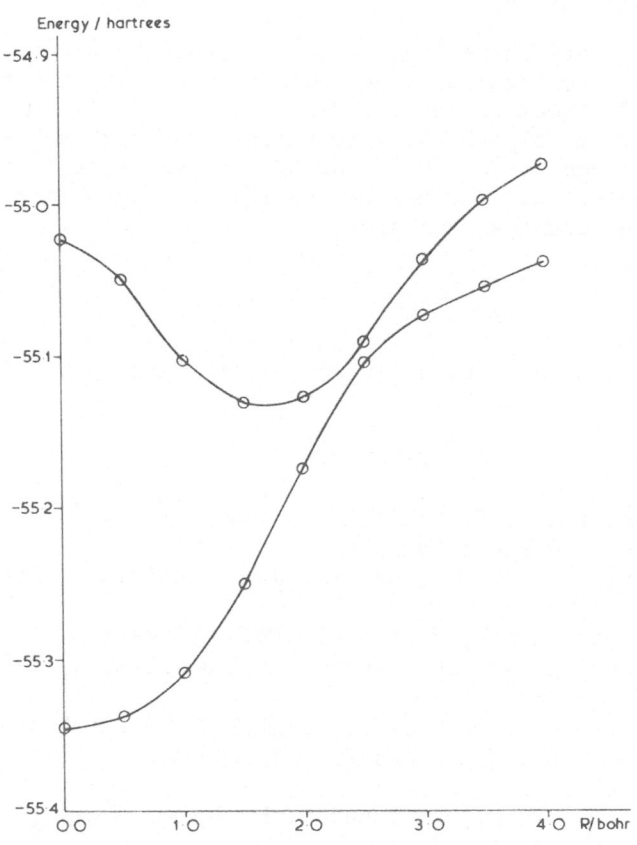

Figure 3. Energy in hartrees of C_s (HNH)$^+$ for r_{HH}=4.0 bohr and $\theta = 80°$.

Thus, provided motion takes place on the lower surface, complex formation is possible. However crude estimates(H-6) of the probability of hopping from the lower to upper surface indicate that, except at the lower translational energies, the hopping probability is high. Thus we can understand why symmetric product distributions are only observed for low energy collisions.

C. Other Approaches to Potential Energy Surfaces

1. The Valence Bond Method

Several groups are concerned with the development of the bond method. In particular Balint-Kurti is interested in potential surfaces,(H-7)and has recently completed some calculations for H_2O^+.

2. Analytic Potential Functions

In recent years Murrell and his co-workers have derived analytic potential surfaces for polyatomic molecules by the superposition of individual asymptotic diatomic potentials and a suitable three body function(H-8). These functions have achieved a broad range of success for molecular systems and have been used in dynamical calculations. It may be possible to extend the approach to ion-molecule potential surfaces.

ACKNOWLEDGEMENT

The results on N^+ + H_2 appeared in Mol. Phys. 35, 1559 (1978).

REFERENCES

H-1. B. H. Mahan and W. E. W. Ruska, J. Chem. Phys. 65, 5044 (1976) and references therein.
H-2. D. J. McLure, C. H. Douglass and W. R. Gentry, J. Chem. Phys. 66, 2079 (1977).
H-3. V. R. Saunders and M. F. Guest, ATMOL 3 Reference Manual, Atlas Computing Division, Rutherford Laboratory, Chilton, Oxon OXII OQ, U.K.
H-4. M. F. Guest and W. R. Rodwell, SPLICE Reference Manual, Atlas Computing Division, Rutherford Laboratory, Chilton, Oxon OXII OQY, U.K.
H-5. M. A. Gittins, D. M. Hirst and M. F. Guest, Faraday Disc. Chem. Soc. 62, 67 (1977).
H-6. G. G. Balint - Kurti and R. N. Yardley, Faraday Disc. Chem. Soc. 62, 77 (1977).

H-7. S. Farantos, E. C. Leisengang, J. N. Murrell, K. S. Sorbie,
J. J. C. Teixeira - Dias and A. J. C. Varandas, Mol. Phys.
34, 947 (1977).

SPIN CONSERVATION IN ELECTRON CAPTURE COLLISIONS - J. J. LEVENTHAL

Experimental studies of energy transfer in collisions between
atomic ions and neutral atoms or molecules were conducted by observing collision-produced luminescence in a crossed beam apparatus
(L-1). For He^{++} -H_2, where double charge exchange to triplet He
states is forbidden according to the spin conservation rule, the
ratio of triplet to singlet signal was roughly 10:1; the cross
sections for production of the diagnostic state, $He(3d^1D)$, was relatively large[1] (~$2Å^2$ at 10 eV). For He^{++}-He collisions, where
triplet formation is similarly forbidden, the triplet: singlet
ratio was about 2:1. In this case however the cross sections were
only about 10^3 $Å^2$ at 200 eV. The apparently anomalous behavior for
He^{++}-He was atributed to the domination of short range effects.

As an example of another system involving few electrons were
studied He^+-Li collisions. In this system there are no spin restrictions on the He atom state formed by single electron capture.
Although the dominant observable inelastic channel is collisional
excitation of the target Li(2s) to Li(2p), charge transfer to excited He(3d) is also observed. In this case only $He(3d^3D)$ is formed
and not $He(3d^1D)$. The preference of this system to form excited
triplet He atoms in charge exchange processes is probably due to
easier access to excited triplet $(HeLi)^+$ states from the repulsive
He^+-Li $^3\Sigma$ state. In order for the system to cross an excited
singlet $(HeLi)^+$ curve the system would have to evolve along the
lower lying $^1\Sigma$ entrance channel with subsequent crossing at small
internuclear separation, i.e. along the repulsive wall.

REFERENCES

L-1. G. D. Meyers, J. G. Ambrose, P. B. James and J. J. Leventhal,
Phys. Rev. A 18, 85 (1978).

ION-DIPOLE COLLISIONS: RECENT THEORETICAL ADVANCES

Walter J. Chesnavich, Timothy Su* and Michael T. Bowers

Department of Chemistry, University of California

Santa Barbara, California 93106 U.S.A.

The calculation of ion-permanent dipole capture rate constants is a problem that has received considerable attention in recent years. This paper summarizes the present state of the theory; particularly, advances occuring since the article of Bowers and Su (1975) was published. Emphasis is placed on a recently developed transition state theory approach which is capable of producing a rigorous upper bound to the true capture rate constant obtained from solution of the classical equations of motion.

The organization of this paper is as follows. Section I provides a brief review of the dynamical and transition state theory formulations of Langevin capture theory. The transition state theory analysis of ion-permanent dipole capture is presented in Section II. Section III summarizes capture theories based on averaging the charge-permanent dipole potential to obtain an effective central potential. In Section IV the classical dynamics of ion-permanent dipole capture is discussed and recently obtained results of trajectory calculations of the thermal capture rate constant are presented. In Section V the various theories are compared to experiment. Some conclusions and a summary of research in progress are given in Section VI.

I. LANGEVIN THEORY

The classical dynamics of a point ion interacting with a point neutral _via_ the charge-induced dipole long range potential has been

---

*Permanent address: Department of Chemistry, Northeastern Massachusetts University, North Dartmouth, Massachusetts 02747 U.S.A.

extensively discussed in the literature (see, for example, Bowers and Su, 1975, and references therein) and will only be summarized here. The transition state theory approach has been discussed by Eyring et al. (1936), Light, Pechukas, and coworkers (see Pechukas, 1976, and references therein) and by Keck (1967).

A. Classical Dynamics

Consider the collision of a point ion with a point neutral interacting via the charge-induced dipole potential. The total energy of the system in the center of mass frame of reference consists only of relative translational energy of the colliding partners, E_{rel}, which can be expressed as

$$E_{rel} = p_r^2/2\mu + V_{eff}(r,L) , \tag{1}$$

where p_r is the radial momentum of the collision partners at separation r, μ is their reduced mass, and the effective potential, V_{eff}, is given by

$$V_{eff}(r,L) = L^2/2\mu r^2 - \alpha q^2/2r^4 , \tag{2}$$

where L is the orbital angular momentum of the system, α is the polarizability of the neutral and q is the charge on the ion. For capture to occur a given trajectory must be able to reach, with $p_r < 0$, the value of r for which V_{eff} is a maximum. Setting $\partial V_{eff}/\partial r = 0$ gives

$$L^2 = 2\mu\alpha q^2/r_c^2 , \tag{3}$$

where $r = r_c$ defines the maximum in V_{eff} for a given L. For a given E_{rel}, the maximum value of L for which capture can occur, L_c, is obtained by setting $p_r = 0$ in Eq.(1) at $r = r_c$. This analysis yields

$$L_c = (8\mu^2 q^2\alpha E_{rel})^{\frac{1}{4}} . \tag{4}$$

The capture rate constant is given by

$$k_c = v_\infty \int_0^{\sigma_c} d\sigma , \tag{5}$$

where $d\sigma = 2\pi b db$ is the differential capture cross section, b is the impact parameter, and v_∞, the relative velocity of the collision partners at infinite separation, is given by $(2E_{rel}/\mu)^{\frac{1}{2}}$. Making these substitutions, along with $L = \mu v_{\infty} b$, into Eq.(5) gives

$$k_c = v_{\infty} \frac{\pi}{2\mu E_{rel}} \int_o^{L_c} 2L dL .$$

(6)

The integration of Eq.(6) gives the well known Langevin-Gioumousis-Stevenson rate constant (Eyring et al., 1936; Vogt and Wannier, 1954; Gioumousis and Stevenson, 1958),

$$k_L = 2\pi q (\alpha/\mu)^{\frac{1}{2}} .$$

(7)

B. Transition State Theory

The transition state theory analysis of capture collisions is based on the fundamental assumption of transition state theory (Eyring, 1935; Wigner, 1937; for recent reviews see Miller, 1976, and Pechukas, 1976) which states that all trajectories crossing the transition state surface from reactants to products (or products to reactants) cross once and only once. It directly follows from this assumption that if the transition state is constructed so that it completely isolates reactant and product regions of the system phase space then the rate constant obtained will be an upper bound to the true result (for further discussion on this point see Pechukas and McLafferty, 1972 and Chapman et al., 1975).

The microcanonical (fixed-energy) transition state theory rate constant for Langevin capture theory can be written as

$$k_{TST}(E_{rel}) = \frac{W^{\ddagger}(E_{rel} - V^{\ddagger})}{h \rho_t(E_{rel})} ,$$

(8)

where $W^{\ddagger}(E_{rel} - V^{\ddagger})$ is the sum of states at the transition state, $V^{\ddagger}$ is the charge-induced dipole potential evaluated at the transition state, $\rho_t(E_{rel})$ is the three-dimensional density of translational states per unit volume of the collision partners (see, for example, Chapter 1 of McQuarrie, 1976), given by

$$\rho_t(E_{rel}) = (2\mu)^{\frac{3}{2}} E^{\frac{1}{2}}/4\pi^2 \hbar^3 ,$$

(9)

and $h = 2\pi\hbar$ is Planck's constant. If L is defined in terms of spherical polar coordinates (see, for example, page 299 of Goldstein, 1950) the sum of states becomes*

---

*Recall that $W^{\ddagger}(E_{rel} - V^{\ddagger})/h$ is the flux through the transition state. The flux through a surface is given by flux = $\int \rho v ds$ where

$$W^{\ddagger}(E_{rel} - V^{\ddagger}) = \left(\frac{1}{2\pi\hbar}\right)^2 \int \delta[r^{\ddagger} - r]\, \delta[E_{rel} - H_{rel}]\, \frac{p_r}{\mu} \cdot$$

$$dr dp_r d\theta\, dp_\theta\, d\phi dp_\phi , \tag{10}$$

where $r^{\ddagger}$ is the location of the transition state, H_{rel} is the system Hamiltonian, given by

$$H_{rel} = p_r^2/2\mu + L^2/2\mu r^2 - \alpha q^2/2r^4 , \tag{11}$$

and L is given by

$$L^2 = p_\theta^2 + p_\phi^2/\sin^2\theta . \tag{12}$$

The integration in Eq.(10) is six-fold and is restricted to the hypersurfaces of constant $r = r^{\ddagger}$ and constant $H_{rel} = E_{rel}$ by the delta functions. Integrating Eq.(10), and combining the result with Eqs.(8) and (9), gives

$$k_{TST}(E_{rel}) = \frac{2\pi}{\sqrt{2\mu E_{rel}}}\, r^{\ddagger 2}\left(E_{rel} + \frac{\alpha q^2}{2r^{\ddagger 4}}\right) . \tag{13}$$

There are two methods by which the final evaluation of Eq.(13) can be performed. According to the dynamical criterion of transition state theory, $r^{\ddagger}$ has the property that all trajectories cross it once and only once. For central potentials it is always possible to find a value of r for which this statement is true. For the charge-induced dipole potential the correct result is given by r_c in Eq.(3). For use in Eq.(13) one must express Eq.(3) as a function of E_{rel}. This is accomplished by setting p_r to zero in Eq.(1) and using Eq.(3) to eliminate L giving

$$r^{\ddagger} = \left(\frac{\alpha q^2}{2E_{rel}}\right)^{\frac{1}{4}} . \tag{14}$$

When Eq.(14) is substituted into Eq.(13) the LGS result, Eq.(7) is obtained.

ρ is the density, v is the velocity normal to the surface and ds is a differential of surface area. Since microcanonical TST is used, the total energy is fixed at E_{rel} and $\rho = {}_n\delta[E_{rel} - H_{rel}]$. The differential surface element is given by $\prod dp_i dq_i\, \delta(r^{\ddagger} - r)/h^n$ where the p's and q's are the momenta and positions of the particles. The delta function $\delta(E_{rel} - H_{rel})$ restricts the integration to surfaces of constant $H_{rel} = E_{rel}$ and $\delta(r^{\ddagger} - r)$ restricts the integration to values of $r = r^{\ddagger}$. When the substitution $v = p_r/\mu$ is made, Eq.(10) results directly.

The fact that the transition state theory rate constant is an upper bound to the true rate constant is the basis of the variational method (Wigner, 1937; Keck, 1967). In this approach $k_{TST}(E_{rel})$ is treated as an explicit function of $r^{\ddagger}$. To determine the best transition state $k_{TST}(E_{rel})$ is varied to determine the value of $r^{\ddagger}$ at which it obtains its minimum value. When the derivative of Eq.(13) with respect to $r^{\ddagger}$ is taken and set equal to zero one finds that the best transition state is again located at $r^{\ddagger}$ given by Eq.(14). The LGS capture rate constant immediately follows. Therefore the variational transition state theory approach is exact for the charge-induced dipole potential. In fact, it is exact for all central potentials for which a capture collision can be defined.

The behavior of Eq.(13) as a function of $r^{\ddagger}$ has the following dynamical interpretation. As $r^{\ddagger} \to \infty$ then $k_{TST}(E_{rel})$ diverges due to contributions from nonreactive trajectories with large L that originate at $r = \infty$, pass through $r = r^{\ddagger}$, reflect off the effective potential, and retreat back to $r = \infty$. As $r^{\ddagger} \to 0$ then $k_{TST}(E_{rel})$ diverges due to contributions from nonreactive trajectories with large L that originate at $r = 0$, pass through $r = r^{\ddagger}$, reflect off the effective potential, and retreat back to $r = 0$. The nonreactive trajectories at large r arise due to the falloff of the centrifugal energy, $L^2/2\mu r^2$, with respect to r whereas those at small r arise due to the falloff of the potential energy. The variational approach seeks a compromise that minimizes the total number of non-reactive trajectories from both sources. For central potentials this compromise is exact.

II. APPLICATION OF TRANSITION STATE THEORY TO ION-PERMANENT DIPOLE COLLISIONS

The long range classical Hamiltonian for the interaction of an ion with a polarizable two-dimensional rigid rotor containing a permanent dipole moment can be written as

$$H = p_r^2/2\mu + L^2/2\mu r^2 + J^2/2I - \alpha q^2/2r^4 - q\mu_D\cos\theta/r^2 , \quad (15)$$

where J, I, and μ_D are the angular momentum, moment of inertia, and dipole moment of the rotor and θ is the relative orientation angle between the dipole and line of centers of the collision. In spherical polar coordinates L, J, and $\cos\theta$ have the following definitions (Goldstein, 1950)

$$L^2 = p_{\theta_1}^2 + p_{\phi_1}^2/\sin^2\theta_1 , \quad (16)$$

$$J^2 = p_{\theta_2}^2 + p_{\phi_2}^2/\sin^2\theta_2 , \quad (17)$$

and, from the law of cosines,

$$\cos\theta = -[\cos(\phi_1 - \phi_2) \sin\theta_1 \sin\theta_2 + \cos\theta_1 \cos\theta_2] . \qquad (18)$$

At total system energy $E = H$ the transition state theory rate constant is

$$k_{TST}(E) = \frac{W^{\ddagger}(E - V^{\ddagger})}{h\rho(E)} , \qquad (19)$$

where $V^{\ddagger}$ is the charge-induced dipole, charge-permanent dipole potential evaluated at the transition state, and $\rho(E)$ is the total density of states per unit volume of the collision partners,

$$\rho(E) = \int_0^E \rho_{rot}(E - E_{rel}) \rho_t(E_{rel}) dE_{rel}$$

$$= (2\mu E)^{\frac{3}{2}} I/3\pi^2\hbar^5 \qquad (20)$$

where $\rho_{rot}(E - E_{rel})$ is the two-dimensional density of rotational states and $\rho_t(E_{rel})$ is given by Eq.(9). When the transition state is chosen at some $r = r^{\ddagger}$, $W(E - V^{\ddagger})$ becomes*

$$W^{\ddagger}(E - V^{\ddagger}) = (\frac{1}{2\pi\hbar})^4 \int \delta[r^{\ddagger} - r]\delta[E - H] \frac{p_r}{\mu} drdp_r \cdot$$

$$\prod_{i=1}^{2} d\theta_i dp_{\theta i} d\phi_i dp_{\phi i} \cdot \qquad (21)$$

Combining Eqs.(19) - (21) and placing the ion on the z-axis gives

$$k_{TST}(E) = \frac{3\pi}{4I(2\mu E)^{\frac{3}{2}}} \int h(E - H^{\ddagger}) \sin\theta d\theta 2LdL2JdJ , \qquad (22)$$

where $\theta = \theta_2$, $h(E - H^{\ddagger})$ is the unit step function, $H^{\ddagger}$ is defined by

$$H^{\ddagger} = L^2/2\mu r^{\ddagger 2} + J^2/2I - \alpha q^2/2r^{\ddagger 4} - q\mu_D \cos\theta /r^{\ddagger 2} , \qquad (23)$$

and the integration is three-fold. The thermal rate constant is obtained by convoluting Eq.(22) with a five-dimensional Boltzmann energy distribution,

$$P(T,E) = \frac{4}{3\sqrt{\pi}} (\frac{1}{kT})^{\frac{5}{2}} E^{\frac{3}{2}} e^{-E/kT} . \qquad (24)$$

Two complimentary transition state theory analyses are presented in the remainder of this section. First, an adiabatic

---

*See footnote on page 3.

approximation, in which L and J are assumed to be independently conserved throughout the collision, is considered. Then, the variational method is applied under the opposite assumption that there is complete energy randomization between radial, orbital, and rotational motion during the collision.

A. Adiabatic Theory

When L and J are assumed to be conserved throughout the course of the collision the analysis used to obtain Eq.(14) can be applied to Eq.(15) giving

$$r^{\ddagger} = [\frac{\alpha q^2}{2(E - J^2/2I)}]^{\frac{1}{4}} . \tag{25}$$

This equation is plotted in Figure 1 for a thermal energy collision of H^- with HCN ($\alpha = 2.59$ Å^3, $\mu_D = 2.98$ Debye). Note that $r^{\ddagger}$ is independent of the product $\mu_D\cos\theta$. This is due to the fact that L^2 and $\mu_D\cos\theta$ both vary as $1/r^2$. Combining Eqs.(22), (23), and (25) under the assumption that $\dot{J} = 0$ gives

$$k_{ADI}(E) = \begin{cases} k_L[\frac{1}{2} + 3\sqrt{\varepsilon}/16 + 3/8\sqrt{\varepsilon}] , & 0 \leq \varepsilon \leq 1 \\ k_L[1 + 1/16 \, \varepsilon^{\frac{3}{2}}] , & 1 \leq \varepsilon \leq \infty \end{cases} \tag{26}$$

where the reduced energy

$$\varepsilon = 2\alpha E/\mu_D^2 , \tag{27}$$

has been introduced. Equation (26) can be convoluted with Eq.(24) to obtain $k_{ADI}(T)$. Note that if the variable change given by Eq.(27) is made, then $k_{ADI}(T)$ depends only on the Langevin rate constant, k_L, and on the reduced temperature T_R defined by

$$T_R = \frac{2\alpha kT}{\mu_D^2} . \tag{28}$$

The result of the convolution, expressed as a function of T_R, is

$$k_{ADI}(T) = k_L[1 + G(T_R)] , \tag{29a}$$

where

$$G(T_R) = \frac{1}{2\sqrt{\pi}} [\frac{1}{\sqrt{T_R}} + \sqrt{T_R}(1 - e^{-1/T_R}) - \sqrt{\pi} \, \text{erf}(1/\sqrt{T_R})] . \tag{29b}$$

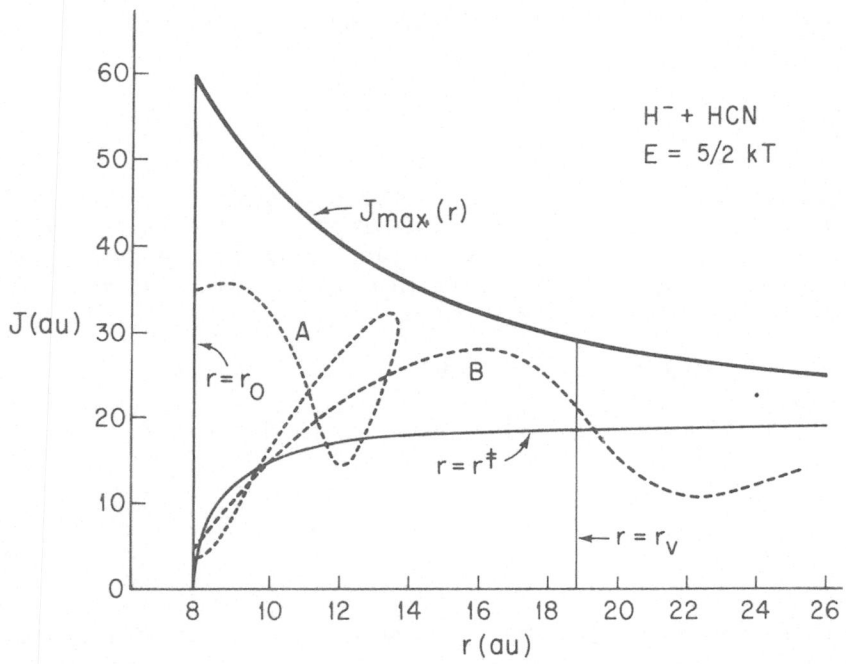

Figure 1. The rJ-plane: $J_{max}(r)$ is obtained by setting p_r and L to zero and $\cos\theta$ to one in Eq.(15): $r = r^{\ddagger}$ is defined by Eq.(25): r_v is the variational theory point of minimum flux: r_0 is the capture radius defined by Eq.(61). The two trajectories, one reactive and one nonreactive, show the large amount of energy transfer between orbital, rotational, and radial motion.

The ratio $k_{ADI}(T)/k_L$ is plotted in Figure 2 as a function of $1/\sqrt{T_R}$.

The adiabatic capture rate constant does not, in general, provide an upper bound to the true capture rate constant because the adiabatic analysis fails to account for the fact that the true rate at which trajectories cross the transition state surface defined by Eq.(25) depends not only on the relative velocity p_r/μ but also on the time rate of change of J, that is, on $\dot{J}$ (or in other words, on the rate of energy flow between orbital, rotational and radial motion). If the adiabatic assumption is relaxed for this surface in order to include the contribution due to $\dot{J}$, one finds that unless the ratio $\mu\mu_D/\alpha I$ is exactly zero the rate constant obtained diverges with respect to $r^{\ddagger}$ due to nonreactive trajectories that cross the surface in the vicinity of $r^{\ddagger} = \infty$. The true rate constant, however, does approach the adiabatic limit from above as T_R approaches infinity since under this condition both results approach the Langevin limit. The range of convergence depends on the ratio μ/I.

Dugan and Magee have proposed a "completely unhindered rotor" model (see, for example, Dugan, 1973, and references therein) which is formally equivalent to the adiabatic theory presented here; the only difference is that Dugan and Magee expressed their result in terms of E_{rel} rather than E. One can obtain Eq.(29) by convoluting $k_{ADI}(E_{rel})$ as given by Dugan and Magee with a three-dimensional thermal distribution in E_{rel}.

B. Variational Theory

In order to perform the variational analysis under the assumption that random energy transfer between radial, orbital, and rotational motion occurs throughout the collision, Eq.(22) must be integrated over L, J and θ before the variation is performed. This procedure gives the following results for the microcanonical capture rate constant;

$$k_{VAR}(E) = \begin{cases} k_L \dfrac{27}{1024\ \varepsilon^{\frac{3}{2}}} [8\varepsilon^2 + 20\varepsilon - 1 + (8\varepsilon + 1)^{\frac{3}{2}}], & 0 \le \varepsilon \le 5/9 \\[2ex] k_L \dfrac{1}{9\ \varepsilon^{\frac{3}{2}}} [2(9\varepsilon^2 + 3\varepsilon + 1)^{\frac{3}{2}} + (3\varepsilon + 2)\cdot \\[1ex] \qquad\qquad (6\varepsilon + 1)(3\varepsilon - 1)]^{\frac{1}{2}}. & 5/9 \le \varepsilon \le \infty \end{cases} \qquad (30)$$

It can be shown that in the limits $T_R \to 0$ and $T_R \to \infty$ that $k_{VAR}(T)$ becomes

$$\lim_{T_R \to 0} k_{VAR}(T) = k_L \frac{9[1 + 2 T_R]}{8\sqrt{\pi T_R}}\ , \qquad\qquad (31)$$

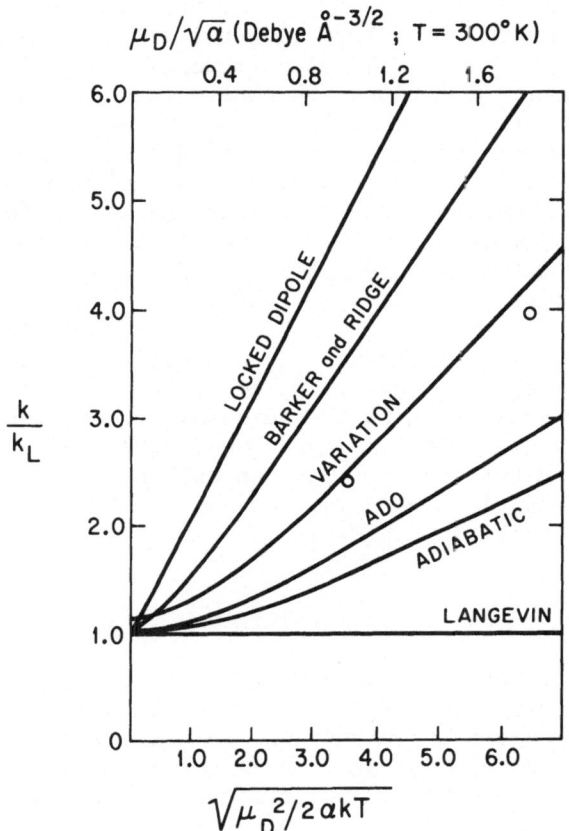

Figure 2. Plot of k/k_L versus $1/\sqrt{T_R}$ for the various theories summarized in the text. The open circles are the results of trajectory calculations of the true capture rate constant on the ion-dipole surface.

$$\lim_{T_R \to \infty} k_{VAR}(T) = \frac{k_L}{\sqrt{3}} [2 + \frac{1}{3T_R}] . \tag{32}$$

The thermal ratio $k_{VAR}(T)/k_L$ is plotted in Figure 2 as a function of $1/\sqrt{T_R}$.

The variational analysis fulfills both criteria necessary to produce a rigorous upper limit to the exact rate constant: every capture trajectory must pass through the transition state surface, and, since the surface is perpendicular to r and parallel to all other coordinates in the system phase space, the rate at which trajectories pass through is given correctly by p_r/μ.

Note that in the limit $T_R \to \infty$ (or $\mu_D \to 0$) the variational rate constant is a factor of $2/\sqrt{3}$ larger than k_L. From a dynamical viewpoint this factor arises because some nonreactive trajectories originating at $r = 0$ and $r = \infty$ pass through the variational surface before being reflected by V_{eff} back to their point of origin. From a thermodynamic viewpoint the factor arises because the variational theory assumes an equilibrium distribution of trajectories in a nonequilibrium situation.

It is well known that the accuracy of transition state theory increases at low energy (Pechukas and McLafferty, 1972; Chapman, et al., 1975; Miller, 1976; Pechukas, 1976). The fact that the reduced temperature that characterizes the variational theory rate constant is proportional to T/μ_D^2 suggests, therefore, that the accuracy of the variational result increases as μ_D approaches infinity. This point is discussed in greater detail in the section on ion-permanent dipole classical dynamics.

III.   THEORIES BASED ON SPHERICAL AVERAGING

Previous approaches to the ion-permanent dipole capture problem have been based on averaging the charge-permanent dipole potential to obtain an effective central potential. The corresponding effective Hamiltonian is given by

$$\bar{H}_{rel} = p_r^2/2\mu + L^2/2\mu r^2 - \alpha q^2/2r^4 - \frac{q\mu_D}{r^2} \overline{\cos\theta}(r) . \tag{33}$$

Since the average potential in $\bar{H}_{rel}$ is spherically symmetric the capture rate constant can be obtained from Eq.(33) by using either the dynamical or the transition state theory procedures of Section I. If the method used to obtain Eq.(13) is applied, the rate constant becomes

$$k_{\overline{\cos}}(E_{rel}) = \frac{2\pi}{\sqrt{2\mu E_{rel}}} r^{\ddagger 2}[E_{rel} + \frac{\alpha q^2}{2r^{\ddagger 4}} + \frac{q\mu_D}{r^{\ddagger 2}} \overline{\cos\theta}(r)] , \tag{34}$$

where, following the procedures used to obtain Eq.(14), the relation between $r^{\ddagger}$ and E_{rel} is given by

$$E_{rel} = \frac{\alpha q^2}{2r^{\ddagger 4}} - \frac{q\mu_D}{2r^{\ddagger}} \left.\frac{\partial \overline{\cos\theta}(r)}{\partial r}\right|_{r=r^{\ddagger}} . \tag{35}$$

Given $\overline{\cos\theta}(r)$ and its derivative, Eqs.(34) and (35) can be solved for $k_{\overline{\cos}}(E_{rel})$. The thermal analogue, $k_{\overline{\cos}}(T)$, is obtained by convoluting $k_{\overline{\cos}}(E_{rel})$ over a thermal distribution in E_{rel}.

Before discussing explicitly the various methods that have been used to evaluate $\overline{\cos\theta}(r)$ it is useful to first consider the partition function for a free-two-dimensional rigid rotor since the averaging methods are closely related to the calculation of this quantity. The Hamiltonian, from Eq.(17), is

$$H_{rot} = \frac{1}{2I} (p_\theta^2 + \frac{p_\phi^2}{\sin^2\theta}) . \tag{36}$$

The semiclassical partition function is (McQuarrie, 1976)

$$Q_{rot} = (\frac{1}{2\pi\hbar})^2 \int d\theta\, dp_\theta\, d\phi\, dp_\phi\, e^{-H_{rot}/kT} , \tag{37}$$

$$= \frac{kT}{B} , \tag{38}$$

where $B = \hbar^2/2I$ is the rotational constant of the rotor. Note that Eq.(38) is also the correct quantum partition function in the limit $kT > B$ (McQuarrie, 1976).

Consider now the calculation of $\overline{\cos\theta}(r)$. Two methods have been proposed in the literature to calculate this quantity, the ADO theory of Su and Bowers (1973) and the theory of Barker and Ridge (1976). Both theories begin with the equilibrium statistical mechanical distribution for a given orientation angle, given by the integrand of Eq.(37),

$$dP(\theta) \propto d\theta\, dp_\theta\, d\phi\, dp_\phi\, e^{-H_{rot}/kT} , \tag{39}$$

and proceed to evaluate this expression using specific dynamical assumptions. Once $P(\theta)$ is obtained the average value of $\cos\theta$ is given by

$$\overline{\cos\theta}(r) = \int \cos\theta\, P(\theta)\, d\theta / \int P(\theta)\, d\theta . \tag{40}$$

The details of the two theories are as follows.

A. The ADO Theory

The ADO theory is based on three main assumptions. The first assumption is that the potential energy released by the ion-permanent dipole interaction goes entirely into rotational motion. That is,

$$E_{rot} = T_{rot} - \frac{q\mu_D}{r^2} \cos\theta , \qquad (41)$$

where the kinetic energy, T_{rot}, is given by H_{rot} in Eq.(36). The second assumption is that E_{rot} is a constant of the motion. An alternate statement of this assumption is that there is no energy transfer between rotational and translational motion. The third assumption of the ADO theory is that E_ϕ, the rotational energy perpendicular to the plane of collision, given by

$$E_\phi = p_\perp^2/\sin^2\theta , \qquad (42)$$

is also a constant of the motion. Since E_{rot} and E_ϕ are assumed to be constants of the motion it follows that their difference, E_θ, given by

$$E_\theta = E_{rot} - E_\phi , \qquad (43)$$

$$= p_\theta^2(r)/2I - q\mu_D \cos\theta /r^2 , \qquad (44)$$

is also a constant of the motion (for notational purposes, Eq.(44) is identical to Eq.(20) of Su and Bowers, 1973). From Eq.(44) one sees, therefore, that the potential energy released by the ion-permanent dipole interaction goes entirely into p_θ, or since $p_\theta = I\dot\theta$, into $\dot\theta$.

Using the above assumptions, Eq.(39) can be evaluated by transferring from p_ϕ to E_ϕ using Eq.(42) and from p_θ to E_θ using Eq.(44). That is,

$$P(\theta) \propto \frac{\partial p_\theta}{\partial E_\theta} \frac{\partial p_\phi}{\partial E_\phi} , \qquad (45)$$

or

$$P(\theta) \propto \frac{\sin\theta}{\dot\theta} . \qquad (46)$$

Note that Eq.(46) is in microcanonical form. To evaluate $\overline{\cos\theta}(r)$ using Eq.(46), $\dot\theta$ is calculated from Eq.(44). Since at infinite separation $E_\theta = p_\theta^2(\infty)/2I$, and since it is assumed that E_θ is a constant of the motion, the thermal average of $\overline{\cos\theta}(r)$ can be obtained by convoluting $\overline{\cos\theta}(r)$ over a one-dimensional thermal rotational energy distribution in E_θ.

The ratio $k_{ADO}(T)/k_L$ is plotted in Figure 2 as a function of $\mu_D/\sqrt{\alpha}$ for $T = 300°K$. Although it has not been proven that this ratio is a function only of T_R there is reason to believe that this is, in fact, true.

It is of interest to note that the ADO capture rate constant lies closer to the adiabatic limit of transition state theory than to the upper bound given by variational theory. The reason for this is that the ADO assumptions that E_{rot}, E_ϕ and, by difference, E_θ are constants of the motion are adiabatic assumptions. The ADO theory may, therefore, be viewed as a semi-adiabatic theory.

Recently, the AADO theory (angular momentum concerned ADO theory) has been introduced by Su et al. (1978) as a modification of the ADO theory. In the AADO theory the second assumption of Su and Bowers has been modified somewhat by assuming that there is angular momentum transfer between p_θ and the ion. The orbital angular momentum now becomes a function of r, assumed to be given by

$$L(r) = \mu V_\infty b - C_L , \tag{47}$$

where C_L, the loss in orbital angular momentum suffered as a function of r, is assumed to be given by

$$C_L = p_\theta^2(r)/2I - \sqrt{IkT} . \tag{48}$$

The quantity $\sqrt{IkT}$ is obtained by noting that $p_\theta^2(\infty)/2I = kT/2$. These assumptions have the effect of increasing the rate constant ($k_{AADO}/k_{ADO} = 1.1$ to 1.3 for most systems) and bring ADO theory into better agreement with transition state theory.

B.   The Barker-Ridge Theory

The Barker-Ridge method of calculating $\overline{\cos\theta}(r)$ may be obtained by first integrating Eq.(39) over $dp_\theta dp_\phi$ at constant J giving

$$dP(\theta) \propto 2JdJ\sin\theta\,d\theta\ e^{-\dfrac{J^2}{2IkT} + \dfrac{q\mu_D}{r^2kT}\cos\theta} . \tag{49}$$

The integral over J produces the partition function given by Eq.(38) which cancels during normalization of $P(\theta)$. Therefore only the integrand for the configuration integral remains, giving

$$P(\theta) \propto \sin\theta \; e^{\dfrac{q\mu_D}{r^2 kT}\cos\theta} \; .$$ (50)

Averaging $\cos\theta$ using Eq.(50) produces the well-known result (see, for example, page 557 of Moore, 1962)

$$\overline{\cos\theta}(r) = \mathcal{L}\left(\frac{q\mu_D}{r^2 kT}\right) \; ,$$ (51)

where $\mathcal{L}$ is the Langevin function.

It is possible to show that the capture rate constant obtained by the Barker-Ridge approach is a function only of the reduced temperature T_R. This is accomplished by replacing r by the reduced distance R, defined by

$$R = \sqrt{\frac{\mu_D}{\alpha q}} \; r \; .$$ (52)

Inserting Eq.(52) into Eq.(33) indicates the Hamiltonian can be written in units of $\mu_D^2/2\alpha$, whereas inserting Eq.(52) into Eq.(51) indicates $\overline{\cos\theta}(R)$ is given by

$$\overline{\cos\theta}(R) = \mathcal{L}\left(\frac{2}{R^2 T_R}\right) \; .$$ (53)

The ratio $k_{BR}(T)/k_L$ is plotted in Figure 2 as a function of $1/\sqrt{T_R}$. It is of interest to note that the Barker-Ridge result is substantially higher than the upper bound given by variational theory. The reasons for this disparity are not yet well understood.

IV. CLASSICAL DYNAMICS OF ION-PERMANENT DIPOLE COLLISIONS

The classical dynamics of ion-permanent dipole collisions can be considered in either Lagrangian form (Dugan and Magee, 1967, and references therein) or Hamiltonian form. In Hamiltonian form the dynamics is governed by Hamilton's equations which, for the conjugate pair $\{q_i, p_i\}$, are

$$\frac{\partial H}{\partial p_i} = \dot{q}_i \; ; \qquad \frac{\partial H}{\partial q_i} = -\dot{p}_i \; .$$ (54)

Hamilton's equations can be expressed in a variety of coordinate systems. In the previous sections spherical polar coordinates proved to be most convenient to work with. In these coordinates the five conjugate pairs which arise due to the five center of

mass degrees of freedom are $\{r,p_r\}$, $\{\theta_i,p_{\theta i}\}$, and $\{\phi_i,p_{\phi i}\}$, where
$i = 1,2$. Other systems are preferable, however, when explicitly
solving the equations of motion. Dugan and Magee used cartesian
coordinates in their trajectory studies of ion-two dimemsional
rotor collisions and later used Euler angles to study ion-symmetric
top collisions. In this section an Euler angle coordinate system,
reduced in order, is used.

In the Euler angle coordinate system the five sets of Hamilton's
equations (54) are reduced to three sets by using conservation of
the magnitude and direction of the total system angular momentum, $\mathscr{J}$
(see, for example, Miller, 1970). The Hamiltonian is

$$H = p_r^2/2\mu + L^2/2\mu r^2 + J^2/2I - \alpha q^2/2r^4 - \frac{q\mu_D}{r^2}\cos\theta , \qquad (55)$$

where

$$\cos\theta = -[\frac{\mathscr{J}^2 - L^2 - J^2}{2LJ}\sin\gamma_1\sin\gamma_2 + \cos\gamma_1\cos\gamma_2] . \qquad (56)$$

In solving Eqs.(54) and (56) the conjugate pairs are $\{r,p_r\}$, $\{\gamma_1,L\}$
and $\{\gamma_2,J\}$. The angles γ_1 and γ_2 describe the orientations of the
ion and dipole in the instantaneous planes of rotation corresponding
to L and J. That is, if $\mu_D = 0$, then $L = \mu r^2\dot{\gamma}_1$ and $J = I\dot{\gamma}_2$. Note
that by using the conservation law

$$\mathscr{J} = \underline{L} + \underline{J} , \qquad (57)$$

to calculate $\mathscr{J} \cdot \mathscr{J}$ one sees that the term

$$-\frac{\mathscr{J}^2 - L^2 - J^2}{2LJ}$$

is the cosine of the instantaneous angle formed by $\underline{L}$ and $\underline{J}$. This
term appears in the potential energy because of the directional
properties of angular momenta.

If all trajectories are initiated at some fixed value of r the
thermal version of the true capture rate constant is

$$k_{CLD}(T) = \frac{1}{2IkT(2\pi\mu kT)^{\frac{3}{2}}} \int \chi(\underline{p},\underline{q})d\gamma_1 \, d\gamma_2 \, dLdJ2\mathscr{J}d\mathscr{J}e^{-E/kT}dE, \qquad (58)$$

where the characteristic function $\chi(\underline{p},\underline{q})$ labels the trajectory
begun with initial conditions $\{\underline{p},\underline{q}\}$ on r as reactive or nonreactive
(see, for example, Miller, 1974). The exact value of χ for a
given trajectory depends on where the trajectories are started;
however, the properties of χ make the rate constant independent of
the starting position.

In the integration of Eq.(58) the nominal bounds on the angles are from zero to 2π, E runs from zero to infinity, and the boundaries on L, J and $\mathcal{J}$ are given by Eq.(57). The integration is restricted by the fact that

$$E - H(p_r = 0) \geq 0 , \tag{59}$$

where $H(p_r = 0)$ is the Hamiltonian evaluated on the initial r with p_r set equal to zero. It is possible to further restrict the upper bound of $\mathcal{J}$ by using variational techniques. By noting that the maximum value of $\mathcal{J}$ according to Eq.(57) is given by L + J, and by using standard techniques (Chesnavich and Bowers, 1977), one can show that

$$\mathcal{J}^2_{max}(r) = (2I + 2\mu r^2) \left(E + \frac{\alpha q^2}{2r^4} + \frac{q\mu_D}{r^2} \right) , \tag{60}$$

where $\mathcal{J}_{max}(r)$ is the maximum value of $\mathcal{J}$ that a trajectory can have as a function of r and E. By setting $\partial\mathcal{J}_{max}(r)/\partial r = 0$, solving for r, and inserting the result back into Eq.(60), one obtains $\mathcal{J}_{max}(E)$, the maximum value of $\mathcal{J}$ that a trajectory with total energy E can have and still pass from $r = \infty$ to $r = 0$.

In the calculation of the true capture rate constant Hamilton's equations must be solved to determine the capture trajectories; that is, the trajectories that originate at $r = \infty$ and reach $r = 0$. This task can be accomplished numerically by first specifying a set of conditions that all capture trajectories must fulfill at some point in time and then solving Hamilton's equations to determine which trajectories fulfill these conditions. For example, Dugan and Magee suggested that any trajectory reaching $r = 2\text{Å}$ must ultimately reach $r = 0$. Although the condition suggested by Dugan and Magee must be fulfilled by all trajectories reaching $r = 0$ from $r = 2\text{Å}$ it is not a sufficient condition since a given trajectory may reach $r = 2\text{Å}$ but still ultimately reflect off the effective potential and retreat to $r = \infty$.

A necessary and sufficient condition that a trajectory must fulfill in order to reach $r = 0$ for the ion-permanent dipole system can be obtained by considering the conditions for which $\dot{p}_r = -\partial H/\partial r$ must be less than zero. It can be shown that if a trajectory with total energy E reaches, with $p_r < 0$, the "capture radius" defined by

$$r_o = \left(\frac{\alpha q^2}{2E} \right)^{\frac{1}{4}} , \tag{61}$$

the trajectory must from that point on accelerate to $r = 0$. Physically the capture radius r_o arises because at small r the charge-induced dipole attractive potential always dominates both the centrifugal potential and the repulsive orientations of the charge-permanent dipole potential. The capture radius does not depend on the dipole moment because both the centrifugal and charge-permanent dipole terms of the Hamiltonian go as $1/r^2$.

One must also specify conditions that a trajectory must fulfill in order to reach $r = \infty$. It is possible, by combining the procedures used to determine r_0 with standard techniques (Chesnavich and Bowers, 1977), to show that if a trajectory with total energy E and total angular momentum $\mathcal{J}$ reaches, with $p_r > 0$, some escape radius $r = r_\infty$ the trajectory must ultimately reach $r = \infty$. Although some r_∞ less than infinity exists and can be calculated for certain values of E and $\mathcal{J}$ it has not been possible to prove that it exists for all E and $\mathcal{J}$.

Numerical calculations of the thermal capture rate constant have been performed for model systems corresponding to $H^- + HCN$ ($\alpha = 2.59 \overset{\circ}{A}^3$, $\mu_D = 2.98$ Debye) and $H_3^+ + NH_3$ ($\alpha = 2.26\ \overset{\circ}{A}^3$, $\mu_D = 1.47$ Debye). The equations of motion were solved using a fourth-order Runge-Kutta procedure and k(T) was obtained by integrating Eq.(58) using a Monte Carlo routine. The trajectories were started on r_0 given by Eq.(61) and were integrated outward to $r = 30$ atomic units. If a trajectory returned to $r = r_0$ the integration was terminated and $\chi = 0$ for that trajectory. If a trajectory reached $r = 30$ atomic units the integration was terminated and $\chi = 1$ for that trajectory. 500 trajectories were run for each system. Two trajectories for $H^- + HCN$ are plotted in Figure 1. The resulting thermal rate constants are plotted in Figure 2. By comparing the exact transition state theory result for r_0 with the transition state theory result obtained from the Monte Carlo integration (i.e. $\chi = 1$ for all trajectories) the error in the numerical results plotted in Figure 2 was estimated to be $\sim 10\%$.

The agreement between the numerical rate constants and the upper bound obtained from the variational transition state theory method is quite satisfactory. It should be kept in mind that the numerical results do not confirm the statement that variational theory provides a rigorous upper bound to the true rate constant. The truth of this statement follows from the definition of a rate constant and from the fundamental assumption of transition state theory. Rather, the numerical results determine the magnitude of the difference between the true rate constant and the variational upper bound.

In Section II it was mentioned that if the variational theory rate constant approaches the true result as T approaches zero it will also approach the true result as μ_D approaches infinity because it is a function only of the reduced temperature T_R. Equations (60) and (61) support, although they do not directly prove, the suggestion that the variational result does approach

the true result as T approaches zero. From Eq.(60) one sees that
when E = 0 then $\mathscr{J}_{max}(r)$ is a monotonically decreasing function of
r. That is, when E = 0 the available phase space expands as r
decreases from infinity opening new $\mathscr{J}$ states. Also, as E → 0 the
capture radius r_c, given by Eq.(61), approaches infinity. To
actually prove this suggestion however, requires a rigorous determination, in the limit E → 0, of the boundaries on the initial
conditions that lead to capture. It may be possible to determine
these boundaries by locating, in the limit E → 0, the conditions
that lead to orbiting (Pollack and Pechukas, 1978). Orbiting
trajectories do exist for the charge-induced dipole, charge-
permanent dipole potential. However, their exact specification
is still a matter of active investigation.

V. COMPARISON WITH EXPERIMENTAL DATA

    Table I compares the experimental values of the total proton
transfer rate constants for the systems H_3^+ + NH_3 and H^- + HCN with
the predictions of the theories discussed in the previous sections.
The locked dipole result, obtained by setting cosθ = 1, is

$$k_{LD} = k_L \left[1 + 2/\sqrt{\pi T_R}\right] .$$
(62)

The highest ratios of experiment to theory occur with the adiabatic
theories. The fact that the ratios are greater than unity suggests
that energy transfer dominates the long range dynamics of these
proton transfer reactions. This suggestion is supported by the
trajectory calculations reported in Section IV which agree quite
well with the variational theory prediction. Both the variational
theory and trajectory calculations predict that the experimental
proton transfer rates of these two reactions go at about eighty to
ninety percent of the capture rate.

    In Figure 3 the variational rate constant is compared to a
number of experimental measurements of thermal proton transfer
rate constants. It should be kept in mind that the variational
theory gives an upper bound to the true capture rate constant for
a model that is based on solution of the classical equations of
motion for the Hamiltonian given in Eq.(15). The agreement is
quite satisfactory and suggests that variational theory adequately
describes the major features of the true long range dynamics of the
model. The few experimental points that exceed predictions of the
variational theory could reflect either experimental error or the
necessity of including additional terms in the potential function.

    Experimental measurements of the temperature dependence of
proton transfer reactions can provide an interesting test of the
variational theory. Possible candidates are the proton transfer

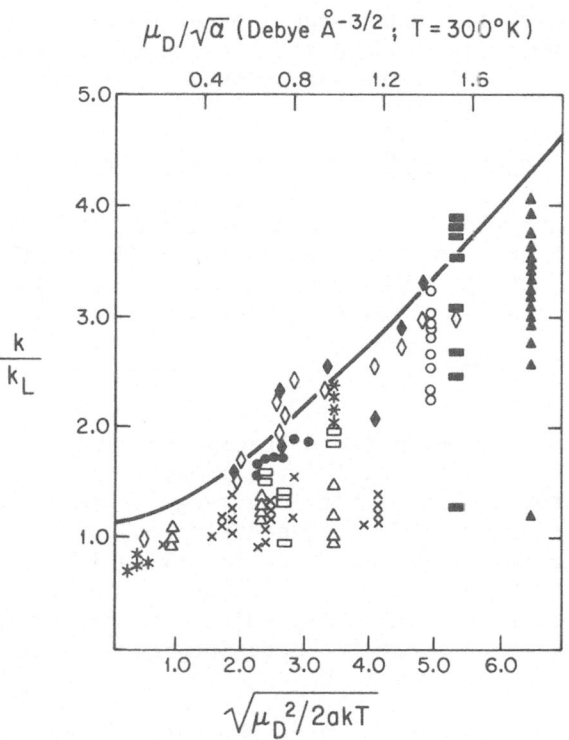

Figure 3. Comparison of k_{VAR}/k_L to k_{EXP}/k_L. ●, Δ, □, X - ICR, Santa Barbara; *, ▲, ■, ◇, ◆ - Flowing Afterglow, York; ○ - High Pressure Mass Spectrometer, Paris.

Table I. Capture Collision Rate Constants

	$H_3^+ + NH_3$	$H^- + HCN$
$T_R(x\ 10^{-2})$	8.62	2.42
k_{EXP}*	4.2	15.0
k_{EXP}/k_L	1.94	3.93
k_{EXP}/k_{LD}	.40	.48
k_{EXP}/k_{BR}	.55	.65
k_{EXP}/k_{AADO}	.93	1.14
k_{EXP}/k_{ADO}	1.06	1.39
k_{EXP}/k_{ADI}	1.25	1.67
k_{EXP}/k_{VAR}	.79	.91
k_{EXP}/k_{CLD}	.80	1.01

---

*Flowing Afterglow - York; x 10^9 cm^3/sec, T = 300°K.

reactions to HCN and CH_3CN. For these systems a temperature range of 200-600°K corresponds to $7.88 \geq 1/\sqrt{T_R} \geq 4.55$. In this region the ratio k_{VAR}/k_L is linear in $1/\sqrt{T_R}$ and, from Eq.(31), has a slope $\sim 9/8\sqrt{\pi}$ with respect to $1/\sqrt{T_R}$. Experimental verification of this prediction relies on the assumption that reaction efficiencies are only slowly varying functions of temperature.

VI. CONCLUSIONS

In this paper a detailed transition state theory analysis of ion-permanent dipole capture collisions was used to obtain an upper bound to the true capture rate constant and to obtain pertinent information on the dynamics of the true capture process. Comparison with preliminary trajectory calculations and with experiment suggests that the upper bound to the capture rate constant obtained from variational theory lies reasonably close to the true result.

There are a number of approaches one can take to establish more precisely the magnitude of the difference between $k_{VAR}(T)$ and $k_{CLD}(T)$. One approach is to perform detailed trajectory calculations of the true capture rate constant. Large savings in computer time can be obtained by beginning the trajectories on the minimum

flux value of r obtained from variational theory since this starting
point samples the minimum total number of trajectories that include
all capture trajectories. However, before meaningful trajectory
calculations can be performed a rigorous criterion for the escape
radius r_∞ must be established which is valid under all conditions.
A second approach is to use transition state theory arguments to
search for surfaces in the system phase space that all trajectories
truly do cross once and only once. These surfaces do exist (Wigner,
1937) and as noted by Pechukas (1976) are defined by the sets of
trajectories that lead to orbiting. If these surfaces can be located then the true capture rate constant can be obtained without
performing expensive trajectory calculations. Both of these
approaches are currently under investigation.

VII.   ACKNOWLEDGEMENT

    We gratefully acknowledge the support of the National Science
Foundation under grant CHE77-15449 and to the Donors of the Petroleum Research Fund, administered by the American Chemical Society,
for support of this research.

REFERENCES

Barker, R. A. and Ridge, D. P. (1976). J. Chem. Phys. 64, 4411.

Bowers, M. T. and Su, T. (1975). Theory of Ion-Polar Molecule
     Collisions in "Interactions Between Ions and Molecules" (Ed.,
     P. Ausloos), Plenum Press, New York.
Chapman, S., Hornstein, S. M. and Miller, W. H. (1975). J. Am.
     Chem. Soc. 97, 892.
Chesnavich, W. J. and Bowers, M. T. (1977). J. Chem. Phys. 66,
     2306.
Dugan, J. V. (1973). Chem. Phys. Lett. 21, 476.
Dugan, J. V. and Magee, J. L. (1967). J. Chem. Phys. 47, 3103.
Eyring, H. (1935). J. Chem. Phys. 3, 107.
Eyring, H. Hirschfelder, J. O. and Taylor, H. S. (1936). J. Chem.
     Phys. 4, 479.
Gioumousis, G. and Stevenson, D. P. (1958). J. Chem. Phys. 29, 294.
Goldstein, H. (1950). Classical Mechanics, Addison-Wesley, Reading,
     Mass.
Keck, J. C. (1967). Adv. Chem. Phys. 13, 85.
McQuarrie, D. A. (1976). Statistical Mechanics, Harper and Row,
     New York.
Miller, W. H. (1970). J. Chem. Phys. 53, 1949.
Miller, W. H. (1974). J. Chem. Phys. 61, 1823.
Miller, W. H. (1976). Accts. Chem. Res. 9, 306.
Moore, W. J. (1962). Physical Chemistry, Prentice-Hall, Englewood
     Cliffs, New Jersey.

Pechukas, P. (1976). "Statistical Approximations in Collision
    Theory" in <u>Dynamics of Molecular Collisions</u> (Ed., W. H. Miller),
    Plenum Press, New York.
Pechukas, P. and McLafferty, F. J. (1972). J. Chem. Phys. 58, 1622.
Pollack, E. and Pechukas, P. (1978). J. Chem. Phys. 69, 1218.
Su, T. and Bowers, M. T. (1973). J. Chem. Phys. 58, 3027.
Su, T., Su, E. C. F. and Bowers, M. T. (1978). J. Chem. Phys. 69,
    2243.
Vogt, E. and Wannier, H. (1954). Phys. Rev. 95, 1190.
Wigner, E. (1937). J. Chem. Phys. 5, 720.

ION MOLECULE COLLISIONS: THEORY AND EXPERIMENT —

SUMMARY OF PANEL DISCUSSION

Douglas P. Ridge

Department of Chemistry, University of Delaware

Newark, Delaware 19711

Douglas P. Ridge: COMMENTS ON INTERMOLECULAR POTENTIALS FOR
POLYATOMIC IONS AND MOLECULES

Over the years the study of collision processes in nonreactive systems has been very useful in characterizing intermolecular
forces. Much of the available information on the importance of
various kinds of forces between neutral molecules, for example,
is the result of studies of diffusion and viscosity in nonreactive
systems (Hirschfelder, Curtiss and Bird, 1954).

Similar information on the forces between ions and molecules
is available from the study of ion mobilities in nonreactive
gases. The importance of the ion induced dipole force and the
relative importance of other forces has been quantitatively
assessed in a variety of systems from such studies (McDaniel and
Mason, 1972). We discuss here what may be learned about intermolecular forces between polyatomic ions and molecules from
studies of mobilities in nonreactive systems.

The drift velocity $\underline{v}_d$ of an ion in a gas in a field $\underline{E}$ is
given by (1). The mobility, K, is related to the momentum

$$\underline{v}_d = K\underline{E} \tag{1}$$

transfer collision frequency, ξ, by (2) and ξ is defined in terms

$$K = \frac{e}{m\xi} \tag{2}$$

55

of the velocity dependent diffusion cross section, $\sigma_d(v)$, by (3)

$$\xi = n \frac{M}{m+M} \langle \sigma_d(v)v \rangle \tag{3}$$

In (3) n is the neutral number density, M is the mass of the neutral, m is the mass of the ion, v is the relative velocity of the ion-neutral pair and the brackets indicate an average over a Boltzmann velocity distribution. If the intermolecular potential is known, then $\sigma_d(v)$ and hence ξ and K may be calculated rigorously. The methods for performing such calculations are well established and described in the literature (McDaniel and Mason, 1972). Langevin calculated K for an $\infty-4$ potential, i.e., a potential that is infinite inside a hard sphere core and attractive and varying with r^{-4} outside the core (Langevin, 1905). Mason and coworkers have calculated K for n-6-4 potentials where n = 8, 12 or 16, i.e., potentials with attractive terms varying as r^{-6} and r^{-4} and a repulsive term varying as r^{-n} where n = 8, 12 or 16 (Viehland, Mason, Morrison, and Flaherty, 1975). Mason, O'Hare and Smith (1972) have calculated K for an accentric potential, i.e., a potential with an attractive term varying as $(r-a)^{-4}$ and a repulsive term varying as $(r-a)^{-12}$. At sufficiently low temperatures ξ and K are dominated by the r^{-4} term in the potential and approach limiting values, $(\xi/n)_p$ and K_p, given by (4) and (5) where α is the neutral polarizability and m_r is

$$(\xi/n)_p = \frac{M}{m+M} 2.210 \ \pi e \left(\frac{\alpha}{m_r}\right)^{1/2} = \frac{M}{m+M} 1.105 \ k_p \tag{4}$$

$$K_p = \frac{1}{n \ 2.210\pi(\alpha m_r)^{1/2}} \tag{5}$$

reduced mass of the ion-neutral pair. We note that in (4) that $(\xi/n)_p$ is simply related to k_p, the familiar orbiting collision rate frequently compared with reaction rate constants. In the ensuing discussion we present K and ξ/n in terms of their ratios to K_p and $(\xi/n)_p$. In Figures 1, 2 and 3 mobilities calculated from an 8-6-4 potential are compared with several mobilities measured over a range of temperatures. The measured mobilities are taken from ion cyclotron resonance linewidth measurements by Buttrill (1973) and Bowers (Bowers, et al., 1977). The theory relating ion cyclotron resonance linewidths to mobilities is well established (Beauchamp, 1967 and Viehland, Mason and Whealton, 1975) as is the agreement between the icr linewidth measurements and drift tube measurements (Ridge and Beauchamp, 1976).

The coefficient of the r^{-4} term in the 8-6-4 potential can be set at $-\alpha e^2/2$, since the ion induced dipole interaction has a

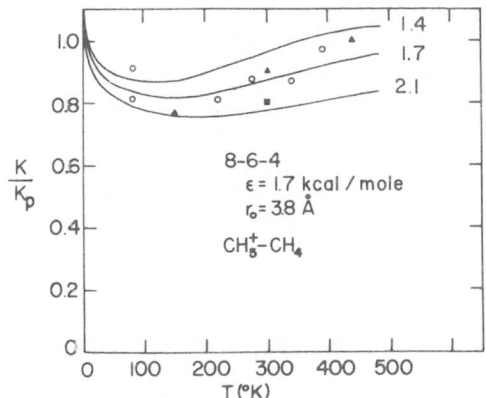

Figure 1. Measured mobilities of CH_5^+ in CH_4 at various temperatures compared with mobilities calculated from an 8-6-4 potential (Viehland, Mason, Morrison and Flannery, 1975) with $r_0 = 3.8$ Å ($V(r_0) = 0$) and well depth $\varepsilon = 1.4$, 1.7 and 2.1 kcal/mole. The open circles are data from Bowers, et al., 1977, the closed triangles are data from Buttrill, 1973, and the closed square is from Ridge and Beauchamp, 1976. The lines represent the calculated values.

firm theoretical basis. That leaves the coefficients of the r^{-8} and r^{-6} terms to be determined by comparing calculated mobilities with experimental ones. In practice an ion-molecule radius, r_0, and a well depth, ε, are chosen. r_0 is defined by the condition that $V(r_0) = 0$. A systematic search reveals the values of r_0 and ε which give calculated mobilities in best agreement with experiment. Comparisons between the measured mobilities and those calculated from the 8-6-4 potential will be the subject of a forthcoming publication (Weddle and Ridge, to be published). The following comments are appropriate, however: (1) Other potentials such as 12-4, 8-4 or 12-6-4 potentials cannot account for the observed mobilities illustrated in Figures 1, 2 and 3 with any reasonable choice of parameters. (2) In particular, no potential that does not include a significant attractive term in addition to the ion-induced dipole term can come close to accounting for the observed mobilities in polyatomic systems (Patterson, 1972, Dymerski, Dunbar and Dugan, 1974, and Ridge and Beauchamp, 1976). In polyatomic systems intermolecular forces depend on many inter-

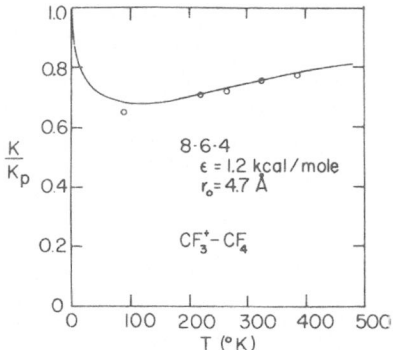

Figure 2. Measured mobilities of CF_3^+ in CF_4 at various temperatures compared with mobilities calculated from an 8-6-4 potential
(Viehland, Mason, Morrison and Flannery, 1975) with r_o = 4.7 A
($V(r_o)$ = 0) and well depth ε = 1.2 kcal/mole. The data are from
Bowers, et al., 1977 and thé line represents the calculated
values.

atomic distances, not just the distance between the two centers of
mass of the interacting molecules. The r^{-6} term compensates for
the failure of a potential dependent only on r to account for the
polyatomic structures of the interacting particles. (See Ridge
and Beauchamp, 1976, for a similar argument.) (3) The potential
parameters obtained are consistent with other data on intermolecular forces. The Lennard-Jones radii of CH_4 and CF_4, 3.8 Å
and 4.7 Å (Hirschfelder, Curtiss and Bird, 1954) respectively,
compare well with the r_o values obtained for CH_5^+-CH_4 and CF_3^+-CF_4.
The well depths are in line with what is known about binding
energies in such species as CH_5^+-methane and $C_2H_5^+$-methane clusters (Hiroaka and Kebarle, 1975). (4) Although the 8-6-4
potential is evidently an improvement on simpler representations,
it is still only an empirically derived approximation to the true
potential surface. The true potential energy depends not only on
r, for example, but also on the structure of the interacting
particles and their precise relative orientation.

Polar Molecules. In cases where the neutral molecule has a
permanent dipole moment a rigorous calculation of the mobility is
extremely complex. An approximation that we have suggested is to
average the ion-dipole energy over a Boltzmann distribution to
obtain a temperature dependent central potential. Orbiting

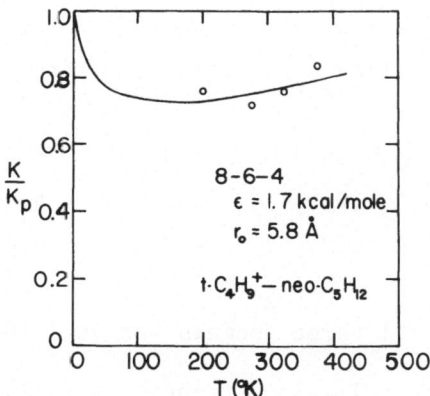

Figure 3. The measured mobilities of t-$C_4H_9^+$ in neo-C_5H_{12} at
various temperatures compared with mobilities calculated from an
8-6-4 potential (Viehland, Mason, Morrison, and Flannery, 1975)
with r_0 = 5.8 Å ($V(r_0)$ = 0) and well depth ε = 1.7 kcal/mole. The
data are from Bowers, et al., 1977, and the line represents the
calculated values.

collision frequencies calculated from this potential agree quite
well with measured momentum transfer collision frequencies of ions
with polar molecules (Barker and Ridge, 1976). While orbiting
collision frequencies approximate momentum transfer collision
frequencies in many simple systems, it would be more satisfactory
to be able to calculate momentum transfer collision frequencies
for polar systems. This may be accomplished by expanding
Boltzmann averaged energy in a power series as indicated in (6)

$$< -\frac{\mu e}{r^2} \cos\theta > = -\frac{\mu e}{r^2} \mathcal{L} \left(\frac{\mu e}{r^2 kT}\right) \cong -\frac{\mu^2 e^2}{3kT} \left(\frac{1}{r}\right)^4 + \frac{\mu^4 e^4}{45k^3 T^3} \left(\frac{1}{r}\right)^8$$

(6)

where μ is the neutral dipole moment, e is the electron charge, θ
is the angle between the dipole and the direction of the approaching charge, the brackets indicate an average over a Boltzmann
distribution, $\mathcal{L}$ (x) is the Langevin function ($\mathcal{L}$ (x) = cotanh (x) −
x^{-1}), and T is the rotational temperature of the neutral. Since

Table I. Momentum Transfer Collision Frequencies for
Dichloroethylenes with Cl^-

	Dipole Moment[a] (D)	$\dfrac{k_{BR}}{(\xi/n)_{exp}}$[b]	r_0[c] (Å)	ε_0[c] (kcal)	$\dfrac{(\xi/n)8\text{-}6\text{-}4}{(\xi/n)_{exp}}$[d]
trans	0.0	.65	4.3	6.0	.95
1,1	1.34	1.04	4.3	6.0	1.07
cis	1.90	1.22	4.3	6.0	1.02

a. Polarizability of all three isomers set at 7.78 Å^3 (Su and Bowers, 1974).

b. k_{BR} is the orbiting collision frequency calculated from the Boltzmann averaged energy (Barker and Ridge, 1976). The experimental momentum transfer collision frequencies $((\xi/n)_{exp})$ are from Su and Bowers, 1974.

c. r_0 and ε_0 are the radius and well depth of the potential in the absence of any ion dipole term.

d. $(\xi/n)8\text{-}6\text{-}4$ is calculated from an 8-6-4 potential with radius r_0 and well depth ε to which is added the r^{-8} and r^{-4} dipole terms in Equation (6). The calculations were done using tabulated collision integrals from Viehland, Mason, Morrison and Flaherty, 1975. The values of $(\xi/n)_{exp}$ come from Su and Bowers, 1974.

the leading terms of the power series depend on r^{-4} and r^{-8} they may be added to the 8-6-4 potential and momentum transfer collision frequencies and mobilities obtained from the available collision integrals for that potential. The results of such calculations for the momentum transfer collision frequencies of Cl^- and the two sets of structural isomers are compared with the experimental collision frequencies (Su and Bowers, 1974) in Tables I and II. Within each set of isomers the potential parameters are held constant except for the dipole moment. The agreement under these circumstances is quite satisfactory, indicating that this is a useful approach to ion polar neutral momentum transfer collision frequencies and mobilities.

Orbiting Cross Sections. The mobility results suggest that a rather large attractive r^{-6} term should be included in the interaction potentials for ion-polyatomic neutral systems. It is of interest to see how such a term in the potential affects the calculated orbiting collision frequencies. The orbiting collision frequencies for the potential given in (7) are plotted in Figure 4. The results are given in terms of the reduced dimensionless

$$V(r) = -\frac{\alpha e^2}{2r^4} - \frac{B}{r^6} - \frac{\mu e}{r^2} \mathcal{L}(\frac{\mu e}{r^2 kT}) \tag{7}$$

parameters P, defined on the figure, and S, defined by (8). The

$$S = B(\frac{kT}{2\alpha^3 e^6})^{1/2} \tag{8}$$

values of B obtained from mobilities in polyatomic systems give values of S at 300°K as large as 1.0. From the Figure it is evident that the r^{-6} term can have a large effect on orbiting cross sections for nonpolar molecules. The more polar the molecule, however, the less important the r^{-6} term.

Table II. Momentum Transfer Collision Frequencies for Difluorobenzenes with Cl⁻

	Dipole Moment[a] (D)	$\dfrac{k_{BR}}{(\xi/n)_{exp}}$ [b]	r_o [c] (Å)	ε_o [c] (kcal)	$\dfrac{(\xi/n)8\text{-}6\text{-}4}{(\xi/n)_{exp}}$ [d]
para	0.0	.57	5.0	6.5	.99
meta	1.58	.95	5.0	6.5	1.07
ortho	2.40	1.01	5.0	6.5	.93

---

a. Polarizability of all three isomers set at 9.77 Å³ (Su and Bowers, 1974).

b. k_{BR} is the orbiting collision frequency calculated from the Boltzmann averaged energy (Barker and Ridge, 1976). The experimental momentum transfer collision frequencies (($\xi/n)_{exp}$) are from Su and Bowers, 1974.

c. r_o and ε_o are the radius and well depth of the potential in the absence of any ion dipole term.

d. $(\xi/n)_{8\text{-}6\text{-}4}$ is calculated from an 8-6-4 potential with radius r_o and well depth ε to which is added the r^{-8} and r^{-4} dipole terms in Equation (6). The calculations were done using tabulated collision integrals from Viehland, Mason, Morrison and Flaherty, 1975. The values of $(\xi/n)_{exp}$ come from Su and Bowers, 1974.

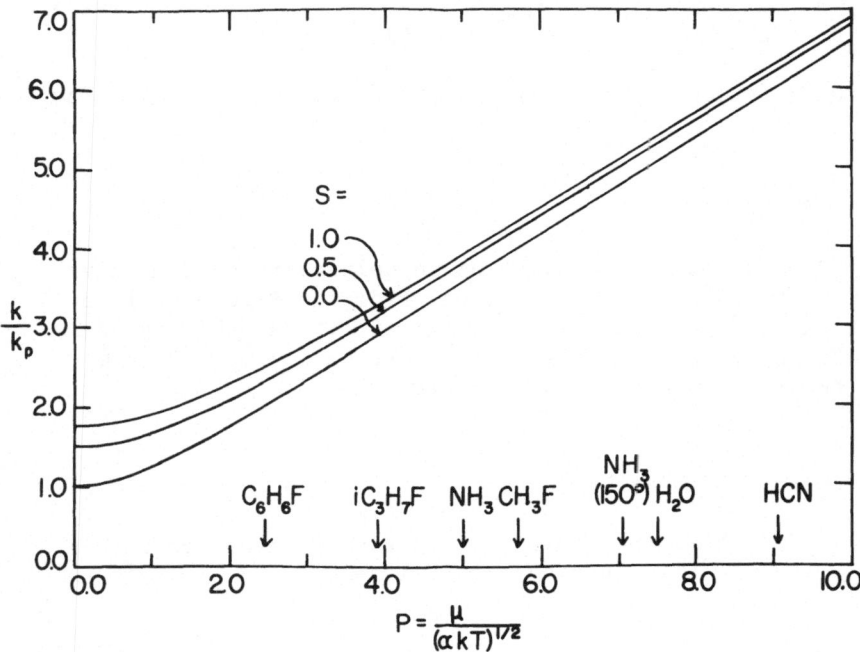

Figure 4.   Orbiting collision frequencies calculated for the potential in Equation (7) as a function of polarity parameter P defined on the figure in terms of molecular dipole moment μ, molecular polarizability α and the temperature.   The parameter S increases with the coefficient of the r^{-6} term in the potential and is defined in Equation (8).   The k in kT is Boltzmann's constant.

Danon and Amdur (1969) have suggested a somewhat different method for "sphericalizing" a non-isotropic potential.   They use a statistical mechanical average free energy rather than a statistical mechanical average energy.   Such a free energy in the case of the ion dipole potential is given in (9) where the integrations are over all angles.   The k in kT is Boltzmann's constant.

$$F = -kT \ln q = -kT \ln \left[\frac{\int e^{\mu e \cos\theta / r^2 kT} \, d\Omega}{\int d\Omega} \right] =$$

$$-kT \ln \left(\sinh\left(\frac{\mu e}{r^2 kT} \right) \right) + kT \ln\left(\frac{\mu e}{r^2 kT} \right) \tag{9}$$

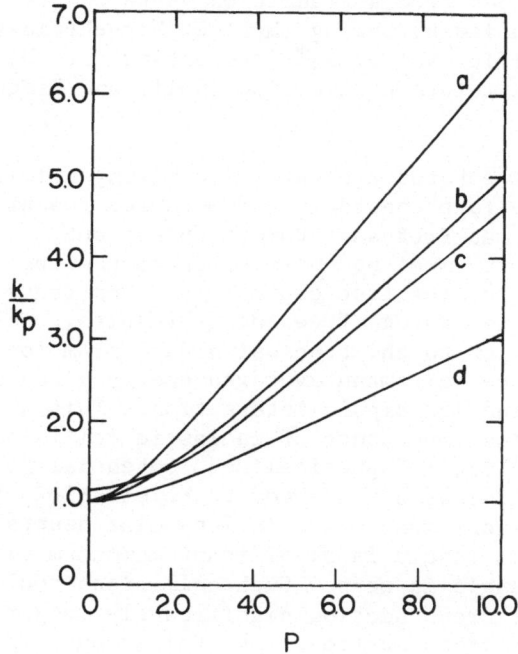

Figure 5. Orbiting or capture collision frequencies for ions
with polar molecules calculated from (a) the Boltzmann averaged
energy (Barker and Ridge, 1976), (b) the average free energy
(see Equation (9)), (c) the transition state theory introduced
in the present volume by Bowers and (d) ADO theory (Su and
Bowers, 1973). The parameter P is defined in Figure 4. Values of P
for the transition state theory were obtained from $\mu/\alpha^{1/2}$ by
setting T = 300°K in the expression for P.

Orbiting collision frequencies calculated using F as a representation of the ion dipole interaction are compared with those determined from the Boltzmann averaged energy (Barker and Ridge, 1972) from ADO theory (Su and Bowers, 1973) and from the new transition state theory (Bowers, present volume) introduced by Professor Bowers in his presentation in Figure 5. The agreement between the free energy approach and the transition state theory approach is noteworthy. It can be shown that the average free energy result can be obtained from a transition state theory approach similar to that applied by Eyring (Eyring, Hirschfelder and Taylor, 1936) for the specific case of H_2^+ interacting with H_2. This will be the subject of a future publication (Celli and Ridge, to be published).

Comparison between the theoretical orbiting collision frequencies or capture rate constants and measured reaction rate constants suggests, as Professor Bowers points out, that the true capture rate constants must be quite close to the transition state theory results. Hence the true capture rate constants must also be quite close to the average free energy results. This is in apparent contradiction to the conclusion made from ion mobility measurements that the Boltzmann averaged energy gives a good representation of the ion dipole interaction. This discrepancy may be related to the importance of inelastic collisions in momentum transfer. In using a "sphericalized" potential to calculate mobilities all collisions are assumed to be elastic. Yet rotational inelastic collisions must occur in ion-polar neutral systems. Such collisions will result in significant momentum transfer even if the scattering angle is zero. Such collisions could result in a momentum transfer cross section significantly larger than the orbiting or capture cross section. The Boltzmann averaged energy may thus give collision frequencies larger than the capture collision frequencies but just large enough to account for effects of inelastic collisions on momentum transfer. The effect of inelastic collisions on momentum transfer in polyatomic gases is the subject of a number of papers in the literature (see McDaniel and Mason, 1972).

In summary three points can be made about capture collision frequencies: (1) The ion induced dipole term may not be the only important term in the potential. An r^{-6} term, for example, may significantly effect the capture collision frequency. (2) The effect of additional terms in the potential diminishes as the polarity of the neutral increases. (3) While the Boltzmann averaged potential seems to give the best momentum transfer collision frequencies, the transition state theory introduced at the meeting by Professor Bowers (or the quite similar average free energy theory described above) seems to give the best capture collision frequencies for polar molecules.

W. Ronald Gentry: COMMENTS ON THE GIOUMOUSIS STEVENSON REACTION
MODEL[*]

The GS model (Gioumousis-Stevenson, 1958) divides trajectories into two categories for an attractive long range potential.
Trajectories with impact parameters less than the orbiting impact
parameter b_o will overcome the centrifugal barrier and penetrate
the region of the potential where shorter range "chemical forces"
become important.   Trajectories with impact parameters less than
b_o do not lead to such close approach encounters.   If the long
range potential is the ion induced dipole potential then the GS
reaction cross section, $\sigma(E)$, is given by (1) where E is the

$$\sigma(E) = P\pi b_o^2 = P\pi e \left(\frac{2\alpha}{E}\right)^{1/2} \tag{1}$$

initial kinetic energy before the collision, α is the neutral
polarizability and P is the average probability of reaction for
collisions at impact parameters between 0 and b_o.   The reaction
rate constant corresponding to this reaction cross section is
given by (2) where m_r is the reduced mass of the ion neutral pair.

$$k = P\ 2\pi e \left(\frac{\alpha}{m_r^2}\right)^{1/2}$$

This model is so appealingly simple that it is often applied
under circumstances where conditions of its validity are not met.
Henchman has given a thorough critique of such abuses which
leads him to state ". . . it can be argued that the field of ion-
molecule rate studies would gain much and lose little if the model
were to be quietly interred. . . " (Henchman, 1972).   Certainly
systems to which it may be applied quantitatively must meet a
variety of conditions.   Specifically, for the GS model to be
quantitatively accurate, the following conditions must be satisfied: (a) Classical mechanics must provide an accurate description of the radial translational motion in the vicinity of the
classical orbiting radius, r_o,   (b) the potential energy near r_o
must be given accurately by the charge induced dipole potential,
(c) the probability of reaction for impact parametric $b > b_o$ must
be zero, and   (d) the probability of reaction averaged over b for
$b < b_o$ must be constant (i.e. P must be independent of E).   By way
of illustration we consider these requirements individually as
applied to several ion atom reactions we have studied using the
merged beam technique.

_____

[*]As summarized by D. P. Ridge who relied on Professor Gentry's
published work, especially McClure, Douglass and Gentry (1977a),
to help refresh his memory of Professor Gentry's comments.

(a)   The GS model is classical and therefore assumes a
continuum of impact parameters or orbital angular momenta.   This
will be a good approximation if the orbital angular momentum
corresponding to the orbiting condition is very much greater than
$\hbar$.   This will generally be true for ion molecule systems at low
energies.   For the $D_2^+ + N$ system, for example, at E = .01 eV the
orbital angular momentum corresponding to the orbiting condition
is ~30 h.   It can be shown that the quantum orbiting cross
section deviates from the classical cross section by an average
of less than 1% under these conditions.

(b)   We have shown that it is necessary in the case of the
$D_2^+ + N$ reaction to modify the basic GS model to include contributions to the potential of the type $-B\ r^{-6}$ (McClure, Douglass
and Gentry, 1977a).   The effect of this contribution to the
potential drops off at lower energies.   It may be neglected at
energies of .01 eV in the $D_2^+ + N$ system without serious error.
It is worthwhile noting that only atoms in S states have perfectly
spherical potentials.   In all other cases the anisotropy of the
potential must be considered.   This is true even of atoms in states
other than S states.   In the case of the $D_2^+ + O(^3P)$ reaction the
oxygen atom has a permanent quadrupole moment characteristic of
atoms in P states.   This gives rise to an r^{-3} term in the potential
(Gentry and Giese, 1977).   The complications which arise in
considering the effect of molecular quadrupole moments on ion
molecule collision trajectories because of the rotational motion
of the molecule are absent in this case.   To the extent that the
Born-Oppenheimer approximation is valid the symmetry axis of the
atom always coincides with the ion-atom axis.   The effects of
this very long range r^{-3} interaction are clearly evident in the
behavior of the reaction cross section at very low energies
(McClure, Douglass and Gentry, 1977b).

(c)   There is no evidence that the low energy ion-atom proton
transfer reactions we have examined involve long range processes.
Such processes would not be energetically feasible.

(d)   It is, of course, possible to discuss measured cross
sections and reaction rates in terms of Equations (1) and (2),
whether or not P depends on E.   In the case that P in (1)
depends on E, then the value of P in (2) will be in general
different from that in (1) and the relationship between the two
values of P will not be simple.   In this case equations (1) and
(2) are primarily useful as a qualitative guide in understanding
the reaction of interest.   In general condition (d) is satisfied
only if reaction probability does not depend on kinetic energy
or the orbital angular momentum of the collision.   It would be
surprising if very low kinetic energies had much effect on
reaction probability for very exothermic reactions.   It is less
clear that effects of orbital angular momentum can be neglected,

since the rotational angular momentum of the reactants is relatively small. The collisional orbital angular momentum will, in fact, impose constraints on some systems such as those in which there is a large change in reduced mass between reactants and products. The $D_2^+ + N$ system is not one to which such circumstances apply. The experimental evidence strongly suggests that P is independent of E up to about E = 1 eV (McClure, Douglass and Gentry, 1977a). Hence the GS model applies to this system.

REFERENCES

Barker, R. A. and Ridge, D. P. (1976), J. Chem. Phys., 64, 4411.
Beauchamp, J. L. (1967), J. Chem. Phys., 46, 1231.
Bowers, M. T. (1979), "Ion Molecule Collisions: Theory and Experiment," this volume.
Bowers, M. T., Neilson, P. V., Kemper, P. R. and Wren, A. G. (1977), Int. J. Mass Spec. and Ion Phys., 25, 103.
Buttrill, S. E., Jr. (1973), J. Chem. Phys., 58, 656.
Celli, F. and Ridge, D. P., to be published.
Danon, F. and Amdur, I. (1969), J. Chem. Phys., 30, 11.
Dymerski, P. P., Dunbar, R. C. and Dugan, J. V. (1974), J. Chem. Phys., 61, 298.
Eyring, H., Hirschfelder, J. O. and Taylor, H. S. (1936), J. Chem. Phys., 4, 479.
Gentry, W. R. and Giese, C. F. (1977), J. Chem. Phys., 67, 2355.
Gioumousis, G. and Stevenson, D. P. (1958), J. Chem. Phys., 29, 294.
Henchman, M. (1972), "Rate Constants and Cross Sections," in Ion Molecule Reactions, Franklin, J. L., Ed., Plenum, New York.
Hiroaka, K. and Kebarle, P. (1975), J. Chem. Phys., 62, 2267.
Hirschfelder, J. O., Curtiss, C. F. and Bird, R. B. (1954), Molecular Theory of Gases and Liquids, Wiley, New York.
Langevin, P. (1905), Ann. Chim. Phys. Ser. 8, 5, 245. Translated in E. W. McDaniel, Collision Phenomena in Ionized Gases (Wiley, New York, 1964), Appendix II.
Mason, E. A., O'Hara, H. and Smith, F. J. (1972), J. Phys. B5, 169.
McClure, D. J., Douglass, C. H. and Gentry, W. R. (1977a), J. Chem. Phys., 66, 2079.
McClure, D. J., Douglass, C. H. and Gentry, W. R. (1977b), J. Chem. Phys., 67, 2362.
McDaniel, E. W. and Mason, E. A. (1973), The Mobility of Ions In Gases, Wiley, New York.
Patterson, P. L. (1972), J. Chem. Phys., 56, 3943.
Ridge, D. P. and Beauchamp, J. L. (1976), J. Chem. Phys., 64, 2735.
Su, T. and Bowers, M. T. (1973), Int. J. Mass Spec. and Ion Phys., 12, 347.

Su, T. and Bowers, M. T. (1974), J. Chem. Phys., 60, 4897.
Viehland, R. A., Mason, E. A., Morrison, W. F., and Flannery, M. R.
    (1975), At. Data and Nuc. Data Tables, 16, 495.
Viehland, R. A., Mason, E. A., and Whealton, J. H. (1975), J. Chem.
    Phys., 62, 4715.
Weddle, G. H. and Ridge, D. P., to be published.

ION-MOLECULE COLLISION COMPLEXES*:

SUMMARY OF THE PANEL DISCUSSION

Cornelius E. Klots

Chemical Physics Section, Health and Safety Research
Division, Oak Ridge National Laboratory,
Oak Ridge, Tennessee 37830

INTRODUCTION

In 1957, Franklin, Field, and Lampe (1) described the ion-molecule reactions which occur in ethylene and acetylene. They noted and commented on the remarkable similarity between, for example, the products of the reaction of $C_2H_4^+$ with C_2H_4 and the fragments produced by the electron-induced ionization of several C_4H_8 isomers. The implication of a common $C_4H_8^+$ intermediate was clear.

In retrospect this work must be recognized as comprising the first "chemical activation" experiments. In the hands of others, the chemical activation technique has made enormous contributions to chemical kinetics. Yet the implied liason between ion-molecule chemistry and unimolecular kinetics has not received comparable attention. It was fitting therefore that a workshop at this conference should be addressed to this subject.

What follows is a summary of the ideas which emerged at this workshop. Many people contributed to this. Especially to be noted are the roles of Professors Fred Lampe and Michael Bowers in leading the discussion of the following two topics.

---

*Research sponsored by the Division of Biomedical and Environmental Research, U.S. Department of Energy under contract W-7405-eng-26 with the Union Carbide Corporation.

DIRECT OBSERVATION OF COLLISION CCMPLEXES

The occurrence of chain-like polymerization reactions and of three-body combination reactions provides compelling, if indirect, evidence that collision complexes may persist long enough to undergo further collisions. Much less common is the direct observation of the collision complex itself. Most of the extant examples involve ion-molecule collisions. In 1959, for example, Pottie and Hamill (2) reported a number of "persistent" collision complexes, exemplary of which is the $(C_2H_5I)_2^+$ ion. Its itensity was proportional to the square of the pressure of the ethyl iodide.

We have collected in Table I a number of examples. It should be recognized that some of these are controversial. Of especial importance then is the recent work of Lampe and his co-workers (12-14) in which care is taken to ensure that the observed complexes are indeed the result of a single "persistent" collision. And, as Lampe showed at this conference, one now can extract directly the lifetime of these complexes as a function of the center-of-mass collision energy.

TABLE I. DIRECT OBSERVATIONS OF COLLISION COMPLEXES

Association Reaction	Reference
$RX^+ + RX$	2, 3
$C_6H_5^+ + C_6H_6$	4, 5
$C_3H_3N^+ + C_3H_3N$	4
$C_2N_2^+ + C_2N_2$	6
$C_2H_4^+ + C_2H_4$	7
$HCO_2^- + HCO_2H$	8
$C_6H_6^+ + C_6H_6$	9, 10
$O^- + C_2H_2$	11
$c-C_5H_{10}^+ + c-C_5H_{10}$	36
$2-C_4H_8^+ + 2-C_4H_8$	37
$SiH_3^+ + C_2H_4$	12, 13
$SiH_3^+, SiH^+, Si^+ + C_6H_6$	14

We have not included in this table the many negative ions
formed by temporary electron-attachment to a molecule; SF_6^- is the
classic example (15). While their exclusion might seem plausible
to the chemist, it should be recognized that the theory of their
lifetime is identical with that pertaining to collision complexes
of a more "chemical" nature.

It is natural to ask why some collision complexes are sufficiently persistent to be observable, while others are not. We
shall see below that theory can offer some guidelines. A common
supposition is that the complex should have a large number of
degrees-of-freedom, that its empirical formula should correspond
to a plausible chemical entity, and that there should exist no
highly exothermic reaction channels. It is by no means obvious
that all the examples in Table I satisfy these criteria.

UNIMOLECULAR REACTIONS OF COLLISION COMPLEXES

These direct observations of collision complexes are of
interest, among other reasons, because of the credibility they
lend to an ancient idea - that the rate of reaction is compounded
from two factors; the rate of formation of a collision complex
and the fraction of these complexes which decomposes to a given set
of products. Implicit in this view is the idea that the collision-
complex "forgets" how it was formed. Its fragmentation into the
several available pathways is then merely a matter of competing
unimolecular reactions; and that is the justification for this
workshop.

It should be emphasized that the direct observation of a
collision complex is by no means the only criterion suggesting the
applicability of unimolecular reaction theory. Isotopic scrambling
and forward-backward symmetry of the reactive scattering in the
center-of-mass coordinate system are also often interpreted as
indicating a "collision-complex." But as Michael Henchman and
Ronald Gentry noted in this workshop, these criteria can be contradictory, ambiguous and misleading. Thus the reaction

$$H_2^+ + H_2 \rightarrow H_3^+ + H \tag{1}$$

exhibits forward-backward symmetry, at low energies, as well as
evidence for isotopic scrambling. Nevertheless, any supposition
of a "collision complex" would be quite incorrect. The symmetry
apparently arises only because of rapid charge exchange. For
example, the reaction

$$D_2^+ + HD \rightarrow D_3^+ + H \tag{2}$$

fails to exhibit this symmetry (16). One might rationalize this result by noting that the H_4^+ ion probably has no bound states. But how then does one understand the remarkable observation that the reaction:

$$H_3^+ + D \rightarrow H_2D^+ + H$$

does exhibit forward-backward symmetry (17)?

If an operational definition of collision complexes is thus by no means simple, we may nevertheless examine what insights into their time-evolution unimolecular reaction theory can offer. Now, the "breakdown pattern" of molecular ions, as a function of energy is a venerable subject. There is, however, an important property of collision complexes which complicates the straightforward application of standard unimolecular fragmentation theory. Let us suppose that that the rate of formation of the collision complex is described by the standard orbiting theory (18,19). It may then be shown that the complexes are formed with a broad range of angular momenta, given by

$$0 \leq J_0^2 \leq (2\mu/m) \; (\alpha\epsilon/a_o^3 \; R)^{1/2} \tag{3}$$

where μ is the reduced-mass of the collision, m is the electron mass, α is the neutral polarizability, a_o the Bohr radius, ϵ the center-of-mass collision energy, and R the Rydberg unit of energy. Thus, in addition to the total energy, it is necessary to establish what effect this angular momentum will have on the decay kinetics of the collision complex.

The usual formulation of unimolecular kinetics, the so-called RRKM formulation, does not lend itself readily to a careful consideration of angular momentum. An alternative formalism, in which the unimolecular reaction is connected via microscopic reversibility to the reaction leading to complex formation, has proven more useful. This approach has come to be known as phase-space theory (20,21).

It has been shown that, when the reactions leading to complex-formation are treated using the standard orbiting model, quite simple expressions can be obtained for unimolecular rate constants and for the kinetic energy spectrum of the reaction products (22-24). Chesnavich and Bowers have examined these topics more rigorously (25) and have further considered such matters as the effect of vibrational energy on ion-molecule rate constants and the threshold behavior of endoergic reactions (26,27).

To illustrate the results which such a treatment can yield, we consider the archetypal example of collision-complex formation:

$$C_2H_4^+ + C_2H_4 \rightarrow C_4H_8^+ \qquad (4)$$

at a center-of-mass collision energy of 2.5 eV. The dominant decay mode is via the reaction:

$$C_4H_8^+ \rightarrow C_3H_5^+ + CH_3 \quad . \qquad (5)$$

Figure 1 illustrates the predicted (28) kinetic energy distribution of the fragments at several of the angular momenta embraced by Eq. (3). These distributions are not (at least as yet) individually observable. What is observable is the average of these distributions over the full range of angular momenta. Good agreement between experiment and theory has been found (23).

As a second example, we consider the competition between reaction (5) and the alternate channel:

$$C_4H_8^+ \rightarrow C_4H_7^+ + H \qquad (6)$$

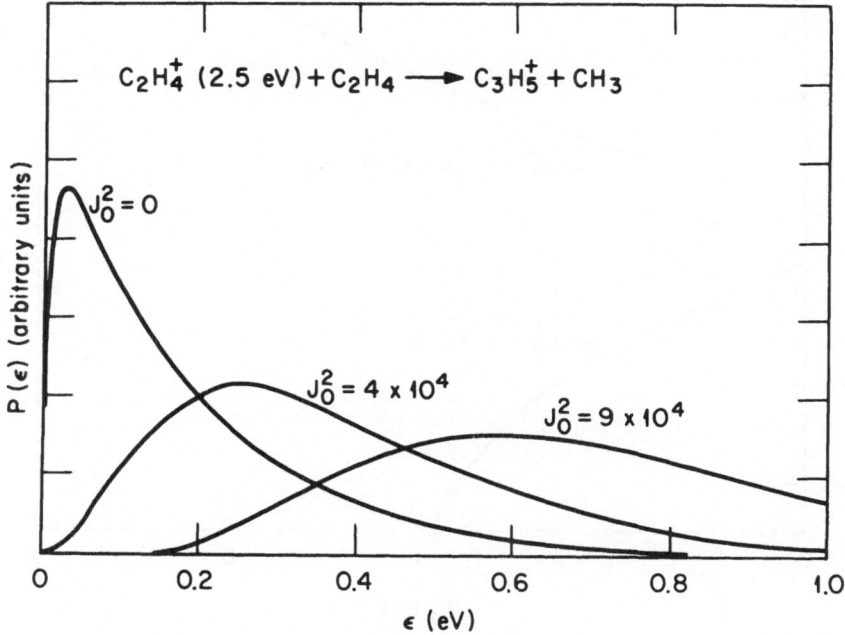

Fig. 1. Predicted kinetic energy spectra for reaction (5) at constant energy but with varying angular momentum.

where the $C_4H_8^+$ has again been formed via reaction (4) at 2.5 eV collision energy. Figure 2 illustrates the rate constants (in arbitrary units) for each channel as a function of angular momentum (28). While the results are fairly sensitive to such parameters as the assumed moments-of-inertia of the final products, it is clear that one cannot assign a unique lifetime or branching ratio to a collision-complex. Any observable will again be an average over the rotational ensemble.

The dramatic increase in rate constant with increasing angular momentum, for reaction (5), has a simple origin. It reflects the decreasing effective well-depth of the complex and so its decreasing density-of-states. Thus unimolecular theory provides a natural explanation for the tendency of ion-molecule reactions to switch from a "complex formation" mechanism to a "direct" mechanism with increasing collision energy. Whether this is the correct explanation for the switch is not clear.

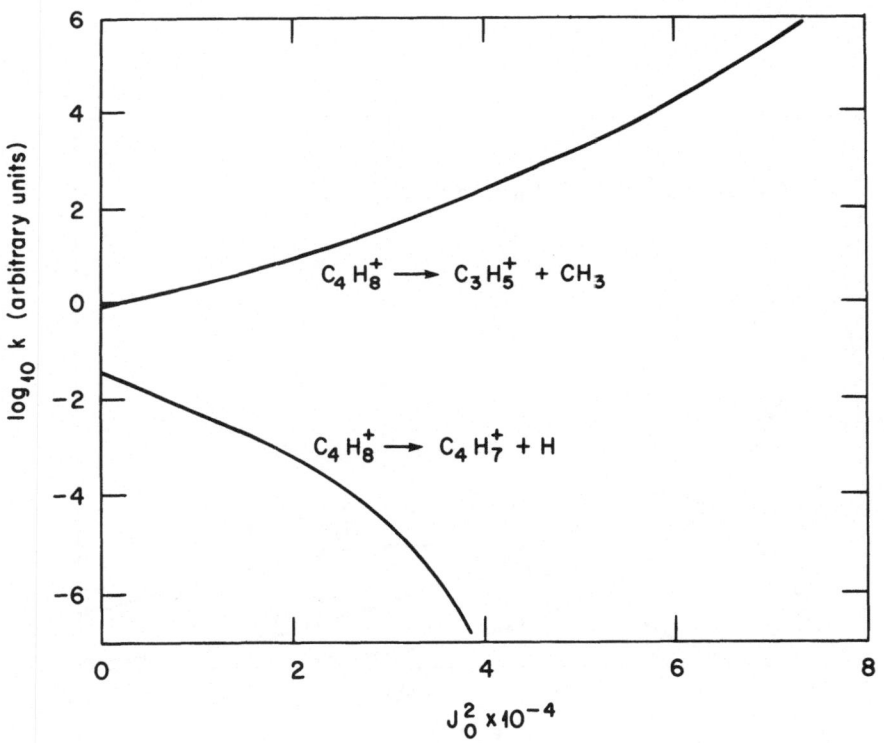

Fig. 2. Decay constants (arbitrary units) at constant energy, as functions of angular momentum, for reactions (5) and (6).

The results of Fig. 2, although model dependent, illustrate what should again be a general result. Increasing angular momentum should tend to mitigate against C-H bond cleavages, compared with C-C bond ruptures, at constant energy. Thus the $C_4H_7^+/C_3H_5^+$ ratio should be smaller in an ion-molecule reaction than following the direct ionization of one of the C_4H_8 isomers, provided the total energy is the same. Experimental evidence directly confirming this prediction was in fact presented by Gerry Meisels at this workshop (29).

DIRECT PREPARATION OF COLLISION COMPLEXES

To those who are interested in unimolecular reactions, collision complexes are unhappily complicated. We have seen how they necessarily comprise a continuum of angular momenta, and thus have no unique lifetime or branching ratio.

There is a second limitation to collision complexes, as customarily generated. Suppose that an exothermic ion-molecule reaction occurs via complex formation; the energy of this complex is necessarily above the threshold for the intended decay channel. The energy regime in which this reaction can be studied is accordingly restricted.

A simple example will illustrate this restriction. The branching ratio for the reactions:

$$H_2^+ + D_2 \rightarrow H_2D^+ + D \tag{7a}$$

$$\rightarrow HD_2^+ + H \tag{7b}$$

is of much interest (16), yet it cannot be obtained closer to threshold than the 1.8 eV exothermicity.

Two methods have been described which avoid these difficulties. Baer and his co-workers have described a photo-ion, photo-electron coincidence technique whereby stable molecules such as C_4H_6 and C_4H_8 may be photoionized, and the decay of their ions studied at well-defined energies (30,31). The relation between these experiments and ion-molecule beam studies has been pointed out (24). Dunbar's method of monochromatic excitation of selected molecular ions (32) constitutes an alternative method for obtaining much the same information.

Nevertheless these techniques do not offer a general method for the direct preparation of collision complexes. They require a starting material of the same empirical formula as the intended complex. But someone interested in reaction (7) will recognize that H_2D_2 is not commercially available.

   It has been realized for some time that the rapid expansion
of a gas through a small nozzle can lead to adiabatic cooling and
to extensive formation of van der Waals polymers of gas. The
reason for this is illustrated in Fig. 3. The boundary between
gas and the condensed state is represented by a Clausius-Clapeyron
line. An isentropic expansion, beginning at some point in the
gas phase, will eventually approach the condensed phase if it
proceeds far enough.

   These van der Waals polymers, either upon ionization or
electron attachment, can undergo what we have called "ion-
molecule half-reactions" (33). They constitute the fragmentation
of collision complexes into ion-molecule reaction channels, but with
the difference that these complexes have been directly prepared,
without the ambiguity or restrictions associated with the first
half of the collision. For example, ionization of an ethylene
dimer has been found (33) to give the reaction:

$$(C_2H_4)_2^+ \rightarrow C_3H_5^+ + CH_3 \tag{8}$$

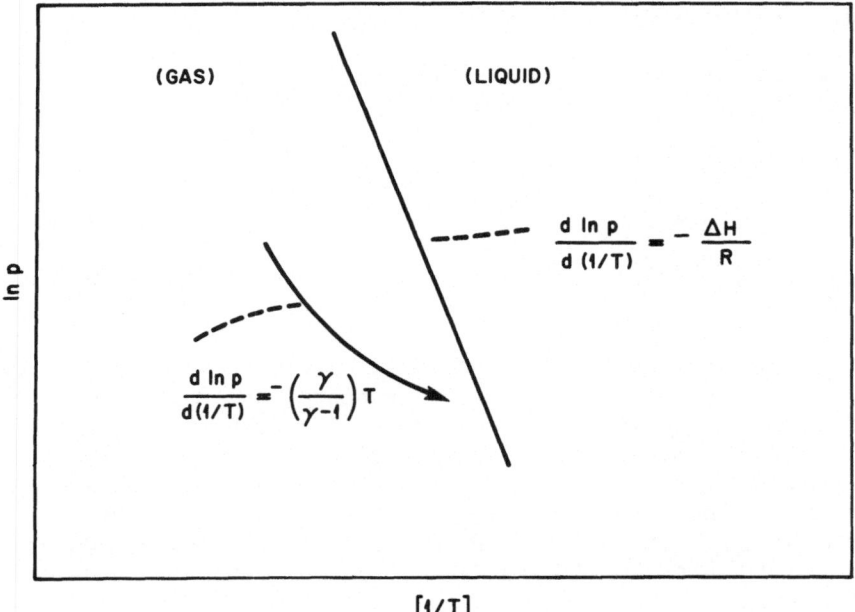

Fig. 3. P-T phase diagram and a gas phase isentrope.

with an appearance potential at the thermochemical threshold.
Likewise, electron attachment to water dimers, at selected energies,
is followed (33) by the reaction:

$$H^-(H_2O) \rightarrow OH^- + H_2 \quad . \tag{9}$$

Reactions (8) and (9) correspond to the second-half of well-
established ion-molecule reactions.

Let us consider one more example. The reaction:

$$CH_4^+ + CH_4 \rightarrow CH_5^+ + CH_3 \tag{10}$$

is a classic. The expansion of CD_4 through a sonic nozzle yields
both the $(CD_4)_2^+$ and the CD_5^+ ions (28), again implicating an ion-
molecule half reaction. But how would the reaction of $SiH_4^+ + SiH_4$
go? It has never been studied because the SiH_4^+ ion itself has
never been isolated.

In an attempt to identify the "half-reaction" analog the
sonic-nozzle expansion of silane was studied (28). Disappointingly,
no ions with the formulas $(SiH_4)_n^+$ or $(SiH_4)_nH^+$ were observed. Thus
neither the ion-molecule half reaction:

$$(SiH_4)_2^+ \rightarrow SiH_5^+ + SiH_3 \tag{11}$$

nor the evaporative stabilization:

$$(SiH_4)_2^+ \rightarrow SiH_4^+ + SiH_4 \tag{12}$$

could be detected.

Instead, we do observe the ions $Si_2H_4^+$ and $Si_2H_5^+$. From their
development as a function of stagnation pressure, these ions
clearly originate with van der Waals dimers generated in the nozzle
expansion. The appearance potentials of the $Si_2H_4^+$ and $Si_2H_5^+$ ions,
as measured with a resolution of ± 0.1 eV, are equal to those of
SiH_2 and SiH_3, respectively, from the silane monomer. We suggest
then that ionization of the silane dimer yields

$$e + (SiH_4)_2 \rightarrow SiH_2^+(SiH_4) + H_2 \tag{13a}$$

$$\rightarrow SiH_3^+(SiH_4) + H \tag{14a}$$

exactly paralleling the lowest energy fragmentations of the silane
monomers. These are apparently then followed by the "half-
reactions":

$$SiH_2^+(SiH_4) \rightarrow Si_2H_4^+ + H_2 \tag{13b}$$

$$SiH_3^+(SiH_4) \rightarrow Si_2H_5^+ + H_2 \tag{14b}$$

analogous to well-established (34) whole reactions. Thus if the original aims of this study were thwarted, we still have some further examples of ion-molecule half-reactions.

Mahan, Lee, and their co-workers (35) have shown how the photoionization of van der Waals molecules can be an important source of thermodynamic information. We suggest that, in conjunction with the photo-ion, photo-electron coincidence technique, they also offer a direct and general route for the preparation of collision complexes, and hence a useful tool for the investigation of ion-molecule dynamics.

REFERENCES

1. F. H. Field, J. L. Franklin, and F. W. Lampe, *J. Am. Chem. Soc.* 79, 2419 (1957); ibid. p. 2665.

2. R. F. Pottie and W. H. Hamill, *J. Phys. Chem.* 63, 877 (1959).

3. L. P. Thread and W. H. Hamill, *J. Am. Chem. Soc.* 84, 1134 (1962).

4. A. Henglein, *Z. Naturforsch.* 17a, 44 (1962).

5. C. Lifshitz and B. G. Reuben, *J. Chem. Phys.* 50, 951 (1969).

6. A. Henglein, G. Jacobs, and G. A. Muccini, *Z. Naturforsch.* 18a, 98 (1963).

7. I. Szabo, *Ark. Fys.* 33, 57 (1966).

8. C. E. Melton, G. A. Ropp, and T. W. Martin, *J. Phys. Chem.* 64, 1577 (1960); but see J.A.D. Stockdale, R. N. Compton, and P. W. Reinhardt, *Phys. Rev.* 184, 81 (1969).

9. J.A.D. Stockdale, *J. Chem. Phys.* 58, 3881 (1973).

10. E. G. Jones, A. K. Bhattacharya, and T. A. Tiernan, *Int. J. Mass Spectrom. Ion Phys.* 17, 147 (1975).

11. J.A.D. Stockdale, R. N. Compton, and P. W. Reinhardt, *Int. J. Mass Spectrom. Ion. Phys.* 4, 401 (1970).

12. T. M. Mayer and F. W. Lampe, *J. Phys. Chem.* 78, 2433 (1974).

13. W. N. Allen and F. W. Lampe, *J. Am. Chem. Soc.* 99, 6816 (1977).

14. W. N. Allen and F. W. Lampe, *J. Chem. Phys.* 65, 3378 (1976);
     J. Am. Chem. Soc. 99, 2943 (1977).

15. W. M. Hickman and R. E. Fox, *J. Chem. Phys.* 25, 642 (1956).

16. C. H. Douglass, D. J. McClure, and W. R. Gentry, *J. Chem. Phys.*
     67, 4931 (1977).

17. C. H. Douglass and W. R. Gentry (to be published).

18. H. Eyring, J. O. Hirschfelder, and H. S. Taylor, *J. Chem. Phys.*
     4, 479 (1936).

19. B. Mahan, *J. Chem. Phys.* 32, 362 (1960).

20. P. Pechukas and J. C. Light, *J. Chem. Phys.* 42, 3281 (1965);
     J. C. Light, *Discuss. Faraday Soc.* 44, 14 (1967).

21. E. E. Nikitin, *Theor. Exp. Chem.* 1, 144 (1965).

22. C. E. Klots, *J. Phys. Chem.* 75, 1526 (1971); *Z. Naturforsch.*
     27a, 553 (1972); *J. Chem. Phys.* 58, 5364 (1973).

23. C. E. Klots, *J. Chem. Phys.* 64, 4269 (1976).

24. C. E. Klots, D. Mintz, and T. Baer, *J. Chem. Phys.* 66, 5100
     (1977).

25. W. J. Chesnavich and M. T. Bowers, *J. Am. Chem. Soc.* 98, 8301
     (1976); *J. Chem. Phys.* 66, 2306 (1977); *J. Am. Chem. Soc.* 99,
     1705 (1977).

26. W. J. Chesnavich and M. T. Bowers, *Chem. Phys. Lett.* 52, 179
     (1977).

27. W. J. Chesnavich and M. T. Bower, *J. Chem. Phys.* 68, 901 (1978).

28. C. E. Klots, unpublished results.

29. G. G. Meisels, G.M.L. Verboom, and M. Weiss (to be published).

30. A. S. Werner and T. S. Baer, *J. Chem. Phys.* 62, 2900 (1975).

31. T. S. Baer, and D. Smith, *Adv. Mass Spectrom.* 7 (1978).

32. M. Riggin, R. Orth, and R. C. Dunbar, *J. Chem. Phys.* 65,
     3365 (1976).

33. C. E. Klots and R. N. Compton, *J. Chem. Phys.* <u>69</u>, 1636, 1644 (1978).

34. F. W. Lampe, *Interactions Between Ions and Molecules*, P. Auloos, Ed., Plenum Press (New York), 1975, p. 445.

35. C. Y. Ng, et al., *J. Chem. Phys.* <u>67</u>, 4235 (1977), and references therein.

36. R. Lesclaux, S. Searles, L. W. Sieck, and P. Ausloos, *J. Chem. Phys.* <u>54</u>, 3411 (1971).

37. L. W. Sieck, S. G. Lias, L. Hellner, and P. Ausloos, *J. Res. Natl. Bur. Stand. (U.S.) A* <u>76</u>, 115 (1972).

MOLECULAR BEAM STUDIES OF ION–MOLECULE REACTIONS

W. Ronald Gentry

Chemical Dynamics Laboratory, Department of Chemistry
University of Minnesota
Minneapolis, MN   55455

INTRODUCTION

In this brief review I will attempt to highlight some of the areas in which molecular beam techniques are contributing to the understanding of ion-molecule reaction dynamics. Because of space limitations, only a few examples can be cited here, out of what has developed over the last decade into a large and vigorous field of research.   Those readers who are interested in a more detailed and comprehensive discussion are referred to a previous review article [1], which includes an extensive bibliography of work published in the period from 1973 through early 1978.

The special role which molecular beam experiments play in the field of reaction kinetics derives from the microscopic nature of the experimental variables which are controlled. For a simple bimolecular reaction

$$A_i + B_j \rightarrow C_m + D_n ,$$ (1)

the microscopic and macroscopic variables of chemical kinetics are, of course, related to each other in a straightforward way, by the expression

$$k(T) = \sum_i \sum_j a_i(T) a_j(T) \sum_m \sum_n \int_0^\infty v f_T(V) \int_0^{2\pi} \int_0^\pi d\sigma_{ij,mn}(v,\theta,\phi) \sin\theta \, d\theta \, d\phi \, dv.$$ (2)

Here, the quantum state-to-state differential reaction cross section $d\sigma_{ij,mn}$ is integrated over polar and azimuthal scattering angles to produce the state-to-state total reaction cross section, which is then averaged over the distribution $f(v)$ of relative collision speed to give the state-to-state reaction rate coefficient, which is then summed over the product quantum states m and n, and averaged over the reactant quantum states i and j (with fractional populations a_i and a_j) to yield finally the macroscopic rate coefficient $k(T)$. Since most chemists think about chemical reactions in microscopic terms (i.e. what will be the outcome of a single molecular collision under specified initial conditions?), it is to be expected that a microscopic experiment will be more sensitive to the features of a conceptual model than will an experiment in which the outcome is more highly averaged over dynamical variables. Specialized molecular beam methods have been developed for controlling or selecting i, j and v, and for measuring m, n, θ, and ϕ. While it remains exceedingly difficult to combine all the various techniques necessary to determine the ultimate microscopic property $d\sigma_{ij,mn}(v,\theta,\phi)$, it is still the goal of most molecular beam experimentalists to push the experiments further toward the microscopic end of the scale, and to extend this sort of measurement to as many different chemical systems as possible.

In fairness to other methods of chemical kinetics, it must be admitted that the principal advantage of the molecular beam method is also its principal disadvantage. The more microscopic are the experimental data, the more pieces of data are required for a complete description of a macroscopic phenomenon. The situation is analogous to that of macroscopic versus microscopic biology. While it is true that examination of the macroscopic habits of a cow will yield little information about the cellular composition of the animal, it is also true that a microscopic examination of individual cells is a very inefficient way of determining, say, the effect of external conditions on its average rate of milk production. Clearly both types of observation are necessary for complete understanding.

There is no such thing as a general-purpose molecular beam apparatus, even for the restricted field of ion-molecule reactions. Instead, a variety of complementary techniques have evolved, each of which has some limited regime of applicability. A comparison of these has been presented elsewhere [1]. Here I will discuss some of the recent advances which have been achieved in three areas: (a) extending the accessible range and resolution of relative kinetic energy, (b) selection of reactant internal states and analysis of product internal states, and (c) the study of new chemical systems and phenomena.

REACTANT KINETIC ENERGY RANGE AND RESOLUTION

Historically, one of the main limitations of ion beam scattering experiments has been a lower bound on the ion beam
laboratory energy of roughly 1 eV, which is imposed by space-
charge and surface-charge effects. In the conventional crossed-
beam and beam-collision cell apparatus geometries, this constraint limits the range of collision energies in the center of
mass (c.m.) system to values which are much higher than the
thermal energy range normally explored by bulk kinetic methods.
This situation has been changed dramatically by the advent of
the merged-beam technique, in which a collinear geometry for the
ion and neutral beams permits the kinetic energy in the c.m.
system to be very small, even though both beams have high
kinetic energies in the laboratory frame. Figure 1 shows how
this is achieved in the University of Minnesota merged-beam
apparatus [2]. A high energy neutral beam is formed by charge
transfer neutralization of a mass-analyzed ion beam from source
A. A second ion beam, from source B, is deflected by 90° in a
magnetic field, and merged with the neutral beam. Parallel
plates at the entrance and exit of a collision cell are used to
accelerate the ions differentially relative to the neutrals, and
thus to establish the desired c.m. kinetic energy in a well-
defined spatial volume. A two-stage electrostatic energy
analyzer serves to separate the product ions from the reactants
and to measure the primary and product ion laboratory energy
distributions with a resolution which may be varied from 0.04%
to 0.8% fwhm. The c.m. kinetic energy resolution is determined
by the spreads in laboratory velocity vectors of both primary
beams. The spread in axial components of velocity is very small
because of the narrowing of the speed distribution Δv which
accompanies electrostatic acceleration of an ion beam:

$$\Delta v = \Delta E/mv , \qquad\qquad (3)$$

where the energy spread ΔE is independent of accelerating voltage.
The transverse components of velocity are minimized by tuning
both beams for the smallest angular divergence within the collision cell. Three x-y scanners are used for this purpose, and
also for measuring the beam density overlap integral, which is
necessary to determine the absolute reaction cross section.

Figure 2 shows an example of the c.m. energy range and
resolution which can be obtained by this method. The actual mean
value of the c.m. kinetic energy and the root-mean-square energy
resolution are shown as functions of the nominal value of kinetic
energy $E_{nom} = 1/2\ \mu g_0^2$, where g_0 is the difference between the
central laboratory speeds of the two beams. These curves were
calculated numerically for an actual experiment [3] on the

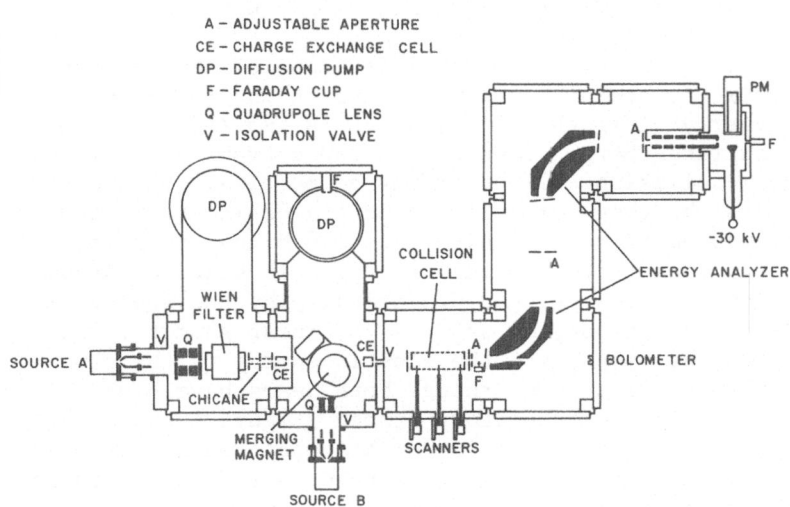

A – ADJUSTABLE APERTURE
CE – CHARGE EXCHANGE CELL
DP – DIFFUSION PUMP
F – FARADAY CUP
Q – QUADRUPOLE LENS
V – ISOLATION VALVE

Fig. 1.　Schematic diagram of the University of Minnesota merged beam apparatus; from Gentry, et al. [2].

reaction

$$D_2^+ + C \rightarrow CD^+ + D \tag{4}$$

by numerically convoluting the two measured primary beam velocity vector distributions at each value of E_{nom}. The useful range of kinetic energy in this experiment extends to less than 0.002 eV, well within the thermal energy range. (The thermal average kinetic energy at 300 K is 0.026 eV.) The experimental values of the absolute total reaction cross section for this system are shown in Fig. 3, over a range of about four orders of magnitude in kinetic energy. Similar data have been obtained for the reactions $D_2^+ + N \rightarrow ND^+ + D$ [4] and $D_2^+ + O \rightarrow OD^+ + D$ [5]. One important feature of these data is that they extend to kinetic energies sufficiently small that the actual reaction cross section is accurately proportional to the "orbiting" or "capture" cross section which can be calculated from the long-range potential function. Thus, a quantitative determination of the low-energy reaction probability per close collision can be made by taking the ratio of the experimental total reaction cross section to the calculated capture cross section. It has been demonstrated [6], however, that the familiar Langevin formalism is not applicable to the long-range interaction of a charged particle with aspherical atoms such as $C(^3P)$ or $O(^3P)$, because the atomic

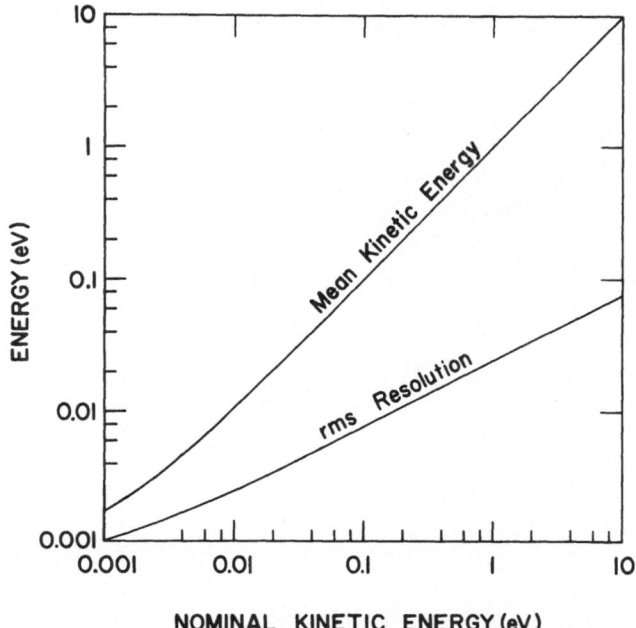

Fig. 2. Actual mean relative kinetic energy, and rms kinetic energy spread, as functions of the nominal c.m. kinetic energy. These results correspond to the measured primary beam energy and angle distributions for a merged beam experiment on the reaction $D_2^+ + C \rightarrow CD^+ + D$; from Schuette and Gentry [3].

quadrupole moment gives rise to a longer-range interaction (r^{-3}) than does the induced dipole moment (r^{-4}). After correctly accounting for the influence of the atomic quadrupole moment and its variation with ion-atom separation caused by spin-orbit coupling effects, the theoretical interpretation of these data indicate that the reaction cross sections for $N(^4S)$ and $O(^3P)$ are much larger than one would predict for an electronically adiabatic reaction mechanism. It is becoming increasingly clear that there will be very many examples of similar nonadiabatic effects in chemical reactions, particularly for reactions of open-shell species. It appears that we must find more effective means of dealing theoretically with the problem of transitions among multiple potential energy surfaces during reactive collisions.

The merged beam technique also yields useful information on the velocity vector distributions of the reaction products. A measurement of the product ion laboratory energy distribution is equivalent to a measurement of the distribution in axial components of the product ion velocity in the c.m. system [4].

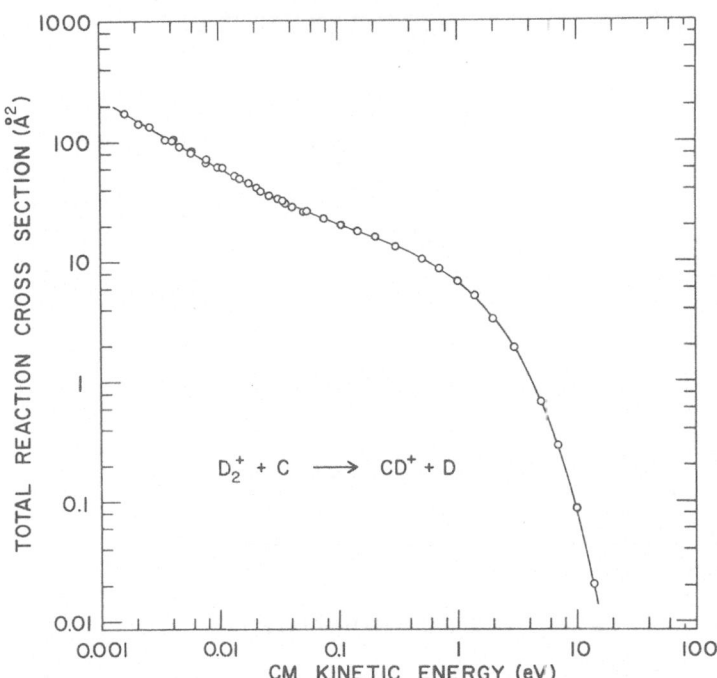

Fig. 3. Energy dependence of the absolute total cross section
for the reaction $D_2^+ + C \rightarrow CD^+ + D$; from Schuette and Gentry [3].

Although a rather elaborate deconvolution procedure [4] is
necessary to extract separate information on the product c.m.
energy and angle distributions, the raw data alone are sufficient
to determine one important feature of the reaction dynamics --
namely, the symmetry of the product velocity distribution about
± 90° in the c.m. system. If the lifetime of the intermediate
"complex" is greater than a few rotational periods of the complex, then the probability of decay of the intermediate will be
equal in all directions in the plane of rotation. In the three-
dimensional distribution of angular momenta appropriate to a
beam experiment, this assumption implies a product distribution
symmetric about the ± 90° plane in product c.m. velocity space
[7]. The observation of such a symmetric distribution does not
prove that the reaction proceeds through a long-lived complex,
because other, direct, mechanisms can also produce symmetric
distributions, e.g. the elastic scattering of hard spheres.
However, the observation of an asymmetric distribution does
disprove a mechanism in which reaction occurs solely through a
rotationally long-lived complex.

It is generally to be expected that the lifetime of a statistical complex will decrease with increasing energy, and that therefore low collision energies will tend to favor a complex reaction mechanism. Since the merged beam experiments extend to such low collision energies, it is worth considering whether the results will be sensitive at all to the detailed reaction dynamics. Will the fact that the collision energy is very much smaller than the potential well depth of the intermediate give the appearance of a complex mechanism, regardless of what actually occurs at small internuclear separations? To answer this question, consider a scattering mechanism which is undeniably direct -- the simple elastic scattering of structureless particles interacting via a potential

$$V(r) = 4\varepsilon [\frac{c}{r^8} - \frac{c}{r^4}] , \qquad (5)$$

which contains a Langevin-like r^{-4} attractive term, and an r^{-8} repulsive core. Orbiting on this potential is possible at all initial kinetic energies $E < \varepsilon/3$, where ε is the well depth. If we look only at collisions with impact parameters less than the value which leads to orbiting, then the results will reflect only those close encounters which we will take to simulate direct, reactive collisions. Figure 4 shows the differential cross sections for these "close collisions", calculated numerically for two values of reduced kinetic energy. At $E = 0.3\varepsilon$, almost at the upper limit for orbiting, we see a substantially higher probability for forward scattering ($0° \leq \theta < 90°$) than for backward scattering ($90° < \theta \leq 180°$), because of the higher weighting of the large impact parameters which give predominantly small-angle scattering. The actual "orbiting" trajectories which give scattering through angles larger than 360° are an extremely small subset of the total. What is perhaps surprising is that if the kinetic energy is lowered to only 1% of the well depth, the angular distribution becomes only slightly less asymmetric. The conclusion is that a well depth much greater than the energy which is available to the separated reactants is not a sufficient condition for a symmetric angular distribution. The essential element is rather the existence of internal degrees of freedom which can remove kinetic energy from the translational coordinate long enough for the complex to rotate several times before decomposing. The merged beam data will be sensitive to this feature of the reaction dynamics even at extremely low kinetic energies. In the $D_2^+ + O$ system, for example, the data show an asymmetric distribution even at $E = 0.015$ eV, with the OD^+ product scattered preferentially in the direction of the incident O atom [5].

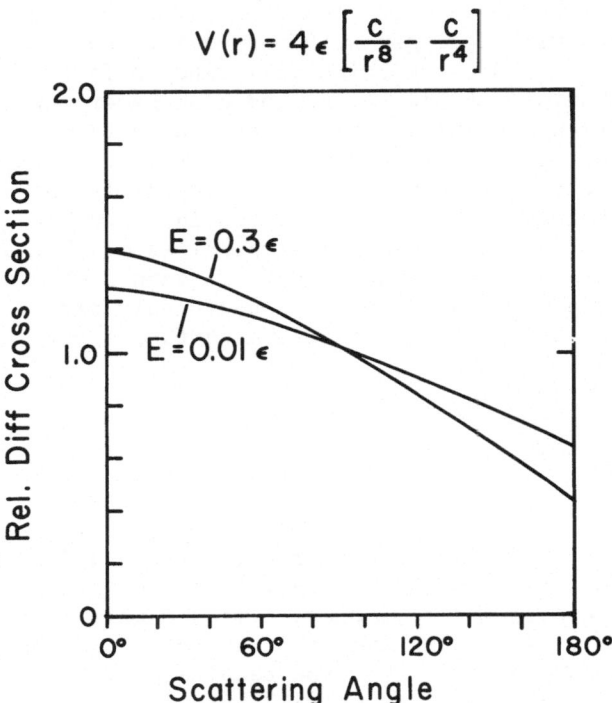

Fig. 4. Differential cross sections for elastic capture collisions on a Lennard-Jones (8-4) potential.

INTERNAL STATE SELECTION AND ANALYSIS

Vibrational-Rotational Excitation

One of the most exciting recent developments in molecular
beam research has been the appearance of inelastic nonreactive
scattering data in which quantum state-to-state transitions are
resolved. These experiments, which are technically more difficult than either elastic or reactive scattering experiments,
come at a particularly propitious time to aid the theoretical
development. Elastic scattering theory, at least for simple
systems, is essentially a solved problem, while reactive scattering in three dimensions is still too complicated for real
systems. The theory of vibrationally and rotationally inelastic
scattering, on the other hand, has reached a point where rigorous
quantum mechanical treatments are just possible for simple systems,
and accurate and efficient approximations in both quantal and
semiclassical frameworks are becoming available for more

complicated systems [8-10]. We are now at just about the same
stage of quantitative understanding of inelastic scattering that
we passed through in the case of elastic scattering in the
early 1960's. The next few years promise to be very fruitful in
this area.

Although the development of new neutral beam techniques
[11,12] may soon change the picture considerably, most of the
state-resolved inelastic scattering data has so far come from
ion beam experiments in which the translational energy defect of
the collision is measured either by electrostatic deflection of
the ions or by time of flight. The two prototype systems, which
have been studied far more thoroughly than any others, are H^+ +
H_2 [13,14] and Li^+ + H_2 [15,16]. The H^+ + H_2 potential energy
surface is reactive, with strongly attractive valence interactions which yield a deeply bound H_3^+ intermediate. The Li^+ + H_2
system is, of course, isoelectronic with He + H_2, and the interaction is mostly repulsive. As one might expect, the mechanisms
of vibrational excitation in these two systems are quite different. In H^+ + H_2 collisions, even large impact parameter,
small scattering angle encounters lead to large stretching forces
between the H_2 nuclei as the passing proton withdraws electron
density from the molecular bond. These long-range interactions
are only weakly anisotropic, and therefore large vibrational
energy transfer occurs without large rotational excitation [17].
In the Li^+ + H_2 system, large vibrational excitation probabilities
are seen only for large scattering angles. Trajectory calculations [18] show that the force acting on the H_2 bond coordinate
is compressive in this case, and is strongly coupled to the molecular rotation, giving rise to multiquantum rotational transitions
concurrent with vibrational excitation.

The techniques for high-resolution ion-molecule scattering
experiments have now been refined to the point where it is
possible to resolve final vibrational states in collisions with
triatomics, such as Li^+ + CO_2 and N_2O [19], although there remain
difficulties with unresolved rotational transitions. At collision
energies of a few eV, these systems show the highest transition
probabilities for excitation of the bending mode vibration. Very
high resolution data on these same systems at kinetic energies in
the range 70-1500 eV have recently been reported by Kobayashi,
et al. [20]. Probably the most microscopically detailed data
currently available for any inelastic scattering process have been
obtained recently by Linder and coworkers, using an apparatus
with a sophisticated tandem electrostatic energy analyzer system
[21]. Figure 5 shows an example of the H^+ + H_2 inelastic scattering data [22] in which baseline resolution of vibrational
transitions was obtained. With an overall energy resolution of
about 0.050 eV, many discrete rotational transitions are resolved

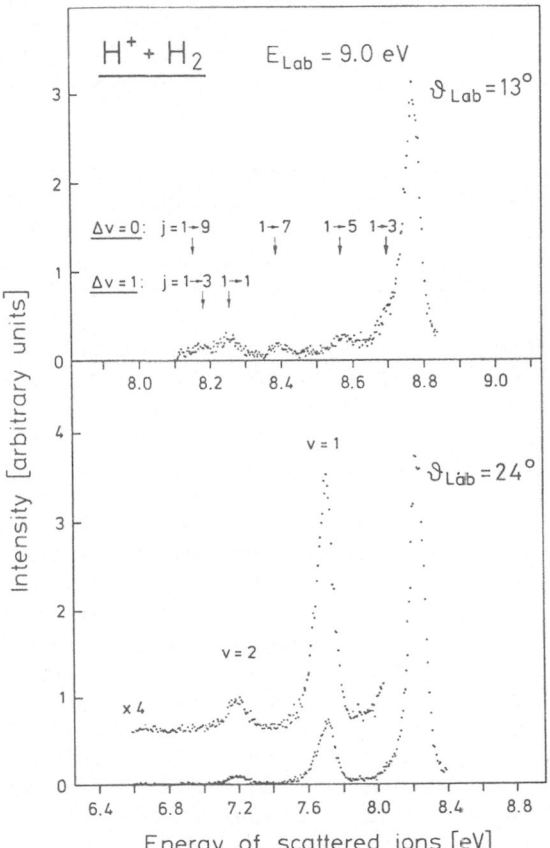

Fig. 5. Proton translational energy spectra for $H^+ + H_2$ scattering, at a H^+ laboratory energy of 9.0 eV, and at laboratory scattering angles of 13° and 24°. The arrows indicate the expected energies for specific vibrational-rotational transitions; from Schmidt, et al. [22].

sufficiently well to permit the discrete state-to-state transition probabilities to be accurately determined by deconvolution.

Reaction Product Vibrational State Analysis

Provided that the product ion analysis system contains provision for mass analysis as well as energy analysis, the same types of apparatus used for the inelastic scattering experiments

discussed above may also be used for reactive scattering experiments. In favorable cases, one can expect to be able to
resolve the reaction product vibrational state distribution, just
as in an inelastic nonreactive collision. A favorable case in
this context means a system in which one of the reaction products has a simple vibrational structure and the other product is
preferably an atom. It is also important that the product
rotational excitation not be large. Vibrational structure has
recently been observed in the product ion energy distribution for
the reaction [23]

$$C^+ + H_2 \rightarrow CH^+ + H \ .\tag{6}$$

With the great improvements in ion beam intensity and energy
resolution which have recently been achieved in several laboratories,
one can be confident that much additional state-resolved data on
ion-molecule reaction products will soon be available.

It is appropriate to mention here also the ion beam experiments of Leventhal, Harris and coworkers and those of Ottinger
and coworkers, in which vibrational and rotational state distributions of products formed in electronically excited states
are observed directly in the fluorescence spectra. These experiments have much better product-state energy resolution than has
ever been achieved in translational energy spectroscopy, but of
course they do not yield differential cross sections, only
integral cross sections for excited electronic states. These
very important results are discussed in detail elsewhere in this
volume.

Effects of Reactant Excitation

Exploration of the effects of reactant internal excitation on
the dynamics of chemical reactions has proven to be somewhat more
difficult experimentally than measuring the distribution in product states. It is especially easy in the case of ion-molecule
reactions to vary the accelerating potential for the ions and
thereby to study reactant translational energy effects. However,
there exists no such general method for selecting specific
internal energy states, and one must therefore rely on the use of
tricks which are specific to particular systems.

One expects electronic excitation of the reactants to cause
the largest effects, because the energy difference between the
ground and the excited state is generally larger than for excitation in other degrees of freedom, and because the shape of
the excited state potential surface will usually be very different

from that of the ground state. Several examples of electronic
excitation effects have now been studied by beam methods. Koski
and coworkers [24,25] have measured cross sections for several
reactions of B^+ in the ground ^3P and excited ^1S states and have
also examined reaction (6) for the ^4P excited state of C^+, which
is exothermic for the excited state and endothermic for the
$C^+(^2P)$ ground state [26]. A particularly striking example of an
electronic excitation effect has been provided by Mahan and
coworkers [27], for the reaction

$$N^+ + H_2 \rightarrow NH^+ + H \quad . \tag{7}$$

Figure 6 shows two velocity vector contour maps for the NH^+ product, measured at similar values of c.m. kinetic energy, but with
different methods of preparation of the N^+ primary beam. N^+

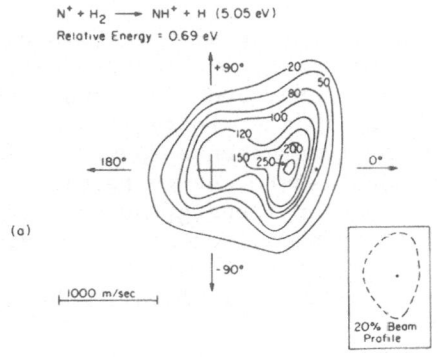

$N^+ + H_2 \longrightarrow NH^+ + H$ (5.05 eV)
Relative Energy = 0.69 eV

(a)

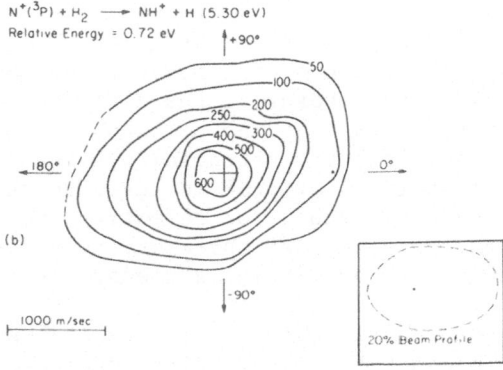

$N^+(^3P) + H_2 \longrightarrow NH^+ + H$ (5.30 eV)
Relative Energy = 0.72 eV

(b)

Fig. 6. Product ion c.m. intensity contour maps for the reaction
$N^+ + H_2 \rightarrow NH^+ + H$, (a) N^+ produced by 160 eV electron impact on
N_2, (b) N^+ formed in a microwave discharge in N_2; from Farrar,
et al. [27].

produced in a microwave discharge, part (b), gives a symmetric product velocity distribution in the c.m. system, but N^+ produced by 160 eV electron bombardment of N_2, part (a), gives a bimodal distribution having one component which resembles that in part (b) and another which is peaked strongly in the direction of the incident N^+ reactant. The symmetric component is due to ground state $N^+(^3P)$, to which the deep ground state potential well of the $NH_2^+(^3B_1 - {}^3\Sigma)$ intermediate is adiabatically accessible. The asymmetric component is probably due to $N^+(^1D)$ excited reactants, which do not correlate with a strongly bound intermediate.

Vibrational energy effects are more subtle, and have to do with the details of the shape of the potential energy surface. Ringer and Gentry [28] have recently studied the influence of reactant vibrational excitation on the dynamics of the endothermic reaction

$$H_3^+ + He \rightarrow HeH^+ + H_2 \quad , \qquad (8)$$

using the merged beam technique. The H_3^+ primary ions are produced from the reaction $H_2^+ + H_2 \rightarrow H_3^+ + H$, which yields highly excited H_3^+. These ions can be rapidly relaxed, however, by proton exchange with cold H_2 in a high-pressure source. At low kinetic energies, one naturally observes a much smaller reaction cross section for H_3^+ produced in a high pressure source than for that produced in a low-pressure source. However, even at kinetic energies much higher than the 2.5 eV reaction endothermicity, differences in the product speed and angle distributions are apparent. Figure 7 shows the HeH^+ laboratory energy distributions for four values of initial c.m. kinetic energy. Products scattered into the forward hemisphere in the c.m. system (0° to 90° relative to the initial He velocity vector) appear to the right of the central vertical line, and those scattered into the backward hemisphere appear to the left. In an endothermic reaction, the reactive trajectories must at some point surmount a repulsive potential energy barrier. To the extent that this barrier extends into the coordinate which is asymptotically translational, the repulsive centrifugal potential of the reactants will impose a constraint on impact parameters which can lead to reaction. Vibrational excitation relaxes this constraint and allows reactions to occur for larger values of impact parameter. This effect is seen in the product energy distributions, which show a much greater extent of scattering into the forward hemisphere when the H_3^+ is more vibrationally excited.

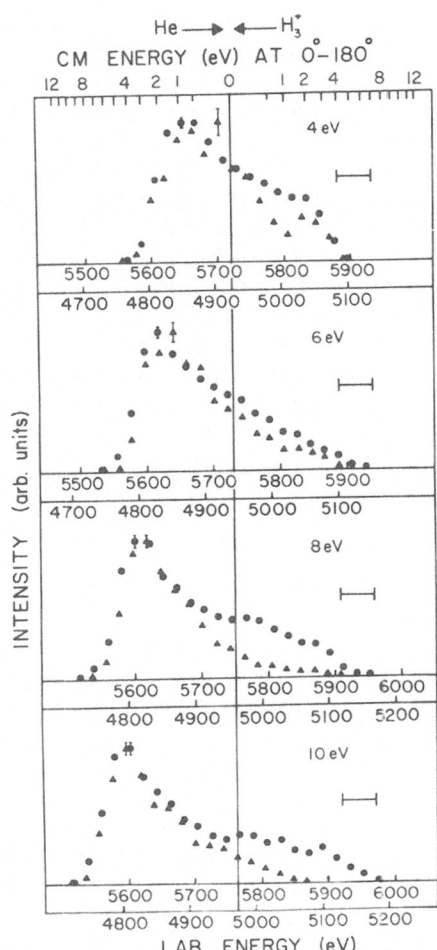

Fig. 7. HeH⁺ product laboratory energy distributions from the
reaction H_3^+ + He → HeH⁺ + H_2, at the indicated values of c.m.
collision energy. The circles correspond to H_3^+ produced in a
low-pressure source, and the triangles to H_3^+ formed in a high-
pressure source; from Ringer and Gentry [28].

NEW CHEMICAL SYSTEMS AND PHENOMENA

It is possible here only to give a few brief examples to
illustrate some of the directions in which molecular beam studies
of ion-molecule reactions are proceeding. These are by no means
an adequate representation of the full range of research activity
in this field.

Even though "simple" single-channel reactions which occur on a single potential energy surface are still far from being understood in detail, many investigators are engaged in studying the more complicated dynamical effects which arise when two or more product channels compete with each other. Not only the overall product branching ratio, but especially the different energy and angle distributions which are observed for different product channels, may prove to be extremely useful in understanding the relationships between the structure of the potential energy surface(s) and the reaction dynamics.

Such effects can occur even among old and familiar reactions which have sometimes been considered to be prototypes for simple models. For example, the reaction

$$Ar^+ + H_2 \rightarrow ArH^+ + H \tag{9}$$

was one of the earliest reactions studied by ion beam techniques, and is often considered to be a typical example of an exothermic abstraction reaction. It occurs with high probability for large impact parameter collisions, giving an ArH^+ product strongly peaked in the direction of the incident Ar^+, and qualitatively obeying "stripping" dynamics. However, recent experiments by Hierl, et al. [29] show that the charge transfer channel

$$Ar^+ + H_2 \rightarrow Ar + H_2^+ \tag{10}$$

also occurs with substantial probability at low collision energies. At a c.m. kinetic energy of 0.13 eV a sharp peak in the H_2^+ distribution is observed at $0°$ and at an energy which corresponds to very small momentum transfer. A peak of comparable intensity is also seen at c.m. angle of $180°$. Despite the large cross section for the rearrangement reaction at this energy, many collisions at both large and small impact parameters lead to charge transfer rather than to atom transfer.

Competition between different rearrangement channels has been examined by Mahan and Schubart [30], for the reactions

$$CO_2^+ + D_2 \rightarrow DCO_2^+ + D \tag{11a}$$

$$\rightarrow DCO^+ + OD . \tag{11b}$$

The product DCO_2^+ corresponds to the type of exothermic atom abstraction process usually associated with large impact parameter collisions and small scattering angles. The data confirm this expectation. The DCO^+ product, on the other hand, can be produced at low kinetic energies only by exchange of an O atom and a D atom. One feels intuitively that a more intimate encounter

will be required in this case than for an abstraction process.
In fact, at low collision energies a symmetric angle distribution
is observed for the DCO$^+$ product -- a result which is consistent
with the formation of a rotationally long-lived intermediate
complex.

Differences in the reaction dynamics for various reaction
mechanisms have also been observed by Douglass and Gentry [31-33]
for reaction in hydrogen systems. Some of the processes studied
include

$$H_3^+ + D_2 \rightarrow D_2H^+ + H_2 \qquad \text{(abstraction)} \qquad (12)$$

$$HD_2^+ + HD \rightarrow D_3^+ + H_2 \qquad \text{(exchange)} \qquad (13)$$

and $\qquad H_3^+ + D \rightarrow H_2D^+ + H \qquad \text{(substitution)}, \qquad (14)$

which provide examples of three distinct reaction mechanisms in
systems for which relatively reliable potential energy surface
calculations are possible. Merged beam measurements of the total
cross sections for the abstraction and exchange reactions (12)
and (13) in H_5^+ systems are compared in Fig. 8. At low kinetic
energies, the exchange process occurs with only about 2% of the
probability for the abstraction process, even though the reaction
energetics differ only in zero-point energy effects. The large
abstraction cross sections imply reaction at large impact parameters, and again this expectation is confirmed by a D_2H^+ product
angle distribution which is strongly peaked in the direction of
the incident D_2 reactant. The exchange process, however, exhibits
more complicated dynamics. Figure 9 shows laboratory energy
distributions for the D_3^+ product. Products to the left of the
central vertical line correspond to scattering in the direction
of the incident HD, and those to the right correspond to scattering of D_3^+ in the direction of the incident HD_2^+ reactant. A
symmetric distribution is seen for a c.m. kinetic energy of 0.002
eV only because of the symmetric distribution of initial relative
velocity vectors for this case [2]. As the collision energy is
increased, however, a real asymmetry in the D_3^+ angular c.m. distribution appears, shifting first in the HD_2^+ direction, then, at
higher energies, in the HD direction. We do not attribute this
behavior to the formation of a long-lived complex, but rather to
a complicated direct mechanism which involves strong coupling
among all five atoms. The complicated change in the exchange
reaction dynamics occurs in the 0.5-5.0 eV energy range, where
the reaction cross section exhibits the unusual feature of drop-
ping, rising, and then falling again with increasing kinetic energy.

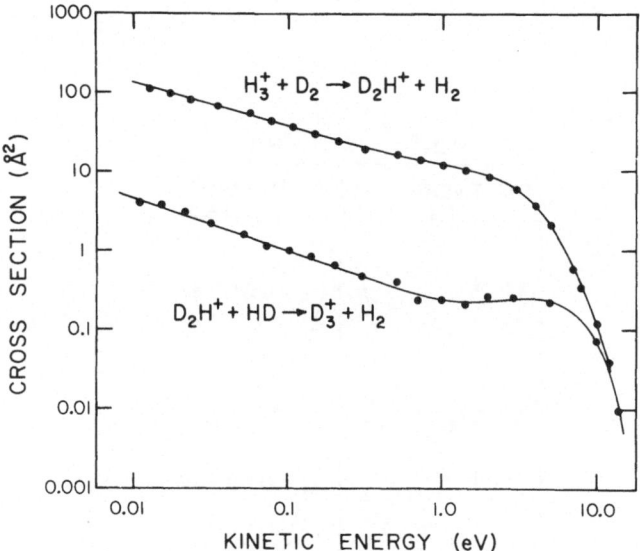

Fig. 8. Energy dependence of the total cross sections for the reactions $H_3^+ + D_2 \rightarrow HD_2^+ + H_2$, and $HD_2^+ + HD \rightarrow D_3^+ + H_2$; from Douglass and Gentry [31,32].

In the limit of high initial kinetic energies, the abstraction reaction cannot take place by a simple stripping mechanism which transfers no momentum to the H_2 product, because to do so would imply an internal energy too large for a bound D_2H^+ product to be observed. A stable D_2H^+ product can only be formed in a more intimate collision which provides a means of dissipating this excess internal energy in either relative translational energy, or internal excitation of the H_2 product. Under these conditions one might expect a statistical distribution of reaction products, which does not favor the abstraction process over the exchange process. The data show that as the reaction cross sections for these two channels become small at high energies, they also become more nearly equal, consistent with a statistical mechanism. The substitution reaction presents an interesting contrast to both of the previous examples. The low-energy reaction probability is large in this case, yet the product velocity vector distributions are perfectly symmetric, within the experimental uncertainty, for collision energies up to 1 eV. The data seem to imply a strongly coupled intermediate complex, but potential surface calculations [34] for H_4^+ yield a binding energy which seems to be much too small for the dissociation lifetime to exceed a rotational period. This appears to be an excellent case for classical trajectory calculations to help elucidate the mechanism.

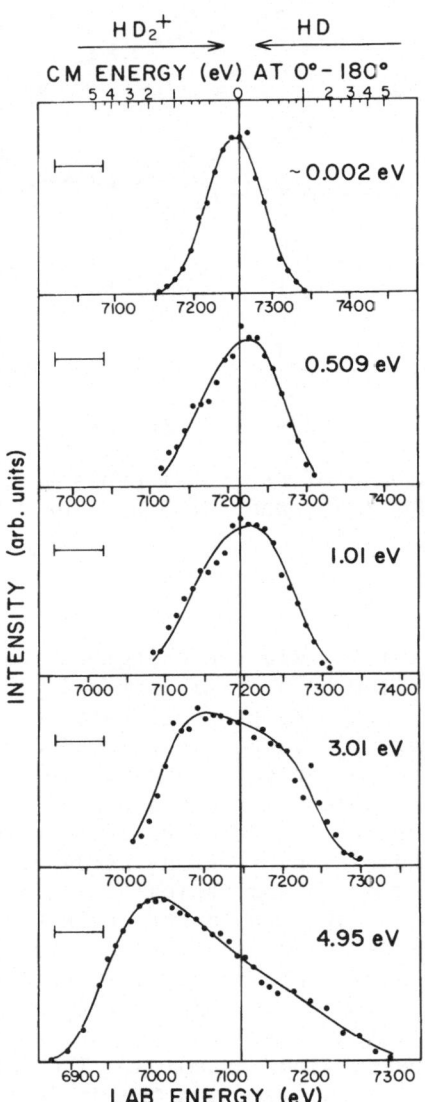

Fig. 9. Product ion laboratory energy distributions for the reaction $HD_2^+ + HD \rightarrow D_3^+ + H_2$, at the indicated initial c.m. kinetic
energies; from Douglass and Gentry [32].

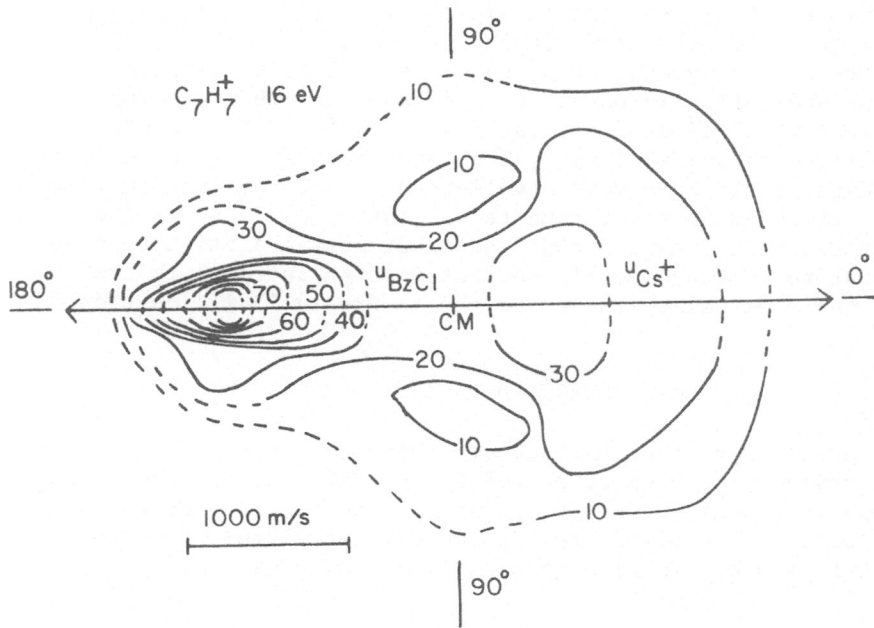

Fig. 10. Product ion c.m. intensity contour map for the reaction $Cs^+ + C_6H_5CH_2Cl \rightarrow C_7H_7^+ + CsCl$, at a Cs^+ laboratory kinetic energy of 16 eV. The quantities u_{Cs^+} and u_{BzCl} are the reactant c.m. velocity vectors; from Safron, et al, [35].

Finally, let me consider the extension of molecular beam techniques to ion-molecule reactions in more complicated chemical systems. An excellent example has been provided by Safron, et al., [35], who used a crossed beam scattering apparatus to study the dynamics of the reactions

$$Cs^+ + C_6H_5CH_2Cl \rightarrow CsCl + C_7H_7^+ \tag{15a}$$

$$\rightarrow CsCl + C_2H_2 + C_5H_5^+ . \tag{15b}$$

Figure 10 shows the $C_7H_7^+$ product velocity vector distribution for an initial Cs^+ laboratory kinetic energy of 16 eV. These workers attribute the sharp forward-scattered peak shown at 180° to a stripping-like process which transfers little momentum to the benzyl ion product. The broader peak at 0° is thought to result from Cs^+ collisions with the benzyl end of the molecule, which convert a large amount of translational energy to internal energy in the process of rearrangement. Some of the benzyl ions receive enough internal excitation to further decompose into $C_5H_5^+ + C_2H_2$.

Since the reactants are both in singlet spin states, it is significant that the only products observed are those whose ground
states are also singlets, even though several triplet-state products are energetically possible. Although reliable a priori
theoretical predictions for this 16 atom, 114 electron system
seem still to be quite distant, the interpretation of the experimental data is not much more complicated than for a much smaller
system. It is extremely fortunate for future progress in the
field of reaction dynamics that the experimental difficulties do
not appear to increase nearly so fast with an increase in the
size of the chemical system as do the theoretical difficulties.

ACKNOWLEDGMENTS

Different aspects of the author's research in the field covered
by this review have been supported by the National Science Foundation (Quantum Chemistry Program), and by the U.S. Department of
Energy (Office of Basic Energy Sciences). An Alfred P. Sloan
Research Fellowship is also gratefully acknowledged.

REFERENCES

1. W. R. Gentry, "Progress in the Application of Molecular Beam
    Techniques to the Study of Ion-Molecule Reactions," in Gas
    Phase Ion Chemistry, Vol. 2, edited by M. T. Bowers (Academic
    Press, New York, in press).

2. W. R. Gentry, D. J. McClure and C. H. Douglass, Rev. Sci.
    Instrum. 46, 367 (1975).

3. G. F. Schuette and W. R. Gentry, "The Dynamics of the Reaction
    $D_2^+ + C \rightarrow CD^+ + D$", (1978) to be published.

4. D. J. McClure, C. H. Douglass, and W. R. Gentry, J. Chem.
    Phys. 66, 2079 (1977).

5. D. J. McClure, C. H. Douglass, and W. R. Gentry, J. Chem.
    Phys. 67, 2362 (1977).

6. W. R. Gentry and C. F. Giese, J. Chem. Phys. 67, 2355 (1977).

7. W. B. Miller, S. A. Safron and D. R. Herschbach, Disc. Faraday
    Soc. 44, 108 (1967).

8. W. R. Gentry, "Vibrational Energy Transfer: Classical and Semiclassical Methods", in <u>Atom-Molecule Collision Theory</u>, <u>a Guide for the Experimentalist</u>, edited by R. B. Bernstein (Plenum, New York, in press).

9. D. J. Kouri, ibid.

10. D. H. Secrest, ibid.

11. W. R. Gentry and C. F. Giese, Rev. Sci. Instrum. <u>49</u>, 595 (1978); J. Chem. Phys. <u>67</u>, 5389 (1977); Phys. Rev. Lett. <u>39</u>, 1259 (1977).

12. U. Buck, F. Huisken, J. Schleusener and H. Pauly, Phys. Rev. Lett. <u>38</u>, 680 (1977); U. Buck, F. Huisken and J. Schleusener, J. Chem. Phys. <u>68</u>, 5854 (1978).

13. H. Udseth, C. F. Giese, and W. R. Gentry, J. Chem. Phys. <u>54</u>, 3642 (1971).

14. H. Udseth, C. F. Giese, and W. R. Gentry, Phys. Rev. A<u>8</u>, 2483 (1973).

15. R. David, M. Faubel, and J. P. Toennies, Chem. Phys. Lett. <u>18</u>, 87 (1973).

16. H. E. van den Bergh, M. Faubel, and J. P. Toennies, Chem. Soc. Faraday Disc. <u>55</u>, 203 (1973).

17. C. F. Giese and W. R. Gentry, Phys. Rev. A<u>10</u>, 2156 (1974).

18. W. R. Gentry and C. F. Giese, J. Chem. Phys. <u>62</u>, 1364 (1975).

19. W. Eastes, U. Ross, and J. P. Toennies, J. Chem. Phys. <u>66</u>, 1919 (1977).

20. N. Kobayashi, Y. Itoh and Y. Kaneko, J. Phys. Soc. Japan <u>45</u>, 617 (1978).

21. V. Hermann, H. Schmidt, and F. Linder, J. Phys. B<u>11</u>, 493 (1978).

22. H. Schmidt, V. Hermann, and F. Linder, (1978) unpublished.

23. C. A. Jones, K. L. Wendell, and W. S. Koski, J. Chem. Phys. <u>66</u>, 5325 (1977).

24. K. Lin, R. J. Cotter, and W. S. Koski, J. Chem. Phys. <u>60</u>, 3412 (1974).

25. K. Lin, H. P. Watkins, R. J. Cotter, and W. S. Koski, J.
     Chem. Phys. 60 , 5134 (1974).

26. C. A. Jones, K. L. Wendell, J. J. Kaufman, and W. S. Koski,
     J. Chem. Phys. 65, 2345 (1976).

27. J. M. Farrar, S. G. Hansen, and B. H. Mahan, J. Chem. Phys.
     65, 2908 (1976).

28. G. Ringer and W. R. Gentry, (1978) to be published.

29. P. M. Hierl, V. Pacak, and Z. Herman, J. Chem. Phys. 67,
     2678 (1977).

30. B. H. Mahan and P. J. Schubart, J. Chem. Phys. 66, 3155 (1977).

31. C. H. Douglass and W. R. Gentry, "The Dynamics of the Reaction
     $H_3^+ + D_2 \rightarrow D_2H^+ + H_2$", (1978) to be published.

32. C. H. Douglass and W. R. Gentry, "A Merged Beam Study of an
     Atomic Exchange Reaction: $HD_2^+ + HD \rightarrow D_3^+ + H_2$", (1978) to
     be published.

33. C. H. Douglass and W. R. Gentry, "A Merged Beam Study of an
     Atomic Substitution Reaction: $H_3^+ + D \rightarrow H_2D^+ + H$", (1978)
     to be published.

34. R. D. Poshusta and D. F. Zetik, J. Chem. Phys. 58, 118
     (1973).

35. S. A. Safron, G. D. Miller, F. A. Rideout, and R. C. Horvat,
     J. Chem. Phys. 64, 5051 (1976).

CHARGE TRANSFERS AT THERMAL ENERGIES: ENERGY DISPOSAL AND REACTION MECHANISMS

R. Marx

Laboratoire de Résonance Electronique et Ionique
(Associé au CNRS) Université de Paris-Sud, Centre
d'Orsay, 91405 Orsay, France

INTRODUCTION

Charge transfer (CT) reactions have been discussed at the last NATO meeting on Ion Molecule Interactions (1) both in a panel devoted to this problem (1a) and as part of different lectures. Special emphasis has been put on low and thermal energy processes considering their importance in lasers, electric discharges,planetary and atmospheric ion chemistry.

Specific features of CT in this energy range, as compared to high energy collisions, were described: - isotropic contour maps for product velocity distributions in low energy beams (1b-1c)- departure of the product vibrational populations from the values corresponding to Franck-Condon Factors (FCF)(1d) indicating short distance effects which do not appear in long distance (near) resonant CT. On the other hand it was outlined that to have efficient CT reactions between A^+ and M: - M^+ has to have an electronic state whose energy is equal to the recombination energy of A^+ (energy resonance) and - the FCF between this state of M^+ and the ground state of M have to be large (vertical transition). The electronic states of M^+ involved may as well be optically allowed states observed by Photo Electron Spectroscopy (PES), as optically forbidden, autoionizing or repulsive states which do not appear in PES and for which FCF are usually unknown (1e).

During these last 4 years a substantial amount of literature appeared, mainly for rare gas ions on molecules, discussing the relative importance of FCF and energy resonance on the magnitude of the rate constant and using rate constant and product distribution measurements to deduce reaction mechanisms (2-5). However

103

it appeared clearly that direct information was needed concerning
the internal and kinetic energy of the products to get more insight
into reaction mechanisms.

1.ENERGY DISPOSAL AND REACTION MECHANISM

In contrast with beam experiments, at thermal energies there
is no kinetic energy (KE) of the reactants to be partly transformed
into internal energy of the products. The maximum energy available
for the reaction is then the recombination energy of the reactant
ion A^+ and of course only exothermic processes may occur.

The reaction exothermicity of $A^+ + M \rightarrow A + M^+$:

$\Delta E = RE(A^+) - IP(M)$, is at most equal to the difference between
the ground state ionization energies of the two reacting species,
but its value for a given reaction depends of course on the internal energy of the reactants and products which may be electronically and (or) vibrationally excited.

In the following discussion we will consider mainly ground
state M and A^+ and very briefly comment on electronically excited A^+.

When M^+ and A are not dissociated, ΔE shows up as recoil kinetic energy of the two products and the internal energy of M^+ can
be deduced from kinetic energy measurements provided $RE(A^+)$ is
known. When dissociation occurs, the balance between internal and
kinetic energy of the products depends on the reaction mechanism.

Two types of mechanisms have been proposed for C.T reactions.

ELECTRON JUMP WITHOUT MOMENTUM TRANSFER: then the CT populates an electronic state of M^+ resonant in energy with RE (A^+)
so that $\Delta E = 0$.

Depending on the interaction between A^+ and M during the
electron jump, M is ionized according to a vertical Franck-Condon
(FC) process or not. F.C populations are always observed in beam
experiments at high kinetic energy, but at low and thermal energies
there are many examples of non FC population. They have been
explained in terms of distortion of M prior to ionization, the
electron jump taking place between a perturbed state of M and the
unperturbed final state of M^+ (6-7).

When an electron jump without momentum transfer populates a
(pre)dissociated state of M^+, the recoil kinetic energy of the
fragments is simply

$$RKE(F) = RE(A^+) - IE(F)$$

where $IE(F)$ is the internal energy of the fragments with respect
to ground-state M.

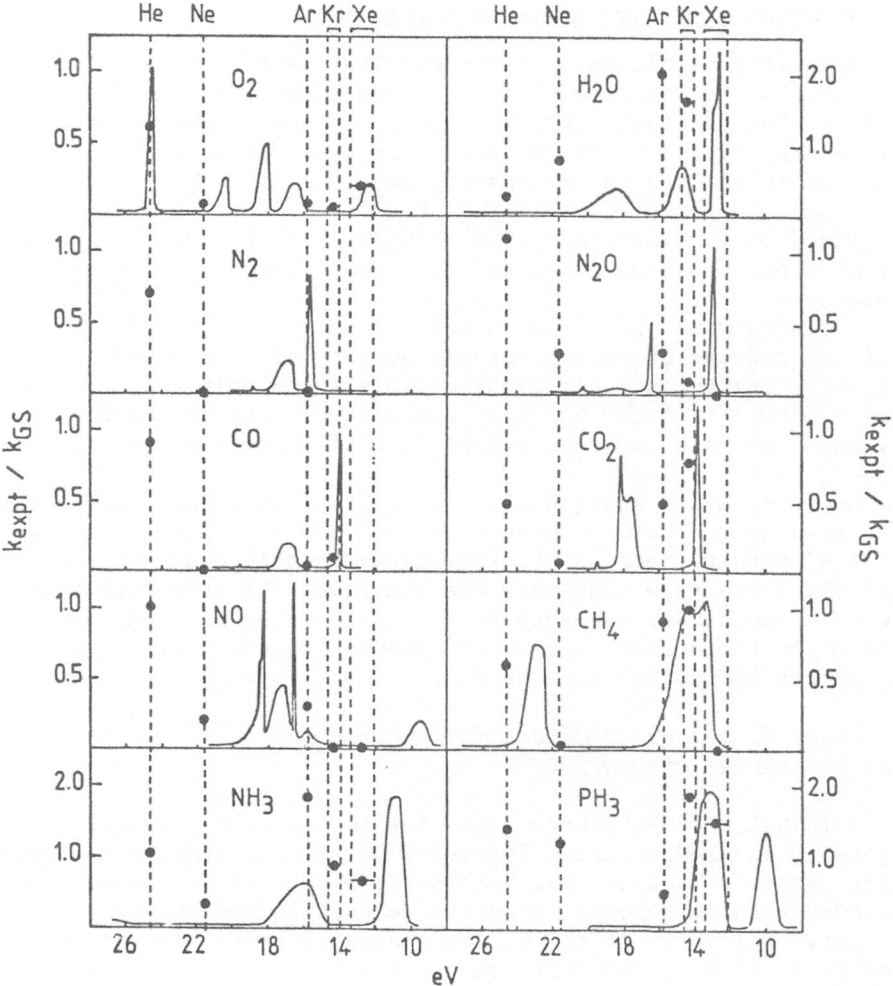

Fig. I: The envelopes of the photo-electron spectra of a few small molecules $O_2, N_2, CO, NO, NH_3, PH_3, H_2O, N_2O, CO_2, CH_4$ are represented in full lines (——). The recombination energies of five rare gas ions are indicated by dotted vertical lines (---) and the corresponding reaction efficiencies (k_{exp}/k_{GS}) by black dots (●).

    To draw the PES envelopes we used the data of : D.W. Turner et al.(Molecular photo-electron spectroscopy, Wiley Interscience N-Y-1970) A.W.Potts et al(Disc.Far.Soc.54, 109,1972) and O.Edquist et al (Phys.Scripta 1, 25,1970).

    The amplitudes are in arbitrary units and are roughly proportional to the FCF for vertical transitions from ground state M to the corresponding M^+. Since they have not been normalized the spectra for different molecules should not be compared.

    The rate constants are taken from ref (2,4,15b) except for Kr^+, H_2O (C.J.Howard et al JCP 53, 3745, 1970).

ELECTRON JUMP WITH MOMENTUM TRANSFER

This is possible only if the electron jump occurs at distances where there is a strong enough interaction between A^+ and M to allow part of the reactant internal energy to be converted into kinetic energy of the products. Such a process may be represented in a schematic way by a curve crossing model.

Electron jump with momentum transfer has been discussed only for CT of atomic ions on atoms and of doubly charged ions on atoms and molecules.

For singly charged ions on molecules it is currently stated that direct conversion of appreciable amounts of electronic energy into kinetic energy is generally inefficient and only *electron jump without momentum transfer* has been considered. This assumption is mainly based on rate constant and product distribution measurements (2-5).

However, as it appears clearly on fig.I,there are many cases where a resonant CT between ground state M and an optically allowed state of M^+ does not account for the measured rate constant. To explain why large rate constants are sometimes observed with reactant ions whose recombination energy falls outside the PES of M, distortion prior to ionization and (or) population of M^+ states which do not appear in the PES have been proposed (2-5).

There is almost no direct information on the M^+ states populated by thermal energy CT.

INTERNAL ENERGY is known only for luminescent products since, to my knowledge, Laser Induced Fluorescence (LIF) experiments used for neutrals have not yet been performed on ion molecule reactions. Moreover,luminescent products account in most cases for only a very small part of the CT process except possibly when electronically excited or molecular ions are used as reactants. However, there is no quantitative luminescence measurement at thermal energies for these systems.

KINETIC ENERGY MEASUREMENTS are currently performed in beam experiments for the ionic products.At thermal energies no measurement existed until recently. A method giving rather crude estimates has been proposed (10): the ions are trapped in the source of a mass spectrometer by the field of the ionizing electron beam and the kinetic energies evaluated through the ion loss as a function of the electron beam intensity, are of the order of a few tenths of an eV. However velocity measurements in low energy beam experiments (11-13) show that intimate collisions, with up to 1 eV KE release in some cases, become increasingly important as collision energy is lowered.

2. REACTION MECHANISMS DEDUCED FROM LUMINESCENCE AND KINETIC ENERGY MEASUREMENTS

To discuss the information on reaction mechanisms which may be deduced from optical emission spectroscopy and kinetic energy measurements of the products I will use the results obtained in our group since this is to my knowledge the only available systematic set of experiments performed at (near) thermal energies.

The experimental procedures using an ICR cell to trap the ions have already been described (14-15) and the kinetic energy measurements will be presented in the panel on experimentation.

2.1. LUMINESCENCE OF (NEAR)THERMAL-ENERGY C.T PRODUCTS

The systems discussed in this section are presented in table 1.

In contrast with high and medium energy beam experiments, very few luminescent molecular ions have been observed.

2.1.1. He^+, N_2 (16): This is the only system where luminescent products (N_2^+) account for $\sim$ 25% of the total CT process and, since one of the luminescent states, N_2^+ ($\tilde{C}$), which has levels very close to RE(He^+) is known to be 90% predissociated, it may be concluded that $\sim$ 70% of the charge transfer is energy resonant. Moreover the unidentified N_2^+(Z) state is also possibly near resonant. Finally $N_2^+(\tilde{B})$, which is a minor product however, may as well result from a higher state also energy resonant with He^+ as from a close encounter. In the last case N_2^+ would take away 13% of the 5 eV KE release.

2.1.2. He^+, N_2O (17): $N_2O^+(\tilde{A}\ ^2\Sigma^+)$ may result from a near resonant electronic state of N_2O^+ ($\tilde{F}\ ^2\Pi$) by a radiative cascade or from a close encounter with momentum transfer. In this case 8% of the 8 eV KE release would be carried by N_2O^+.

$N_2^+(\tilde{B})$ also correlates according to ref (18) to the ($\tilde{F}\ ^2\Pi$) state of N_2O^+. The most abundant fragments, which do not radiate, may all be correlated to the ($\tilde{E}\ ^2\Pi$) state of N_2O^+ which is $\sim$ 2 eV off resonance. Kinetic energy measurements could not be performed in this system because of experimental difficulties (too many different product ions).

Finally it is interesting to notice that Ne^+ at thermal energies does not procuce $N_2O^+(\tilde{A})$ and $N_2^+(\tilde{B})$ as it does at $\sim$ 1 eV in beam experiments (19). This could indicate that thermal energy He^+ populates state(s) above $\sim$ 22 eV.

2.1.3. He^+, H_2O (20): OH^+ ($\tilde{A}\ ^3\Pi$) correlates to the $^2A_1$ state of H_2O too high in energy ($\sim$ 27 eV) to be populated by thermal energy He^+ (24.6 eV). Distortion of H_2O^+ from C_{2V} to near C_s

symmetry may lower the required energy and explain the observed fluorescence.

The most abundant ionic fragments: H^+ and OH^+ (ground $X^3\Sigma^-$ or metastable $a^1\Delta$) correlate to the $\tilde{a}^4B_1$ and ^2B states of H_2O^+ which are very near resonant with He^+. The KE measurement confirms this interpretation as we will see later.

TABLE 1: Fluorescence and Product Distributions

Ref (16)	$He^+ + N_2$: $k_t = 1.2\times10^{-9} cm^3 s^{-1}$				
Obs.Prod.	N^+	$N_2^+ \; (\tilde{C} \to \tilde{X})$ v' not F.C	$N_2^+ (Z \to \tilde{X} \text{ or } \tilde{A})$	$N_2^+ (\tilde{B} \to \tilde{X})$ v' not F.C	$N_2^+(?)$
% of k_t	60	$\leqslant 6$	~ 20	~ 1	10–20
(M^+)	$N_2^+(\tilde{C})$		$N_2^+(\text{near } \tilde{C})$	$N_2^+(\tilde{B}$ or near $\tilde{C})$	$N_2^+(?)$
ΔE_{eV}	$(v=3,4) \leqslant 0.3$		~ 0	~ 5 or ~ 0	?

Ref(17)	$He^+ + N_2O$: $k_t = 2.3\times10^{-9} cm^3 s^{-1}$						
Obs.Prod.	$N_2O^+ (\tilde{A} \to \tilde{X})$ v' not F.C	$N_2^+ \; (B \to X)$	N_2O^+	N_2^+	NO	O^+	N^+
% of k_t	0.2 – 1	1 – 5	< 1	54	21	12	13
(M^+)	$N_2O^+(\tilde{F}$ or $\tilde{A})$	$N_2O^+(\tilde{F})$	$N_2O^+ \; (\tilde{E})$?				
ΔE_{eV}	0.5 or 8.2	~ 0.5	2.				

Ref (20)	$He^+ + H_2O$: $k_t = 0.5\times10^{-9} cm^3 s^{-1}$			
Obs.Prod.	H_2O^+	OH^+	$OH^+(A \to X)$	H^+
% k_t	10	52	3 – 4	37
(M^+)	?	$\tilde{a}^4B_1$	$^2A_1$	$^2B_1$
ΔE_{eV}	$\sim 6eV$	0.5	~ -3	~ -1

Continued on next page →

Ref(21)	$He^+ + C_2H_2$: $k_t = 3\ 10^{-9} cm^3 s^{-1}$				
Obs.Prod.	$C_2H_2^+$	C_2H^+	C_2^+	CH^+	$\underline{CH^+(\tilde{A} \to \tilde{X})} - \underline{CH(\tilde{A} \to \tilde{X})} - \underline{CH(B \to X)}$,
% of k_t	7	25	46	22	0,15
ΔE eV	1.05–13.2	4.3–6.8	0–1.3	0.1–4	
(M^+)	$C_2H_2^+$ ($\tilde{C} - \tilde{B} - \tilde{A}$ or $\tilde{X}$)				?
ΔE eV	1.05 – 6.2 – 8.2 or 13.2			1.1	1.2 0.9

Ref(22)	$Ar^+ + H_2O$: $k_t = 1.9\ 10^{-9} cm^3 s^{-1}$	
Obs.Prod.	H_2O^+ ($\tilde{A} \to \tilde{X}$), v' = 9 →18	H_2O
% k_t	≲ 3	97%
(M^+)	H_2O^+ ($\tilde{A}$)	$H_2O^+(\tilde{X})$
v'	9–10–11–12–13–14–15–16–17–18	(12–13)
Obs.popul.	5– 8–11–17–14–12–14– 9– 6– 4	
F.C.popul.	17–16–16–14–12– 9– 7– 5– 3– 1	
ΔE $Ar^+\ ^2P_{1/2}$	0.04	
	1 ± 0.1	1 ± 0.1
$Ar^+\ ^2P_{3/2}$	0.13	

Table 1: This table summarizes the observations concerning fluorescence of reaction products and product distribution for a few CT reactions.

Obs.Prod.: refers to the ions observed by ICR. The fluorescent species are underlined.

% k_t : is the percentage of the total rate constant corresponding to each ion.

(M^+) : is the electronic state of M^+ from which the observed products most probably originate (see discussion in the text).

ΔE eV : is the energy difference between (M^+) and the recombination energy of the reactant ion. In (Ar^+, H_2O) ΔE is given for the two states of $Ar^+(^2P_{1/2}$ and $^2P_{3/2})$ populating the lowest and highest vibrational levels observed in $H_2O^+(\tilde{A})$ and the levels most probably populated in $H_2O^+(\tilde{X})$ according to KE measurements (see table 2).

2.1.4.$\underline{He^+,C_2H_2}$ (21): We do not know to which state(s) of $C_2H_2^+$ correlate the observed $CH^+(\tilde{A}^1\Pi)$, $CH(\tilde{A}^2\Delta)$ and $CH(\tilde{B}^2\Sigma^+)$ fluorescent fragments. However the adiabatic appearance potentials of these species are: 23.46, 23.4 and 23.72 eV respectively (table 1 ref 21). The excess energy is ~ 1 eV or even less if the products have a few tenths of an eV vibrational energy.

Due to the lack of correlation diagram, nothing can be concluded concerning the most abundant product ions: $C_2H^+-C_2^+-C_2H_2^+$ and ground state CH^+. Kinetic energy measurements would not be very helpful since, even if these fragments result from resonant states of $C_2H_2^+$, the excess energy will be carried by the neutral H, except for $C_2H_2^+ \rightarrow CH^+(X)+CH(X)$ where half of the 4 eV excess energy would be carried by CH^+.

2.1.5.$\underline{Ar^+,H_2O}$ (22): Since $RE(Ar^+)$ is well below the first dissociation limit of H_2O^+, only molecular ions may be formed in the CT reaction. However luminescent $H_2O^+(\tilde{A}^2A_1)$ is in this case also a minor product: 97 % of the CT produces $H_2O^+(\tilde{X}^2B_1)$. For $H_2O^+(\tilde{A}^2A_1)$ vibrational levels from $v' = 18$ (near resonant with $Ar^{+2}P_{1/2}$) to $v' = 9$ (corresponding to 1 eV excess energy) are observed and their population is not F.C. This clearly indicates an intimate collision with a broad distribution of internal and kinetic energies. Kinetic energy measurements show that most H_2O^+ ions have a mean kinetic energy of 1 eV.

This is the first example of a fast CT reaction which does not proceed via a long distance electron jump.

2.2. KINETIC ENERGY OF (NEAR) THERMAL ENERGY CT PRODUCT

Kinetic energy measurements have been performed as well on fragments as on molecular product ions and, depending on the system, different types of information could be deduced from these measurements (tables 2 and 3).

2.2.1. FRAGMENT IONS IN POLYATOMIC MOLECULES: (He^+,CH_4) AND (He^+,H_2O).

In both cases H^+ fragment ions with large KE have been observed (15) which were not seen in standard ICR experiments (23).

Reaction mechanism cannot be deduced from the KE of the ionic fragment only. However it may be determined if electron jump without momentum transfer is a possible mechanism i.e consistent with the measured KE.

With CH_4, fragmentation into ground state H^+ and CH_3 requires 18.1 eV and the total kinetic energy release is 3.4 eV. If the corresponding electronic state of CH_4^+ is populated via a resonant CT, the excess energy (24.6 – 21.5=3.1 eV) must be used as vibrational excitation of CH_3.

With H_2O, fragmentation into H^+ and ground state OH requires 18.7 eV and the total KE release is 4.4 eV. This leaves: 24.6 - 23.1 = 1.5 eV as vibrational energy for OH.

In both cases we find that (near) resonant CT with a reasonable amount of vibration in the neutral fragment is a likely mechanism. Moreover in both cases there are electronic states of the molecular ions located around the recombination energy of He^+ and correlated to the observed fragments: $CH_4^+(\tilde{A}^2A_1)$ and $H_2O^+(^2B_1$ or $\tilde{A}^2A_1)$ respectively.

2.2.2. FRAGMENT IONS IN DIATOMIC MOLECULES

In diatomic molecules potential energy curves are reasonably well known and there is of course no vibration in the atomic fragments. This allows somewhat more reliable conclusions concerning reaction mechanisms deduced from KE measurements. As an example in He^+, O_2 (24) there are two types of O^+ ions having 1.8 and 0.3 eV KE respectively. In a (near)resonant CT process this corresponds to 3.6 and 0.6 eV KE release in the O^+ and O fragments which leaves 21 eV and 24 eV respectively as internal energy of the products ($O^+ + O$) with respect to ground state O_2.

Dissociation limits of O_2^+ close to these values (within the experimental uncertainties) are known: $O^+(^4S°) + O(^1D)$ at 20.7 eV, $O^+(^2P) + O(^3P)$ and $O^+(^2D) + O(^1D)$ at 23.75 and 24.02 eV. The fast (1.8 eV) $O^+(^4S°)$ ions result probably from predissociation of O_2^+ ($^4\Sigma_u^-$, v=0) which is very near resonant with RE(He^+) while the slow (0.3 eV) $O^+(^2P$ or $^2D)$ ions are possibly formed via a dissociative continuum also near resonant with RE(He^+).

It is interesting to notice that dissociative ionization process in O_2 seems to be very energy dependent: in beam experiments at low energies different fragmentation patterns have been observed (25) and in threshold photoion - photoelectron coincidence experiments (26) drastic changes in fragmentation are found when the photon energy is varied from 24.6 to 24.5 eV.

2.2.3. MOLECULAR IONS

As already discussed, observation of kinetically excited molecular ions in a CT reaction is a direct evidence of "close encounter".

This mechanism is usually supposed to give small rate constants at least for singly charged ions on molecules.

However this is not always true as shown on table 3 : small and large rate constants are observed with kinetic energy releases $KE(M^+ + A)$ up to 1.8 eV.

We will briefly discuss the three following cases:

In Ar^+, O_2 (27) since 1.8 eV goes into KE, we are left with 14 eV internal energy for O_2^+. This corresponds, within the experimental uncertainty of the KE measurements, to $O_2^+(X^2\Pi_g$ v=8-10). i.e $\Delta E_{vib} = 1.9 \pm 0.2$ eV.

TABLES 2 and 3: Product internal energy and reaction mechanisms deduced from product ion KE measurements (see text)

TABLE 2: Fragments from He^+ + M						
M	Fragment	$\overline{KE}_{eV}$	AP_{eV}	$\overline{\Delta E}_{vib}$	(M^+)	ΔE
CH_4	H^+	3.2±0.5	21.5	CH_3:3.1 eV	$\tilde{A}^2A_1$	near resonant states
H_2O	H^+	4.2±0.4	23.1	OH:1.5 eV	$^2B_1$ or $\tilde{A}^2A_1$ continuum	
M	Fragment	$\overline{KE}_{eV}$	$IE(O^++O)_{eV}$	Dissociation limits $_{eV}$		(M^+)
O_2	O^+	0.3	24	$O^+(^2D)+O(^1D)$:24.02 $O^+(^2P)+O(^3P)$:23.75		$^4\Pi$dissoc. continuum
		1.8	21	$O^+(^4S°)+O(^1D)$:20.7		$^4\Sigma_u^-$

TABLE 3: Molecular ions from Ar^+ + M → M^+ +Ar						
M	k_t $10^{-9}cm^3s^{-1}$	$KE(M^+)$	$KE(M^++A)$	(M^+)	ΔE_{vib}	$R_{c_\circ}$ Å
O_2	0.06	1	1.8	X	1.9	1.7
H_2O	1.6	0.6_5	0.9_5	$\tilde{X}$ and $\tilde{A}$	2.2	2.2
N_2O	0.31	0.6	1.2_5	$\tilde{X}$	1.7	-

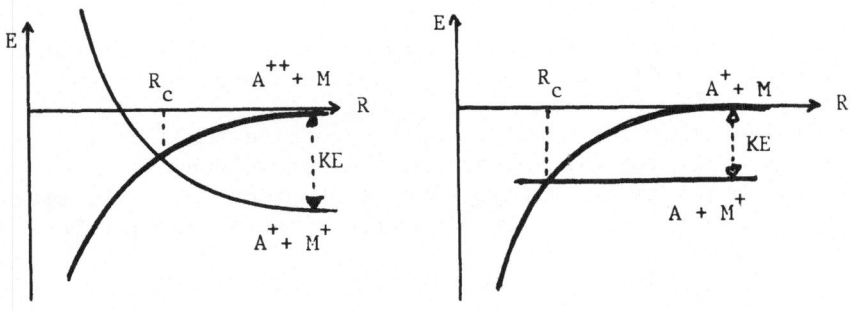

a) Doubly charged ions   b) Singly charged ions

Fig. 2

Franck-Condon Factors for vertical ionization from ground state O_2 to these levels of O_2^+ are very small, but in close encounters undisturbed FCF are not expected to play any role.

    In Ar$^+$, H$_2$O(22) if 0.9 eV goes into KE, H$_2$O$^+$ may be ($\tilde{X}^2B_1$ v=12.13) or (and)($\tilde{A}^2A_1$ v=9.10). Fluorescence experiments (see above 2.1.5) indicate that ($\tilde{A}^2A_1$ v=9.10) accounts for less than 0.4% of the total CT. So ($\tilde{X}^2B_1$ v=12.13) must be the dominant product. In this case $\Delta E_{vib} = 2.2 \pm 0.2$ eV.

    Finally in Ar$^+$, N$_2$O, the fact that no dissociation is observed means that Ar$^+$ populate states below 15 eV (17). Our KE measurement leads to 14.6±0.1 eV internal energy i.e ΔE_{vib}=1.7±0.2 eV. KE estimated in ref 10 would leave N$_2$O$^+$ with internal energies higher than 15.5 eV.

    As shown on table 3   a fairly large amount of the reaction exothermicity goes into vibration of the molecular ion products. Unfortunately we have no information on the vibrational populations since the kinetic energy measurements give only mean values (27b). In N$_2$O$^+$ however, the absence of dissociation indicates a rather narrow distribution around the mean value 1.4 eV.

    Although KE measurements have not yet been performed on thermal energy CT reaction products of doubly charged ions on molecules, recoil energies of several eV have been predicted on the basis of a curve crossing model (9). Simple ion-dipole attractive potential for the reactants (A^{++}+M) and ion-ion repulsive potential for the products (A$^+$+M$^+$) were used to calculate crossing distances R_c ranging from 2.5 to 14.4 Å and KE from 5.77 to 1 eV. The maximum reaction efficiency: (k/k$_L$) for these systems (Mg^{2+} on 7 different molecules) is found for $R_c \sim$ 4-6 Å and KE $\sim$ 2-3 eV. Moreover reaction efficiencies on atoms are three orders of magnitude lower than on molecules with similar values of R_c.
    No theoretical interpretation of these results (especially of the much larger rate constants for molecules than for atoms) could be proposed and the applicability of simple Landau-Zener calculations at thermal energies and for molecules is severely questioned.

    A similar treatment may be applied to CT of singly charged ions on molecules. Using ion-dipole and (or) ion-induced dipole potential for the reactants (A$^+$+M), distance independent energy for the products (A+M$^+$) and the total kinetic energy release as energy difference between reactants and products at infinite distance (see fig. 2 b) gives:

$$R_c = 2.2 \text{ Å for (Ar}^+,H_2O) \text{ and } R_c = 1.7 \text{ Å for (Ar}^+,O_2)$$

This difference at least qualitatively explains the difference in reaction efficiencies.

However: i–The potential curves used in these calculations are very crude approximations since the molecule is considered as a dipole with fixed orientation.

ii– There are certainly more than one state of M^+ populated in the charge transfer so that there must be several exit channels with different crossing probabilities to be considered.

It is very unfortunate that no progress has been made these last years concerning ion molecule interaction potential and curve crossing calculations.

3. CT FROM MOLECULAR AND (OR) ELECTRONICALLY EXCITED IONS

This topic will not be discussed in detail here since there are very few experimental results in the thermal energy range.

Rate constant measurements indicate in most cases large reaction efficiencies as well for molecular ions (27–28) as for electronically excited molecular or atomic ions (24c–29). The only luminescence experiment at thermal energies has been performed with He_2^+ with CO and N_2 (27). A wide range of product ion states have been observed with nearly F.C vibrational populations. This is attributed to the fact that, He_2^+ neutralization being dissociative, there is a wide range of recombination energies so that the resonance condition: $RE(He_2^+) = IP(M)$, is less restrictive than for atomic ions. The large rate constants are thought to be a consequence of long distance electron jump.

Luminescence of CT products of electronically excited ions $(N^+-C^+-O^+)$ on molecules has been studied recently in low energy beam experiments (30–32) and are discussed in another lecture (33). Energy resonance and large FCF seem to play an important role indicating that an electron jump takes place at large distances. There is no obvious reason why CT in these systems should be very different at thermal energies.

CONCLUSION

Most of the results available to date on thermal energy CT reactions concern ground state atomic ions (mainly rare gases) on molecules.

Recent experiments on luminescence spectra and kinetic energy of CT products showed that both long distance energy resonant and short distance non resonant processes may occur.

Electron jump without momentum transfer does not populate, in most cases, the electronic state(s) of M^+ according to vertical F.C rule. This is qualitatively accounted for by a distortion of the electronic state of M by the approaching ion A^+.

Electron jump with momentum transfer leading to 1-2 eV kinetic energy release and ~ 2 eV vibrational excitation has been observed when the CT is not dissociative. This indicates that CT may be an efficient way to produce vibrationally excited molecular ions.

For dissociative CT it is impossible, except in a few cases, to deduce mechanisms from the experimental results available to date: both resonant or a few eV exothermic processes are consistent with the data.

It appeared clearly in this discussion that more information is needed in the following topics:

i- <u>Electronic states of molecular ions and their relaxation channels</u>. One of the most promising methods seems to be threshold Photo-Electron Photo-Ion coincidence measurements (TPES). However, they do not solve the problem of forbidden states which may be populated through charge transfer. Discussion of the $(He^+, O_2)CT(24c)$ clearly illustrates the contribution and limits of TPES results.

ii- <u>Internal states of ground and metastable ions</u>. In all cases, except N_2^+, electronically excited luminescent products are in very small quantity, which means that most products are in ground or metastable states. A possible method to study them is Laser Induced Fluorescence which has been used for neutrals and for ions produced by electrons or discharges but not yet for products of ion-molecule reactions.

iii- <u>Theoretical calculations</u>. As already outlined, our theoretical background to interpret the experimental observations on Charge Transfer (which is apparently the simplest ion-molecule reaction) is very poor.

- Short range effects in resonant charge transfer are only qualitatively accounted for. The recent more elaborated calculations (7) hold only for medium or low collision energies; no extension at thermal energies has been made.

- To interpret CT reactions with momentum transfer we need: good ion-molecule interaction potentials - curve crossing models which would be valid for systems other than atomic ions on atoms. These data do not exist even for simple rare gas atomic ions on diatomic molecules other than He^+ on H_2.

Finally, new developments in the field of charge transfer reactions may be expected mainly for electronically excited and doubly charged ions whose behavior is apparently different from singly charged ground state ions (33). This will be a good topic for the next summer school on ion-molecule reactions.

BIBLIOGRAPHY

1. NATO Advanced study institutes series B.6 Interaction between ions and molecules 1975. Edited by P.Ausloos

   1a) Charge transfer.Panel discussion led by J.Durup p.619-633

   1b) J.H.Futrell in 1a p.627

   1c) K.Birkinshaw, V.Pačak, Z.Herman p.95

1d) R.Marx p.563

1e) W.T.Huntress p.622

2. J.B.Laudenslager, W.T.Huntress Jr, M.T.Bowers, J.C.P. $\underline{61}$, 4600 (1974)

3. P.Ausloos, J.R.Eyler, S.G.Lias, Chem.Phys.Lett. $\underline{30}$, 21 (1975)

4. M.Chau, M.T.Bowers Chem.Phys.Let., $\underline{44}$, 490 (1976)

5. V.G.Anicich, J.B.Laudenslager, W.T.Huntress Jr, J.H.Futrell, J.C.P., $\underline{67}$, 4340 (1977)

6. M.Lipeles, J.C.P., $\underline{51}$, 1252 (1969)

7. J.D.Kelley, G.H.Bearman, H.H.Harris, J.J.Leventhal, Chem.Phys. Let., $\underline{50}$, 295, (1977)

8. J.M.Green, C.E.Webb, J.Phys.B, $\underline{7}$, 1698 (1974)

9. K.G.Spears, F.C.Fehsenfeld, M.McFarland E.E.Ferguson, J.C.P.$\underline{56}$ 2562 (1972)

10. P.W.Harland, K.R.Ryan, Int. J. Mass Spectrom. Ion Phys., $\underline{18}$ 215 (1975)

11. G.H.Rayborn, T.L.Bailey, J.C.P., $\underline{60}$, 813 (1974)

12. P.M. Hierl, V.Pačak, Z.Herman, J.C.P., 67, 2678 (1977)

13. Z.Herman, Private communication

14. G.Mauclaire, R.Marx, C.Sourisseau, C.Van de Runstraat, S.Fenistein, Int.J.Mass Spectrom.Ion Phys., $\underline{22}$, 339 (1976)

15. a) G.Mauclaire, R.Derai, R.Marx, Dynamic Mass Spectrometry, $\underline{5}$, 139 (1978)
    b) G.Mauclaire, R.Derai, R.Marx, Int.J. Mass Spectrom. Ion Phys. $\underline{26}$, 289 (1978)

16. T.R.Govers, M.Gérard, G.Mauclaire, R.Marx, Chem. Phys. $\underline{23}$, 411 (1977)

17. M.Gérard, T.R.Govers, R.Marx, Submitted to Chem. Phys.

18. J.C.Lorquet, C.Cadet, Int.J.Mass Spectrom. Ion Phys., $\underline{7}$, 245(1971)

19. J.J.Leventhal, Private communication

20. T.R.Govers, M.Gérard, R.Marx, Chem. Phys., $\underline{15}$, 185 (1976)

21. M.Gérard, T.R.Govers, R.Marx, Chem.Phys.,$\underline{30}$, 75 (1978)

22. R.Marx,R.Derai, M.Gérard, T.R.Govers, G.Mauclaire, C.Z.Profous, Proc. 26th An.Conf.Mass Spect. and Allied Topics St Louis(1978) in press

23. J.K. Kim, W. T. Huntress Jr., Int. J. Mass Spectrom. Ion Phys., $\underline{16}$, 451 (1975)

24. G.Mauclaire, R.Derai, S.Fenistein, R.Marx
    a) Proc.26th An.Conf.Mass Spect. and Allied Topics St Louis (1978)
       in press
    b) Ne^+ and Ar^+ on O_2 submitted to J.C.P
    c) He^+ on O_2 submitted to J.C.P

25. a) H.Udseth, G.F. Giese, W.R.Gentry,VIIIth ICPEAC 1973
       abstract of papers p. 109

    b) W.R.Gentry this meeting

26. P.M.Guyon, T.Baer, L.F.A.Ferreira, I.Nenner, A.Tabché-Fouhaillé,
    R.Botter, T.R.Govers, J.Phys.B11 L 141 (1978)

27. L.G.Piper, L.Grundel, J.E.Velasco, D.W.Setser, JCP, 62, 3883
    (1975)

28. C.B.Collins, F.W.Lee, JCP, 68, 1391 (1978)

29. J.Glosik, A.B.Rakshit, N.D.Twiddy, N.G.Adams, D.Smith, J.Phys.
    B, 11, 1 (1978)

30. J.A.Rutherford, D.A.Vroom, JCP, 62, 1460 (1975)

31. T.F.Moran, J.B.Wilcox, JCP, 68, 2855 (1978)

32. Ch.Ottinger, J.Simonis, Chem. Phys., 28, 97 (1978)

33. See W. Lindinger and P. Hierl in the Panel on Internal Energy
    Partitioning.

ENERGY DEPENDENCES OF ION-NEUTRAL REACTIONS STUDIED IN DRIFT TUBES

Daniel Lee Albritton

Aeronomy Laboratory, Environmental Research Laboratories,
National Oceanic and Atmospheric Administration, Boulder,
Colorado 80303, U.S.A.

INTRODUCTION

All of us are aware, particularly when we try to keep up with
the literature, of the tremendous growth rate of the field of ion-
neutral reactions. Eighteen years ago, Lampe, Franklin, and Field
(1961) compiled the ion-neutral rate constants that were known at
the time. There were a little over a hundred; now there must be
thousands (e.g., see Ferguson, 1973; Sieck and Lias, 1976; Huntress,
1977; Albritton, 1978; and Sieck, 1979).

Although this growth rate would seem to suggest that the field
is still perhaps in its heady adolescence, there are presently indications of the maturity and sophistication of, say, young adult-
hood, at least. Namely, fifteen to twenty years ago, the basic need
associated with the application of ion-neutral reaction-rate data
was simply knowing the approximate magnitude of the thermal-energy
rate constant: was it large or was it small? For example, because
of the scarcity of even such qualitative data, ionospheric modelers
of that time were forced to make the now-antiquated assumption that
all exothermic ion-neutral reactions proceeded at their gas-kinetic
rate. In contrast to those times, rate constants are now available
for the reactions of a wide variety of ions and neutrals. This
ready availability of a rather large data set has had payoffs in
many areas. For example, when it began to appear that ion-neutral
reactions could possibly be responsible for the formation of interstellar molecules, there were enough of the relevant rate constants
already available to demonstrate the attractiveness of this scheme
and to justify mounting an effort to study the missing links, as is
discussed by D. Smith elsewhere in these proceedings.

Furthermore, it is a clear sign of maturity of the field that, in addition to pursuing simply the thermal-energy rate constant for the reactions of ground-state ions with "gas-bottle" neutrals, more detailed and sophisticated questions are now being asked about ion-neutral reactions and they are often being answered! Some of these new areas of endeavor are (a) the reactions of electronically excited ions and neutrals, (b) the energy dependence of rate constants, (c) the branching ratios of multiple product ions, (d) the possible excited states of these products, and (e) the reaction mechanisms wherewith to rationalize all of these observations. One has only to glance down the titles of the contributions to these proceedings to confirm that these subjects are among the topics being addressed. The particular one on which the present contribution will concentrate is the energy dependence of ion-neutral reaction rate constants.

Although "room temperature" is an energy that is near and dear to us all, most ionic processes occur at energies that are vastly different from the small range in which we just happen to live. If even a small fraction of these processes are to be understood or utilized, it is essential to have data on the energy dependences of the rate constants involved.

Yet, there have been long-standing "gaps" in the rate-constant data that are available as a function of energy. For example, most of the experimental studies of ion-neutral reactions have been made either near room temperature or at energies above a few electron volts. This is because, on the one hand, it is very difficult to operate a typical many-collision "rate-constant" apparatus, such as a flowing afterglow, above about 900 K and, on the other hand, it is equally difficult to operate a typical single-collision "cross-section" apparatus, such as crossed ion and neutral beams, at relative kinetic energies below 1 eV. Interpolation across this energy gap is fraught with pitfalls because, as will be noted below, the energy dependence of rate constants often changes drastically in just this energy range. This energy gap is a particularly awkward one for many fields. For example, in aeronomy, it includes the ionospherically important temperature range of 200 to 2000 K.

A number of new experimental techniques have been devised for the express purpose of providing data in this energy range, e.g. merged beams, which W. R. Gentry discusses elsewhere in these proceedings. Another technique that is being used increasingly is the drift tube, which is, of course, not a new device, having been employed since the beginning of this century to study the non-reactive interactions, or transport properties, of ions in neutral gases (McDaniel and Mason, 1973). The drift tube has been pressed into service in the study of reactive ion-neutral interactions since, simply by changing the drift potential, the average energy in the ion-reactant center of mass can be varied from room temperature

to a few eV. This energy variability, combined with mechanical
simplicity and chemical versatility, gives the technique its appeal.

However, the drift tube is a <u>nonthermal</u> swarm technique and the
characterization of the motion of ions through a gas under the
influence of an electric field is a nontrivial enterprise. Because
of this, the energy dependences of rate constants measured in drift
tubes have many subtle features that should influence their interpretation and application. Only in the last few years has there
been significant progress in understanding the details of the field-
induced, drifting motion of ions. It is appropriate to note here
that it has been <u>ion-neutral</u> <u>reactions</u> that have been used to test
much of this understanding.

In the present contribution, the energy dependences of ion-
neutral reactions are examined as seen through the sampling aperture
of a drift tube. Hopefully, this discussion will answer two complementary questions: (a) what information have drift tubes provided us
about the energy dependences of ion-neutral reactions and (b) what
information has the energy dependences of ion-neutral reactions
provided us about drift tubes? Although these goals may sound
rather limited, they are nevertheless appropriate here for several
reasons. First, other contributions discuss energy dependences from
the viewpoints of different experimental techniques and from the
viewpoint of theory. Secondly, drift tubes of various types are
being used in several ion-neutral reaction-kinetics laboratories
around the world and others are now being constructed or planned.
Finally, even if one is strictly a beam or ICR devotee, one is at
least honor-bound to compare one's results to those that may be
available from drift tubes. Thus, there may be some general interest in the question: what is it that drift tubes <u>really</u> measure?

The steps taken here to answer that question, insofar as now is
possible, are the following. After a very short description of the
basic relevant features of a typical drift tube (just to establish a
common starting point), the three energy characteristics of an ion-
neutral reaction are addressed in terms of drift tubes: (a) what is
the mean relative ion-reactant <u>kinetic</u> energy in a drift-tube reaction-kinetics study, (b) what is the <u>internal</u> energy of the reactant
neutral in such a study, and (c) what is the <u>internal</u> energy of the
drifting reactant ion? Since these three quantities are the energy
parameters of a rate constant, it is important to know how well each
of them is characterized in a drift-tube ion-neutral reaction
investigation.

DRIFT TUBE FEATURES

Figure 1 schematically illustrates the key features of a typical drift tube. Ions are created and are caused to enter a drift

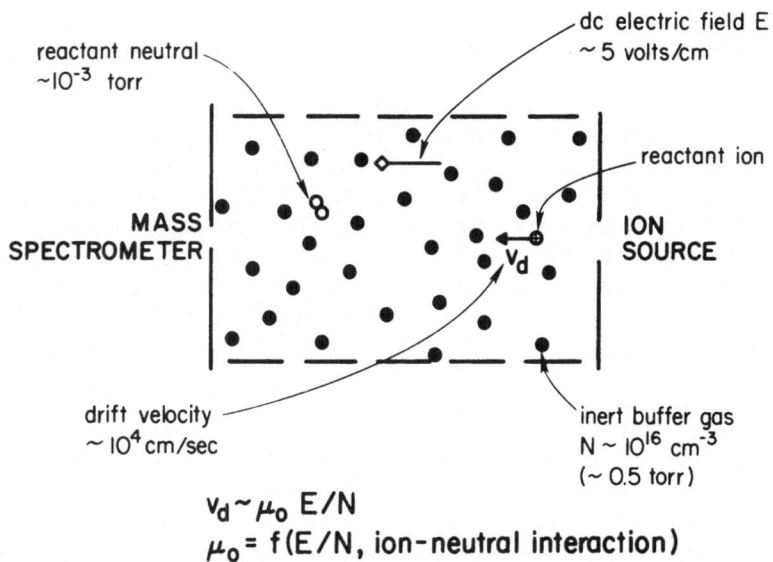

$$v_d \sim \mu_0 \, E/N$$
$$\mu_0 = f(E/N, \text{ion-neutral interaction})$$

Figure 1.   Schematic drift tube and basic definitions.

region that has two characteristics that are of interest to the ions. First, there is an inert buffer gas with a typical number density of 10^{16} cm^{-3} (or about 0.5 torr). Secondly, there is a weak electric field, a few V/cm, directed down the tube. The motions of the ions rapidly adjust to these two parameters. Namely, superimposed on the usual random thermal motion in the gas is a drift motion directed along the electric field. This mean drift velocity v_d is a constant for a given set of conditions; that is, a drift tube is not an accelerator. A typical drift velocity is 10^4 cm/sec. Its actual magnitude depends, first of all, on the driving force, E/N, and then on a factor that describes how large is the response to a given driving force. As shown, this factor, the ion mobility μ_0, depends itself on E/N (therefore, the drift velocity is not linearly proportional to E/N) and the scattering potential between the particular ion and the particular buffer-gas neutral.

Thus, as E/N is increased, the drift velocity increases higher and higher above the mean thermal velocity. In this way, the average kinetic energy of the ions can be varied from room temperature (that is, zero field) to a few eV. Consequently, when one adds small traces of a reactant neutral, reactions occur between the ions and these neutrals and, because of the drift velocity of the ions, these reactions occur at elevated kinetic energies. Consequently, the kinetic energy dependence of a reaction can be studied simply by varying the electric field strength, as has now been demonstrated in several laboratories using different types of drift tubes (e.g., see

Kaneko et al., 1966; Miller et al., 1968; Hiemerl et al., 1969;
Kaneko et al., 1970; McFarland et al., 1973; Elford and Milloy,
1974; and Lindinger et al., 1977).

However, the picture is not as rosy as this brief description
might imply. Several interpretive problems exist, but by far the
classic problem that has been associated with drift tubes throughout
their many-decade history is that they are a nonthermal technique.
What is meant by this is the following. As the ions make their
tortuous way down the drift tube, on the one hand being urged on by
the electric field and on the other hand being held back by scattering off of the buffer gas neutrals, they assume a distribution of
speeds. This speed distribution is an equilibrium distribution;
i.e., it is constant with time. However, the problem is that the
speed distribution of drifting ions is generally non-Maxwellian and
unknown. This shortcoming has several important implications in
terms of using drift tubes to study the kinetic energy dependences
of ion-neutral reaction rate constants.

ION-REACTANT RELATIVE KINETIC ENERGY

To be sure, it is clear that, by increasing E/N and thereby
causing the ions to drift faster, the mean kinetic energy of the
ions has indeed been increased. However, quantitatively, what is
that mean energy?

Determination of the Mean Energy

It has been pointed out that the basic quantity in the motion
of ions in electric fields is E/N. It is straightforward to measure
reaction rate constants as a function of E/N, but this is not a
useful stopping point. For example, rate constants as a function of
E/N cannot be directly compared to rate constants obtained as a
function of temperature or beam center-of-mass energies. What is
required here is the mean ion kinetic energy that corresponds to
each E/N. Even non-Maxwellian distributions must have means and
clearly the mean of a distribution, i.e. its first moment, should be
the easiest of the many moments to estimate. Therefore, estimating
the mean kinetic energy associated with ions drifting at a given E/N
is logically the first step in relating drift-tube rate constants to
the needs of many applications or in comparing with the results from
other types of measurements.

Events have moved so fast in this area recently that one tends
to forget that, just a few years ago, the mean energy of drifting
ions was a topic that was seldom encountered numerically. This
reluctance to quote a mean energy occurred in spite of the fact
that, 25 years ago, theory had provided a simple expression that
related mean energy and E/N:

$$\langle KE_{lab} \rangle \equiv \frac{1}{2}m_i \langle v_i^2 \rangle = \frac{3}{2}kT + \frac{1}{2}m_i v_d^2 + \frac{1}{2}M_b v_d^2. \tag{1}$$

It was derived by Wannier (1953) and consists of three components. The first is easily recognizable as simply the thermal energy of the ion. The second is the kinetic energy associated with the drift motion, since m_i is the mass of the ion and v_d is its drift velocity. The third term, which is the most unusual one since it contains the mass of the buffer gas neutral M_b, is the kinetic energy associated with the scattered, or random, motion of the ion. This simple expression yields the mean kinetic energy as a function of E/N, since the drift velocity can be expressed as a function of E/N (Fig. 1).

However, the Wannier expression had always seemed a bit <u>too</u> simple. The theory on which it was based was only the point-charge, induced-dipole interaction, i.e., the familiar polarization attraction. Since one believes that this is certainly not the only interaction between ions and neutrals, confidence in the general applicability of the Wannier expression has never been very widespread. An example of the reasoning behind this lack of confidence is given in Fig. 2.

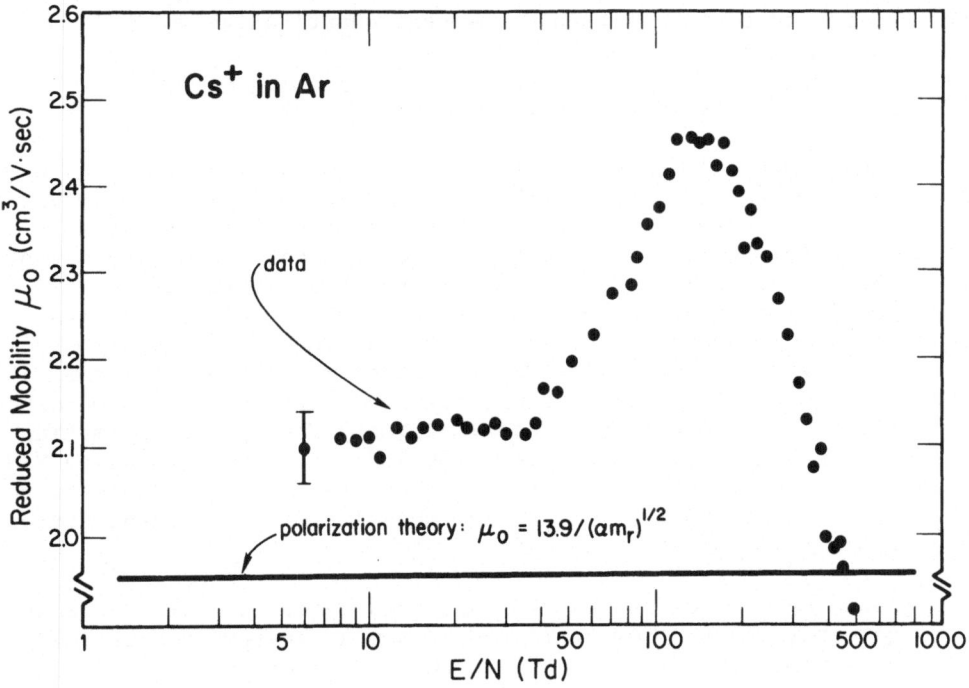

Figure 2. Experimental and polarization-theory mobilities for Cs^+ in argon at 300 K as a function of E/N (1 Td = 10^{-17} V·cm²).

The circles represent the mobility data obtained for Cs^+ in an argon buffer gas by Gatland et al. (1978). For those unfamiliar with the units, a mobility of 2 or 3 cm^2/V·sec is a rather slow mobility, an E/N of 6 Td is a small driving force, and an E/N of 500 Td is a very large driving force. The solid line is the mobility calculated with polarization theory, which expresses the mobility in terms of the polarization α of the buffer gas neutral and the reduced mass M_r of the ion-neutral pair (McDaniel and Mason, 1973). At low E/N, the polarization model does not do too badly; there is only a 7% difference between theory and experiment. Argon is a fairly polarizable atom, so at the low energies, the polarization attraction could well be the dominant interaction. These rather "sticky" collisions are no doubt what gives this somewhat low mobility for this pair. However, as indicated, the polarization mobility is independent of E/N and experiment clearly shows that the real mobility is not. This variation is attributed to the influence of other interactions, such as chemical forces, coming into play at these higher energies. From comparisons like this one, it was thought that if the mobility deduced from polarization theory was inadequate, then the mean energy deduced from the same theoretical framework is probably also inadequate. One almost never, for example, saw mean-energy scales on mobility plots, where they often would be useful, say, in understanding the reasons for the maximum in the mobilities in Fig. 2.

In the last few years, however, the situation has completely reversed. Using Monte Carlo techniques, Skullerud (1973) simulated the drift motion of ions for a wide variety of ion-neutral interaction potentials. In each case, it was found that the numerically obtained mean energy agreed with that given by the Wannier expression within 10%. A little later, Viehland et al. (1974), working with the Boltzmann transport equation, found that a derivative of the observed mobility as a function of E/N can be used to correct even those few percent error in the Wannier expression, all of this, of course, sending the confidence in the Wannier expression soaring almost overnight.

In hindsight, it is clear why this simple expression works so well. If the drift velocity computed from the theoretical mobility in Fig. 2 is used as input in Eq. 1, i.e. to consistently use polarization theory throughout, then the Wannier expression would give poor results. However, use of the experimental drift velocity as input, which is the customary approach, corrects for almost all of the deficiencies in the polarization-model framework.

To this impressive theoretical development, one can add the experimental evidence, also gathered in the last few years, that also demonstrates the reliability of the Wannier expression. Figure 3 shows the rate constants measured for the reaction of N^+ with O_2 using a variety of experimental techniques. The X denotes a room-

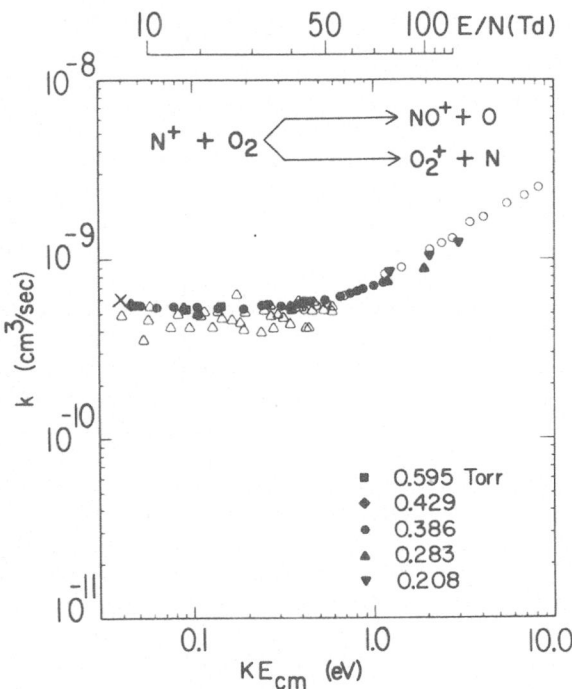

Figure 3. Rate constants for the reactions of N^+ with O_2 as a
    function of mean relative kinetic energy.

temperature flowing-afterglow value obtained by Dunkin et al. (1968),
which is virtually identical with the value obtained recently at the
same temperature by Smith et al. (1978) in a selected-ion flow tube.
The open triangles are the static drift-tube data of Johnsen et al.
(1970) and the solid figures are flow-drift tube data of McFarland
et al. (1973). These drift-tube data were, of course, taken as a
function of E/N, as indicated by the upper scale. The Wannier
expression was used to establish the lower energy scale, which is
the mean energy in the N^+-O_2 center of mass. Although the agreement
at the lower energies between the data of the flow tubes and the
drift tubes is indeed gratifying to all involved, it cannot be used
to test the Wannier-established energy scale. At zero field, the
Wannier expression collapses to only the thermal term and at the
higher energies, uncertainties in the Wannier expression would not
cause the two sets of drift tube data to differ, since this expression was used to plot both sets. However, the open circles are the
crossed beam data of Rutherford and Vroom (1971) extending down to
energies of 1 eV, which was impressively low for such data. The
energies quoted with these crossed-beam data have nothing to do with
the Wannier expreession, of course, since they were established by
distances and voltages. Therefore, the agreement between these data
and the drift-tube data supports the reasonableness of the Wannier-

established energy scale. Other such comparisons (McFarland et al., 1973) provide additional support.

Consequently, on the basis of the transport theory and the drift tube and beam rate-constant comparisons, the simple Wannier expression is now believed to give a mean energy for drifting ions that is accurate to a few percent, which is certainly sufficient for the purpose of expressing the average kinetic energy at which a drift-tube rate constant was obtained.

Determination of Speed Distributions

However, despite the fact that the mean energy can be calculated very easily and very accurately, it corresponds to only one moment of the squared-velocity distribution. Two speed distributions can have the same mean and can still be quite different, say, in their high-energy tails. Equation 2 indicates why this is important:

$$k(E/N) = \int_0^\infty \sigma(v) \, f_{DT}(v) \, v \, dv. \qquad (2)$$

The drift tube can easily provide a rate constant as a function of E/N, having done for us this familiar integration, where $\sigma(v)$ is the energy-dependent reaction cross section, $f_{DT}(v)$ is the speed distribution present under the particular drift-tube conditions, and v is the ion-reactant relative speed. For different conditions, say different buffer gases, f_{DT} can be substantially different; hence, the observed drift-tube rate constant k(E/N) can be different for the same reaction even for the same mean energy in both cases. Thus, it is clear that a drift-tube rate constant is, strictly speaking, a phenomenological quantity. What one ideally wants is the reaction cross section; i.e., to be able to unfold the reaction cross section as a function of energy from the drift-tube rate constants as a function of E/N. In principle, this can be done if the complete drift-tube speed distributions are known, but that is an "if" of considerable magnitude.

Again, theory has made dramatic breakthroughs recently in this area. Lin and Bardsley (1977) have extended the Monte Carlo techniques of Skullerud in the following way. Beginning with a trial ion-neutral scattering potential for an ion-buffer pair of interest, mobilities appropriate to this potential were calculated and compared to those from experiment. The potential was then suitably modified to yield mobilities that better agree with experiment and this procedure was iterated until a satisfactory match was obtained. One then has not only a scattering potential at lower energies than most beams can probe, but also, and more importantly for this discussion, one has the speed distributions that correspond to this

potential for a range of E/N values. With these, the integral in
Eq. 2 can be inverted.

However, it would be reassuring if there were some independent
way to test these theoretical speed distributions, namely, to compare theory and experiment in this regard. Understandably, it is
extraordinarily difficult to <u>directly</u> measure these speed distributions, since the distribution is easily alterable by the measurement process. Such direct measurements have failed to agree with
theory (Lin and Bardsley, 1975).

There is, however, an indirect way to make this test, a way
that is close to the interests of these proceedings, namely, to use
the energy dependence of ion-neutral reactions. If a reaction is
studied under different drift-tube conditions, say different buffer
gases, one can ask that the observed difference in rate constants be
correctly predicted by theory, that is, to deliberately exploit the
phenomenological nature of a drift-tube rate constant. The requirements for a sensitive test are the following: (1) the ion-
neutral reaction selected should have a rate constant that increases
steeply with increasing energy in order to amplify small differences
in the speed distributions, (2) there should be reasons to believe
that the speed distributions of the ion in the two selected buffer
gases should be quite different, and (3) it should be an <u>atomic</u> ion
drifting in <u>atomic</u> buffer gases (no need to complicate things with
structure).

The two reactions selected are

$$O^+ + O_2 \rightarrow O_2^+ + O + 1.6 \text{ eV} \tag{3}$$

$$O^+ + N_2 \rightarrow NO^+ + N + 1.1 \text{ eV.} \tag{4}$$

Both of these exothermic reactions are known to be slow at room
temperature, about 10^{-11} and 10^{-12} cm^3/sec, respectively, and fairly
fast at elevated energies, 10^{-10} cm^3/sec at about 2 eV. Thus, both
rate constants increase rapidly with increasing energy, particularly
the latter, which increases over two orders of magnitude.

Figure 4 shows the observed (solid figures) and polarization-
theory (μ_L) mobilities for O^+ ions in helium (Lindinger and Albritton, 1975) and argon (Dotan et al., 1976). These data present
strong indications that the O^+ speed distributions are different in
these two buffer gases. In helium, O^+ is a heavy ion in a light
buffer gas and, in argon, O^+ is a light ion in a heavy buffer gas.
The magnitude of the mobilities are quite different: O^+ in He is
fast and O^+ in Ar is slow. Furthermore, the two E/N dependences are
dissimilar: O^+ in He rises to a pronounced maximum and O^+ in Ar
declines to a weak minimum.

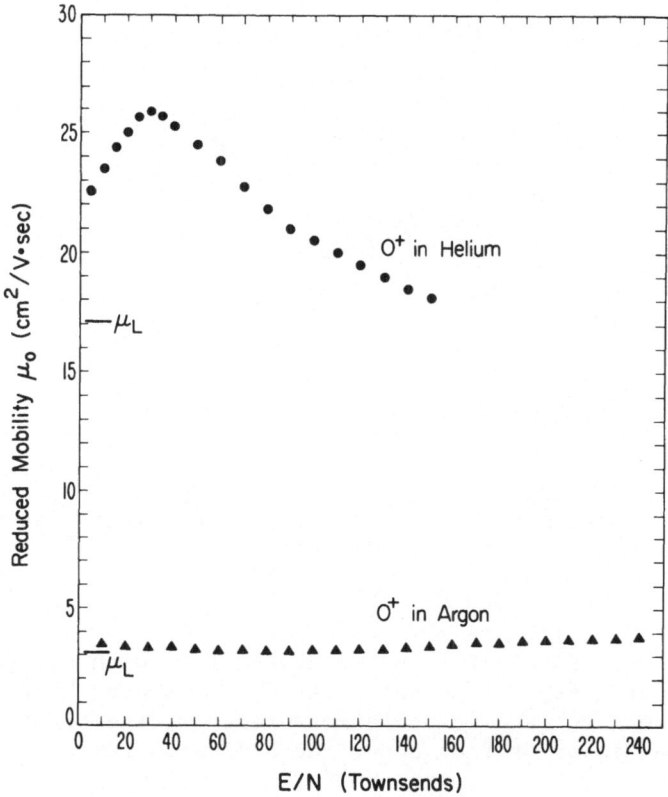

Figure 4. Experimental (solid figures) and polarization-theory (μ_L) mobilities of O^+ ions in helium and argon at 300 K as a function of E/N (1 Townsend = 10^{-17} V·cm^2).

Finally, O^+, He, and Ar are all atomic species, meeting the last of the above three requirements for a sensitive test. Thus, an examination of the reactions of O^+ with O_2 and N_2 in He and Ar buffer gases separately should provide experimental evidence of the effects of ion speed distributions.

Figure 5 shows the experimental results obtained for the reaction of O^+ with O_2. The solid figures represent the rate constants measured in a helium buffer and the open figures represent the values obtained in an argon buffer (Albritton et al., 1976). At low O^+-O_2 energies, there is good agreement with other thermal-energy data (Dunkin et al., 1968, and Smith et al., 1978) and there is very little difference between the data taken in the two buffer gases, as expected, since the two speed distributions should differ very little from a Maxwellian at these low E/N. However, at the higher E/N, where the speed distributions could be different, there is a difference that is much larger than the experimental uncertainties, the argon-buffered data lying distinctly higher. Figure 6

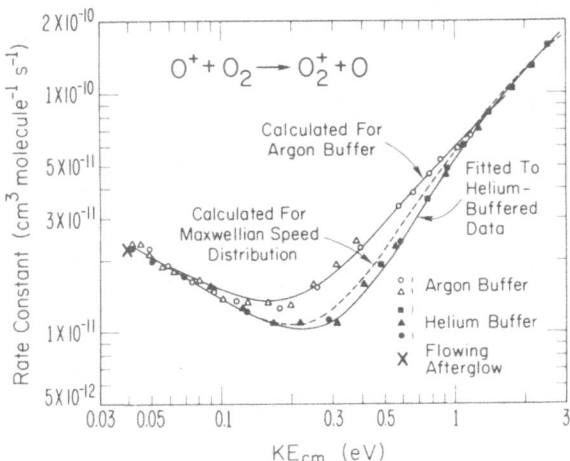

Figure. 5. Rate constants for the reaction of O^+ with O_2 as a function of mean relative kinetic energy.

shows the same type of data for the reaction of O^+ with N_2, which has a much steeper energy dependence. At 0.2 eV where the O^+-in-Ar signal got too weak to make reliable measurements, the argon-buffered data lie a factor of four above the helium-buffered data.

If these differences are due to different O^+ speed distributions in helium and argon, it implies that the O^+-in-Ar distribution is probing much more deeply into the steeply-rising energy-dependent reaction cross section than is the O^+-in-He distribution. That is, the O^+-in-Ar distribution must have a <u>larger</u> high-energy tail. Lin and Bardsley's (1977) Monte-Carlo calculations for O^+ in argon and in helium do indeed show just this, as Fig. 7 demonstrates. Compared here are three O^+ speed distributions, all of which have the same mean energy, 0.64 eV. The dashed line is a Maxwellian distribution for reference. The solid line is the calculated O^+ distribution in helium, which in this case, turns out to be fairly close to the Maxwellian. The dotted line is the calculated O^+ distribution in an argon buffer, which, as can be noted, has a distinctly larger high-energy tail, just as expected from the data.

However, one can make a much more <u>quantitative</u> test of theory and experiment than this, as is shown by the solid curves in Figs. 5 and 6, which were computed in the following way. First, for each reaction, the helium-buffered data and the Monte Carlo O^+-in-He speed distributions were used together to unfold a reaction cross section via Eq. 2. As a check on this unfolding process, the cross section and the speed distributions were simply folded back together to yield calculated rate constants, which are given by the lower

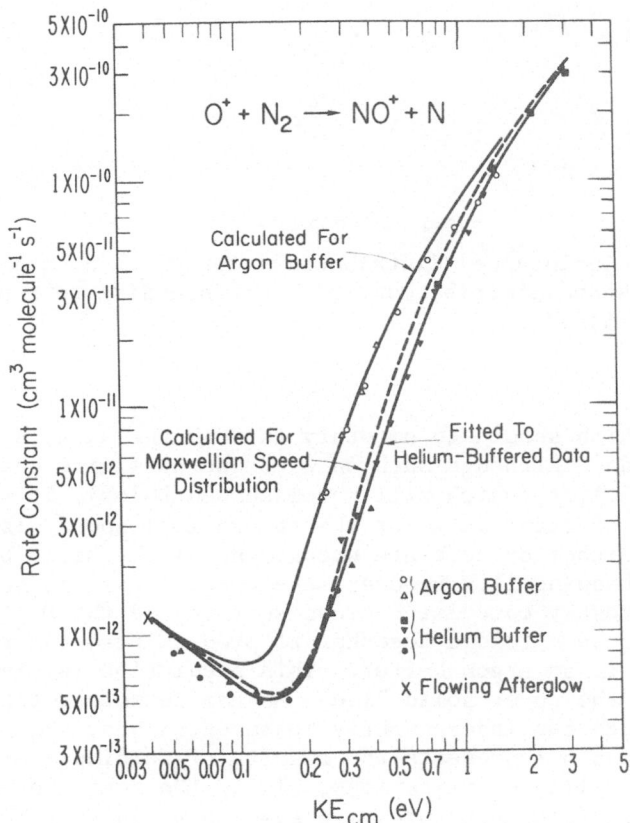

Figure 6.   Rate constants for the reaction of O^+ with N_2 as a function of mean relative kinetic energy.

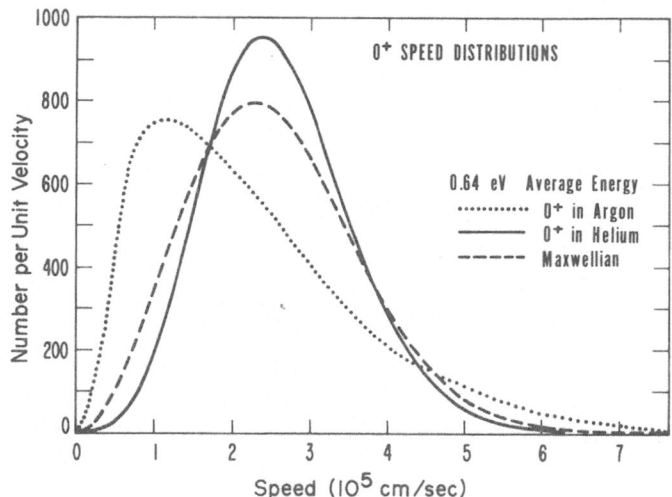

Figure 7. Monte Carlo speed distributions for O^+ in He and Ar and a
    Maxwellian speed distribution, all corresponding to a mean
    energy of 0.64 eV.

solid line and which should go smoothly through the data, as Figs. 5
and 6 show they do. Although such an exercise does give one a
reaction cross section (which will be discussed below), it tests
neither the rate-constant data nor the theoretical speed distributions, since if either or both are wrong, one simply has a bad cross
section without knowing it. However, the crucial test is now
possible. The freshly calculated cross section and the O^+-in-Ar
speed distribution are folded together to predict what the rate constants should be in an argon buffer. This prediction is given in
Figs. 5 and 6 by the upper solid line. As can be noted, the predictions track through the independently measured values. The dashed
line in Figs. 5 and 6 represents the results of folding the cross
section into Maxwellian distributions, giving the rate constant as
a function of kinetic temperature, expressed here in eV. These
values lie very close to the helium-buffered results, which corresponds to our earlier observation that the O^+-in-He speed distribution was fairly close to a Maxwellian. (This is, of course, a
feature of the O^+-He interaction potential in particular and would
not be true for helium buffer gases in general).

    Figures 8 and 9 give these cross sections for the reactions of
O^+ with O_2 and N_2. Both cross sections have a pronounced minimum
and then increase steeply with increasing energy. At the higher
energies, the drift-tube cross sections agree well with those determined independently by crossed beams (Stebbings et al., 1966, and
Rutherford and Vroom, 1971). It is the almost-vertical part of the

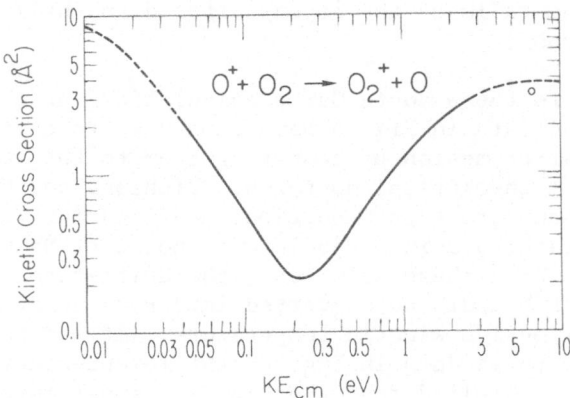

Figure 8. Kinetic cross section deduced for the reaction of O^+
with O_2 from the helium-buffered data in Fig. 5 and Monte
Carlo O^+-in-He speed distributions. The open circle is a
crossed-beam measurement.

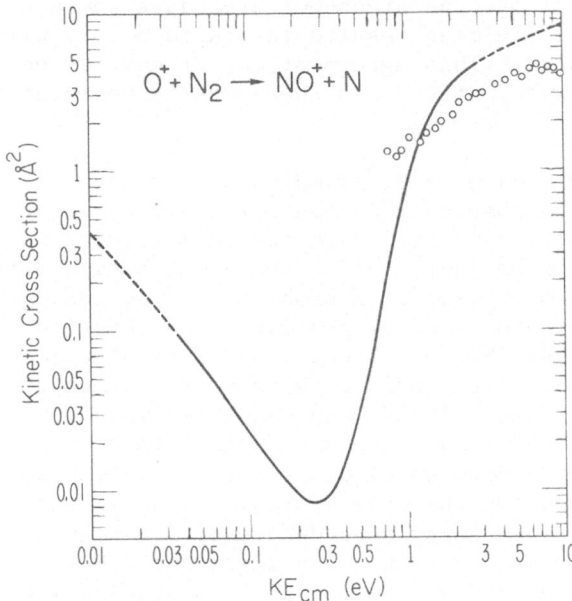

Figure 9. Kinetic cross section deduced for the reaction of O^+
with N_2 from the helium-buffered data in Fig. 6 and Monte
Carlo O^+-in-He speed distributions. The open circles are
crossed-beam measurements.

$O^+ + N_2$ cross section, for example, that effectively amplifies the differences in the tails of the argon-buffered and helium-buffered speed distributions.

In addition to these Monte Carlo calculations of the O^+ speed distributions, the data in Fig. 6 have been used to test the understanding of ion drift motion by comparing them to the results of an entirely different theoretical approach. Viehland and Mason (1977) solved the Boltzmann transport equation, modified for reactive loss, in an iterative fashion using a judicious choice of Maxwellian-like basis functions. With these solutions, the drift-tube rate constants as a function of E/N could be converted into rate constants as a function of an effective kinetic temperature. It was found that, for the helium-buffered data in Fig. 6, the lowest-order solution was adequate, which implied that the kinetic temperature associated with each E/N was given simply by $\frac{3}{2}kT_{eff} = KE_{cm}$, where KE_{cm} is the mean relative kinetic energy based on the Wannier expression. This simplicity corresponds to the finding in the Monte Carlo study that the O^+-in-He speed distribution closely resembled a Maxwellian at all E/N values investigated. For the argon buffer, however, Viehland and Mason found that higher-order solutions were required and hence the relation between T_{eff} and KE_{cm} was more complicated. Using the higher-order relations, the argon-buffered data could be plotted as a function of T_{eff} yielding results that agreed very well with the helium-buffered data. This agreement is, of course, equivalent to the agreement between Monte Carlo theory and experiment shown in Fig. 6.

These $O^+ + O_2$ and $O^+ + N_2$ examples were reactions that were, as mentioned above, deliberately chosen for their hyperactive energy dependences. Most ion-neutral reaction rate constants do not vary this dramatically with energy and speed-distribution effects would be expected to be much less of a problem. Thus, not only do such quantitative atomic-ion studies provide a "calibration" for the magnitude of speed-distribution effects, but they also demonstrate the utility of using different buffer gases to experimentally test for the magnitude of these effects when quantitative calculations cannot be made. From all of this, one can conclude that the long-standing problem of non-Maxwellian speed distribution of ions in drift tubes has now been solved for the case of atomic ions in rare-gas buffers; or alternatively, the energy dependences of ion-molecule reactions obtained in drift tubes for the very important 0.04-3 eV energy range need no longer suffer from possible obscuring speed-distribution effects and therefore can be used, without reservation, to deduce properties of the reactive collision itself.

REACTANT-NEUTRAL INTERNAL ENERGY

The internal energy of the reactant neutral can also play a role in an ion-neutral reaction. What does the electric field of the drift tube do to neutral reactants? The answer is, of course, nothing. The electric field "heats" only the ions; therefore, it is important to remember that, in drift-tube studies of ion-neutral reaction energy dependences, the reactant neutral is, in general, at room temperature.

Consequently, for the reaction of O^+ with N_2, for example, this drift-tube property implies that the helium-buffered rate constant data shown in Fig. 6 <u>cannot</u> be used directly in an application where the O^+ ions and N_2 neutrals are in thermal equilibrium at high temperatures, despite the fact that the O^+-in-He kinetic speed distributions have been shown to be very nearly Maxwellian. The reason is that this reaction is known to be separately sensitive to the vibrational temperature of N_2. Schmeltekopf <u>et al</u>. (1968) varied the N_2 vibrational temperature from 300 to 6000 K in their flowing-afterglow study, in which the ion-reactant <u>kinetic</u> energy was 300 K throughout, and found that the reaction rate constant increased a factor of 50 over that range. The drift-tube study of the kinetic-energy dependence of this reaction as depicted in Fig. 6 was, of course, blind to this vibrational effect, which is, in many ways, a scientific blessing. This "blindness" has permitted the kinetic and vibrational effects in this important reaction (see the aeronomic application discussed by E. E. Ferguson in these proceedings) to be examined <u>separately</u>, a point of no little value in unraveling the reaction mechanism involved.

Another example of how this ostensible shortcoming can be, in reality, an advantage is the reaction of Ne^+ with N_2, which is a second example of a reaction known to be very sensitive to the vibrational temperature of N_2 (Albritton <u>et al</u>., 1973). The reaction is immeasurably slow at room temperature, but increases dramatically when the N_2 vibrational temperature is increased. Since the Ne^+-N_2 kinetic energy in this flowing-afterglow study was, like $O^+ + N_2$ above, near 300 K throughout, it raises the question whether it too will be sensitive to kinetic energy as well. In an unpublished flow-drift tube study in our laboratory, it was found that the rate constant remained less than 10^{-12} cm^3/sec over the mean relative kinetic-energy range 0.04 to 2 eV. Such negative evidence is, of course, nevertheless useful, since it identifies the reaction of Ne^+ with N_2 as one that, in contrast to the reaction of O^+ with N_2 above, is sensitive to <u>only</u> the internal energy of the reactant neutral.

The fact that the reactant neutrals are at room temperature in drift-tube studies must be borne in mind when comparing drift-tube

energy dependences to those expected from collision theories.   For
example, the averaged-dipole-orientation (ADO) theory (Su and Bowers,
1973a and 1973b) was designed for situations where the ions and
reactants are equilibrated at one temperature.   For comparison to
drift-tube energy dependences, however, one would want a "two-temp-
erature" ADO theory in which one temperature describes the rotation
of the neutral reactant, which always corresponds to room temperature, and another temperature describing the ion-reactant relative
kinetic energy, which is, of course, dependent on E/N.

REACTANT-ION INTERNAL ENERGY

An ion-neutral reaction can also be sensitive to the internal
energy of the reactant ion.   Therefore, it is important to ask how
well this energy parameter is characterized in drift-tube reaction-
kinetics studies.   In contrast to the room-temperature neutral, the
ion, being urged onward by the electric field, is colliding its way
down the tube through the buffer gas.   A fraction of the drift-
induced kinetic energy could possibly be transferred, via the ion-
buffer collisions, into ionic internal energy.   Hence, it is possible
that drifting non-atomic ions have both suprathermal kinetic energy
and suprathermal internal energy.   Clearly, if drift tubes are to be
used to examine the energy dependence of ion-neutral reactions up to
a few eV, it is important to be able to assess, at least qualitatively at first, the role of possible ionic internal energy.   Some
progress has been made regarding internal excitation of drifting
ions and its effect on ion-neutral reactions, but it is by no means
as quantitative yet as the situation described above for atomic-ion
speed distributions.   The remainder of this contribution will summarize this progress.

The observations of collision-induced dissociation of polyatomic positive and negative ions at high E/N (e.g. Albritton et
al., 1968, and Dotan et al., 1976) strongly suggest large internal
excitation.   In addition, the energy dependence of the rate constants of the forward and reverse directions of nearly thermoneutral
proton-transfer reactions indicate the occurrence of internal
excitation, which enhances the rate constants of the endothermic
direction and inhibits those of the exothermic direction (Lindinger
et al., 1975).   Furthermore, laser photodetachment studies (Cosby et
al., 1975) of drifting negative ions have reported changes in the
detachment cross section at high E/N that can be interpreted as
being due to increased internal excitation.   To these studies, one
can now add recent drift-tube investigations that have correlated
changes in internal-energy-sensitive rate constants with the addition of a trace gas that rapidly quenches the ionic internal energy.

Figure 10 shows the rate constants for the reaction of O_2^+ with
CH_4 measured as a function of mean energy in helium and argon buffer

gases separately (Dotan et al., 1978). At thermal energies, this
reaction is very slow, 7×10^{-12} cm^3/sec, in agreement with earlier
room-temperature studies. As indicated, the reaction has proven to
be quite sensitive to increasing energy, the rate constant increasing a factor of 15 by 1 eV. Furthermore, the increase in the argon-
buffered measurements is considerably larger than that in the helium-
buffered measurements and the difference is much larger than one
would expect on the basis of speed distributions alone. One strongly
suspects that the O_2^+ ions drifting in the heavier argon buffer,
where the transfer of kinetic to internal energy is much more efficient, are internally excited and that this excitation is promoting the reaction over and above what just the kinetic energy alone
does. This suspicion was confirmed by the recent results shown in
Fig. 11 (Albritton et al., 1979).

The solid lines are smooth representations of the argon- and
helium-buffered data in Fig. 10. The data points are two additional
sets of measurements made in the following way. A small amount of
extra oxygen is added to the argon buffer gas to quench the O_2^+
internal excitation via resonant charge transfer, but not enough
additional oxygen to alter the O_2^+-in-Ar speed distributions. As
Fig. 11 shows, when the O_2 concentration is increased up to 3% in

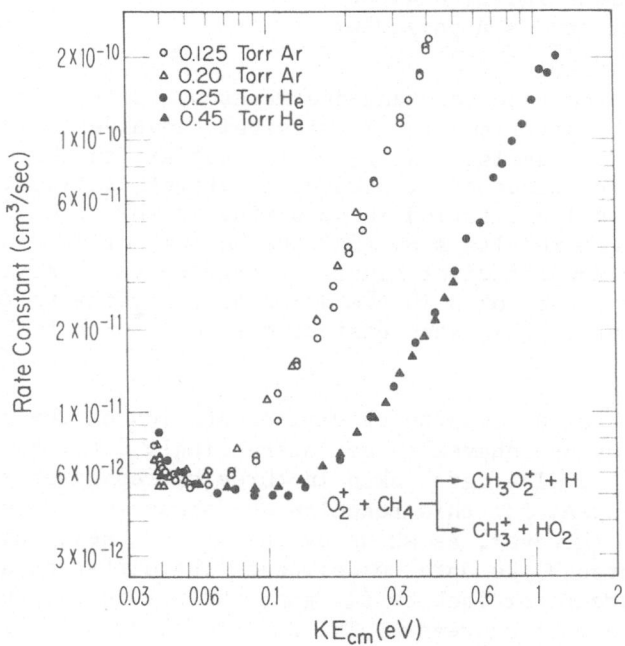

Figure 10. Rate constants for the reaction of O_2^+ with CH_4 as a
    function of mean relative kinetic energy.

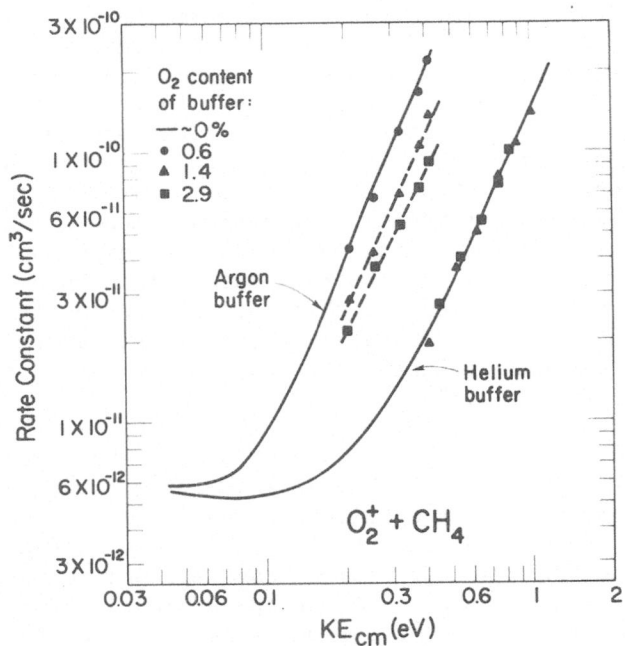

Figure 11.   Rate constants for the reaction of O_2^+ with CH_4 as a
      function of mean relative kinetic energy, with the O_2 content
      of the buffer gas as a parameter.

the argon buffer, the rate constant decreases steadily, since the
O_2^+ internal-energy distribution progressively involves only the
lower, less-reactive levels.   In the helium buffer, on the other
hand, the same percentages of O_2 produce no effect, indicating
little or no internal excitation is occurring in the lighter helium
buffer.   Thus, these results show that the helium-buffered data are
exhibiting purely the effect of kinetic energy on the reaction and
the additional increase found in the argon buffer reflects the
further enhancement of the rate constant due to O_2^+ internal excitation.

    Figure 13 shows the results of similar studies on the reaction
of CO_2^+ with H_2, which behaves in an interestingly different, but
consistent, fashion.   The data taken in three buffer gases all show
that the rate constant for this reaction decreases as the collision
energy increases.   However, as shown by the solid lines, which are
smoothly drawn through the data to delineate the differences, the
decrease is different for each buffer gas, the heavier argon buffer
gas giving the steepest decrease.   The additional data in Fig. 13
show that these differences are due to internal excitation of the
CO_2^+ ions in the heavier buffer gases and its suppression of the
reaction rate contant.   The solid lines in Fig. 13 are the smooth
representations of the helium- and argon-buffered data in Fig. 12.

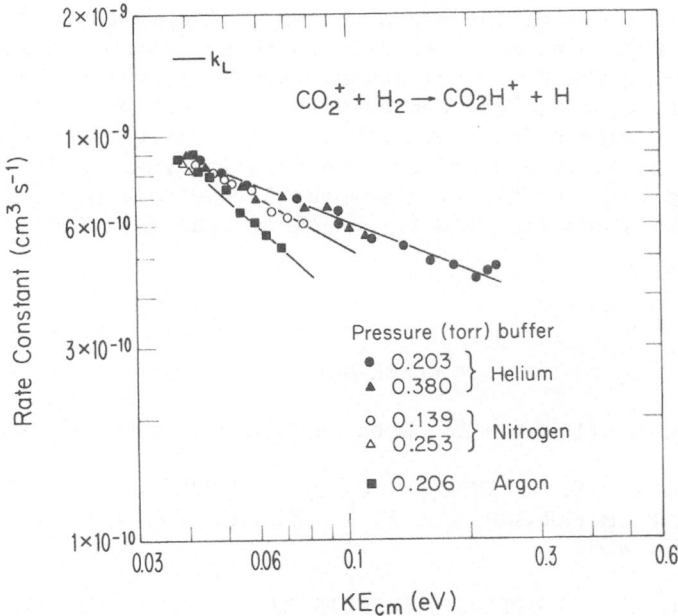

Figure 12. Rate constants for the reaction of CO_2^+ with H_2 as a function of mean relative kinetic energy.

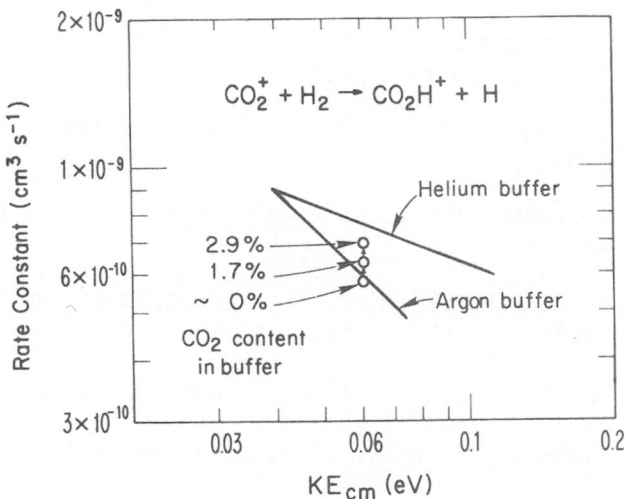

Figure 13. Rate constants for the reaction of CO_2^+ with H_2 as a function of mean relative kinetic energy, with the CO_2 content of the buffer gas as a parameter.

When the CO_2 content of the argon buffer is increased to a few
percent, the rate constants steadily increase, since the additional
CO_2 is quenching the CO_2^+ excitation, thereby removing the inhibition
on the reaction. While one can only thus far estimate the extent of
excitation, perhaps a few vibrational levels, these data provide hard
evidence for the existence of considerable internal excitation of
ions drifting in heavy buffer gases and demonstrate how drift tubes
can be used to study its effects on ion-neutral reactions.

REFERENCES

ALBRITTON, D. L. (1978). At. Data Nucl. Data Tables 22, 1.

ALBRITTON, D. L., Y. A. BUSH, F. C. FEHSENFELD, E. E. FERGUSON, T. M.
    GOVERS, M. McFARLAND, and A. L. SCHMELTEKOPF (1973). J. Chem.
    Phys. 58, 4036.

ALBRITTON, D. L., I. DOTAN, W. LINDINGER, M. McFARLAND, J. TELLING-
    HUISEN, AND F. C. FEHSENFELD (1976). J. Chem. Phys. 66, 410.

ALBRITTON, D. L., W. LINDINGER AND F. C. FEHSENFELD (1979). J.
    Chem. Phys. (to be submitted).

ALBRITTON, D. L., T. M. MILLER, D. W. MARTIN, AND E. W. McDANIEL
    (1968). Phys. Rev. 171, 94.

COSBY, P. C., R. A. BENNETT, J. R. PETERSON, AND J. T. MOSELEY (1975).
    J. Chem. Phys. 63, 1612.

DOTAN, I., F. C. FEHSENFELD, AND D. L. ALBRITTON (1978). J. Chem.
    Phys. 68, 5665.

DOTAN, I., W. LINDINGER, AND D. L. ALBRITTON (1976). J. Chem.
    Phys. 64, 4544.

DUNKIN, D. B., F. C. FEHSENFELD, A. L. SCHMELTEKOPF, AND E. E.
    FERGUSON (1968). J. Chem. Phys. 49, 1365.

ELFORD, M. T., AND H. B. MILLOY (1974). Aust. J. Phys. 27, 795.

FERGUSON E. E. (1973). At. Data. Nucl. Data Tables 12, 159.

GATLAND, I. R., M. G. THACKSTON, W. M. POPE, F. L. EISELE, H. W.
    ELLIS, AND E. W. McDANIEL (1978). J. Chem. Phys. 68, 2775.

HEIMERL, J., R. JOHNSEN, AND M. A. BIONDI (1969). J. Chem. Phys.
    51, 5041.

HUNTRESS, W. T. (1977). Astrophys. J. Suppl. 33, 495.

JOHNSEN, R., H. L. BROWN, AND M. A. BIONDI (1970). J. Chem. Phys. 52, 5080.

KANEKO, Y., N. KOBAYASHI, AND I. KANOMATA (1970). Mass Spectrosc. 18, 920.

KANEKO, Y., L. R. MEGILL, AND J. B. HASTED (1966). J. Chem. Phys. 45, 3741.

LAMPE, F. W., J. L. FRANKLIN, AND F. H. FIELD (1961). Prog. React. Kinet. 1, 67.

LIN, S. L., AND J. N. BARDSLEY (1975). J. Phys. B 8, L461.

LIN, S. L., AND J. N. BARDSLEY (1977). J. Chem. Phys. 66, 435.

LINDINGER, W., AND D. L. ALBRITTON (1975). J. Chem. Phys. 62, 3517.

LINDINGER, W., E. ALGE, H. STÖRI, M. PAHL, AND R. N. VARNEY (1977). J. Chem. Phys. 67, 3495.

LINDINGER, W., M. McFARLAND, F. C. FEHSENFELD, D. L. ALBRITTON, A. L. SCHMELTEKOPF, AND E. E. FERGUSON (1975). J. Chem. Phys. 63, 2175.

MILLER, T. M., J. T. MOSELEY, D. W. MARTIN, AND E. W. McDANIEL (1968). Phys. Rev. 173, 115.

McDANIEL, E. W., AND E. A. MASON (1973). The Mobility and Diffusion of Ions in Gases (John Wiley and Sons, Inc., New York).

McFARLAND, M., D. L. ALBRITTON, F. C. FEHSENFELD, E. E. FERGUSON, AND A. L. SCHMELTEKOPF (1973). J. Chem. Phys. 59, 6610, 6620, and 6629.

RUTHERFORD, J. A., AND D. A. VROOM (1971). J. Chem. Phys. 55, 5622.

SCHMELTEKOPF, A. L., E. E. FERGUSON, AND F. C. FEHSENFELD (1968). J. Chem. Phys. 48, 2966.

SIECK, L. W. (1979). NSRDS-NBS 64, U. S. Printing Office, Washington D.C. (in press).

SIECK, L. W., AND S. G. LIAS (1976). J. Phys. Chem. Ref. Data 5, 1123.

SKULLERUD, H. R. (1973). J. Phys. B 6, 728.

SMITH, D., N. G. ADAMS, AND T. M. MILLER (1978). J. Chem. Phys.
     69, 308.

STEBBINGS, R. F., B. R. TURNER, AND J. A. RUTHERFORD (1966). J.
     Geophys. Res. 71, 771.

SU, T., AND M. T. BOWERS (1973a). J. Am. Chem. Soc. 95, 1370.

SU, T., AND M. T. BOWERS (1973b). Int. J. Mass Spectrom. Ion
     Phys. 12, 347.

WANNIER, G. H. (1953). Bell Syst. Tech. J. 32, 170.

VIEHLAND, L. A., AND E. A. MASON (1977). J. Chem. Phys. 66, 422.

VIEHLAND, L. A., E. A. MASON, AND J. H. WHEALTON (1974). J. Phys.
     B 7, 2433.

INTERNAL ENERGY PARTITIONING:

SUMMARY OF THE PANEL DISCUSSION

J. J. Leventhal

Department of Physics, University of Missouri

St. Louis, Missouri, U.S.A. 63121

From the standpoint of relevance to this panel the term "internal energy partitioning" was interpreted very broadly. Thus, contributions to the session included discussion of the effects of reactant internal energy as well as the distribution of internal energy among accessible product final states. All participants reported experimental results, and a wide range of techniques were represented. For experiments in which reactant internal energy effects were studied flow drift tubes or selected ion flow tubes (SIFT) were the primary sources of data. Also included in this discussion was a report of ion fragmentation threshold measurements using photoionization. Various beam techniques were employed for study of product internal energy partitioning. These included kinematic analysis in beam gas and crossed beam apparatuses and detection of collision produced luminescence in a beam-gas setup.

The popularity of the subject led to an unusually large number of panelists (eleven), so that the discussion was extended an additional hour making it a four hour session.

The following summary of the separate contributions is divided into two categories: Effects of Reactant Internal Energy and Product Internal Energy Distributions. Inevitably, some of the contributions would be appropriate in either category.

A. EFFECTS OF REACTANT INTERNAL ENERGY
    N. D. Twiddy: EXCITED ION REACTIONS

Reaction rates of ground and metastable states of O^+, NO^+, O_2^+, C^+, N^+, S^+ and N_2^+ with H_2, CO, NO, O_2, CO_2, H_2O, NH_2 and CH_4 at 300K were measured and product ion distributions determined. The apparatus

used was the Aberystwyth SIFT outfitted with a low pressure electron
impact source. This source was capable of producing the above ions
with 10-45% in metastable states. The monitor ion method in which
a monitor gas M is introduced and reacts only with excited species
was used to distinguish between ground state and excited state reactions. In the case of NO^{+*}, S^{+*} and N_2^{+*} reactions collisional
quenching often accounts for half of the loss of metastable ions.
For S^+ and N_2^+ it was possible to distinguish two metastable states
through their markedly different reaction rate coefficients.

D. Smith: REACTIONS OF $Ar^{++}(^3P, {}^1S_0)$ AND $Xe^{++}(^3P, {}^1D_2, {}^1S_0)$

This work was done in collaboration with N. G. Adams.

Measurements were reported of the reactions of the ground (^3P)
and metastable electronically excited $(^1D_2$ and $^1S_0)$ states of Xe^{++}
ions with several atoms (Ar and Xe) and molecules (H_2, N_2, O_2 and CO_2)
at 300K (N. G. Adams et.al., 1979). The data were obtained using a
Selected Ion Flow Tube (SIFT) apparatus. State selection was achieved
by introducing suitable "chemical filter" gases into the flow tube.
The rate coefficients for the reactions with the molecules were large
with the exception of some of the reactions of the $^1S_0$ state. Conversely, the atom reactions were generally slow except for the ^3P
state reaction with Ar in which a favorable pseudo energy resonance
must occur. All the molecular reactions proceeded via single charge
transfer and in the $^1D_2$ reactions with molecules dissociation of the
product ion was usually observed. Significant quenching of the $^1D_2$
state was evident in the reactions with all of the molecules studied.

I. Dotan: ON THE PRODUCT RATIOS IN Ar^+-NO AND
            $NO^+(a^3\Sigma^+)$-Ar REACTIONS AT 300K

This work was done in collaboration with D. L. Albritton and
F. C. Fehsenfeld.

The reaction of Ar^+ with NO yield to the possible ground-state
$NO^+(X^1\Sigma^+)$ and metastable-state $NO^+(a^3\Sigma^+)$ product ions has been
studied near 300K in a flow-tube using helium and argon buffer gases
separately.

$$Ar^+ + NO \rightarrow NO^+(X^1\Sigma^+) + Ar + 6.496 \text{ eV}$$

$$\rightarrow NO^+(a^3\Sigma^+) + Ar + 0.092 \text{ eV}$$

The "apparent" rate constant obtained in the argon buffer is an
order of magnitude smaller than the rate constant obtained in the
helium buffer. The difference in the rate constants is due to the
"back" reaction of the $NO^+(a^3\Sigma^+)$ product ions with the argon buffer-
bas atoms to yield the possible product ions $NO^+(X^1\Sigma^+)$ and Ar^+.

$$NO^+(a^3\Sigma^+) + Ar \rightarrow NO^+(X^1\Sigma^+) + Ar + 6.404 \text{ eV}$$

$$\rightarrow Ar^+ + NO - 0.092 \text{ eV}$$

The ratio of the rate constants places precise limits on the product ratios in both cases; namely, the Ar^+ + NO reaction yields greater than 90% $NO^+(a^3\Sigma^+)$ ions, and/or the $NO^+(a^3\Sigma^+)$+Ar reaction yields greater than 90% Ar^+ ions.

These are the first reported results of the branching ratios of the two reactions measured at rcom temperature.

W. Lindinger: REACTIONS OF DOUBLY CHARGED IONS

Data on the energy dependence of reactions of state selected $Ar^{++}(^3P)$ and $Ar^{++}(^1S)$ ions with various neutral species were presented. Reactions of $Ar^{++}(^1S)$ with N_2, O_2, CO_2, C_2H_2 and CH_4 at thermal energies proceed with rates that are the order of Langevin. A similar situation exists for $Ar^{++}(^3P)$ with CH_4, C_2H_2 and NO_2. It is found that cross sections are generally largest if the energy defect is 3-5 eV. A curve crossing model was shown to be consistent with the observations in all cases involving single electron transfer.

H. M. Rosenstock: PRECISION MEASUREMENTS OF ION FRAGMENTATION
        THRESHOLDS

This work was done in collaboration with K. E. McCulloh and R. L. Stockbauer.

Ion fragmentation threshold measurements have been used for many years to obtain ion energetics information. Early work with electron impact techniques suffered from poor accuracy and precision because of the energy spread of the electron beam and contact potentials. More recently the use of photoionization has eliminated these problems and permitted one to concentrate on the fragmentation processes themselves. Even for small molecules, at finite temperature the fragmentation "thresholds" are not well defined because of thermal rotational population effects. For large molecules, in addition, one has to consider vibrational hot bands and also the unimolecular kinetics of the fragmentation process. Recent work using variable temperature photoionization and time-delayed photoion photoelectron coincidence techniques is described. It is possible to obtain quantitative results on rotational effects on small ion fragmentation thresholds. In addition for large molecules it is possible to demonstrate the shift in fragmentation behavior due to unimolecular kinetics ("kinetic shift"). A number of accurate fragmentation threshold values are obtained, corresponding to processes at absolute zero temperature.

B.   PRODUCT INTERNAL ENERGY DISTRIBUTIONS

P. M. Hierl: KINEMATIC MEASUREMENTS OF PRODUCT EXCITATION

This work was done in collaboration with V. Pacak and Z. Herman.

Product angular and velocity vector distributions have been measured in a cross beam experiment for the charge transfer process $Ar^+ + H_2 \rightarrow Ar + H_2^+$ at relative collision energies of 0.13, 0.48, and 3.44 eV. We have found (P. M. Hierl et.al., 1977) that the process proceeds in a direct manner rather than by the formation of a long-lived intermediate complex over the energy range studied, and that two distinct reaction mechanisms are operative: (a) an electron jump at large separations occurs at all collision energies, and (b) an intimate collision mechanism becomes increasingly important as the collision energy is lowered.   Secondly, the results show that electron transfer competes effectively with atomic rearrangement in small impact parameter collisions, particularly at the lower collision energies.   Thirdly, energy disposal depends upon the reaction mechanism: the electron-jump process produces a rather narrow distribution of internal states of H_2^+, with the most highly populated internal states of the product ion being those most nearly resonant with the recombination energy of the reactant ion.   The intimate collisions, on the other hand, produce H_2^+ in a broad range of vibrational states, the most probable channels being those which are slightly endothermic.

We have also performed kinematic studies of partial charge transfer reactions of the general type $A^{++} + B \rightarrow A^+ + B^+$.   Conservation of energy permits one to deduce the ionization potential of B from a knowledge of the recombination energy of the reactant ion and measurements of the initial and final relative translational energies, E_T and E_T', respectively:

$$IP(B) = RE(A^{++}) - (E_T' - E_T).$$

In agreement with Lindinger's results (this panel), we find that the most probable reaction products are those which have an energy defect (at infinite separation) of 2-5 eV.   For example, the reaction of Ar^{++} with CH_4 produces Ar^+ ions whose velocity distribution is found to consist of two clearly resolved components, indicating the occurrence of two different reaction mechanisms.   One component (I) was characterized by the release of considerable translational energy in the products, the most probable value of the translational exoergicity, $Q = E_T' - E_T$, being about 4 eV.   The second peak in the Ar^+ velocity distribution corresponded to a net release of about 1.5 eV of translational energy.   We have interpreted Process I as being the reaction

$$Ar^{++}(^3P) + CH_4 \rightarrow Ar^+(^2P) + CH_4^+(2a_1)^{-1}$$

where the term $(2a_1)^{-1}$ signifies that ionization occurs by the loss of an electron from the $2a_1$ molecular orbital of CH_4. This represents formation of the CH_4^+ ion in its first electronically excited state. We ascribe Process II to the reaction

$$Ar^{++}(^3P) + CH_4 \rightarrow Ar^+(^2S) + CH_4^+ (1t_2)^{-1}$$

which produces CH_4^+ in its ground electronic state but leaves the Ar^+ ion in its first electronically excited state. We saw no evidence for the occurrence of a reaction channel which simultaneously produced both product ions in their ground electronic states, a process for which the energy defect would have been about 14.6 eV.

S. Zimmermann: PRODUCT TRANSLATIONAL ENERGY SPECTRA FROM THE REACTION N^++CO

This work was done in collaboration with Ch. Ottinger.

In a tandem mass spectrometer the products NO^+, CN^+, O^+, C^+ and CO^+ from N^+-CO collisions were observed in the forward direction and their kinetic energy distributions determined. At 5 and 8 eV (lab) projectile energy the NO^+ product is observed at a kinetic energy consistent with the formation of an intermediate NOC^+ complex. At 10 and 14 eV the peak of the NO^+ distribution shifts to a position consistent with a stripping process. From 20 to 60 eV the Q value becomes less than for stripping, peaking at a constant value of about 10 eV, with a secondary peak at 20 eV. Tentatively, this points to electronic excitation of the NO^+ and, for the secondary peak, of the C atom as well. Furthermore, from 30 eV up to 200 eV N^+ energy, a peak at about zero lab energy is observed in the NO^+ kinetic energy distribution. This peak is explained by a collinear, hard-sphere like exchange reaction $N^+ + CO \rightarrow NO^+ + C$. The CN^+ distributions show very similar features. O^+, C^+ and CO^+ are predominantly observed at very small kinetic energies.

J. J. Kaufman: DETERMINATION OF ION PRODUCT STATES FROM THE LIMITING VALUES OF TRANSLATIONAL EXOERGICITY

This work was done in collaboration with W. S. Koski.

In the general reaction

$$A^+ + BC \rightarrow AB^+ + C$$

the conservation of energy permits the following relation for Q:

$$T_R + U_R - \Delta H = T_p + U_p$$

$$Q = U_R - U_p - \Delta H$$

T and U represent the translational and internal energies of the reactants (R) and the products (P), respectively, and ΔH is the heat of reaction. When U_p reaches the product dissociation energy, Q reaches a plateau in the plot of Q vs projectile kinetic energy. This limiting value of Q, designated as Q_{MIN} has been used to determine product ion electronic states. However, examination of the literature shows that frequently the reported value of Q_{MIN} does not correspond to that expected from the thermodynamic data. We have examined the factors that influence the measurement of Q_{MIN} and found that the most important single factor is instrumental resolution. Most of the older data reported in the literature have been obtained with instruments with low energy resolution and in general gave values of Q_{MIN} that were more positive than expected by as much as 1 eV. We have reexamined a number of these cases and found that if good energy resolution was used (FWHM of 0.1 eV) that the Q_{MIN} obtained was consistent with thermodynamic expectations. The reactions that were reexamined are: C^+ (^{2}P) [D_2,D] CD^+, C^+ (O_2,0) CO^+, F^+ (D_2,D) FD^+ etc.

If a diatomic ion has rotational excitation a dynamic rotational barrier to dissociation exists and the dissociation energy differs from the value expected for the rotationless state, thus shifting Q_{MIN} to a more negative value. Furthermore, if the product ion is a diatomic ion containing hydrogen one can expect further complications due to tunneling; such effects (C. A. Jones, et.al., 1977) have been found in the reaction, F^+ (H_2,H) FH^+. Another reaction that reflects the influence of a rotational barrier on Q_{MIN} is $C^+(^2P) + O_2(^3\Sigma_g)$ $\rightarrow CO^+(A^2\Pi) + O(^3P)$; the Q_{MIN} value for this reaction was previously measured and reported to be -4 eV (J. J. Leventhal, 1971). The thermodynamically expected value is -5 eV, and when we remeasured this reaction under good instrumental resolution we obtained a Q_{MIN} of -5.5 eV. This was interpreted as indicating the presence of a barrier of 0.5 eV to dissociation in the product diatomic suggesting considerable rotational excitation in the CO^+ product. This is consistent with chemiluminescence studies (Ch. Ottinger and J. Simonis, 1975) in which it was reported that the CO^+ from this reaction was in the $A^2\Pi$ state and was highly rotationally excited.

W. R. Gentry: INTERNAL ENERGY PARTITIONING IN THE DISSOCIATIVE CHARGE TRANSFER REACTION $He^+ + O_2 \rightarrow He + O + O^+$

This work was done in collaboration with H. Udseth and C. F. Giese.

Molecular beam experiments on this dissociative charge transfer reaction reveal a particularly simple dynamical mechanism for this process, from which it is possible to deduce the relative probabilities of each pair of final atomic ion + atomic neutral states from measurements of the O^+ energy and angle distributions. The apparatus

used for this study (H. Udseth et.al., 1973) permits the collision
cell to be cooled with liquid nitrogen to decrease the spread in
target molecule internal and translational energies. The O^+ product energy distributions at fixed scattering angle show four distinct peaks, at energies which are, within experimental error,
independent of laboratory scattering angle, implying that the O^+
product energy distribution is approximately symmetric about the
laboratory, not the c.m., origin. Furthermore, the O^+ peak energies
are also virtually independent of He^+ kinetic energy below about
3 eV. Taken together, these data imply that the charge transfer
reaction occurs with very little concomitant momentum transfer to
the oxygen species from the incoming He^+. To a first approximation,
the O_2 molecule transfers its electron to He^+ at a sufficiently
large separation that the interaction potential is small at this
point. The resulting O_2^+ ion is left in a dissociative state about
24.58 eV above the O_2 ground state. As the O_2^+ explodes about its
own center of mass (which is nearly stationary in laboratory coordinates), the O atom and O^+ ion each receive half of whatever reaction exoergicity is available to translation. The O^+ energies
which are actually observed correlate well with the energies predicted by this simple mechanism.

The energetically accessible states of $(O + O^+)$ and the relative probabilities observed for each are:

a. $O(^3P) + O^+(^4S)$ 59%

b. $O(^1D) + O^+(^4S)$ 27%

c. $O(^3P) + O^+(^2D)$ 11%

d. $O(^1S) + O^+(^4S)$ absent

e. $O(^3P) + O^+(^2P)$

f. $O(^1D) + O^+(^2D)$ 4% (unresolved)

Despite a careful search, no O_2^+ products were observed.

One can imagine either of two charge transfer mechanisms to be
responsible for the observed reaction dynamics. At the moment of
charge transfer the O_2 molecule could make a transition directly to
one of several repulsive O_2^+ potential energy functions which cross
the resonance energy (24.58 eV above the O_2 ground state) at separations near the O_2 potential minimum. Alternatively, the transition
could occur to a bound excited state of O_2^+ which has a short predissociation lifetime, or which radiates to a repulsive or predissociated state. While other simultaneous processes are not completely
ruled out, the preponderance of the evidence indicates that the

dominant reaction mechanism is charge transfer to a $^2\Pi_u$ state of
O_2^+, which predissociates to the observed products in a time comparable to the vibrational period.

J. J. Leventhal: STUDIES OF VIBRONIC STATE DISTRIBUTIONS BY
    DETECTION OF LUMINESCENCE

    This work was done in collaboration with J. D. Kelley,
G. H. Bearman and H. H. Harris.

    Observation and analysis of collision-produced luminescence
can yield precise data on product internal energy partitioning
among accessible vibronic levels. Using this technique it has
been shown that A^+-BC charge transfer excitation (CTE) processes
at low kinetic energy yield decidedly non-Franck-Condon (FC) distributions in the product BC^+. It has been suggested that (M.
Lipeles, 1969) polarization distortion of the BC molecule by the
approaching A^+ ion can account for this behavior. To test this
model we have performed a series of experiments on BC^+-A collisions
(J. D. Kelley et.al., 1977 and 1978). By observing luminescence
from excited BC^+ resulting from collision-induced excitation (CIE)
state distributions from such processes can also be determined.
The results showed that here again non-FC distributions were prevalent at low kinetic energies (down to ~10 eV), and, as was the
case for CTE, the distributions tended toward the FC picture as the
kinetic energy was increased. Since polarization-distortion by an
incoming charge in BC^+-A collisions cannot account for these results
the need for more general consideration of the problem is established.

    The model that we propose is applicable to either CTE or CIE
processes and is based on the assumption that deviations from FC
behavior are caused by the short-range repulsion between the target
and projectile. It has been applied specifically to N_2^+-He CIE
processes, but is in fact quite general and applicable to neutral-
neutral reactions as well. In this model the deviations from FC
behavior are the result of T-V energy exchange on both the ground
and excited-state potential energy surfaces. These calculations
qualitatively reproduced the observed bahavior, that is, non-FC
distributions at low energy with a tendency toward the FC picture
as the energy is increased. Recently the model has been extended
(J. D. Kelley and H. H. Harris, 1979) using Monte Carlo-selected
trajectories on two intersecting diabatic potential surfaces.
Again, the results were consistent with the experimental observations.

REFERENCES

Adams, N. G., Smith, D., and Grief, D. (1979) J. Phys. B.
  (in press).
Hierl, P. M., Pacak, and Herman, Z. (1977) J. Chem. Phys. $\underline{67}$,
  2678.
Jones, C. A., Wendell, K. L., and Koski, W. S. (1977) J. Chem.
  Phys. $\underline{67}$, 4917.
Kelley, J. D., Bearman, G. H., Harris, H. H., and Leventhal J. J.
  (1977) Chem. Phys. Lett. $\underline{50}$, 295.
Kelley, J. D., Bearman, G. H., Harris, H. H., and Leventhal J. J.
  (1978) J. Chem. Phys. $\underline{68}$, 3345.
Kelley, J. D. and Harris, H. H., J. Phys. Chem. (submitted).
Leventhal, J. J. (1971) J. Chem. Phys. $\underline{55}$, 4654.
Lipeles, M. (1969) J. Chem. Phys. 51, 1252.
Ottinger, Ch. and Simonis, J. (1975) Phys. Rev. Lett. $\underline{35}$, 924.
Udseth, H., Giese, C. F. and Gentry, W. R. (1973) Phys. Rev. A
  $\underline{8}$, 2483.

FACTORS INFLUENCING THERMAL ION-MOLECULE RATE CONSTANTS

John I. Brauman

Department of Chemistry, Stanford University

Stanford, California, 94305, U. S. A.

INTRODUCTION

This talk (and chapter) should more properly be called "some factors influencing some thermal ion-molecule rate constants." Much of this work has already appeared in print, so this chapter contains only the essential arguments necessary to understand some of our (and others') current work in this area.

It is well known that many exoergic ion-molecule reactions occur on essentially every collision. This is a consequence of long range attractive forces often coupled with deep wells which together can mask any consequences of details of the potential surface. The reactants, once having entered some interaction region (a capture radius, say) are efficiently converted to products. On the other hand, it had previously been shown by us (1), by Bohme (2), and many others as well, that many reactions, especially "organic" ones, occurred considerably slower than the encounter rate. Thus, our motivation in studying these slow reactions is to try and learn why they are slow, and how their rates can be related to structure.

The class of reactions discussed here is <u>assumed</u> to be thermal, to involve long lived complexes, and to be governed by statistical theory. By this I mean that conversion of reactants to products takes place in a way such that energy is equilibrated statistically among all of the available vibrational modes and that the other usual postulates of transition state theory also apply (3,4). Effects of angular momentum can be treated in a variety of ways; we will not consider this here, however. The reactions to be discussed are (a) three-body association, (b) proton transfer and nucleophilic displacement, and (c) carbonyl addition reactions.

THREE-BODY ASSOCIATION

The general model for an ion-molecule reaction involving a complex is:

$$A^+ + B \underset{k_b}{\overset{k_c}{\rightleftharpoons}} X^{+*} \underset{}{\overset{k_p}{\rightleftharpoons}} C^+ + D. \tag{1}$$

If conversion to $C^+ + D$ is slow (k_p), then presumably reaction to give back A^+ and B (k_b) must be faster. Thus, we are going to have to model two reactions. We recognize that if there is no exit channel we can "measure" k_b directly via the familiar three body association reaction:

$$A^+ + B \underset{k_b}{\overset{k_c}{\rightleftharpoons}} AB^{+*} \overset{k_s[M]}{\longrightarrow} AB^+. \tag{2}$$

Since k_c is a capture rate constant calculable from Langevin or ADO theory (5), and k_s is also, if we assume that every collision removes enough energy to stabilize the complex, then the forward rate constant k_f can be seen to depend structurally on k_b.

$$k_f = \frac{k_c k_s[M]}{k_b + k_s[M]} \tag{3}$$

That is, k_c and k_s are "physical" quantities while k_b depends on the "chemical" or structural properties of the complex. In fact, this familiar Lindemann mechanism is not correct, because k_b is formally pressure dependent. At low pressure most of the complexes decompose (to reactants) and the average lifetime of those which decompose is long (k_b is small). At higher pressure, more complexes are collisionally stabilized; thus the average lifetime of those which escape collision is shorter (k_b is larger). This problem is accommodated in unimolecular reaction rate theory by integrating the microscopic rate constant for dissociation, $k(E)$, over the appropriate distribution function $F(E)$. An important consequence

$$\frac{d[AB^+]}{dt} = \frac{d[AB]^{+*}}{dt} \cdot \int_0^\infty \frac{k_s[M]}{k(E) + k_s[M]} F(E) dE \tag{4}$$

of this is that a plot of $1/k_f$ vs. $1/P$ should not be linear and one should not attempt to derive k_b from such plots.

For many years it has been believed that this model for association reactions should be valid, but attempts to calculate association rates quantitatively have not been very successful. In principle one calculates the sum of states $G(E^\dagger)$ for the dissociation transition state A (activated complex), and the density of states, $N(E^*)$, in the complex X, see Figure 1, and determines

$$k(E) = \frac{G(E^\dagger)}{hN(E^*)}.$$ (5)

Information about the complex is derived from the measured thermodynamic quantities ΔH_f^o and ΔS_f^o. The well depth is determined by ΔH_f^o; the vibrational frequencies of the complex are chosen by analogy with known compounds and must be consistent with ΔS_f^o. The transition state is chosen to be at the point where the sum of the centrifugal and potential energies is at a maximum. At this distance (~ 7 Å) the ion and neutral are taken to be freely rotating and to otherwise have the properties of the isolated ion and neutral.

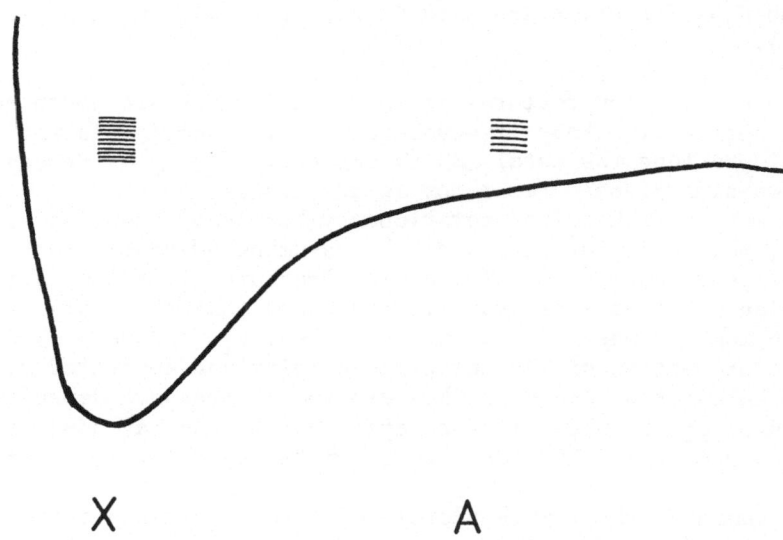

X   A

Figure 1.   Energy levels in complex X and transition state A.

Thus, the rotational quantum states of the transition state are very dense, which is consistent with the large impact parameters typically found for ion molecule reactions. It is immediately apparent from this model that essentially all of the (negative) temperature dependence is in k_b (k_c and k_s are at best weakly temperature dependent); since k_b will increase with increasing temperature, the overall forward rate constant k_f will decrease. There is no obvious functional form of the dependence and one should probably be wary of interpreting the slopes of plots of log k vs. log T.

As indicated above, this general model has been previously suggested by many workers to account for three-body associations, but it appeared to be quantitatively inadequate. Based on our recent results (6,7), I believe that this is a consequence of the approximations used to evaluate the rate constants. In our work we have used an efficient, accurate algorithm developed by Stein and Rabinovitch (8) to calculate $G(E^\dagger)$ and $N(E^*)$. Then, using the correct distribution for this type of (thermal) reaction as shown by Forst (4), we do a numerical integration to determine $\langle k_b \rangle$. The effect of angular momentum is accounted for using the approximations of Waage and Rabinovitch (9)(recently discussed extensively by Troe (10)). The results of our work on alkyl amine-ammonium ion dimers have been published; the agreement between experimental and calculated results is quite good (6). More recent calculations on $H_5O_2^+$ and $H_5S_2^+$ formation are also in good agreement with experiment (7).

There are a few features of this model which are worth noting. (a) The choice of unknown frequencies for the complex is not critical provided the total ΔS^o is correct. (b) It is necessary to use a two-dimensional free rotor in the transition state. (c) The correct method of handling rotational-vibrational coupling remains unclear, but the error introduced by assuming adiabatic rotations does not seem severe. (d) In the low pressure limit the choice of transition state is irrelevant in the sense that the system is in "equilibrium" between $A^+ + B$ and AB^{+*}. Thus k_f depends on k_c/k_b and only the properties of the energized complex and separated reactants. (e) The calculated values of $\langle k_b \rangle$ are indeed pressure dependent (factors of ca. 5-10). (f) For large systems one may need to go to quite low pressure (10^{-3} torr) to be in the true low pressure regime.

In summary, the simple picture of three body association reactions of "large" systems appears to be consistent with the experiment when the model is evaluated properly. No adjustable parameters are required to predict rate constants accurately; only thermodynamic and spectroscopic data are used. Predicting thermodynamic data from rate constants is not sufficiently sensitive to be useful (7).

SLOW REACTIONS: PROTON TRANSFER AND NUCLEOPHILIC DISPLACEMENT

One of the major interesting features of ion-molecule reactions is that many exoergic reactions occur on essentially every colli- sion. Indeed, one occasionally sees statements to the effect that all such reactions occur on every collision, although a good many counter-examples are known. Two kinds of reactions which are clearly slow are nucleophilic displacement at saturated carbon, (S_N2),

$$X^- + RY \longrightarrow RX + Y^- \tag{6}$$

and proton transfer reactions in which the negative ion is delocal- ized. For example:

$$CD_3\overset{O}{\overset{\|}{C}}CD_2^- + CH_3\overset{O}{\overset{\|}{C}}CH_3 \longrightarrow CD_3\overset{O}{\overset{\|}{C}}CD_2H + CH_3\overset{O}{\overset{\|}{C}}CH_2^- \tag{7}$$

These reactions reveal a number of interesting structural character- istics. In the proton transfer reactions (11), the rate constant depends on the exothermicity, and the slowness is clearly associated with the delocalized carbanion system. Phenoxide anions appear to react rapidly. We have associated the slowness with a loss of resonance stabilization in the transition state. The displacement reactions (12) are not well correlated with exothermicity and depend in detail on the specific nucleophile X^- and leaving group Y^-. Thus, the slowness in these reactions appears to be connected with the intimate interaction of X, R, and Y in the transition state. What is the nature of the potential surface which causes these reactions to be slow? It is important to keep in mind that these reactions are really slow in that there is only one available thermal product channel. Most of the collisions simply result in return to reactants. In fact, as Dr. Albritton has pointed out in his talk (13), slow reactions often (always?) have rate constants which decrease with increasing energy. We have tried to develop a general model which explains how a thermal ion-molecule reaction can be inefficient. I think this is a quite general model which will be applicable to many thermal, statistical ion-molecule reac- tions. It has the properties of explaining qualitatively the slow rates, predicting a general decrease in rate for increasing tempera- ture or energy, and being capable of quantitative evaluation. First, we should recognize that it is convenient to speak of a reaction efficiency which represents the fraction of encounters which result in product formation (this eliminates the differences in encounter rates which are "physical" rather than chemical in origin). We expect that a potential surface with a single minimum should lead to efficiencies which correspond roughly to the ratio of densities of states in reactants vs. products. For a thermoneutral reaction,

for example isotope exchange, the efficiency will be one-half.
Obviously, then, the reactions $Cl^- + CH_3Cl \rightarrow$, or $CH_3COCH_2^- +$
$CD_3COCD_3 \rightarrow$, because they are slow, must occur on a more complex
surface. The least complicated surface which will suffice to
explain this behavior is one containing two minima. Appealingly,
the surface is nominally symmetrical and is consistent with microscopic reversibility. We write the reaction phenomenologically as
formation of a reactant encounter complex, conversion (over a
barrier) to a product encounter complex, and dissociation to
products:

$$A^+ + B \underset{k_{-1}}{\overset{k_1}{\rightleftharpoons}} AB^+ \underset{k_{-p}}{\overset{k_p}{\rightleftharpoons}} CD^+ \underset{k_2}{\overset{k_{-2}}{\rightleftharpoons}} C^+ + D. \qquad (8)$$

The efficiency of the reaction depends upon the relative values of
k_{-1}, k_p, k_{-p}, and k_{-2}. Under the collisionless conditions of low
pressure ion-molecule reactions, the complexes AB^+ and CD^+ are chemically activated, containing the thermal and kinetic energy associated with the reactants. Thus the rate constants k_{-1}, k_p etc., are
not thermally averaged rate constants but rather microscopic rate
constants averaged over an appropriate energy distribution function.
The potential surface appears as shown in Figure 2. The first

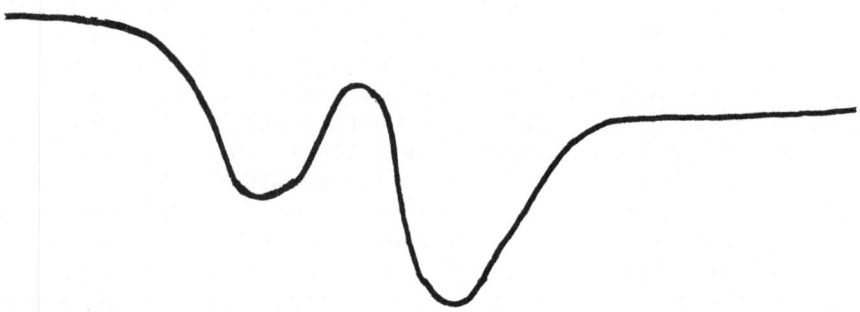

Figure 2. Potential surface (Energy <u>vs</u>. Reaction Coordinate).

obvious question is how can such a surface give rise to an inefficient reaction, since there is clearly always enough energy to cross the barrier associated with k_p, while there is only just enough energy to cross back via k_{-1}. The answer lies in what the kineticist recognizes as a very large difference in entropies of activation or pre-exponential factors. The forward reaction k_p is, in effect, a unimolecular isomerization with a comparatively "tight" transition state (low A-factor). The back dissociation, k_{-1}, is a fragmentation reaction with an exceptionally "loose" transition state involving essentially free rotations of the product fragments (high A-factor). (This is the same transition state we used earlier to describe the decomposition of the complex in the association reactions.) Thus, although "energy" favors k_p, "entropy" favors k_{-1}. The efficiency of the reaction thus results from a trade-off between these two effects. In more formal terms, it is the available sum of states (weighted by the appropriate energy distribution function) which determines the branching ratio between k_{-1} and k_p. The density of states is determined both by the energy above the critical energy (E) and also the frequencies of the oscillators which affect the A-factor.

(It is easy to see how this model can be extended to more complex situations in which there is more than one product channel. Since in general the transition state for the particular reaction will reflect factors in addition to the overall exothermicity one should not expect that the product branching ratios will follow the exothermicities of the products.)

In favorable cases, where one can make a reasonable estimate of the structure and vibrational frequencies of the transition state, one can make a calculation of the branching ratio k_{-1}/k_p without knowing any of the properties of the intermediate complex AB^+. Although the properties of the complex are required in order to calculate absolute rate constants, the branching ratio can be determined just from the properties of the two transition states. The only unknown quantity is the critical energy for k_p. This is obtained by finding the value required to fit the experimental data. This method has a long history in neutral chemical activation studies. When one applies it to the reaction of Cl^- + $CH_3Br \rightarrow CH_3Cl + Br^-$, it is found that the energy of the transition state is only slightly lower than that of the reactants, suggesting that most of the activation energy in S_N2 reactions is due to desolvation effects (12). The reader should be aware that the model is crude in that the frequencies of the activated complex are not known exactly, but the barrier height estimates are not especially sensitive to their choice.

There are some interesting consequences of this model which can be tested experimentally. For example, it predicts that slow

reactions should become slower with increasing energy (negative
temperature dependence). Eventually, of course, the mechanism will
change and the reaction rate may increase. It is clearly predicted
that polyatomic nucleophiles will react more slowly than atomic
nucleophiles with the same barrier height due to losses in ΔS_{rot}.
In general, we have found polyatomics to be slower, but the correct,
rigorous test remains to be done. This effect suggests that
apparent steric effects will not be simply interpretable, since
for equal barrier heights larger nucleophiles react more slowly
anyway. Finally, some further interesting effects will occur as a
function of structure. In comparing the (calculated) reactions of
Cl⁻ with methyl and n-butyl bromides we find that at high barrier
heights (slow reaction) n-butyl bromide is slower--largely due to
the greater density of states in the separated species. However,
for small barrier heights (fast reactions), n-butyl bromide becomes
faster, because the density of states in the reaction transition
state grows in very rapidly once one gets very far above the critical energy.

ADDITION TO CARBONYL CENTERS

Reactions of carbonyl groups play a central role in organic
chemistry. Condensations, esterification, saponification, etc. are
all of great synthetic importance. Among the major mechanistic
problems in this area is the question of the "tetrahedral intermediate". Do nucleophilic additions to carbonyl centers involve
stable tetravalent structures, 1, nominally similar to conventional
alkoxide ions? Theoretical analyses suggest that indeed such

$$
\underset{\text{1}}{
\begin{array}{c}
O \\
\parallel \\
RC\text{-}X + Y^- \longrightarrow R\text{-}\overset{O^-}{\underset{Y}{\overset{|}{C}}}\text{-}X \longrightarrow RC\text{-}Y + X^-
\end{array}} \qquad (9)
$$

intermediates are stable and that they have no activation barriers
to decomposition in addition to the thermodynamic ones (14). That
is, the potential surface for this reaction has a single minimum.
The analysis given previously suggests that if the surface has only
a single minimum, carbonyl addition reaction will be very efficient.
Experimentally this does not appear to be the case. Riveros (15)
and Comisarow (16) have shown that when other channels are accessible,
carbonyl addition reactions may not occur even though the reactions
are energetically favorable. Thus, for example

$$CH_3O^- + CH_3COF \longrightarrow CH_2COF^- + CH_3OH \qquad \Delta H^o = -18.4 \text{ kcal/mole}$$

$$\xrightarrow{\ \ //\ \ } F^- + CH_3COOCH_3 \qquad \Delta H^o = -19.9 \text{ kcal/mole}$$

In addition, it is known that the competing reactions themselves may have barriers. Consistent with this behavior of discrimination against addition, we have found that the thermoneutral isotope exchange reaction:

$$\overset{O}{\overset{\|}{R C}}Cl + Cl^- \longrightarrow \overset{O}{\overset{\|}{R C}}Cl + Cl^- \qquad (10)$$

is less than 50% efficient even when there is no other available exothermic reaction channel (17). Thus, these reactions must also have barriers on their surfaces as do the displacement and proton transfer reactions. If the potential surface calculations are correct in the sense that the tetrahedral intermediate has no activation energy associated with its formation and decomposition, how can the reaction be slow? As we will see, the slowness of the reaction arises from the existence of other, more stable, complexes which may have essentially nothing structurally to do with progress along the reaction coordinate.

What can we say about the complex between $Cl^- + CH_3COCl$? We have been able to show that the reaction:

$$CH_3BrCl^- + CH_3COCl \rightleftharpoons CH_3Br + CH_3COCl_2^- \qquad (11)$$

has $K \approx 1$, indicating a heat of reaction of ~ 11 kcal/mole for the binding of Cl^- to CH_3COCl. Binding energies of this size are suggestive of a loose rather nonspecific complex of the type frequently observed by Kebarle (18) and also by Dougherty (19). In contrast to the S_N2 complexes, however, these carbonyl complexes are able to exchange halide ions as evidenced by further reactions (20).

If we assume the transition state in the reaction to be a tetrahedral adduct (and use vibrational frequencies appropriate for this structure) we require the top of the barrier to be about 7 kcal/mole below the energy of the reactants to account for the rate. (The method for determining this is the same as that described previously.) Thus, the secondary minimum associated with this structure is actually a saddle point. The interaction of the ion and molecule is seen now to result first in a loose complex which then either dissociates to reactants or passes through a tetravalent structure and goes on ultimately to products. This picture shows that the dynamics cannot be understood or predicted by calculating a single trajectory between reactants and products, even if there is no activation energy. Indeed, one may actually have to know the appearance of the entire surface.

There is a final feature of the carbonyl addition reaction which deserves comment. If the "transition state" is a tetrahedral intermediate, and if the bonding in this structure is normal, then its energy seems remarkably low. If the alkoxy radical derived by removing an electron:

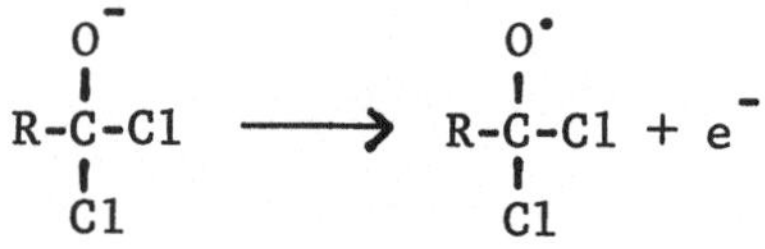

$$\begin{array}{ccc} \overset{-}{O} & & \overset{\bullet}{O} \\ | & & | \\ R-\overset{|}{C}-Cl & \longrightarrow & R-\overset{|}{C}-Cl + e^- \\ | & & | \\ Cl & & Cl \end{array}$$

has a heat of formation consistent with group additivity, then the electron affinity of this alkoxy radical would be ca. 92 kcal/mole, an unusually high value for an alkoxy radical (21). This value can be found approximately by noting that the C-Cl bond formed is about as strong as the C=O π bond broken, the exothermicity of the reaction is ~ 7 kcal/mole, and the electron affinity of Cl is ~ 83 kcal/mole.

CONCLUSION

The model for reactions described here seems to be generally applicable to a large number of reactions. It explains qualitatively how exoergic reactions can be slow, and predicts correctly the energy dependence of these slow reactions.

ACKNOWLEDGMENTS

I would like to thank my colleagues who have carried out work in this area, particularly C. A. Lieder, W. E. Farneth, W. N. Olmstead, O. I. Asubiojo, J. M. Jasinski, and R. N. Rosenfeld, as well as M. Lev-On at SRI International. I would like especially to thank David Golden who was and continues to be our source of knowledge regarding unimolecular reaction rate theory and its applications, and also for his other special contributions.

I am grateful for research support from the National Science Foundation and the donors of the Petroleum Research Fund, administered by the American Chemical Society.

REFERENCES

(1) J. I. Brauman, W. N. Olmstead, and C. A. Lieder, J. Am. Chem. Soc., 96, 4030 (1974).

(2) D. K. Bohme, G. I. Mackay, and J. D. Payzant, J. Am. Chem. Soc., 96, 4027 (1974). K. Tanaka, G. I. Mackay, J. D. Payzant, and D. K. Bohme, Can. J. Chem., 54, 1643 (1976) and earlier references cited therein.

(3) P. J. Robinson and K. A. Holbrook, "Unimolecular Reactions," Wiley-Interscience, London, 1972.

(4) W. Forst, "Theory of Unimolecular Reactions," Academic Press, New York, N. Y., 1973.

(5) T. Su and M. T. Bowers, J. Chem. Phys., 58, 3027 (1973); Int. J. Mass Spectrom. Ion Phys., 12, 347 (1973). L. Bass, T. Su, W. J. Chesnavich, and M. T. Bowers, Chem. Phys. Lett., 34, 119 (1975).

(6) W. N. Olmstead, M. Lev-On, D. M. Golden, and J. I. Brauman, J. Am. Chem. Soc., 99, 992 (1977).

(7) J. M. Jasinski, R. N. Rosenfeld, D. M. Golden, and J. I. Brauman (submitted for publication).

(8) S. E. Stein and B. S. Rabinovitch, J. Chem. Phys., 58, 2438 (1973).

(9) E. V. Waage and B. S. Rabinovitch, Chem. Rev., 70, 377 (1970).

(10) J. Troe, J. Chem. Phys., 66, 4745, 4758 (1977).

(11) W. E. Farneth and J. I. Brauman, J. Am. Chem. Soc., 98, 5546 (1976).

(12) W. N. Olmstead and J. I. Brauman, J. Am. Chem. Soc., 99, 4219 (1977).

(13) D. L. Albritton, paper presented at this NATO Institute.

(14) For example: H. B. Bürgi, J. M. Lehn, and G. Wipff, J. Am. Chem. Soc., 96, 1956 (1974). S. Scheiner, W. N. Lipscomb, and D. A. Kleier, ibid., 98, 4770 (1976). G. Alazoma, E. Scrocco, and J. Tomasi, ibid., 97, 6976 (1975).

(15) S. M. José and J. M. Riveros, Nouv. J. Chim., 1, 113 (1977) and references therein.

(16)   M. Comisarow, Can. J. Chem., 55, 171 (1977).

(17)   O. I. Asubiojo and J. I. Brauman, submitted for publication.

(18)   R. Yamdagni and P. Kebarle, J. Am. Chem. Soc., 93, 7139
       (1971).

(19)   R. C. Dougherty and J. D. Roberts, Org. Mass. Spectrom., 8,
       81 (1974).

(20)   O. I. Asubiojo, L. K. Blair, and J. I. Brauman, J. Am. Chem.
       Soc., 97, 6685 (1975).

(21)   B. K. Janousek, A. H. Zimmerman, K. J. Reed, and J. I.
       Brauman, J. Am. Chem. Soc., 100, 6142 (1978).

MECHANISTIC ASPECTS OF ION-MOLECULE REACTIONS

N.M.M. Nibbering

Laboratory of Organic Chemistry
University of Amsterdam
1018 WS Amsterdam, The Netherlands

INTRODUCTION

Ion-molecule reactions in the gas phase have been the subject of numerous studies and the interest in this field has grown steadily since the NATO Advanced Study Institute on Ion-Molecule Interactions held in Biarritz, France, in 1974 (Ausloos, 1975). This assertion is born out by the appearance of several books (Lias and Ausloos, 1975; Lehman and Bursey, 1976; Bowers, 1979) and reviews on proton-transfer equilibria (Taft and Arnett, 1975) and on ion thermochemistry and solvation from gas phase ion equilibria (Kebarle, 1977). The field of negative chemical ionization mass spectrometry has grown rapidly (Jennings, 1977) and developments in the study of ion-molecule reactions have recently been reviewed (Jennings, 1978). It is hoped that this paper will supplement the previous review of the reaction mechanisms of organic and inorganic ions in the gas phase (Beauchamp, 1975) by presenting a survey covering the mechanistic aspects of data which has been obtained in the intervening years by the ion cyclotron resonance and flowing afterglow techniques (Jennings, 1978). Unless stated otherwise, the coventions of physical organic chemistry textbooks will be used (Lowry and Richardson, 1976). Reactions of both positive and negative ions with neutral species will be described and it is hoped that the paper will be as comprehensive as is possible within the space allotted. Only organic systems will be included since the chemis-

try of inorganic systems will be described in the Chapter by J.L. Beauchamp. Attention will be paid to the
following topics:
(i)   nucleophilic displacement reactions at saturated
        carbon centers,
(ii)   tetrahedral intermediates,
(iii)   addition-elimination reactions,
(iv)   cycloaddition reactions,
(v)   hydrogen-deuterium exchange reactions of carban-
        ions and
(vi)   formation of $(M-H_2)^{-\cdot}$ ions from reactions of $O^{-\cdot}$
        with some organic substrates.
Where possible isotopic labelling studies are described
for these complex ion-molecule reactions since they are
capable of giving considerable insight into the detailed
mechanisms involved. It is hoped that this approach will
contribute to other branches of the field, such as kinetic studies and studies, designed to make comparisons
between gas phase and solution chemistry.

NUCLEOPHILIC DISPLACEMENT
REACTIONS AT SATURATED CARBON CENTERS

     This type of reaction may occur during encounters
between neutral nucleophiles with positive ions and between anionic nucleophiles with neutral molecules as
defined by the overall reactions 1 and 2:

$$X + RY^+ \longrightarrow Y + RX^+ \tag{1}$$

$$X^- + RY \longrightarrow Y^- + RX \tag{2}$$

The simple rules governing the occurrence of reaction 1
(Beauchamp, 1975) are also applicable to reaction 2:
(i) the reactions must be exothermic
(ii) there should be no facile proton transfer process
     from RY$^+$ to X or from RY to X$^-$.

     Examples of process 1 are the reactions of methoxy-
methyl cations with acetaldehyde and acetone, studied by
ICR (van Doorn and Nibbering, 1978). They react with the
expulsion of formaldehyde:

$$CH_3OCH_2^+ + CH_3CRO \longrightarrow CH_3O\overset{+}{C}RCH_3 + CH_2O \tag{3}$$
$$(R = H, CH_3)$$

^2H and ^{18}O labelling (i.e. acetone-^{18}O) indicate that
reaction 3 corresponds formally to the transfer of the

original methyl group of the methoxymethyl cation to
the carbonyl compound when the ion is generated from
dimethyl ether at 13 eV. This observation suggests the
intermediacy of collision complex I -a methyl cation
bonded complex of two n-donor bases- in reaction 3.
However, in view of the ambident nature of the methoxymethyl cation it is also possible that its methylene
group is attacked by the lone pair electrons of the
carbonyl group so as to give collision complex II.

$$CH_2=O----\overset{\overset{H\ H}{\diagup}}{\underset{H}{C^+}}----O=C\underset{R}{\overset{CH_3}{\diagdown}} \qquad CH_3OCH_2\overset{+}{O}=C\underset{R}{\overset{CH_3}{\diagdown}} \qquad (\ R=H\ or\ CH_3\ )$$

$$\text{I} \qquad\qquad\qquad \text{II}$$

Such an attack has been reported for the reaction of
labelled methoxymethyl cations with toluene (Dunbar,
Shen, Melby and Olah, 1973):

$$CD_3OCH_2^+ \ + \ C_6H_5CH_3 \longrightarrow C_8H_9^+ \ + \ CD_3OH \qquad\qquad (4)$$

The behaviour of collision complex II for R = H is
known from an electron impact study of methoxymethyl
isopropyl ether (Schoemaker, Nibbering and Cooks,
1975). Combined ^{2}H and ^{18}O labelling of this compound
has shown that its $(M-CH_3)^+$ ions, which are similar to
collision complex II for R = H, eliminate formaldehyde
via two reaction channels leading to the same product
ion, namely by methoxyl migration and by methyl migration as indicated in reactions 5a and 5b.

$$\overset{O}{\underset{CH_2\text{---}OCH_3}{|}}\diagdown\overset{+}{\underset{}{C}}\overset{CH_3}{\diagup}\overset{}{\diagdown H} \xrightarrow{-CH_2O} CH_3\text{---}\overset{*}{\underset{+}{O}}=C\overset{CH_3}{\diagup}\diagdown H \qquad (5a)$$

$$CH_2\text{---}\overset{+}{O}=C\overset{CH_3}{\diagup}\diagdown H \quad\overset{*}{\underset{O\text{---}CH_3}{|}} \xrightarrow{-CH_2O} CH_3\text{---}\overset{*}{O}=C\overset{CH_3}{\diagup}\diagdown H \qquad (5b)$$

Channel 5a requires about 9 kcal/mole more activation
energy than channel 5b (van Doorn and Nibbering, 1978).
An appropriate reaction coordinate diagram for reaction
3, going via collision complex II, is shown in Figure 1.

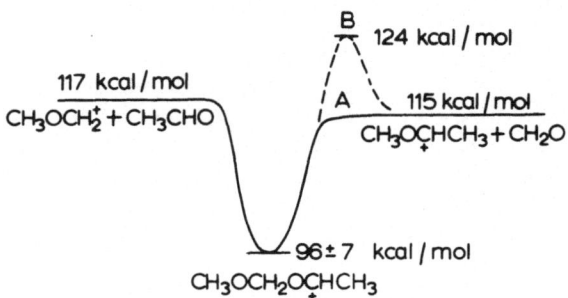

<u>Fig. 1</u>. Reaction coordinate diagram illustrating the energetics of the reaction between the methoxymethyl cation and acetaldehyde. Route A corresponds to a methyl shift (reaction 5b) and route B corresponds to a methoxyl shift (reaction 5a) in the supposed collision complex prior to fragmentation to the products.

From Fig. 1 it can be seen that the reaction between the methoxymethyl cation and acetaldehyde is exothermic by 2 kcal/mole, but route B (reaction 5a) requires an activation energy whereas route A (reaction 5b) does not. A similar diagram can be constructed for the reaction of the methoxymethyl cation with acetone.
Channel B in Fig. 1 will not be accessible to methoxymethyl cations which are in their ground state. However, it is known that they can accomodate 83 kcal/mole internal energy before fragmentation by methane loss occurs, though they suffer from 1,3 hydrogen shifts leading to hydrogen scrambling at the lower internal energy of 58 kcal/mole (Hvistendahl and Williams, 1975). The latter value is in excellent agreement with the threshold value of 53 kcal/mole for 1,3 hydrogen shifts in methoxymethyl cations, now determined by ICR (van Doorn and Nibbering, 1978). Methoxymethyl cations with this amount of internal energy are found to react with acetone so as to expel molecules of formaldehyde which contain the original randomised hydrogen/deuterium atoms of the methoxymethyl cations. Therefore, reaction channel B should become accessible when higher electron energies are used in the generation of the methoxymethyl cations. However, methoxymethyl cations generated over an electron energy range from 12 to 50 eV always react with acetone-^{18}O so as to produce unlabelled formaldehyde. This may be taken to be evidence against the involvement of collision complex II in reaction 3. Thus it seems that reaction 3 proceeds

through a mechanistically simple nucleophilic displacement reaction involving collision complex I.

Very detailed studies carried out in the last few
years have shown that collision complexes of type I are
also involved in the general reaction 2.
In solution chemistry reaction 2 is well known as an
"S_N2 displacement reaction", where the essential feature is that the formation of the new bond occurs syn-
chronously with cleavage of the old bond with the result that inversion of configuration occurs at the site
of attack (Walden inversion). This implies backside attack at the saturated carbon center as represented pic-
torially in reaction 6 (Lowry and Richardson, 1976):

$$X^- + \quad \overset{|}{\underset{/}{C}}{-}Y \longrightarrow \left[X{-}{-}{-}\overset{|}{\underset{\wedge}{C}}{-}{-}{-}Y \right]^- \longrightarrow X{-}\overset{|}{C}{\diagdown} + Y^- \tag{6}$$

In a very elegant study, Cl$^-$ has been allowed to react
with cis- and trans-4-bromocyclohexanols in the cell of
a pulsed ICR spectrometer. The vacuum can was isolated
from the pumping system so that the neutral products
might be accumulated (Lieder and Brauman, 1974; Lieder
and Brauman, 1975). After an appropriate time the contents of the vacuum can were analyzed in the positive
ion mode. The stereochemistries of the 4-chlorocyclo-
hexanol products were determined by measurement of
their $[M-H_2O]^{+\cdot}/[M^{+\cdot}]$ ratios, which were compared with
the $[M-H_2O]^{+\cdot}/[M^{+\cdot}]$ ratios of authentic samples of cis-
and trans-4-chlorocyclohexanol. In this way the neutral
product from reaction of Cl$^-$ with cis-4-bromocyclohexa-
nol has been determined to be 0.92 $\pm$ 0.04 mole fraction
of trans-4-chlorocyclohexanol and from reaction of Cl$^-$
with trans-4-bromocyclohexanol 0.87 $\pm$ 0.07 mole fraction of cis-4-chlorocyclohexanol. Thus the mechanisms
of these reactions do indeed involve Walden inversion
and hence backside attack at the reacting carbon centers. The belief that backside attack is impossible for
bridgehead halides is substantiated by the finding that
1-chloro- and 1-bromo-adamantane do not show ionic
reaction products following attack by F$^-$ (Lieder and
Brauman, 1975).
More recently rate constants for reaction 2, in which
X = F, Cl, Br, CN, OH, OCH$_3$, (CH$_3$)$_3$O, CH$_3$S etc. and
Y = Cl, Br, CF$_3$CO$_2$, C$_6$H$_5$O, CH$_3$CO$_2$ etc. have been measured both by the flowing afterglow technique (Tanaka,
Mackay, Payzant and Bohme, 1976) and by pulsed ICR
spectroscopy (Olmstead and Brauman, 1977). Values of

the rate constant for reaction 2 spanning the range
from almost collision controlled down to too slow to be
observed have been found as the nucleophile (X^-), the
leaving group (Y^-) and the alkyl substrate (R) are altered. A discussion of the factors influencing ion-molecule reaction rate constants is to be found in the
Chapter by J.I. Brauman. Here it should be mentioned
that the equation

$$k_{exptl} = k_c \cdot \exp(-E_a/RT) \qquad (7)$$

has been used to determine the apparent Arrhenius activation energy E_a for a number of gas phase S_N2 reactions (Tanaka, Mackay, Payzant and Bohme, 1976), where
k_{exptl} = experimentally determined rate constant and
k_c = capture rate constant, based on the average dipole
orientation theory (Su and Bowers, 1973a and 1973b). In
general E_a appears to cover the range of 0.1-5
kcal/mole, although there are exceptions such as the
reaction of CN^- with CH_3F which has an $E_a \geq 17$
kcal/mole. At present it is not known whether or not
some of these E_a values correspond to activation energy
barriers higher than the potential energies of either
the separated reactants or products. Recent determinations of translational energy dependences of reactive
scattering cross sections (so-called "Excitation functions") have shown for example that there is an activation barrier of the order of $\simeq$ 10 kcal/mole for the
proton transfer reaction 8, even though it is exothermic by 10.3 kcal/mole (Lifshitz, Wu and Tiernan, 1978).

$$CH_2=CH-CH_2^- + C_6H_5CH_3 \longrightarrow C_6H_5CH_2^- + CH_2=CH-CH_3 \qquad (8)$$

Perhaps, as more data becomes available, rule (i) given
at the beginning of this section will need to be quali-
fied by the addition of "with no activation energy barriers higher than the potential energies of either the
separated reactants or products."
On the other hand, an alternative explanation for the
great spread of rate constant values has been proposed
which involves a double potential well model (Fig. 2)
(Olmstead and Brauman, 1977). RRKM calculations show
that variation of the height of the central barrier in
Fig. 2 (which is assumed to be lower than the potential
energies of either the separated reactants or products)
can result in variation of the overall rate constant.
The height of the central barrier will of course depend
upon the nature of X^-, Y^- and R.

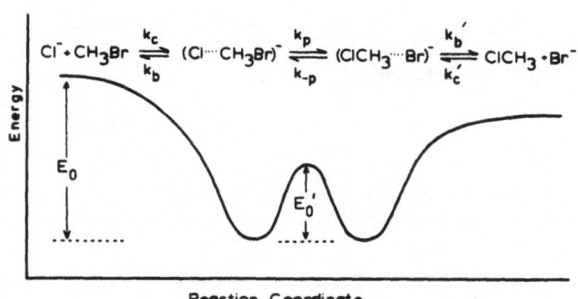

Fig. 2. Mechanism and potential energy-reaction coordinate diagram for a representative S_N2 reaction in the gas phase (Olmstead and Brauman, 1977). (By courtesy of the American Chemical Society.)

The concept of a double-well potential is further supported by the following experimental observations:
1°. Stable adducts between anions and alkyl halides have been observed in a high-pressure mass spectrometer, varying in stability from 8.6 to 14.4 kcal/mole. "Backside" attachments of the anions to the carbon atoms is suggested by the failure to observe adducts between halide ions and bridgehead halides (Dougherty, Dalton and Roberts, 1974; Dougherty and Roberts, 1974; Dougherty, 1974).
2°. The adduct of chloride ion with methyl bromide has been made by reaction 9 to 11:

$$CF_3Cl^{-\bullet} + CH_3Br \longrightarrow ClCH_3Br^- + {}^{\bullet}CF_3 \qquad (9)$$

$$Cl_2^{-\bullet} + CH_3Br \longrightarrow ClCH_3Br^- + Cl^{\bullet} \qquad (10)$$

$$COCl^- + CH_3Br \longrightarrow ClCH_3Br^- + CO \qquad (11)$$

These adducts are able to transfer Cl^- but not Br^- to neutral acetonitrile and 1,1-difluoroethane (reactions 12 and 13). This indicates that there is a barrier to exchange between the halide atoms in $ClCH_3Br^-$:

$$ClCH_3Br^- + CH_3CN \nearrow \begin{array}{l} ClCH_3CN^- + CH_3Br \;, \Delta H° = -2.5 \text{ kcal/mole} \\[2em] BrCH_3CN^- + CH_3Cl \;, \Delta H° = -10.2 \text{ kcal/mole} \end{array} \qquad (12)$$

$$ClCH_3Br^- + CH_3CHF_2 \nearrow \begin{matrix} ClCH_3CHF_2^- + CH_3Br \\ \\ \cancel{\searrow} \\ BrCH_3CHF_2^- + CH_3Cl \end{matrix} \qquad (13)$$

Unfortunately, attempts to prepare the adduct of bromide ion with methyl chloride have so far been unsuccessful (Olmstead and Brauman, 1977).
Nevertheless, it is clear that the $ClCH_2Br^-$ adducts, generated in reactions 9 to 11, are best described as a Cl^- ion solvated methyl bromide:

$$Cl^- \text{---} CH_3 \text{---} Br$$

This is corroborated by observations summarized in reactions 14 and 15 (Riveros, Breda and Blair, 1974):

$$CO^{35}Cl^- + CH_3{}^{37}Cl \longrightarrow {}^{35}Cl^- \text{----} CH_3{}^{37}Cl \qquad (14)$$

$${}^{35}Cl^- \text{---} CH_3{}^{37}Cl + CH_3CF_3 \nearrow \begin{matrix} {}^{35}ClCH_3CF_3^- + CH_3{}^{37}Cl \\ \\ \cancel{\searrow} \\ {}^{37}ClCH_3CF_3^- + CH_3{}^{35}Cl \end{matrix} \qquad (15)$$

In this model the collision complex in reaction 6 and collision complex I should be regarded as representing transition state collision complexes, i.e. at the top of the central barrier.

TETRAHEDRAL INTERMEDIATES

In solution many carbonyl compounds react with nucleophiles by forming initially an adduct (III) which is known as a "tetrahedral intermediate" (Lowry and Richardson, 1976):

$$\begin{matrix} \diagdown & \diagup O^- \\ & C & \quad III \\ \diagup & \diagdown Nuc \end{matrix}$$

Under suitable conditions such intermediates may be sufficiently stable to be observed by NMR as in the case of the reaction of an alkoxide ion with an amide (Fraenkel and Watson, 1975).
Various attempts have been made to show that neutral nucleophiles may react with positive ions and anionic

nucleophiles with neutral molecules through such intermediates in gas phase ion-molecule reactions:

$$Nuc + \underset{X}{\overset{R_1}{>}}C = \overset{+}{O}R_2 \longrightarrow \left[R_1 - \underset{Nuc^+}{\overset{O-R_2}{\underset{|}{C}}} X \right] \longrightarrow \underset{Nuc^+}{\overset{R_1}{>}}C = O + R_2X \quad (16)$$

$$Nuc^- + \underset{X}{\overset{R_1}{>}}C = O \longrightarrow \left[R_1 - \underset{X}{\overset{O^-}{\underset{|}{C}}} - Nuc \right] \longrightarrow \underset{Nuc}{\overset{R_1}{>}}C = O + X^- \quad (17)$$

The adducts in reactions 16 and 17 are much more tight
than the loose collision complexes described in the
previous section, because now a real chemical bond is
formed.

   An example of reaction 16 is the elimination of
methanol from the collision complex of the methoxymethyl cation with 2-methoxy-ethanol, for which the
isotopic labelling results (as summarized in reaction
18) strongly support the intermediacy of a tetrahedral
intermediate (Pau, Kim and Caserio, 1974 and 1978a):

$$CH_3OCH_2CH_2\bar{O}H + CH_2 = \overset{18+}{O} - CH_3$$

$$\downarrow$$

$$\left[CH_3OCH_2CH_2 - \underset{+}{\overset{H}{O}} - \underset{H}{\overset{H}{C}} - {}^{18}OCH_3 \right]^* \quad (18)$$

$$\downarrow$$

$$CH_2 = \overset{+}{O} - CH_2CH_2OCH_3 + CH_3{}^{18}OH$$

Note here that the nucleophile attacks the methylene
carbon atom of the methoxymethyl cation in contrast to
the reactions of this ion with carbonyl compounds (see
previous section).
In principle protonated esters, aldehydes and ketones
may also react with alcohols via reaction 16. However,
it has been shown that these ions play the role of pro-
tic acids rather than electrophiles while the alcohols
function as bases rather than as nucleophiles (Pau, Kim
and Caserio, 1974 and 1978b).

   The same is true for the anionic ion molecule reac-

tion 17: if a facile proton transfer can occur from the
neutral to Nuc⁻, the formation of a tetrahedral intermediate is largely suppressed. Evidence for this has
followed from reactions of $^{18}O^{-\bullet}$ and $^{18}OH^-$ with acetone
and dimethylsulphoxide, where the abundant product
ions: $(M-H)^-$, $(M-H_2)^{-\bullet}$ and $(M-H_2-CH_3)^-$ do not incorporate the ^{18}O label at all. The ions CH_3COO^- and CH_3SOO^-,
generated from $O^{-\bullet}$ and these compounds presumably via
tetrahedral intermediates, represent only about 1% of
the overall yield (Dawson and Nibbering, 1978a). Nevertheless, various clear-cut examples of anionic tetrahedral intermediates and adducts have been discovered by
a judicious choice of Nuc⁻ and substrate. $Cl_2^{-\bullet}$ and
$COCl^-$ react with acetyl chloride, loosing $Cl^\bullet$ and CO
respectively, to give the adduct $CH_3COCl_2^-$:

$$Cl_2^{-\bullet}/COCl^- \; + \; CH_3COCl \longrightarrow CH_3COCl_2^- \; + \; Cl^\bullet/CO \qquad\qquad (19)$$

The $CH_3COCl_2^-$ ions thus formed are then found to transfer either of the chlorines with equal probability as
Cl^- to CF_3COCl. One may thus deduce that the $CH_3COCl_2^-$
ions have structurally equivalent chlorines (Asubiojo,
Blair and Brauman, 1975):

$$CH_3-\underset{\underset{Cl}{|}}{\overset{\overset{O^-}{|}}{C}}-Cl \quad \mathbf{IV}$$

Similar 1:1 adducts (but in this case collisionally
stabilized) have been reported for reactions of the
perfluoroacetate anion with perfluoroacetic anhydride
(Bowie and Williams, 1974) and of the acetate anion
with acetic anhydride (Wilson and Bowie, 1975). In the
latter case the adduct may decompose further by the loss
of ketene. By working with acetic anhydride (α,α'-d_2) it
has been shown, that in the loss of CH_2CO and $CHDCO$ from
the corresponding adduct (V and Va) a kinetic isotope
effect is operative, which varies from 1.2 to 2.4 as the
ion transit time through the ICR cell is varied from
10^{-3}-10^{-4} s to 10^{-1}-10^{-2} s.

The intermediacy of tetrahedral intermediates has also been suggested for the reactions of F^- and OH^- with alkyl formates (Faigle, Isolani and Riveros, 1976). Both ions react exothermally to generate $HCOO^-$:

$$F^- + HCOOCH_3 \longrightarrow HCOO^- + CH_3F \, , \, \Delta H° = -26 \text{ kcal/mole} \tag{20}$$

$$OH^- + HCOOCH_3 \longrightarrow HCOO^- + CH_3OH \, , \, \Delta H° = -45 \text{ kcal/mole} \tag{21}$$

As noted by the authors, a tetrahedral intermediate in reaction 21 could explain the formation of $HCOO^-$ if one assumes that the acidic hydrogen is then transferred to the least acidic moiety at fragmentation. However, they felt that such an intermediate in reaction 20 would be faced with a rather akward CH_3F elimination. The formation of $HCOO^-$ was not observed when nucleophiles like RO^-, RS^-, SH^- or NH_2^- were allowed to react with alkyl formates. Reaction 20 might therefore proceed through an S_N2 type collision complex rather than through a tetrahedral intermediate. Changing from the methyl to the ethyl ester of formic acid causes the rate constant for the reaction of F^- producing $HCOO^-$ to increase several--fold and is therefore at variance with the expectations of an S_N2 mechanism. Neutral product analysis was not inconsistent with the idea that the other reaction products were $C_2H_4 + HF$ and not C_2H_5F. The importance of the β-hydrogen atoms in accelerating the formation of $HCOO^-$ was seen more emphatically in reactions of F^- with the butyl formates, where the rate constants were found to follow the order tert-butyl > sec-butyl, n-butyl > iso-butyl. The authors rejected an explanation for these observations based on the notion of a base-induced elimination reaction (Ridge and Bauchamp, 1974), because this mechanism would also fail to explain why stronger bases like alkoxide ions are unable to promote this elimination:

$$F^- + \quad \longrightarrow \quad \longrightarrow HF + HCOO^- + C_2H_4 \tag{22}$$

Therefore, they have proposed the mechanism of reaction 23 involving a tetrahedral intermediate for the formation of $HCOO^-$ from ethyl and higher alkyl formates containing β hydrogen atoms:

$$F^- + \quad \longrightarrow \quad \longrightarrow HCOO^- + HF + C_2H_4 \tag{23}$$

Reactions of $^{18}OH^-$ and $^{18}O^{-\bullet}$ with methyl formate have shown about 80% and 60% incorporation of the label in formate ions (Dawson and Nibbering, 1978a). Incorporation of ^{18}O into unlabelled formate ions by reaction with $H_2^{18}O$ was not found:

$$HCOO^- + H_2^{18}O \not\rightleftharpoons HCO^{18}O^- + H_2O \qquad (24)$$

These observations taken together point to two mechanisms for the formation of $HCOO^-$ from methyl formate by OH^- and $O^{-\bullet}$, one via a tetrahedral intermediate and the other via an S_N2 type complex:

$$\qquad (25a)$$

$$\qquad (25b)$$

Methyl formate has been used as a reagent to distinguish between different structures of the $(M-H)^-$ and $(M-H_2)^{-\bullet}$ ions generated from propyne by reaction with $O^{-\bullet}$ (Dawson, Kaandorp and Nibbering, 1977). These ions appear to exist in two pairs of stable non-isomerising structures – $CH_3-C\equiv C^-$ & $^-CH_2-C\equiv CH$ and $^\bullet CH=C=CH^-$ & $^\bullet CH_2-C\equiv C^-$ – which reacted with methyl formate as shown in reactions 26 to 29:

$$\qquad (26)$$

$$\qquad (27a)$$

$$k_{27a}/k_{27b} \approx 3$$

$$\qquad (27b)$$

$$\underset{m/e\ 40}{\dot{C}D_2-C\equiv C^-} + HCOOCH_3 \longrightarrow \text{no products observed} \qquad (28)$$

$$\underset{m/e\ 39}{\begin{matrix}\dot{C}D=C=\bar{C}H\\ \updownarrow\\ \bar{C}D=C=\dot{C}H\end{matrix}} + HCOOCH_3 \longrightarrow \left[\begin{matrix}&&OCH_3\\ CD\equiv C&-\underset{(D)}{\bar{C}H}&-\overset{|}{\underset{|}{C}}-O\cdot\\ (H)&&H\end{matrix}\right]^* \longrightarrow \underset{(H)}{CD\equiv C}-\underset{(D)}{CH}=CH-O^- + CH_3O\cdot \quad (29)$$

$$\text{(} m/e\ 68 \text{)}$$

Recently it has been found that NH_2^- reacts with hexafluoroacetone to generate the ions $C_2F_3HNO^-$ and NCO^- (reaction 30a) and that OH^- reacts to form just $C_2F_3O_2^-$ (reaction 30b) CF_3H being expelled in all cases (Noést and Nibbering, 1978a). The involvement of tetrahedral intermediates in these reactions is the only rational explanation possible. Multiple elimination of CF_3H may occur in a stepwise or concerted manner.

$$NH_2^- + \underset{CF_3}{\overset{CF_3}{C}}=O \longrightarrow \left[\underset{CF_3}{HN-\overset{H\ CF_3}{C}-O^-}\right]^* \xrightarrow{-CF_3H} \underset{CF_3}{NH=C-O^-} \xrightarrow{-CF_3H} NCO^- \qquad (30a)$$

$$OH^- + \underset{CF_3}{\overset{CF_3}{C}}=O \longrightarrow \left[\underset{CF_3}{O-\overset{H\ CF_3}{C}-O^-}\right]^* \xrightarrow{-CF_3H} CF_3-\overset{O}{\underset{O}{C}}- \qquad (30b)$$

It is of interest to note that the $(M-H)^-$ ions of CF_3COCH_2COR, generated by electron capture in a double focusing mass spectrometer, have also been found to eliminate CF_3H (Gregor, Jennings and Sen Sharma, 1977).

ADDITION-ELIMINATION REACTIONS

Olefinic Compounds

In solution chemistry it is well known that carbon--carbon multiple bonds can be destroyed and formed by addition and elimination reactions respectively. Both electrophiles and nucleophiles can add to carbon-carbon multiple bonds, especially when these are substituted respectively by strong electron-donating or electron--withdrawing groups (Lowry and Richardson, 1976):

$$E^+ \; + \; \Large{>}\normalsize C{=}C\Large{<} \; \longrightarrow \; \Large{>}\normalsize \overset{}{C}{-}\overset{+}{C}\Large{<} \qquad (31)$$
$$\qquad\qquad\qquad\qquad\qquad\quad |$$
$$\qquad\qquad\qquad\qquad\qquad\; E$$

$$Nuc^- \; + \; \Large{>}\normalsize C{=}C\Large{<} \; \longrightarrow \; \Large{>}\normalsize \overset{}{C}{-}\overset{\cdot}{C}\Large{<} \qquad (32)$$
$$\qquad\qquad\qquad\qquad\qquad\qquad |$$
$$\qquad\qquad\qquad\qquad\qquad\; Nuc$$

An example of reaction 31 is the addition of the methoxymethyl cation to olefins (van Doorn and Nibbering, 1978). The collision complex formed with propene eliminates both methanol and formaldehyde through the mechanisms shown in reactions 33 and 34 (elucidated by deuterium labelling experiments):

$$CH_3OCH_2^{\cdot} \; + \; CH_2{=}CH{-}CH_3 \; \longrightarrow \; \left[\quad \right]^* \xrightarrow{-CH_3OH} CH_3{-}CH{-}CH{-}CH_2 \;\; ,\Delta H^\circ{=}{-}12\,kcal/mole \qquad (33)$$

$$k_{33}\,/\,k_{34} \; = 1.8$$

$$CH_3OCH_2^{\cdot} \; + \; CH_2{=}CH{-}CH_3 \; \longrightarrow \; \left[\quad \right]^* \xrightarrow{-CH_2O} n{-}C_4H_9^{\cdot} \;\; ,\Delta H^\circ{=}{-}6\,kcal/mole \qquad (34)$$

Similar mechanisms have been reported for the unimolecular decompositions of corresponding $C_4H_9O^+$ ions (Bowen and Williams, 1977).

Reaction 32 is exemplified by interactions of strongly basic anions with fluorethylenes; CF_2CH_2 will be taken as an example. Reactions 35 to 38 have been observed when this compound is attacked by CD_3O^- (Riveros and Takashima, 1976; Sullivan and Beauchamp, 1977):

$$CF_2CH_2 \; + \; CD_3O^- \left\{ \begin{array}{l} \xrightarrow{0.09} CD_3OH \; + \; CF_2CH^- \qquad (35) \\[4pt] \xrightarrow{0.30} CFCH \; + \; CD_3OHF^- \qquad (36) \\[4pt] \xrightarrow{0.08} CD_3OH \; + \; CFCH + F^- \qquad (37) \\[4pt] \xrightarrow{0.53} CD_3F \; + \; CH_2CFO^- \qquad (38) \end{array} \right.$$

Reactions 35 and 36 correspond to proton transfer and elimination of HF via the chemically activated intermediate VI:

CD$_3$O$^-$ ... H ... C=C with F, F   VI

The generation of F$^-$ by reaction 37 may result from the
breakup either of the intermediate VI or of the ionic
product of reaction 36. Reaction 38 corresponds to a
nucleophilic addition followed by the loss of methyl
fluoride. CH$_2$CFO$^-$ (reaction 38) is the only product generated when alkoxide ions weaker than CH$_3$O$^-$ are used
(Sullivan and Beauchamp, 1977). Reaction 38 is calculated to be exothermic by 48.5 kcal/mole, when the recently determined fluoride affinity of ketene (McMahon and
Northcott, 1978) and heats of formation of CF$_2$CH$_2$ (Sullivan and Beauchamp, 1977), CH$_3$O$^-$ (Sullivan and Beauchamp, 1976) and of CH$_2$CO and CH$_3$F (Rosenstock, Draxl,
Steiner and Herron, 1977) are used. Reaction 38 becomes
less exothermic as the base becomes weaker CH$_3$O$^-$ >
CH$_3$CH$_2$O$^-$ > (CH$_3$)$_2$CHO$^-$ > (CH$_3$)$_3$CO$^-$. This is most probably
the reason why reactions 35 to 37 do not compete effec-
tively with reaction 38 for the last three bases. This
is also shown by the influence of base strength on the
reactivity of the related fluoroethane compounds (Sullivan and Beauchamp, 1976). In reaction 38 the CH$_3$O$^-$ ion
will attack the electropositive geminal difluoro carbon
atom. This leads to the formation of the intermediate
VII which may then eliminate methyl fluoride via a four-
-center rearrangement mechanism as indicated (Sullivan
and Beauchamp, 1977) or by an internal S$_N$2 type displacement reaction (Riveros and Takashima, 1976) possibly
via a solvated F$^-$ complex ion (DePuy, 1977):

$^-$CH$_2$—C(F, F)—CH$_3$ with O   VII

The rate of reaction 38 increases significantly when the
bases C$_2$H$_5$O$^-$ and (CH$_3$)$_2$CHO$^-$ are used. This has been explained by a six-center rearrangement of a β-hydrogen
atom to one of the fluorine atoms with concomitant loss
of HF and an olefin (Sullivan and Beauchamp, 1976; cf.
reaction 23). However, the rate of the reaction with
(CH$_3$)$_3$CO$^-$ which has nine β-hydrogens drops sharply from
that of (CH$_3$)$_2$CHO$^-$, so that it would be helpful to have
analysis of the neutral product(s) which could be either
C$_n$H$_{2n+1}$F or C$_n$H$_{2n}$+HF (Method of Lieder and Brauman, 1974;
Lieder and Brauman, 1975).

A remarkable difference has been observed between
cis-CFHCFH and trans-CFHCFH in reactions with alkoxide
ions: the former exhibits reactions 35, 36 and 38, but
the latter shows only reaction 35, even though the heats
of formation of these fluoroethylenes differ by as
little as 1 kcal/mole. It has been argued that the explanation may be that fluorine trans to the site of base
attack has a larger portion of the distributed negative
charge, making it more labile than a cis fluorine (Sullivan and Beauchamp, 1977).

Addition-elimination reactions have also been ob-
served in reactions of $O^{-\cdot}$ with α,β unsaturated nitriles, as exemplified in reaction 39 (Dawson and Nibbering, 1978b):

$$O^{-\cdot} + CH_3\overset{\beta}{C}H=\overset{\alpha}{C}HCN$$

$$\left[{}^{\bullet}O-\underset{H}{\overset{CH_3}{C}}-\bar{C}HCN \right]^{*} \tag{39}$$

$$\underset{\overset{|}{O=\overset{CH_3}{C}-\bar{C}HCN}}{\xleftarrow{-H^{\bullet}}} \quad \underset{O=\overset{}{\underset{H}{C}}-\bar{C}HCN}{\xleftarrow{-|^{\bullet}CH_3}} \quad \underset{{}^{\bullet}\bar{C}HCN}{\xrightarrow{-CH_3CHO}}$$

These three reactions clearly demonstrate the attack of
$O^{-\cdot}$ upon the β-carbon atom, a process known in solution
chemistry as "Michael Addition" (Lowry and Richardson,
1976). A similar attack of $O^{-\cdot}$ has been shown to occur
with maleic anhydride, which is followed by a successive
loss of $H^{\bullet}$, CO and CO_2. Reactions with ${}^{18}O^{-\cdot}$ have shown
that the reactant oxygen is fully retained in the
$(M-H-CO)^{-}$ ions, but that complete oxygen equilibration
then occurs prior to CO_2 loss (Dawson and Nibbering,
1978b).
It is interesting to note that the type of attack in
reaction 39 has recently been demonstrated to occur in
the tetracyanoethylene system, where the corresponding
$(M-CN)^{-}$ ions induce anionic polymerization reactions,
giving:

(Bowie, 1977)

Aromatic Compounds

Both electrophiles and nucleophiles react with aromatic compounds. Electrophilic attack on the aromatic ring has been demonstrated in reactions of methoxymethyl cations with benzene and toluene (Dunbar, Shen, Melby and Olah, 1973) as exemplified by reactions 40 and 41:

$$CD_3OCH_2^+ + C_6H_6 \longrightarrow C_7H_7^+ + CD_3OH \tag{40a}$$

$$CH_3OCD_2^+ + C_6H_6 \longrightarrow C_7H_5D_2^+ + CH_3OH \tag{40b}$$

$$CD_3OCH_2^+ + C_7H_8 \longrightarrow C_8H_9^+ + CD_3OH \tag{41a}$$

$$CH_3OCD_2^+ + C_7H_8 \longrightarrow C_8H_7D_2^+ + CH_3OH \tag{41b}$$

$$CH_3OCH_2^+ + C_6H_5CD_3 \longrightarrow C_8H_6D_3^+ + CH_3OH \tag{41c}$$

No positional selectivity has been found in the reactions of o-, m- and p-toluenes-d$_1$ with $CH_2OCH_2^+$. It has been suggested, that these results are not inconsistent with the solution-phase mechanistic concepts involving initial formation of π-complex aronium ions followed by conversion into σ-complexes (arenium ions):

$$\tag{42}$$

Within this picture, the lack of positional selectivity can be understood either as reflecting statistical formation of σ-complexes or ring hydrogen scrambling in the σ-complex. A similar attack has been proposed for $C_7H_7^+$ ions on toluene, which on the basis of deuterium labelling experiments can best be formulated as involving a benzyl cation transferring its methylene group to toluene (Shen, Dunbar and Olah, 1974):

$$\phi CH_2^+ \ + \ \text{(toluene)} \longrightarrow [\ \text{(σ-complex)}\ ] \longrightarrow \text{(product)} \ + \ \phi H \qquad (43)$$

Since then, ICR photodissociation experiments have shown, that $C_7H_7^+$ ions do exist in reactive and unreactive forms, presumably benzyl and tropylium respectively (Dunbar, 1976). This is in agreement with a recent ICR study of $C_7H_7^+$ ions (Jackson, Lias and Ausloos, 1977) and with previous ICR experiments, which have shown that the $C_7H_7^+$ ions from 7-methoxycycloheptatriene are unreactive towards dimethylamine (Venema and Nibbering, 1974) whereas the $C_7H_7^+$ ions from benzyl methyl ether do react with dimethylamine (Bruins and Nibbering, 1974). Similar conclusions have been arrived at from "Collisional Activation" experiments on $C_7H_7^+$ ions in a double focusing mass spectrometer with a reversed geometry (McLafferty and Winkler, 1974). Moreover, a recent study has shown that benzyl and tropylium ions are generated by attack of $C_3H_5^+$ on benzene (Houriet, Elwood and Futrell, 1978).
With regard to the π-complex in reaction 42 it may be noted that a consideration of measurements of the rate of acetylation of substituted anisoles and phenols, especially with the acetylating agent $CH_3COCO(COCH_3)CH_3^+$, has led to the postulation that π-complexes are involved in the course of these ion-molecule reactions (Chatfield and Bursey, 1976).

Recently reactions have been reported which demonstrate nucleophilic attack upon aromatic rings. Alkoxide anions react with fluorobenzene to give F^- (reaction 44), with difluorobenzenes to give (M-H)$^-$ and fluorophenoxide anions, with 1,3,5-trifluorobenzene, 1,2,3,4--tetrafluorobenzene and pentafluorobenzene to give exclusively (M-H)$^-$ ions and with hexafluorobenzene to give pentafluorophenoxide anions (reaction 45) (Briscese and Riveros, 1975):

$$RO^- \ + \ C_6H_5F \ \longrightarrow \ F^- \ + \ C_6H_5OR \tag{44}$$

$$RO^- \ + \ C_6F_6 \ \longrightarrow \ C_6F_5O^- + \ RF \tag{45}$$

The $C_6H_5O^-$ ions in reaction 45 can also be generated by attack of F^- ions upon pentafluoroanisole with expulsion of methyl fluoride (Sullivan and Beauchamp, 1977). Both reactions 44 and 45 can best be rationalized by postulating the Meisenheimer-type intermediate VIII which is well-known in solution chemistry (Lowry and Richardson, 1976):

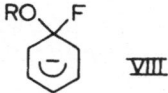

The observations listed above indicate, that these complexes are formed only if the acidity of the aromatic compounds is less than or comparable with that of the alcohol. $(M+NO_2)^-$ adducts have been observed in the ICR spectra of o-, m- and p-dinitrobenzenes and these may also be regarded as Meisenheimer-type complexes. In this case the formation of neutral trinitrobenzenes from reaction of NO_2^- and o-dinitrobenzene and of neutral chloronitrobenzene and chlorodinitrobenzene from the interaction of Cl^- with o-dinitrobenzene was reported (Bowie and Stapleton, 1977).
Probably complexes similar to VIII are also involved in reactions of $O^{-\cdot}$ ions with benzene, naphthalene, pyridine, alkylbenzenes, methylpyridines and fluorotoluenes where $(M-H+O)^-$ ions are generated by displacement of a ring hydrogen atom. Positional selectivity has been found in the reaction of $O^{-\cdot}$ with pyridine where deuterium labelling has demonstrated a preferential attack at the 4-position (Bruins, Ferrer-Correia, Harrison, Jennings and Mitchum, 1978).

Cycloaddition Reactions

In the chemistry of neutral unsaturated molecules it is a commonplace that two or more of such molecules may condense to form a ring by transferring electrons from π bonds to new σ bonds. Some typical examples,

characterized by the number of π electrons in the reacting components, are [2+2] and [2+4] cycloadditions giving four- and six-membered ring systems. The new σ bonds may be formed in a simultaneous way or in a stepwise manner. This is largely determined by the allowed or forbidden character of the cycloaddition in terms of orbital symmetry considerations. If the reaction has a forbidden character, cycloaddition may nevertheless occur via the intervention of biradical or zwitterionic intermediates. For more details the reader is referred to the literature (Lowry and Richardson, 1976).
Both positive and negative ions derived from π electron systems may in principle react with neutral π electron systems in the gas phase through cycloaddition type reactions. However, from the few studies performed so far it is not clear whether or not orbital symmetry rules are applicable. This is particular true for reactions of open shell systems (radical cations or anions). Most of the work done so far has been devoted to determining whether or not the supposed ionic cycloadditions do actually occur, an essential prerequisite to further work.

A clear example of a [2+2] cycloaddition (the numbers in brackets denote in this case and in following cases the number of carbon atoms involved) is the reaction between ionized ethylene and 1,1-difluoroethylene as summarized in equation 46:

$$C_2D_4^{+\bullet} + CH_2CF_2 \rightleftharpoons \left[\begin{array}{c} CD_2-CD_2 \\ | \quad | \\ CH_2-CF_2 \end{array} \right]^{+\bullet *} \longrightarrow C_2D_2F_2^{+\bullet} + C_2H_2D_2 \qquad (46)$$

(Ferrer-Correia and Jennings, 1973; Anicich and Bowers, 1974). It is interesting to note that in the collision complex of reaction 46 no H/D interchange takes place. Similar [2+2] cycloadditions have been shown to occur in ion-molecule reactions of monofluoropropenes (Drewery, Goode and Jennings, 1976a), in mixtures of fluorinated ethylenes and propylenes (Drewery, Goode and Jennings, 1976b) and in the vinyl methyl ether system (Drewery and Jennings, 1976). The latter molecule has proved to be a useful CI reagent gas since its molecular ion undergoes reactions similar to equation 46 with several classes of olefins. Such reactions can then be used to infer the location of the double bond in unknown compounds (Ferrer-Correia, Jennings and Sen Sharma, 1975 and 1976).

The first example of a [2+4] cycloaddition was discovered many years ago in an ion-molecule reaction of
styrene. Its radical cation reacts with neutral styrene
to produce a "dimeric" complex which subsequently loses
the elements of benzene. Such behaviour did not occur
with the isomeric cyclooctatetraene system, which also
turned out to be unreactive towards both ionized and
neutral styrene. Moreover, a [2+2] cycloaddition between
the olefinic bonds of ionized styrene and neutral styrene was ruled out, because the resulting trans- or cis-
-1,2-diphenylcyclobutanes were not expected to lose
benzene, as indicated by the 19 eV electron impact spectra of these model compounds. 1-Phenyltetralin however
eliminates benzene upon electron impact and deuterium
labelling shows this to be by a 1,4 elimination. An
identical behaviour was found for the "dimeric" complex
formed from appropriately deuterated ionized and neutral
styrene, thus showing the occurrence of a [2+4] cycloaddition pictured in reaction 47 (Wilkins and Gross,
1971):

$$\tag{47}$$

Recently, reactions of the 1,3-butadiene radical
cation with some olefinic compounds have been studied to
see whether they proceed as [2+4] cycloadditions. In
reaction with ethylene the collision complex eliminates
a methyl radical, which contains the hydrogen atoms from
the reagent and substrate at random whereas the carbon
atom originates exclusively from one of the terminal positions of the butadiene radical cation (shown by deuterium and ^{13}C-labelling experiments) (Gross, Russell,
Phongbetchara and Lin, 1978). These observations do not
fit in with an ionized cyclohexene type of intermediate
collision complex since field ionization kinetic studies
on ionized cyclohexene show that it rapidly ($\sim 10^{-11}$ s)
undergoes hydrogen scrambling by double-bond isomerization prior to methyl loss (Derrick, Falick and Burlingame, 1972). This would result in a statistical participa-
tion of all carbon atoms in the methyl loss. Therefore,
it has been postulated that the reaction of interest
proceeds (48) either through an acyclic intermediate or
through a [3+2] cycloaddition rather than through a
[4+2] cycloaddition:

$$(48)$$

However, the 1,3-butadiene/vinyl methyl ether system where either species may be ionized reacts via a [2+4] cycloaddition to form an ion m/e 80 (van Doorn, Nibbering, Ferrer-Correia and Jennings, 1978). The evidence for this is as follows:

1. The collision complex of 1,3-butadiene-1,1,4,4-d_4 and vinyl methyl ether loses exclusively CH_3OD.
2. There is a close similarity between the 15 eV electron impact spectrum of 4-methoxycyclohexene and the ICR spectrum of the 1,3-butadiene/vinyl methyl ether system.
3. The methanol lost in the electron impact spectrum of 4-methoxycyclohexene-2,6,6-d_3 is almost exclusively CH_3OD.
4. Collisional activation has shown that the product ions of m/e 80, generated in the ion-molecule reaction of 1,3-butadiene and vinyl methyl ether, have a cyclic structure, although the position of the double bond is not known exactly as indicated by the dotted lines (reaction 49).

$$(49)$$

By way of comparison, it is interesting to note that on the basis of deuterium and ^{13}C-labelling experiments arguments have been put forward for a [2+4] cycloaddition between the radical cation of furan, which serves as the dienophile, and 1,3-butadiene as summarized in reaction 50 (Gross, Russell, Phongbetchara and Lin, 1978):

$$(50)$$

Anionic cycloadditions have only recently come to receive some attention. It has been found that 2-cyano-allyl anions (generated for example by the reaction of $O^{-\cdot}$ ions with isocrotonitrile) react readily with tetra-fluoroethylene so as to expel two molecules of HF from the collision complex. The process probably involves two consecutive losses, although the product ions resulting from loss of one molecule of HF from the collision complexes have not been observed (Dawson and Nibbering, 1977). The reaction might proceed as a [2+3] cycloaddition ultimately yielding a substituted cyclopentadienyl anion (in terms of electrons this is an allowed [2+4] cycloaddition). This system is currently being studied by deuterium labelling in the hope of obtaining definitive results. It may be noted that the $(M-H)^-$ ions of cyclopropyl cyanide, obtained by abstraction of the α-proton by $O^{-\cdot}$ (shown by deuterium labelling) do not show any loss of 2xHF in reaction with tetrafluoroethylene. This observation taken together with the fact that the $(M-H)^-$ ions of cyclopropyl cyanide are more basic than the isomeric 2-cyanoallyl anions and the fact that the latter ions exchange their hydrogens with D_2O whereas the former do not indicates that these isomeric $C_3H_4CN^-$ ions are distinct stable species in the gas phase as in solution (Dawson and Nibbering, 1978b).

Reactions of allyl anions with tetrafluoroethylene have been studied in which 2xHF are also eliminated from the collision complex. Labelled allyl anions could easily be generated from labelled propenes but first it had to be proved that these anions do not suffer from internal hydrogen scrambling. This has been demonstrated by the following experiments (Dawson, Noest and Nibbering, 1978a): allyl anions, deuterated in the central position and derived from propene-2-d_1 by reaction with $O^{-\cdot}$, do not exchange their deuterium atom with water. The same conclusion was reached independently in a flowing after-glow study (Stewart, Shapiro, DePuy and Bierbaum, 1977). Allyl anions, fully deuterated in one of the terminal positions and derived either from $CH_3CH=CD_2$ or from $CD_3CH=CH_2$ by reaction with $O^{-\cdot}$, react with N_2O to expel mainly either H_2O or D_2O, but little HDO. This observation precludes internal H/D scrambling in the corresponding allyl anions and proves at the same time, that the N_2O reaction product is the 1-diazoprop-2-enyl anion (reaction 51) (Bierbaum, DePuy and Shapiro, 1977) rather than the pyrazole anion as had been suggested (reaction 52) (Smit and Field, 1977).

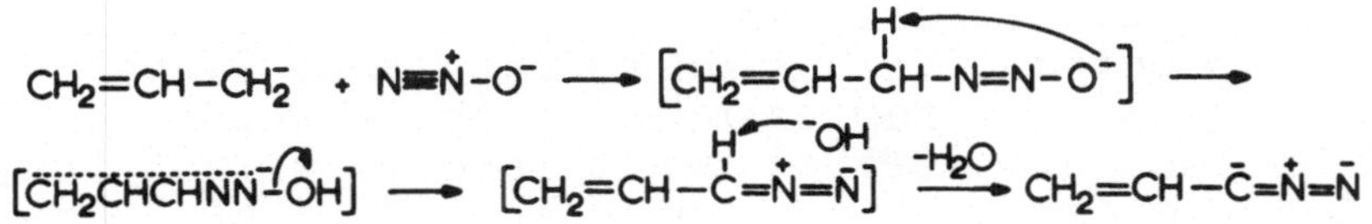

$$CH_2=CH-CH_2^- \ + \ N\equiv\overset{+}{N}-O^- \ \longrightarrow \ [CH_2=CH-\overset{H}{\underset{|}{C}}H-N=N-O^-] \ \longrightarrow \tag{51}$$

$$[\overset{\frown}{C}H_2CHCHNN-OH] \ \longrightarrow \ [CH_2=CH-\overset{H}{\underset{}{C}}=\overset{+}{N}=\overset{..}{N}] \ \xrightarrow{-H_2O} \ CH_2=CH-\overset{..}{C}=\overset{..}{N}=\overset{..}{N}$$

m/e 41

m/e 67

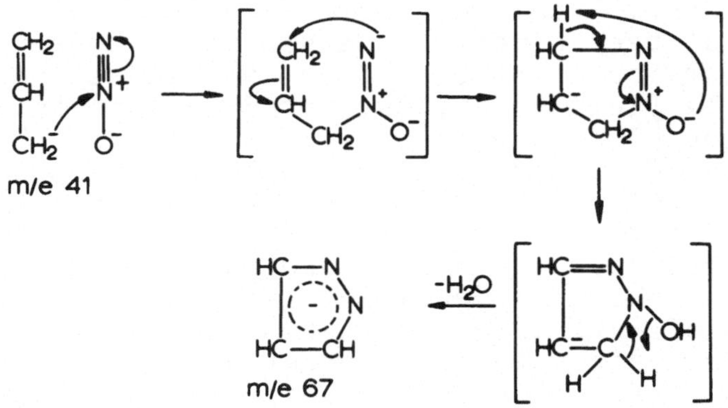

$$(52)$$

More definite evidence for the absence of internal H/D scrambling in $CH_2=CH-CD_2^-$ ions has been found from their reaction with F_2CO, where either 2xHF or 2xDF, but not HF+DF, are eliminated from the collision complex (Dawson, Noest and Nibbering, 1978).

Preliminary results, obtained for reactions of $CH_2=CD-CH_2^-$ and $CH_2=CH-CD_2^-$ anions with tetrafluoro-ethylene, have indicated that about 65% linear addition (reaction 53), 20% [2+2] cycloaddition (reaction 54) and only 15% [2+3] cycloaddition (reaction 55) occurs (Noest, Dawson and Nibbering, 1978):

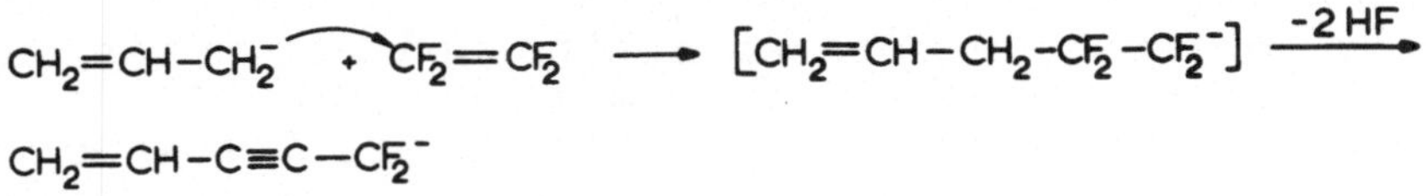

$$CH_2=CH-CH_2^- \ \overset{\frown}{\cdot} \ CF_2=CF_2 \ \longrightarrow \ [CH_2=CH-CH_2-CF_2-CF_2^-] \ \xrightarrow{-2HF}$$

$$CH_2=CH-C\equiv C-CF_2^- \tag{53}$$

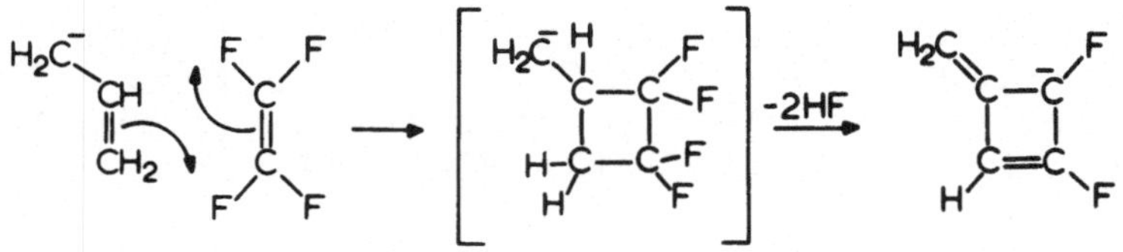

$$(54)$$

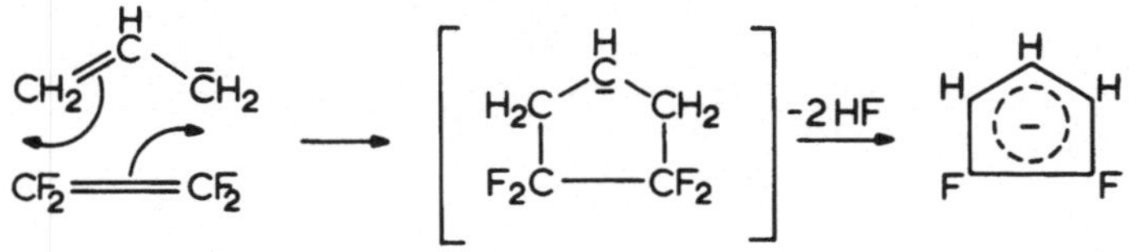

$$(55)$$

It should be noted, that the precise mechanistic details
of the loss of HF molecules from collision complexes are
not known. It is quite likely that they occur in a highly asymmetric or even a stepwise fashion in which fluoride anion solvated complexes may play a role. For example, although many apparent 1,2 elimination are known,
the $CD_3C \equiv C^-$ anions generated from reaction of $O^{-\cdot}$ with
$CD_3C \equiv CH$ react with F_2CO and C_2F_4 to expel 2xDF. In these
expected collision complexes the fluorine atoms cannot
come into close proximity with the trideuteromethyl
group because of the triple bond, unless the C-F bond is
considerably stretched (Noest and Nibbering, 1978b).

HYDROGEN-DEUTERIUM EXCHANGE REACTIONS OF CARBANIONS

Recently it has been discovered that many hydrogen
containing anions undergo bimolecular gas phase hydrogen
exchange reactions with weak acids such as water and alcohols (Stewart, Shapiro, DePuy and Bierbaum, 1977; Dawson, Kaandorp and Nibbering, 1977; MacKay, Lien, Hopkinson and Bohme, 1978; DePuy, Bierbaum, King and Shapiro,
1978; Hunt, Sethi and Shabanowitz, 1978).
A general picture is now beginning to emerge in which it
is seen that the exchange processes, which must occur in
an intermediate reaction complex, only become favourable
when the acidity of the exchange reagent is just a
little below the relevant acidity of the parent neutral
of the anion. This is in reassuring agreement with the
interpretation given for variations of the rate constants of proton transfer reactions involving delocalized negative ions. It has been proposed that these reactions occur in a three-step process corresponding to a
double-well potential with a small central barrier (cf.
Fig. 2) (Farneth and Brauman, 1976). This is summarized
in equation 56:

$$AH + B^- \underset{k_b}{\overset{k_c}{\rightleftharpoons}} [AH \cdots B^-] \overset{(A^- \cdots H^+ \cdots B)^{\ddagger}}{\underset{}{\overset{k_p}{\rightleftharpoons}}} [A^- \cdots HB] \underset{k_{c'}}{\overset{k_{b'}}{\rightleftharpoons}} A^- + HB \qquad (56)$$

(A similar process has been proposed for proton transfer
between neutral bases (Olmstead, Lev-On, Golden and
Brauman, 1977)).

If reaction 56 is slightly endothermic in the forward
direction, the initially formed $[AH \cdot \cdot B^-]$ may convert to
$[A^- \cdot \cdot HB]$ over a small central barrier involving the ac-

tivated complex $[A^-\cdot\cdot H^+\cdot\cdot B^-]^{\ddagger}$, which may then revert
(but now involving the shift of a proton originally be-
longing either to A^- or to B^-) and eventually dissociate
into the initial products. The exchange of hydrogens is
thus effected by interconversion between the solvated
$[AH\cdot\cdot B^-]$ and $[A^-\cdot\cdot HB]$ complexes. The exchanged B^- ions
may subsequently meet again another AH molecule, so that
multiple exchange of hydrogens becomes possible. Of
course, one should be able to write the structure of the
anion in such a way that the charge becomes localized at
a site bearing the hydrogen atom(s) which are to be exchanged. For example, the $(M-H)^-$ ion of ethyl acetate
($^-CH_2COOC_2H_5$) can exchange two of its hydrogen atoms
with D_2O, but the $(M-H)^-$ ion of methanol can not exchange at all (Stewart, Shapiro, DePuy and Bierbaum, 1977).
As noted above, the acidities of AH and BH should not
differ too much, otherwise exchange of hydrogens will
not occur. This is well demonstrated by the observation
that the $(M-H)^-$ ions from acetone and acetaldehyde do
not exchange with D_2O, but give up to five and two ex-
changes respectively with CH_3OD (DePuy, Bierbaum, King
and Shapiro, 1978). In this way the number of exchange-
able hydrogen atoms in anions can be counted, if suitable exchange reagents are chosen (Hunt, Sethi and Shabanowitz, 1978). This is a very promising method for
distinguishing isomeric anions, for example the acetate
anion does not exchange with D_2O, but the $(M-H)^-$ ion of
glycol aldehyde exhibits the exchange of one hydrogen
atom with D_2O. The technique may provide mechanistic details of an ion molecule reaction through product ion
identification as was the case with proton abstraction
from dihydropyran (reaction 57):

(57a)

or

(57b)

This reaction might have proceed either as an allylic
proton abstraction (reaction 57a) or as a β-proton abstraction with concomitant ring opening to generate an
enolate anion (reaction 57b). The product ion was not
found to undergo exchange (with D_2O) thus favouring the
enolate ion formation via reaction 57b (Stewart, Shapiro, DePuy and Bierbaum, 1977).

Further details on proton transfer reactions in the gas phase are to be found in the Chapter by R.W. Taft.

FORMATION OF $(M-H_2)^{-\cdot}$ IONS FROM REACTIONS OF $O^{-\cdot}$ WITH SOME ORGANIC SUBSTRATES

In addition to the expected $H^{\cdot}$ and H^+ abstractions, a large number of aliphatic compounds undergo $H_2^{+\cdot}$ abstraction in reaction with $O^{-\cdot}$ ions to give $(M-H_2)^{-\cdot}$ ions.
A well-known example is the rapid reaction of $O^{-\cdot}$ with ethylene, where deuterium labelling has established that the two hydrogen atoms are abstracted from the same carbon atom to produce the $H_2C=C^{-\cdot}$ ion (Goode and Jennings, 1974). This ion reacts with N_2O to form CH_2CN^- and ^{15}N labelling has shown that reaction takes place at the terminal nitrogen (Dawson and Nibbering, 1978c). Here there is a <u>solid</u> phase analogy in that the $H_2C=C^{-\cdot}$ ion has been generated from interaction between ethylene and O^- centres on magnesium oxide surfaces and has recently been identified by its electron spin resonance spectrum (Taarit, Symons and Tench, 1977). The abstraction of $H_2^{+\cdot}$ from a single carbon atom in compounds, containing a suitably activated methylene group, has been strongly indicated by reactions of $O^{-\cdot}$ with saturated aliphatic nitriles, fluorinated ethylenes and propenes, acrylonitrile, vinyl methyl ether (Dawson and Jennings, 1976), methyl alkyl ketones and cyclic ketones (Harrison and Jennings, 1976), which all show the production of $(M-H_2)^{-\cdot}$ ions. This is further corroborated by the observation that for example isopropyl cyanide and di-isopropyl ketone do not generate $(M-H_2)^{-\cdot}$ ions by reaction with $O^{-\cdot}$. It is in this respect truly amazing to note, that $O^{-\cdot}$ ions abstract $H_2^{+\cdot}$ exclusively from one corner of the ring of cyclopropyl cyanide, whereas H^+ is abstracted exclusively from the α-position (shown by deuterium labelling experiments) (Dawson and Nibbering, 1978b). Additional arguments for the $H_2^{+\cdot}$ abstraction from a single carbon atom follow from further decomposition of $(M-H_2)^{-\cdot}$ ions. For example, $(M-H_2)^{-\cdot}$ ions of n-butyronitrile and of acetone eliminate a methyl radical as rationalized in equations 58 and 59 (Dawson and Jennings, 1976; Harrison and Jennings, 1976):

$$CH_3-CH_2-C^{-\cdot}-CN \longrightarrow CH_2=\bar{C}-CN \;+\; {}^{\cdot}CH_3 \qquad (58)$$

$$CH_3-C-CH \longrightarrow HCCO^- + {}^{\bullet}CH_3 \qquad (59)$$

The evidence for the structure of the $(M-H_2)^{-\bullet}$ ion of acetone as shown in reaction 59 is strong on the basis of methyl expulsion. However, in view of the presence of active groups at both sides of the carbonyl group there remains the possibility that some of the non-decomposing ions may have the alternative structure generated via the mechanism given in reaction 60:

$$O^{-\bullet} + CH_3CCH_3 \longrightarrow \left[\begin{array}{c} H-CH_2 \\ -O \quad C=O \\ H--CH_2 \end{array} \right]^{*}$$

$$H_2O + CH_2-C-CH_2 \longleftarrow \left[\begin{array}{c} H \quad CH_2 \\ -O \quad C-O^{\bullet} \\ H--CH_2 \end{array} \right]^{*} \qquad (60)$$

A recent "stop-press" study of the reaction of $O^{-\bullet}$ with CH_3COCD_3 has shown, that about half of the $(M-H_2)^{-\bullet}$ ions of acetone are formed through reaction 60. These $(M-H_2)^{-\bullet}$ ions do not lose a methyl radical whereas the remaining lose a methyl radical according to reaction 59 without any deviance from the expected labelling specificity (Dawson, Noest and Nibbering, 1978b).
Deuterium labelling has uncovered a similar mechanism for the formation of $(M-H_2)^{-\bullet}$ ions from m-xylene, where each of the methyl groups furnish one of their hydrogen atoms to form water with $O^{-\bullet}$ (Bruins, Ferrer-Correia, Harrison, Jennings and Mitchum, 1978).

CONCLUSION

This review has sought to show that mechanisms of ion molecule reactions of organic species in the gas phase can be classified in a systematic way. Although an impressive body of data has become available within the past few years, the field is still in its infancy compared to the state of knowledge about organic chemistry in solution. Nevertheless, the field is blossoming and may be expected to have implications for chemists working in different media. This is already demonstrated by a recent publication on hydride transfer between ions and neutral molecules in the gas phase, a proces which is directly related to reductions in solution by hydride donors (DePuy, Bierbaum, Schmitt and Shapiro, 1978).

ACKNOWLEDGEMENTS

The author wishes to express his most sincere gratitude to Dr. J.H.J. Dawson, Drs. R. van Doorn, Drs. A.J. Noest and all the other members of his group for their enthusiasm and motivation to work in the field of ion molecule reactions and mass spectrometry. He also wishes to thank Prof. M.L. Gross (University of Nebraska, USA), Prof. K.R. Jennings (University of Warwick, England) and Prof. C.H. DePuy and members of his group (University of Colorado, Boulder, USA) for stimulating discussions and for preprints of their most recent papers. His thanks extend to all workers in the field, who have published their work, without which the present survey could not have been written. Finally the author would like to thank the Netherlands Organisation for Pure Research (SON/ZWO) for financial assistance in providing the basic research tools necessary in this field of activity.

REFERENCES

Anicich, V.G. and Bowers, M.T. (1974). Int. J. Mass Spectrom. Ion Phys. 13, 359.
Ausloos, P. (Ed.) (1975). Interactions between Ions and Molecules, Plenum Press, New York, N.Y.
Asubiojo, O.I., Blair, L.K. and Brauman, J.I. (1975). J. Am. Chem. Soc. 97, 6685.
Beauchamp, J.L. (1975). In "Interactions between Ions and Molecules", P. Ausloos Ed., Plenum Press, New York, N.Y., p. 413.
Bierbaum, V.M., DePuy, C.H. and Shapiro, R.H. (1977). J. Am. Chem. Soc. 99, 5800.
Bowen, R.D. and Williams, D.H. (1977). J. Am. Chem. Soc. 99, 6822.
Bowers, M.T. (Ed.) (1979). Gas Phase Ion Chemistry, Academic Press, New York, N.Y.
Bowie, J.H. (1977). Aust. J. Chem. 30, 2161.
Bowie, J.H. and Stapleton, B.J. (1977). Aust. J. Chem. 30, 795.
Bowie, J.H. and Williams, B.D. (1974). Aust. J. Chem. 27, 1923.

Briscese S.M.J. and Riveros, J.M. (1975). J. Am. Chem.
Soc. 97, 230.
Bruins, A.P., Ferrer-Correia, A.J.V., Harrison, A.G.,
Jennings, K.R. and Mitchum, R.K. (1978). Advan. Mass.
Spectrom. 7A, 355.
Bruins, A.P. and Nibbering, N.M.M. (1974). Tetrahedron
Lett. 2677.
Chatfield, D.A. and Bursey, M.M. (1976). J. Am. Chem.
Soc. 98, 6492.
Dawson, J.H.J. and Jennings, K.R. (1976). J.C.S. Farady
Trans. II, 72, 700.
Dawson, J.H.J., Kaandorp, Th.A.M. and Nibbering, N.M.M.
(1977). Org. Mass Spectrom. 12, 332.
Dawson, J.H.J. and Nibbering, N.M.M. (1978a). In "Lec-
ture Notes in Chemistry, Vol. 7: Ion Cyclotron Resonance
Spectrometry", H. Hartmann and K.-P. Wanczek, Eds.,
Springer Verlag, Berlin, p. 146.
Dawson, J.H.J. and Nibbering, N.M.M. (1978b). To be pu-
blished.
Dawson, J.H.J. and Nibbering, N.M.M. (1977). Prelimina-
ry reported at the Ninth British Mass Spectroscopy Group
meeting at the University of Swansea, England, 27-29
September, paper no. 20.
Dawson, J.H.J. and Nibbering, N.M.M. (1978c). J. Am.
Chem. Soc. 100, 1928.
Dawson, J.H.J., Noest, A.J. and Nibbering, N.M.M.
(1978a). Int. J. Mass Spectrom. Ion Phys., in press.
Dawson, J.H.J., Noest, A.J. and Nibbering, N.M.M.
(1978b). Int. J. Mass Spectrom. Ion Phys., in press.
DePuy, C.H. et al. (1977). Private discussion.
DePuy, C.H., Bierbaum, V.M., King, G.K. and Shapiro,
R.H. (1978). J. Am. Chem. Soc. 100, 2921.
DePuy, C.H., Bierbaum, V.M., Schmitt, R.J. and Shapiro,
R.H. (1978). J. Am. Chem. Soc. 100, 2920.
Derrick, P.J., Falick, A.M. and Burlingame, A.L. (1972).
J. Am. Chem. Soc. 94, 6794.
Doorn, R. van and Nibbering N.M.M. (1978). Org. Mass.
Spectrom. 13, 527.
Doorn, R. van, Nibbering, N.M.M., Ferrer-Correia, A.J.V.
and Jennings, K.R. (1978). Org. Mass Spectrom, in press.
Dougherty, R.C. (1974). Org. Mass Spectrom. 8, 85.
Dougherty, R.C., Dalton, J. and Roberts, J.D. (1974).
Org. Mass Spectrom. 8, 77.
Dougherty, R.C. and Roberts, J.D. (1974). Org. Mass.
Spectrom. 8, 81.
Drewery, C.J., Goode, G.C. and Jennings, K.R. (1976a).
Int. J. Mass Spectrom. Ion Phys. 20, 403.
Drewery, C.J., Goode, G.C. and Jennings, K.R. (1976b).
Int. J. Mass Spectrom. Ion Phys. 22, 211.
Drewery, C.J. and Jennings, K.R. (1976). Int. J. Mass

Spectrom. Ion Phys. 19, 287.
Dunbar, R.C. (1976). J. Am. Chem. Soc. 97, 1382.
Dunbar, R.C., Shen, J., Melby, E. and Olah, G.A. (1973).
J. Am. Chem. Soc. 95, 7200.
Faigle, J.C.G., Isolani, P.C. and Riveros, J.M. (1976).
J. Am. Chem. Soc. 98, 2049. See also: José, S.M. and Ri-
veros, J.M. (1977). Nouveau Journal de Chimie 1, 113.
Farneth, W.E. and Brauman, J.I. (1976). J. Am. Chem.
Soc. 98, 7891.
Ferrer-Correia, A.J.V. and Jennings, K.R. (1973). Int.
J. Mass Spectrom. Ion Phys. 11, 111.
Ferrer-Correia, A.J.V., Jennings, K.R. and Sen Sharma,
D.K. (1975). J.C.S. Chem. Comm. 973. and (1976). Org.
Mass Spectrom. 11, 867.
Fraenkel, G. and Watson, D. (1975). J. Am. Chem. Soc.
97, 231.
Goode, G.C. and Jennings, K.R. (1974). Advan. Mass Spec-
trom. 6, 797.
Gregor, I.K., Jennings, K.R. and Sen Sharma, D.K.
(1977). Org. Mass Spectrom. 12, 93.
Gross, M.L., Russell, D.H., Phongbetchara, R. and Lin,
P.-H. (1978). Advan. Mass Spectrom. 7A, 129.
Harrison, A.G. and Jennings, K.R. (1976). J.C.S. Fara-
day Trans. I 72, 1601.
Houriet, R., Elwood, T.A. and Futrell, J.H. (1978). J.
Am. Chem. Soc. 100, 2320.
Hunt, D.F., Sethi, S.K. and Shabanowiz, J. (1978). 26th
Annual Conference on Mass Spectrometry and Allied To-
pics, St. Louis, Missouri, May 28-June 2, paper no. MF1.
Hvistendahl, G. and Williams, D.H. (1975). J. Am. Chem.
Soc. 97, 3097.
Jackson, J.A., Lias, S.G. and Ausloos, P. (1977). J. Am.
Chem. Soc. 99, 7515.
Jennings, K.R. (1977). In "Mass Spectrometry-Vol. 4,
Specialist Periodical Reports", R.A.W. Johnstone, Ed.,
The Chemical Society, London, Chapter 9.
Jennings, K.R. (1978). Advan. Mass Spectrom. 7A, 209.
Kebarle, P. (1977). Ann. Rev. Phys. Chem. 28, 445.
Lehman, T.A. and Bursey, M.M. (1976). Ion Cyclotron Re-
sonance Spectrometry, Wiley-Interscience, New York.
Lias, S.G. and Ausloos, P. (1975). Ion-molecule reac-
tions - their role in radiation chemistry, American
Chemical Society, Washington, D.C. 20036.
Lieder, C.A. and Brauman, J.I. (1974). J. Am. Chem. Soc.
96, 4028.
Lieder, C.A. and Brauman, J.I. (1975). Int. J. Mass
Spectrom. Ion Phys. 16, 307.
Lifshitz, C., Wu, R.L.C. and Tiernan, T.O. (1978). J.
Am. Chem. Soc. 100, 2040.
Lowry, T.H. and Richardson, K.S. (1976). Mechanism and

Theory in Organic Chemistry, Harper & Row, Publishers,
Inc., New York.
Mackay, G.I., Lien, M.H., Hopkinson, A.C. and Bohme,
D.K. (1978). Can. J. Chem. 56, 131.
McLafferty, F.W. and Winkler, J. (1974). J. Am. Chem.
Soc. 96, 5182.
McMahon, T.B. and Northcott, C.J. (1978). Can. J. Chem.
56, 1069.
Noest, A.J., Dawson, J.H.J. and Nibbering, N.M.M. (1978).
To be published.
Noest, A.J. and Nibbering, N.M.M. (1978a). Unpublished
results.
Noest, A.J. and Nibbering, N.M.M. (1978b). Unpublished
results.
Olmstead, W.N. and Brauman,J.I. (1977). J. Am. Chem. Soc.
99, 4219.
Olmstead, W.N., Lev-On, M., Golden, D.M. and Brauman,
J.I. (1977). J. Am. Chem. Soc. 99, 992.
Pau, J.K., Kim, J.K. and Caserio, M.C. (1974). J. Chem.
Soc. Chem. Comm. 120.
Pau, J.K., Kim, J.K. and Caserio, M.C. (1978a). J. Am.
Chem. Soc. 100, 3838.
Pau, J.K., Kim, J.K. and Caserio, M.C. (1978b). J. Am.
Chem. Soc. 100, 3831.
Ridge, D.P. and Beauchamp, J.L. (1974). J. Am. Chem. Soc.
96, 3595.
Riveros, J.M., Breda, A.C. and Blair, L.K. (1974). J. Am.
Chem. Soc. 96, 4030.
Riveros, J.M. and Takashima, K. (1976). Can. J. Chem. 54,
1839.
Rosenstock, H.M., Draxl, K., Steiner, B.W. and Herron,
J.T. (1977). J. Phys. Chem. Ref. Data, Vol. 6, suppl. 1.
Schoemaker, H.E., Nibbering, N.M.M. and Cooks, R.G.
(1975). J. Am. Chem. Soc. 97, 4415.
Shen, J., Dunbar, R.C. and Olah, G.A. (1974). J. Am.
Chem. Soc. 96, 6227.
Smit, A.L.C. and Field, F.H. (1977). J. Am. Chem. Soc.
99, 6471.
Stewart, J.H., Shapiro, R.H., DePuy, C.H. and Bierbaum,
V.M. (1977). J. Am. Chem. Soc. 99, 7650.
Su, T. and Bowers, M.T. (1973a). J. Chem. Phys. 58, 3027.
Su, T. and Bowers, M.T. (1973b). Int. J. Mass Spectrom.
Ion Phys. 12, 347.
Sullivan, S.A. and Beauchamp, J.L. (1976). J. Am. Chem.
Soc. 98, 1160.
Sullivan, S.A. and Beauchamp, J.L. (1977). J. Am. Chem.
Soc. 99, 5017.
Taarit, Y.B., Symons, M.C.R. and Tenck, A.J. (1977). J.C.
S. Faraday Trans. I 73, 1149.
Taft, R.W. and Arnett, E.M. (1975). In "Proton-Transfer

Reactions", E. Caldin and V. Gold Eds., Chapman and Hall, London.

Tanaka, K., Mackay, G.I., Payzant, J.D. and Bohme, D.K. (1976). Can. J. Chem. 54, 1643.

Venema, A. and Nibbering, N.M.M. (1974). Tetrahedron Lett. 3013.

Wilkins, C.L. and Gross, M.L. (1971). J. Am. Chem. Soc. 93, 895.

Wilson, J.C. and Bowie, J.H. (1975). Aust. J. Chem. 28, 1993.

INTRAMOLECULAR SELECTIVITY, STEREOCHEMICAL AND STERIC ASPECTS OF ION-MOLECULE REACTIONS

Fulvio Cacace

University of Rome

00100 Rome, Italy

INTRODUCTION

Over the past few years, the impact of gas-phase ionic chemistry upon major fields of the chemical research has been steadily increasing. In particular, organic applications of icr (Beauchamp, 1975) and high pressure mass spectrometry (Kebarle 1975, 1977) have provided kinetic and thermochemical results of great value to physical organic chemistry.

By comparison, the mechanistic impact has been undoubtedly less significant, especially in view of the scarce and largely indirect information provided by most gas-phase experiments on basic features of organic reactions, such as intramolecular selectivity and orientation, stereochemistry, steric effects, etc. that are mechanistically fundamental and are easily deduced from product analysis in solution-chemistry studies.

Such limitations arise from the recognized lack of structural and stereochemical discrimination typical of mass spectrometry, the technique that has played an almost exclusive role in the study of ion-molecule interactions. Structural discrimination of gaseous ions represents today the most serious problem in the mechanistic study of ion-molecule interactions, and despite its current sophistication and ingenious experiments (Lieder and Brauman, 1975), a purely mass spectrometric approach can hardly provide a general and direct answer to mechanistically fundamental problems of structural and stereochemical nature. These limitations have renewed interest in alternative experimental approaches, specifically designed to extend to gas-phase ionic processes the old chemical practice based

on the actual isolation of the reaction products. After a brief
outline of the experimental techniques, this presentation will
illustrate a few representative applications concerning problems of
intramolecular selectivity, stereochemical and steric features of
typical gas-phase ion-molecule reactions.

EXPERIMENTAL TECHNIQUES

The principles, and the early applications of the two major experimental approaches have been previously illustrated, in particular at the Biarritz NATO Institute (Cacace, 1975 a, Lias, 1975).
Their common feature is the introduction of ionic reactants of
precisely defined nature into gaseous systems at any convenient
pressure (typically from a few Torr to several atmospheres),
followed by isolation of the neutral end products from the ion-
molecule reactions. The actual isolation of products allows direct
determination of their molecular structure, isomeric and isotopic
composition, stereochemical features, etc., which in turn provides
valuable information on their ionic precursors, much in the same
way as in conventional solution-chemistry studies. Both techniques
can be in fact regarded as a direct extension to the dilute gas
state of the classical solution-chemistry approach, and utilize all
its mechanistic and kinetic tools, including competition experiments,
pressure-dependence analysis, radical scavengers and ionic inter-
ceptors, isotopic labeling, etc.

Radiolytic Technique

In this approach, pioneered by Ausloos (Ausloos, 1966, 1969,
Ausloos and Lias 1972, Lias and Ausloos, 1975), the irradiation of
the bulk constituent of a carefully selected system gives a low concentration of the ionic reagent(s), whose nature and yields must be
precisely known from preliminary mass spectrometric and radiolytic
studies.

The reagent, usually formed in an excited state from the radiolysis, must undergo many unreactive collisions with the bulk component of the system in order to be thermalyzed before reacting with
the neutral substrate of interest, present in the gas at very low
concentrations. The neutral end products from the ion-molecule reactions are eventually analyzed by suitable techniques, such as glc
and/or mass spectrometry.

The ionic nature of the processes responsible for the formation
of the products is ensured by appropriate radical scavengers that
remove reactive neutral species from the radiolysis. Furthermore,
independent confirmation is usually obtained using ionic intercep-
tors (e.g. NH_3 for H_3^+, CH_5^+, and other gaseous acids) that cause

a decrease in the rate of formation of the products from ion-molecule reactions.

Decay Technique

In the decay technique (Cacace, 1964, 1970, 1975b), the charged reactant is obtained from the spontaneous β decay of a precursor containing several T atoms in equivalent positions of the same molecule, e.g.

$$CT_4 \xrightarrow{\text{beta decay}} CT_3^+ + {}^3He + \beta^- \tag{1}$$

which is contained in the gas at extremely low (tracer) concentrations. The reactions of the decay ions give eventually tritiated neutral products, whose nature and yields can be measured by suitable radiometric techniques, e.g. radio glc.

Of course, the nature and the abundance of the ionic reagent must be precisely known by preliminary mass spectrometric experiments designed to determine the decay-induced fragmentation pattern of the precursor (Wexler, 1965). It should be noted that, since the charged reactant is obtained via a nuclear process, entirely insensitive to the environmental factors, the decay approach can be used in dense gases, or even in condensed phases (Cacace and Giacomello, 1977).

COMPETITION AMONG π-TYPE CENTERS. ALKYLATION OF ARENES BY GASEOUS CARBOCATIONS

Carbenium ions can be conveniently generated in the dilute gas state with the radiolytic and the decay techniques. For instance, thermal t-butyl cations can be obtained from the radiolysis of neopentane

$$neo\text{-}C_5H_{12} \longrightarrow [t\text{-}C_4H_9^+]_{exc} + CH_3 + e \tag{2}$$

$$[t\text{-}C_4H_9^+]_{exc} \xrightarrow[\text{collisions}]{\text{Many}} t\text{-}C_4H_9^+ \tag{3}$$

i-propyl ions from propane or i-butane, ethyl ions from methane, etc. Gaseous carbenium ions have been also obtained from the decay of tritiated alkanes, including CT_3^+ ions from CT_4, equation (1), $C_2H_4T^+$ ions from $C_2H_4T_2$, $C_3H_6T^+$ ions from $C_3H_6T_2$, etc.

Alkyl cations in general react efficiently with arenes, present in small concentration in a gaseous alkane, yielding the correspondent alkylbenzenes, whose isomeric composition provides otherwise inaccessible information on the intramolecular selectivity of
the attack.

Alkylation of Arenes by Carbenium Ions

Irradiation of gaseous neopentane, containing typically 0.5
mol% toluene, 1.5 mol% O_2 (as a radical scavenger) and traces of
gaseous bases (EtOH or NH_3) yields exclusively para and meta-t-
butyltoluenes, whose proportions depend on the pressure of the
system. No ortho substitution is observed, and increasing concentrations of NH_3 depress, and eventually suppress the formation
of alkylated products, consistent with the ionic nature of the t-
butylation (Cacace and Giacomello, 1973).

In the high-pressure range (500 to 720 Torr) a constant isomeric composition is observed, corresponding to 95% para and 5%
meta substitution. As the pressure is lowered, orientation gradually
shifts in favour of the meta isomer, and at the lowest pressure reported (8.7 Torr) 37% para and 63% meta-t-butyltoluene are formed.
On the grounds of kinetic and chemical ionization evidence (Munson
and Field, 1967), the results have been interpreted with a general
scheme that applies to other alkylations as well. The t-butyl
ions from the radiolysis (eqs. 2 and 3) condense with the arene
yielding as primary intermediates isomeric arenium ions (in the
specific case of t-butylation, essentially the para isomer) excited
by the exothermicity of the reaction, e.g.

Effective collisional stabilization, followed by proton loss to a gaseous base (eq. 5) eventually gives neutral alkylated products whose isomeric composition reflects that of the arenium ions formed in the kinetically controlled step (4).

On the other hand, as the pressure is lowered, the efficiency of collisional stabilization decreases, allowing increasing isomerization toward more stable arenium ions, e.g. (II), whose deprotonation yields neutral products characterized by an increasing thermodynamic control of isomeric composition. As a general remark, since isomerization cannot be entirely excluded even at the highest pressures, especially for the most exothermic substitutions, the intramolecular selectivity observed can be regarded as a lower limit. Anyhow, these results characterize the free t-butyl ion as a remarkably selective electrophile, as shown by the high para:meta ratio, which actually exceeds that measured in solution.

Isopropylation by free $i-C_3H_7^+$ cations is considerably less selective than t-butylation. Only at high pressure, under conditions favouring kinetic control of products, a typically electrophilic orientation pattern emerges (43% ortho-, 32% meta-, 25% para-i-propyltoluene), characterized by a para:meta ratio significantly higher than the statistical value (Cacace and Possagno, 1973, Cacace et al., 1974). At lower pressures, the orientation rapidly shifts toward thermodynamically more stable arenium ions, and in fact the isomeric distribution measured at 100 Torr (44.2% ortho, 40.4% meta and 15.4% para-i-propyltoluene) was unduly regarded (Takamuku et al., 1971) as evidence for the lack of positional selectivity in gas-phase ionic alkylations. There is no steric hindrance to ortho substitution, as shown by the fact that at high pressures the ortho:para ratio approaches the statistical value.

Ethylation is still more unselective, as shown by the isomeric composition of ethyltoluenes from the attack of ethyl ions (produced from the radiolysis of methane) on toluene highly diluted in methane gas at 720 Torr, namely 43% ortho, 34% meta and 22% para (Cacace et al., 1974). Of course, taking into account the higher exothermicity of the process with respect to other alkylations, isomerization of the primary arenium ions cannot be excluded, even in methane at atmospheric pressure.

The interplay of electronic and steric effects is particularly evident in the gas-phase alkylation of the xylenes by free t-butyl ions (Giacomello and Cacace, 1975, Giacomello and Cacace, 1976). p-Xylene is entirely unreactive, giving no alkylated products, which shows that the steric hindrance of an ortho methyl group outweights its activating effect, consistent with the lack of ortho-t-butylation of toluene. o-Xylene is alkylated only at position 4, the activating effect of one methyl group being again insufficient to overcome its steric hindrance, thus preventing substitution at position 3.

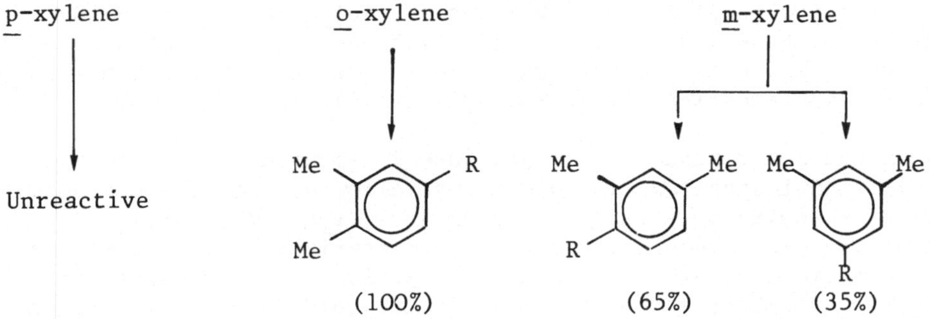

p-xylene   o-xylene   m-xylene

Unreactive

(100%)   (65%)   (35%)

On the other hand, meta xylene is alkylated predominantly (65%) at position 4, ortho to one methyl group and para to the other, showing that the combined activating effects of the two methyl groups overcome the steric hindrance of one, but not two, methyl groups. The latter point is clearly demonstrated by the lack of alkylation at position 2, ortho to both methyl groups. All these data refer to high-pressure systems, under conditions favouring kinetic control of products. At lower pressures, the orientation shifts in favour of the most stable symmetrically-substituted arenium ion showing that secondary isomerization is occurring.

Isopropylation provides an interesting contrast with t-butylation of the xylenes. The composition of the alkylated products in propane at 720 Torr (Attinà et al., 1977 a) shows that contrary to early solution-chemistry evidence,

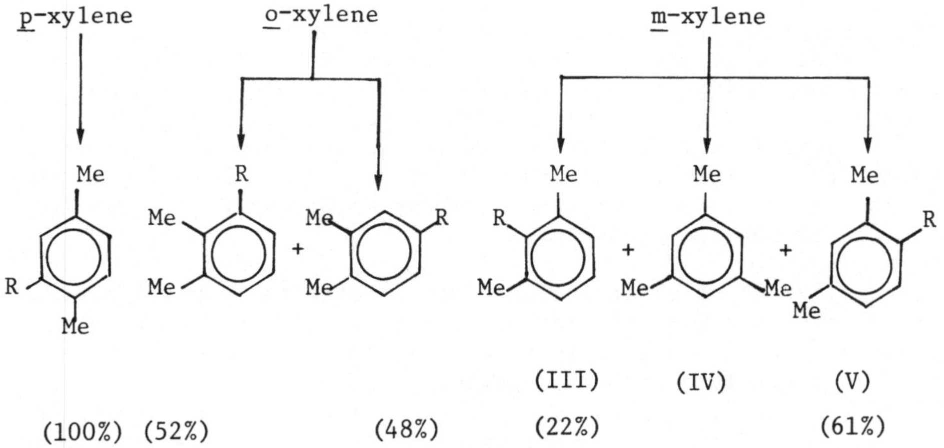

p-xylene   o-xylene   m-xylene

(III)   (IV)   (V)

(100%) (52%)   (48%)   (22%)   (61%)

an <u>ortho</u> methyl group causes no steric hindrance at all to a free,
gaseous i-propyl cation. Such conclusion is supported by the predominant <u>ortho</u> orientation in <u>o</u>-xylene, the efficient alkylation
of <u>p</u>-xylene (unreactive toward t-C$_4$H$_9$$^+$) and especially the high
reactivity of the position 2 of <u>m</u>-xylene, despite the steric hindrance of <u>two</u> methyl groups.

At lower pressures, the isomeric composition changes in favour
of the most stable arenium ions, <u>e.g.</u> the "symmetrically" substituted one in the case of <u>m</u>-xylene. Thus, at 20 Torr, the lowest pressure investigated, the isomeric composition corresponds to 8.4%
(III), 54.3% (IV) and 37.3% (V), consistent with the expected highest
stability of the "symmetric" arenium ion (VI).

(VI)

Gas-phase methylation of toluene has been invesitgated with
the decay technique (Cacace and Giacomello, 1978) using free CT$_3$$^+$
ions from the decay of methane-T$_4$ (eq. 1). The apparent intramolecular selectivity of the methyl ion, deduced from the isomeric
composition of the xylenes formed in the pressure range 36-350 Torr,
<u>i.e.</u> 19.5% <u>ortho</u>, 68.5% <u>meta</u>, 12.0% <u>para</u>, is characterized by pre-
dominant <u>meta</u> orientation that, <u>prima facie</u>, contrasts with the obvious electrophilic character of the attack. However, one has to
take into account the extremely high exothermicity of the methylation
process, <u>e.g.</u>

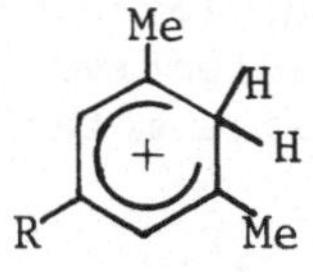

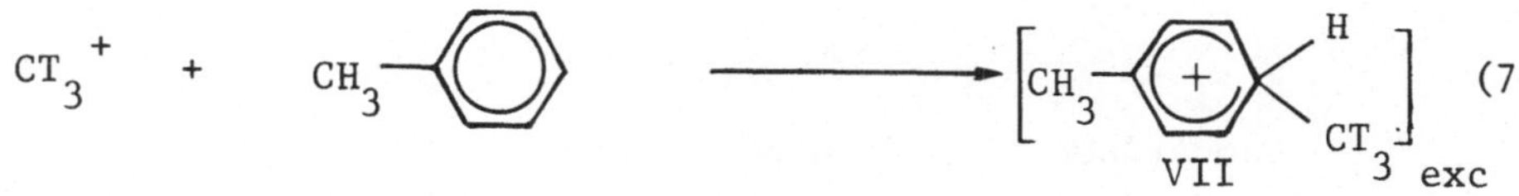

(7)

estimated around 90 kcal mol^{-1}, as compared to ca. 50 kcal mol^{-1}
for ethylation, 30 kcal mol^{-1} for i-propylation, and 10 kcal mol^{-1}
for t-butylation. The high excitation level of the primary arenium
ions is expected to promote their isomerization into the most stable
structure, <u>i.e.</u> the 2,4-dimethylbenzenium ion

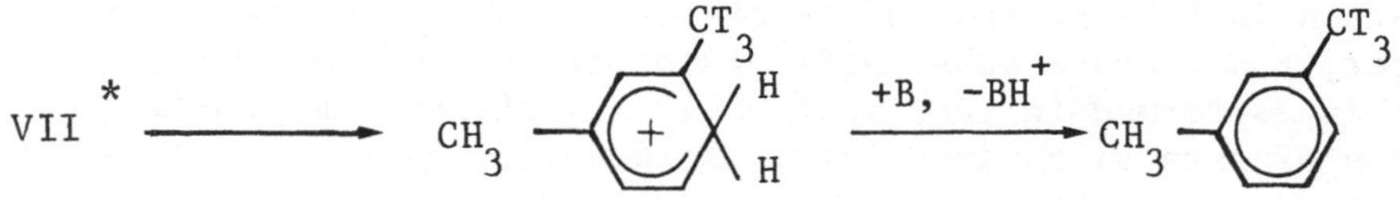

(8)

which accounts for the predominant <u>meta</u> orientation. Clearly, even
at 350 Torr, collisional deactivation cannot prevent extensive isomerization, and therefore the observed composition of xylenes does
not reflect the initial selectivity of the attack, being largely
thermodynamically controlled. This view is supported by the results
of liquid-phase methylation of toluene with free CT_3^+ ions generated
in <u>situ</u> by the decay of dissolved CT_4 (Cacace and Giacomello, 1977).
The much higher efficiency of collisional stabilization in the liquid
phase reduces, if not entirely suppresses, secondary isomerization,
and the isomeric composition of xylenes, 40% <u>ortho</u>, 27% <u>meta</u>, and
33% <u>para</u>, reflects more faithfully the relative abundances of the
primary arenium ions, showing the electrophilic character of the
reagent.

Similar results have been reported (Giacomello and Schüller,
(1977), for the reaction of t-butylbenzene with CT_3^+ decay ions.
The extremely exothermic and unselective character of the alkylation is underlined by the extensive methyl de-t-butylation observed

$$CT_3^+ + C_6H_5\text{-}t\text{-}C_4H_9 \rightarrow C_6H_5CT_3 + t\text{-}C_4H_9^+ \tag{9}$$

and by the appreciable attack to the side-chain, giving amylbenzenes.
The predominant <u>meta</u> orientation (56%) has been traced again to secondary isomerization of the highly excited primary arenium ions.
In fact, when CT_3^+ decay ions are allowed to react with <u>liquid</u>
t-butylbenzene, where collisional deactivation is of course considerably enhanced, the isomeric composition of the t-butyltoluenes
is 25% <u>ortho</u>, 37% <u>meta</u>, and 38% <u>para</u>, consistent with the predominant
<u>ortho-para</u> orientation expected for a charged electrophile. While
occurrence of secondary isomerization cannot be certainly excluded,
even in the liquid phase, for such an extremely exothermic reaction,
the 1/2 <u>ortho:para</u> ratio below 0.3 clearly reveals the effects of
steric hindrance to methylation <u>ortho</u> to the bulky t-butyl substituent.

Comparison with Solution-Chemistry Data

As a whole, aromatic substitution by free carbenium ions can
be regarded, from the standpoint of intramolecular selectivity, as
a homogeneous class of reactions, following the inverse reactivity-
selectivity correlation long established in solution chemistry.
A possible exception is alkylation by free methyl ions, whose peculiar
orientation is however reasonably traced to its extremely exothermic
character, that causes substantial isomerization of the charged
intermediates formed in the kinetically significant step, even at
pressures of several hundred Torr. In general, gas-phase orientation
measured under conditions favouring kinetic control of products displays no substantial differences with that measured in aprotic

solvents. Indeed, gas-phase experiments provide the first direct
data on the <u>intrinsic</u> steric requirements of carbenium ions, independent of their solvation state and specific counterion, and
will prove to be of value for comparison with the results of theoretical calculations, where alkyl cations are necessarily treated
as free, unsolvated species.

COMPETITION OF π-TYPE AND n-TYPE CENTERS FOR GASEOUS BRØNSTED ACIDS PROTONATION OF HALOBENZENES

Combined application of icr and chemical ionization mass spectrometry with neutral products analysis has allowed identification
of the competitive reaction channels promoted by the gas-phase
attack of strong Brønsted acids (HeT$^+$, D$_2$T$^+$, CH$_5^+$ and C$_2$H$_5^+$) to
halo- and dihalobenzenes, typical multidentate substrates containing
both π-type and n-type nucleophilic centers.

The first reagent used was HeT$^+$, the labeled counterpart of
HeH$^+$, an extremely strong Brønsted acid (ΔH_f^0 = 323 kcal mol^{-1}, PA of
He $\sim$ 40 kcal mol^{-1}) that can be conveniently obtained in the dilute
gas state from the decay of molecular tritium

$$T_2 \xrightarrow{\text{beta decay}} HeT^+ + \beta^- \qquad (10)$$
$$(95\%)$$

Its gas-phase reaction with halobenzenes yields (Cacace and
Perez, 1970) tritiated products from two distinct mechanisms, <u>i.e.</u>
T-for-H substitution (tritiodeprotonation) and T-for-halogen substitution (tritiodehalogenation). The relative rate of the latter process, insignificant for fluorobenzene, increases for chlorobenzene
and bromobenzene, as shown by the absolute yields of the tritiated
products identified, < 0.2% C$_6$H$_5$T from C$_6$H$_5$F, 3.7% C$_6$H$_5$T from
C$_6$H$_5$Cl, 8.1% C$_6$H$_5$T from C$_6$H$_5$Br.

The D$_2$T$^+$ ion (which is the counterpart of H$_3^+$, a reagent
frequently employed in chemical ionization) is a strong Brønsted
acid that can be conveniently obtained in the dilute gas state from
the radiolysis of D$_2$/DT mixtures. Its reaction with a number of
halo- and dihalobenzenes has been investigated in D$_2$ gas at 720 Torr
(Cacace and Speranza, 1976a, 1976b). The end products analysis, and
independent evidence from the icr study of the D$_3^+$ attack to halobenzenes (Speranza <u>et al</u>., 1977) suggest the following general
reaction pattern, where X = D,T

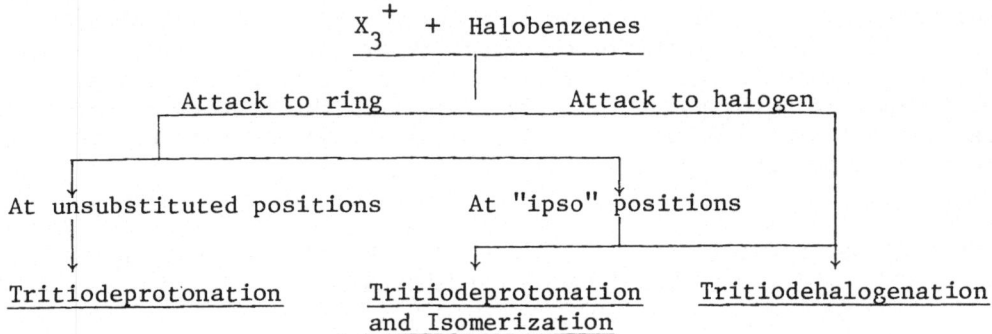

Tritiodeprotonation involves attack to an unsubstituted ring position, giving an arenium ion, e.g.

$$C_6H_5 F + X_3^+ \longrightarrow \text{[arenium ion with H, X, F]} + X_2 \tag{11}$$

whose deprotonation leads to the observed T-for-H substitution.

Attack to the ring position bearing the substitutent ("ipso" attack) occurring directly or via fast intramolecular proton shifts can promote isomerization, e.g.

$$\tag{12}$$
$$\tag{13}$$

The observed migratory ability of the halogens (F<<Cl<Br) reflects their different ability to assume the positive polarization that characterizes the postulated transition state for the $1 \rightarrow 2$ halonium ion shift, e.g.

Protonation at the "ipso" position can also cause dehalogenation via intermolecular halonium ion shifts to any nucleophile N contained in the gas, including the substrate itself, e.g.

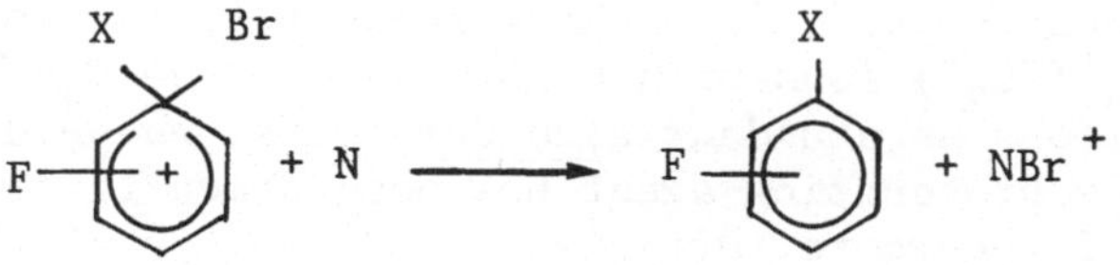

$$F \longleftarrow \overset{X}{\underset{+}{\bigcirc}} \overset{Br}{} + N \longrightarrow F \longleftarrow \overset{X}{\bigcirc} + NBr^+$$

(14)

According to this mechanism, however, the dehalogenation rate should parallel that of isomerization, F<<Cl<Br. Analysis of neutral products and icr evidence shows that this order is actually reversed, suggesting intervention of an alternative dehalogenation route, promoted by attack to the halogen atom, causing direct or indirect halide-ion abstraction, e.g.

$$C_6H_5F + X_3^+ \quad \begin{cases} \longrightarrow C_6H_5^+ + X_2 + XF & (15a) \\ \longrightarrow X_2 + (C_6H_5FX)^+ \rightarrow C_6H_5^+ + XF & (15b) \end{cases}$$

The phenylium ions formed react efficiently with the (labeled) deuterium gas, giving tritiated arenium ions, e.g.

$$C_6H_5^+ + X_2 \rightarrow \overset{X}{\underset{X}{\bigcirc{+}}}$$

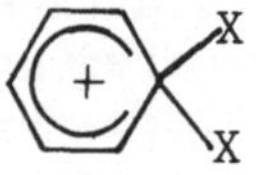

(16)

whose deprotonation yields the observed tritiodehalogenated products. The rather unusual reaction (16), that involves electrophilic attack of a carbocation to molecular hydrogen, has been unequivocally demonstrated by icr and chemical ionization studies (Speranza et al., 1977, Leung et al., 1978), and finds analogies in processes occurring in cold superacid solutions.

The mechanism deduced from neutral products analysis has found independent support from icr and chemical ionization mass spectrometric studies. All the charged intermediates postulated have been observed in the H_2 and CH_4 chemical ionization spectra of halobenzenes. Initially, one-step mechanisms

$$H_3^+ + C_6H_5F \rightarrow HF + C_6H_7^+$$

(17a)

$$CH_5^+ + C_6H_5F \rightarrow HF + \overset{H}{\underset{CH_3}{\bigcirc{+}}}$$

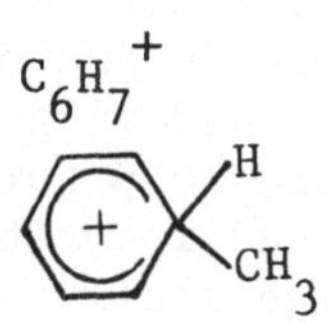

(17b)

were suggested (Harrison and Li, 1975) instead of the two-step
mechanisms described in eqs. 15-16. However, subsequent icr
measurements (Speranza et al., 1976) have shown that the reactions
of $H_3^+(D_3^+)$ and $CH_5^+(CD_5^+)$ ions with halobenzenes give $C_6H_5^+$ (presumably phenyl) cations as abundant secondary ions, whose direct
correlation with the protonating agent has been demonstrated by
double resonance. Furthermore, efficient condensation of the phenylium ions with $H_2(D_2)$ and $CH_4(CD_4)$ has been demonstrated by icr
spectrometry, providing independent evidence for both steps of the
dehalogenation mechanism. Finally, a detailed reinvestigation of
the problem by chemical ionization mass spectrometry has led Harrison
and coworkers to concur in the conclusion that two-step dehalogenation is indeed the only process operative in $H_2(D_2)$, and a major
channel in methane as well.

The current understanding of the rich ionic chemistry initiated
by gas-phase protonation of halo- and dihalobenzenes provides a unified picture of the reactivity of different Brønsted acids, HeT^+,
D_2T^+, and $C_nH_5^+$. All these reactants attack the halogen substi-
tuent(s) and the ring competitively. However, in the case of HeT^+
the composition of the system prevents formation of labeled, and
therefore detectable, organic products from the attack to the halogen. Consequently, only the dehalogenation channel promoted by
attack to the ring can be observed, which accounts for its increase
in the order F<<Cl<Br.

In the case of D_2T^+, attack both to the ring and to the halogen
give detectable dehalogenated products, that however are the same
and cannot be distinguished. Use of CH_5^+ and $C_2H_5^+$ in methane allows
resolution of the two dehalogenation routes, namely protodehalogena-
tion (promoted by ring attack) increasing in the order F<<Cl<Br
and methyldehalogenation (following halogen attack) increasing in
the opposite order.

Concerning intramolecular selectivity, the most significant
feature is undoubtedly the much greater rate of fluoride ion
abstraction in comparison with the correspondent dechlorination and
debromination processes.

COMPETITION OF π-TYPE AND n-TYPE CENTERS FOR GASEOUS LEWIS ACIDS
ALKYLATION OF PHENOL, ANISOLE AND HALOBENZENES

Carbenium ions can react in the gas phase with phenol and anisole both as Brønsted acids and as Lewis acids, yielding in the
latter case O-alkylated and ring-alkylated products. The competition
between the two channels is determined by the nature of the reagent
and of the substrate, and is critically affected by the pressure of
the system. When energetically allowed, proton transfer is generally

the only reaction pathway observed in the mass spectrometer under typical chemical ionization conditions. However, at the substantially higher pressures allowed by application of the radiolytic or the decay technique, alkylation becomes a major reaction channel, and analysis of the neutral alkylated products has provided otherwise inaccessible information on the intramolecular selectivity of the attack.

Free t-$C_4H_9^+$ reacts with phenol in the dilute gas state essentially as a Lewis acid, giving high yields of alkylated products. In this respect, its reactivity is simpler than those of the lower carbocations, that in addition to alkylation promote protonation of the substrate. The gas-phase reaction of phenol with t-butyl ions, obtained from the radiolysis of neopentane, has been investigated by Attinà et al. (1976, 1977b).

The results have been rationalized assuming competitive attack of the carbenium ion (R^+) both to the oxygen atom of phenol, giving an excited alkylaryloxonium ion

$$R^+ \;+\; C_6H_5OH \;\rightarrow\; \left[C_6H_5 - \overset{+}{O} \overset{H}{\underset{R}{\diagdown}} \right]_{exc} \tag{18}$$
$$(VIII)*$$

and to the ring, giving an excited arenium ion

$$R^+ \;+\; C_6H_5OH \;\rightarrow\; \left[\begin{array}{c} R \\ H \end{array} \diagup\!\!\!\bigoplus\!\!\!\diagdown^{OH} \right]_{exc} \tag{19}$$
$$(IX)*$$

The reaction is kinetically biased in favour of the attack to oxygen (eq. 18), but the fate of the oxonium ion, the major primary intermediate, depends on the efficiency of collisional stabilization. At high pressures, effective stabilization and deprotonation of (VIII) lead to the observed high yields of t-butyl phenyl ether

while at low pressure extensive isomerization to the more stable arenium ions takes place. In the same way, isomerization of the

ortho-substituted arenium ion into the para-alkylated one is responsible for the shift in the isomeric composition of the ring-
substituted products observed at low pressures.

From the standpoint of intramolecular selectivity, the major
features of the t-butylation of phenol can be summarized as follows:

i.   There is a strong kinetic bias for attack to oxygen. The result
      is even more significant if one considers that the observed
      selectivity ratio (O-alkylation: ring alkylation >5) is only
      a lower limit, since isomerization of (VIII)* into (IX) cannot
      be entirely excluded even at 720 Torr.

ii.   The higher stability of arenium ions (IX), in particular the
      para-substituted one, with respect to the isomeric oxonium ion
      is clearly suggested by the pressure dependence of the products
      composition, and agrees with independent mass spectrometric and
      theoretical evidence.

The gas-phase i-propylation of C_6H_5OH by radiolytically formed
$i-C_3H_7^+$ ions (Attinà et al., 1978) provides an interesting contrast
with t-butylation in that the ratio of O-alkylation to ring-alkyla-
tion is actually reversed, passing from 5.5 for $t-C_4H_9^+$ to 0.3 for
$i-C_3H_7^+$. Furthermore, the absolute yields of i-propylated products
are significantly lower than those of t-butylated ones, under
otherwise identical conditions. The striking discrepancy has been
rationalized on energetic grounds, in that proton transfer from
the carbenium ion to the oxygen atom of phenol, that is slightly
endothermic, or at best thermoneutral from $t-C_4H_9^+$, is certainly
exothermic from $i-C_3H_7^+$. Since exothermic proton transfer to n-
type nucleophiles is generally recognized as an extremely fast
process, it is likely that the electrophilic attack to oxygen,
kinetically predominant as in t-butylation, promotes protonation of
the n-type center which competes successfully with the relatively
slower alkylation thus reducing the yields of O-alkylated products
in comparison with t-butylation, where protonation at oxygen is
endothermic.

The intramolecular selectivity of $C_3H_7^+$ is characterized by
the high extent of ortho substitution, with a ortho:para ratio
of 5.9 at 720 Torr, well in excess of the statistical value.

In solution, predominant ortho orientation is usually explained
with a mechanistic hypothesis based on the initial attack of the
electrophile to the n-electrons of the substitutent, giving a
n-complex, or onium ion (e.g. oxonium, halonium ion, etc.) that re-
arranges intramolecularly by a formal 1,3 sigmatropic shift to the
correspondent ortho-substituted arenium ion. Indeed, disproportion-
ately high amounts of ortho products have become an informal criterion

for the intermediacy of onium ions in the reaction pathway (see
for instance Taylor and McKillop, 1970). This time-honored mechanistic hypothesis deserves special consideration in the present case,
owing to the unusually pronounced ortho orientation and to the
evidence provided by the pressure-dependence analysis for the isomerization of the oxonium ion.

The gas-phase alkylation of anisole by i-$C_3H_7^+$ and t-$C_4H_9^+$ ions,
radiolytically formed, respectively, in C_3H_8 and neo-C_5H_{12}, gives
predominant ortho substitution at 720 Torr, which gradually changes
in favour of the para isomer at lower pressures (Attinà et al.,
1977c, 1978). The product distribution, and the kinetic features
of the alkylation are consistent with a mechanism involving kinetically predominant formation of a dialkylaryloxonium ion

$$R^+ \; + \; C_6H_5OMe \;\rightarrow\; \left[C_6H_5 - \overset{+}{O} \underset{R}{\overset{Me}{\diagup\kern-0.6em\diagdown}} \right]_{exc} \qquad (22)$$

(X)*

that however, contrary to the secondary alkylaryloxonium ions and
from phenol, cannot undergo direct deprotonation to stable neutral
products. The tertiary ion (X) is consequently long-lived, and re-
arranges into isomeric arenium ions, in particular the ortho-
substituted one

(X)* ───────────────→ [arenium ion, H, R, OMe] $\xrightarrow{+B, \; -BH^+}$ [substituted ring, R, OMe] (23)

whose deprotonation leads to the observed alkylphenols.

The most relevant feature of the intramolecular selectivity
of i-$C_3H_7^+$ and t-$C_4H_9^+$ in their attack to anisole is the pronounced
bias for ortho substitution, especially evident in i-propylation,
where the ortho isomer reaches the unusual proportion of 90% under
conditions favouring kinetic control of products.

The General Kinetic Bias for the Attack to n-Type Substitutents

A general feature of the results illustrated in the previous
paragraphs, concerning alkylation of phenol and anisole is the
significant kinetic bias for attack to the n-type substitutent of
the substrate, to form a gaseous onium ion as the major primary
intermediate, despite the higher stability of the isomeric arenium
ions.

The bias, directly measurable in the case of phenol and deduced from kinetic features, and especially from the disporportionately high extent of ortho substitution in the case of anisole and also chlorobenzene (Attinà and Giacomello, 1977) finds a close analogy with the intramolecular selectivity in the protonation of halobenzenes, illustrated in a previous section. It should be noted that the kinetically predominant formation of the onium ions occurs independently of the activating, or deactivating properties of the substituent. Several reasonable explanations can be suggested to rationalize the general kinetic predominance of the attack of gaseous carbenium ions, and other gaseous cations as well, to strongly electronegative n-type substituents such as O, F and Cl.

We speculate that the kinetic bias arises from the early formation (a kind of clustering) of some directional electrostatic bonding between the n-type substitutent and a hydrogen atom of the electrophile, carrying a fractional positive charge. This early adduct can subsequently collapse either via proton transfer, when energetically allowed, or via a condensation-type process, leading to the formation of a geasoues onium ion. In addition to accounting for the higher reactivity of the electronegative substituents with respect to the ring, the model provides some clue to the otherwise unexpected preference for ortho orientation. In fact, formation of the early electrostatic adduct can conceivably entail some degree of regioselectivity in the approach of the electrophile to the ring, as the rearrangement of the intermediate onium ion is likely to favour ortho positions over the equally activated para position.

COMPETITIVE ACETYLATION OF π-TYPE AND n-TYPE CENTERS

Gaseous $C_2H_3O^+$ species, generally identified as acetylium (CH_3CO^+) ions, have been long observed in the mass spectrometer, and their reactivity toward various substrates has been investigated by icr and high-pressure mass spectrometry. In particular, Bursey and coworkers have studied the condensation of acetylium ions and other acetylating species, obtained by electron impact on acetone and 2,3-butanedione, with ambident aromatic substrates, including phenol, anisole, and their ring-substituted derivatives.

These studies are of interest to the present review, since they represent one of the very few cases where a purely mass spectrometric approach has given a clue to the intramolecular selectivity of an ion-molecule reaction. In fact, from the decrease of the condensation rate caused by the presence of bulky substituents ortho to the hydroxyl, or methoxyl, group (Benezra and Bursey, 1972), and from a comparative sutdy of the condensation rate of substituted anisoles (Chatfield and Bursey, 1976), kinetically predominant attack of the acetylating reagent to the oxygen atom of the substrate was inferred.

This background stimulated application of both the decay and the radiolytic technique to probe the structure and the reactivity of the gaseous species conventionally identified as acetylium cations (Giacomello and Speranza, 1977). The reagent was prepared with the decay technique, alkylating carbon monoxide with CT_3^+ cations from the decay of CT_4 (eq. 1)

$$CT_3^+ + CO \rightarrow [CT_3CO^+]_{exc} \xrightarrow{+M, -M^*} CT_3CO^+ \qquad (24)$$

The excited acetylium ions from the strongly exothermic alkylation process were allowed to undergo a large number of unreactive collisions with the large excess of carbon monoxide present in the gas, before attacking the aromatic substrate.

The radiolytic approach, based on the radiolysis of CH_3F-CO mixtures, represents an extension to much higher pressures of the nucleophilic displacement process

$$CO + CH_3M^+ \rightarrow CH_3CO^+ + M \quad (M = HF, CH_3F) \qquad (25)$$

studied by Holtz and Beauchamp (1971) with icr techniques. Irrespective of the preparation method, acetylium ions react efficiently with n-type nucleophiles, giving almost quantitative yields of acetylated products For instance, both decay-formed CT_3CO^+ and radiolytically formed CH_3CO^+ ions react with methanol, ethanol, n- and i-propanol, etc., giving the correspondent acetates, which, incidentally, supports the acetylium structure assigned to the $C_2H_3O^+$ species. On the other hand, acetylium cations are entirely unreactive toward benzene and toluene, consistent with mass spectrometric results obtained at much lower pressures. Acetylation of phenol takes place essentially at the oxygen atom, giving high yields of phenyl acetate

that represents the major product (84%), while ring acetylation
gives a smaller yield (16%) of hydroxyacetophenone. Thus, two
independent approaches, based on neutral product analysis, have
largely confirmed the conclusion inferred by Bursey and coworkers
on the grounds of indirect mass spectrometric evidence, in that it
has been conclusively shown that the CH_3CO^+ cation displays a sharp
(over 5:1) preference for attack to oxygen.

However, the superior structural discrimination allowed by
actual isolation of products, has revealed another interesting
facet of the acetylium ion reactivity, namely its ability to promote
ring substitution of highly activated aromatic substrates, e.g.
phenol and anisole (eq. 27). The intramolecular selectivity of the
acetylation as deduced from the isomeric composition of products
formed at 720 Torr

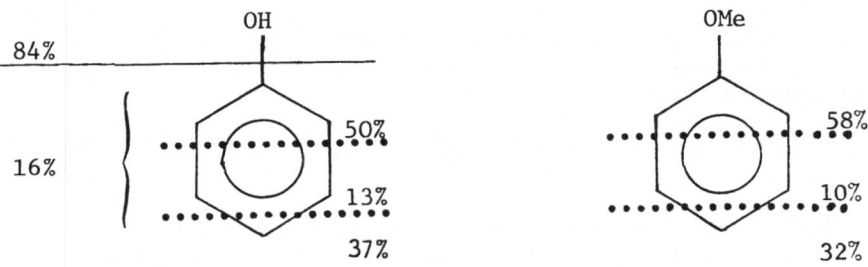

characterizes CH_3CO^+ as a remarkably selective, mild electrophile,
whose reactivity is similar to that of t-butyl cation. In fact,
t-butylation of phenol leads to a ratio of O-alkylation to ring
alkylation of ca. 8:1, as compared to the 5.3:1 ratio obtained from
acetylation. Furthermore, the isomeric composition of ring-substituted products from the two reagents is very close at 720 Torr.

As a whole, the study of gas-phase acetylation provides further
support to the inverse reactivity-selectivity correlation already
noted for other gaseous cations.

STEREOCHEMICAL ASPECTS OF ION-MOLECULE REACTIONS

One of the most serious limitations of the current studies of
ion-molecule reactions is the almost complete lack of information
on the stereochemical features of the processes investigated and their
products, despite their obvious mechanistic interest.

The very few data available have been obtained from the application of different experimental techniques whose common features is
the analysis of the neutral products (Cacace and Speranza, 1972,
Lieder and Brauman, 1975).

Recent extension of the stereochemical studies, and especially the successful application of the radiolytic technique by Angelini and Speranza (1978a, 1978b) have proved remarkably fruitful, providing new and significant data on different ion-molecule reactions.

Among the processes investigated, protonation of halogenated derivatives with $C_nH_5^+$ ions from the radiolysis of methane provided the first experimental evidence for inversion caused by gas-phase SN 2 attack. As an example, the halonium ion (XI) from the protonation of a chlorocyclohexane, e.g.

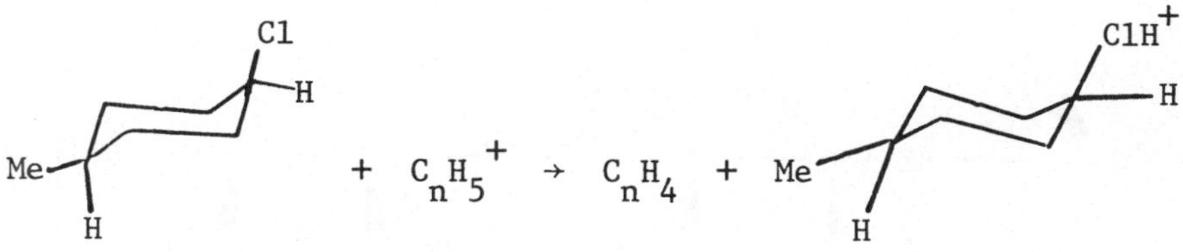

$$+ \ C_nH_5^+ \ \rightarrow \ C_nH_4 \ + \qquad\qquad\qquad (28)$$

(XI)

undergoes nucleophilic attack by water

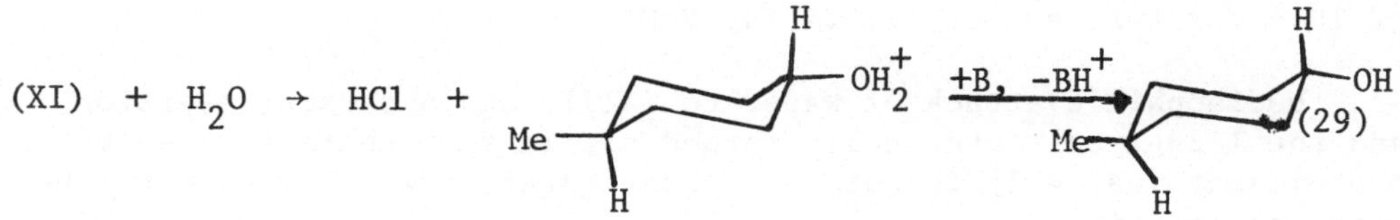

$$(XI) \ + \ H_2O \ \rightarrow \ HCl \ + \qquad \xrightarrow{+B, \ -BH^+} \qquad (29)$$

yielding the corresponding cyclohexanols, in a predominantly (3:1) inverted configuration. Inversion following SN 2 attack is quite general in the gas phase, as shown by methylation of chlorocyclohexanes by dimethylfluoronium ions

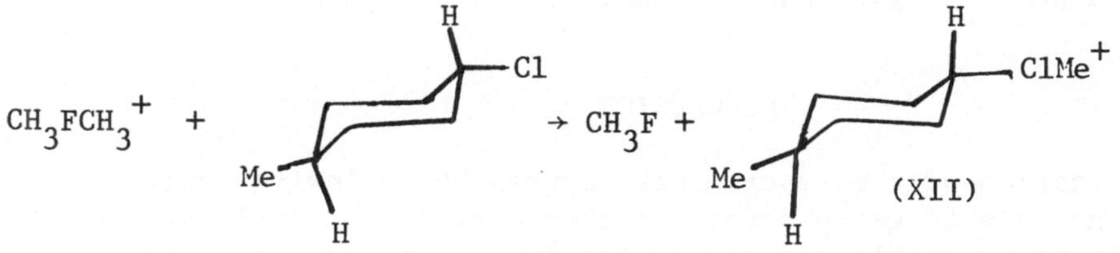

$$CH_3FCH_3^+ \ + \qquad\qquad \rightarrow \ CH_3F \ + \qquad\qquad\qquad (30)$$

(XII)

$$(XII) \ + \ H_2O \ \rightarrow \qquad \xrightarrow{+B, \ -BH^+} \qquad (31)$$

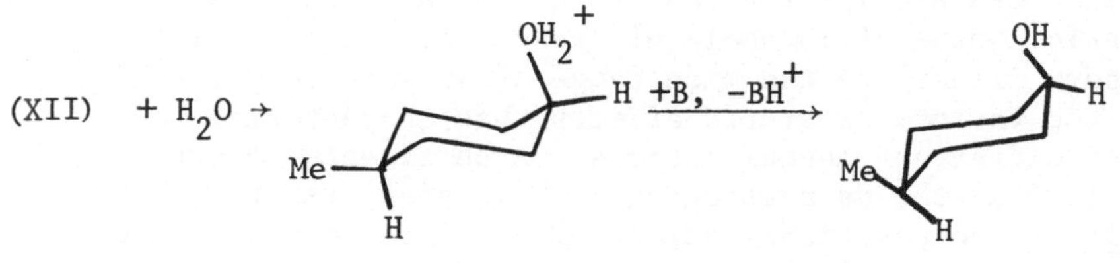

The lower exothermicity of the methylation leads in general to a higher stereospecificity, and the cyclohexanols formed are characterized by an inversion: retention ratio of 19:1. The SN i process, also investigated by Angelini and Speranza (1978 a), is characterized by predominant inversion. Thus, the reaction sequence

$$\text{(32)}$$

leads to formation of epoxides with an inversion: retention ratio of 10:1 for X=F, and of 12.5:1 for X=Cl.

Nucleophilic attack of water to (XIII) again causes inversion, and the 1,2-glycols eventually formed are in fact characterized by predominant (ca. 4:1) <u>retention</u> of configuration with respect to the starting material.

The few examples, chosen among the numerous systems studied by Angelini and Speranza (1978b), provide ample demonstration of the great potential value of the radiolytic techniques as a direct and elegant tool for investigating otherwise inaccessible stereochemical aspects of gas-phase ion-molecule reactions.

CONCLUSION

Application of the experimental approaches allowing neutral product analysis to gas-phase ionic chemistry is currently providing previously inaccessible information on the intramolecular selectivity and steric course of ion-molecule reactions. The results discussed above illustrate the wide range of intramolecular selectivity, and the variety of steric effects that characterize the reactivity of different gaseous cations. A pertinent, general observation is that the data measured at high pressures represent a <u>lower limit</u> of the positional selectivity, since secondary isomerization cannot be <u>a priori</u> ruled out, especially for the most energic electrophiles. However, whenever the orientation is kinetically controlled, all typical features that characterize intramolecular selectivity in solution can be easily identified, if to different extents, in the gas phase as well. Thus, the inverse reactivity-selectivity correlation emerges from a comparison of the reactivity of homologous carbenium ions, intervention of steric hindrance is clearly revealed by the typical orientation of bulky cations, e.g. t-$C_4H_9^+$. In conclusion it appears that such fundamental facets of chemical reactivity as intramolecular selectivity and orientation, steric effects and stereochemical features measured in the gas phase reveal no basic departures from more familiar solution-chemistry patterns, if allowance is made for the different energetics, and especially for the lack of the complicating phenomena of solvation and ion pairing.

Such a conclusion underlines the fundamental interest of gas-phase mechanistic studies, whose extension and refinement will, it is hoped, define intrinsic and unified reactivity parameters of free ionic species, independent of the specific reaction environment, and directly comparable with the results of theoretical approaches.

Acknowledgments: The author wishes to express his deep personal appreciation for the substantial contribution made by his colleagues M. Attinà, G. Ciranni, and P. Giacomello to much of the work reported in the present review.

REFERENCES

Angelini, G., and Speranza, M., J. Chem. Soc. Chem. Commun., 213 (1978a).

Angelini, G., and Speranza, M., unpublished results (1978b).

Attinà, M., Cacace, F., Ciranni, G., and Giacomello, P., J. Chem. Soc. Chem. Commun., 466 (1976).

Attinà, M., and Giacomello, P., Tetrahedron Letters, 2373 (1977).

Attinà, M., Cacace, F., Ciranni, G., and Giacomello, P., J. Am. Chem. Soc. 99, 2611 (1977a).

Attinà, M., Cacace, F., Ciranni, G., and Giacomello, P., J. Am. Chem. Soc. 99, 5022 (1977b).

Attinà, M., Cacace, F., Ciranni, G., and Giacomello, P., J. Am. Chem. Soc. 99, 4101 (1977c).

Attinà, M., Cacace, F., Ciranni, G., and Giacomello, P., J. Chem. Soc. Perkin II, in press (1978).

Ausloos, P., Ann. Rev. Phys. Chem. 17, 205 (1966).

Ausloos, P., Progr. Reaction Kinetics 5, 113 (1969).

Ausloos, P., and Lias, S. G., Ion-Molecule Reactions, J. L. Franklin, editor, Plenum Press, New York, (1972).

Beauchamp, J. L., Interactions Between Ions and Molecules, P. Ausloos, editor, Plenum Press, New York, (1975).

Benezra, S. A. and Bursey, M. M., J. Am. Chem. Soc. 94, 1024 (1972).

Cacace, F., Proceedings of the Conference on the Methods of Preparing and Storing Marked Molecules, Euratom, Bruxelles, (1964).

Cacace, F., Adv. Phys. Org. Chem. 8, 79 (1970).

Cacace, F., Interactions between Ions and Molecules, P. Ausloos, editor, Plenum Press, New York, (1975a).

Cacace, F., Hot Atom Chemistry Status Report, IAEA, Vienna, (1975b).

Cacace, F., and Caronna, S., J. Am. Chem. Soc. 89, 6840 (1967).

Cacace, F., and Perez, G., J. Chem. Soc. (B), 2086 (1971).

Cacace, F., and Speranza, M., J. Am. Chem. Soc. 94, 4447 (1972).

Cacace, F., and Stöcklin, G., J. Am. Chem. Soc. 94, 5818 (1972).

Cacace, F., Cipollini, R., and Occhiucci, G., J. Chem. Soc. Perkin II, 84 (1972).

Cacace, F., and Giacomello, P., J. Am. Chem. Soc. 95, 5851 (1973).

Cacace, F., and Possagno, E., J. Am. Chem. Soc. 45, 3397 (1973).

Cacace, F., Cipollini, R., Giacomello, P., and Possagno, E., Gazz. Chim. 104, 977 (1974).

Cacace, F., and Speranza, M., J. Am. Chem. Soc. 98, 7229 (1976a).

Cacace, F., and Speranza, M., J. Am. Chem. Soc. 98, 7305 (1976b).

Cacace, F., and Giacomello, P., J. Am. Chem. Soc. 99, 5477 (1977).

Cacace, F., and Giacomello, P., J. Chem. Soc. Perkin II, 652 (1978).

Chatfield, D. A., and Bursey, M. M., J. Am. Chem. Soc. 98, 6492 (1976).

Cipollini, R., Lilla, G., Pepe, N., and Speranza, M., J. Phys. Chem., in press (1978).

Giacomello, P., and Cacace, F., J. Chem. Soc. Chem. Commun., 379 (1975).

Giacomello, P., and Cacace, F., J. Am. Chem. Soc. 98, 1823 (1976).

Giacomello, P., and Schüller, M., Radiochimica Acta 24, 111 (1977).

Giacomello, P., and Speranza, M., J. Am. Chem. Soc. 99, 7818 (1977).

Giacomello, P., unpublished results (1978).

Harrison, A. G., and Lin, P. H., Can. J. Chem. 53, 1314 (1975).

Holtz, D., and Beauchamp, J. L., Nature (London) Phys. Sci. 231, 204 (1971).

Kebarle, P., Interactions between Ions and Molecules, P. Ausloos, editor, Plenum Press, New York, (1975).

Kebarle, P., Ann. Rev. Phys. Chem. 28, 445 (1977).

Leung, H. W., Ichikawa, H., Li, Y. H., and Harrison, A. G., J. Am. Chem. Soc. 100, 2479 (1978).

Lias, S. G., Rebbert, R. E., and Ausloos, P., J. Chem. Phys. 57, 2080 (1972), and references therein.

Lias, S. G., Interactions between Ions and Molecules, P. Ausloos, editor, Plenum Press, New York (1975).

Lias, S. G., and Ausloos, P., Ion-Molecule Reactions; Their Role in Radiation Chemistry, ACS/ERDA Monograph, Washington, DC (1975).

Lieder, C. A., and Brauman, J. L., Int. J. Mass Spectrom. Ion Phys. 16, 307 (1975).

Munson, M. S. B., and Field, F. H., J. Am. Chem. Soc. 89, 1047 (1967).

Speranza, M., and Cacace, F., J. Am. Chem. Soc. 99, 3051 (1977).

Speranza, M., Henis, J. M. S., Sefcik, M. D., and Gaspar, P. P., J. Am. Chem. Soc. 99, 5583 (1977).

Speranza, M., Pepe, N., and Cipollini, R., J. Chem. Soc. Perkin II, in press (1978).

Takamuku, S., Iseda, K., and Sakurai, H., J. Am. Chem. Soc. 93, 2420 (1971).

Taylor, E. C., and McKillop, A., Accounts Chem. Res. 3, 345 (1970), and references therein.

Wexler, S., Actions Chimiques et Biologiques des Radiations, M. Haissinsky, editor, Volume VII, Masson, Paris (1965).

THERMOCHEMISTRY OF POLYATOMIC CATIONS

Sharon G. Lias

National Bureau of Standards

Washington, DC, 20234, U.S.A.

The enthalpy of formation of a chemical species is defined as the difference between the enthalpy of the compound and the sum of the enthalpies of the elements of which it is composed. The definition of the enthalpy of formation of a cation must also include the energy required to remove an electron, an "ionization potential";

$$M \rightarrow M^+ + e^- \tag{1}$$

Adiabatic ionization potentials are rigorously equal to the difference in the energies or enthalpies of formation of the ion and the molecule or radical at zero degrees Kelvin. Therefore, in order to arrive at values for enthalpies or energies of formation of an ion at temperature T, account must be taken of the integrated heat capacities of all the species in reaction 1:

$$\Delta H_f(M^+)_T = IP_0 + \Delta H_f(M)_T + \left[\int_0^T C_p(M^+)dT - \int_0^T C_p(M)dT + \int_0^T C_p(e^-)dT \right] \tag{2}$$

(where IP_0 is the adiabatic ionization potential and the term in the brackets corrects the ionization potential to the enthalpy of ionization at temperature T). In practice, for most ions the assumption is made that:

$$\int_0^T C_p(M^+)dT \sim \int_0^T C_p(M)dT \tag{3}$$

Current usage includes two equally arbitrary ways of treating the term $\int_0^T C_p(e^-)dT$ in equation 2, a fact of which most ion chemists

have until now apparently been unaware. Consequently, there are
in the literature numerous examples of thermochemical quantities
calculated using conflicting conventions in a single equation.
Specifically, in calculating values for proton affinities (defined
below in equation 12), workers should be cautioned that the 298 K
enthalpy of formation of the proton listed in thermodynamic compilations[1], 367.186 $\pm$ 0.01 kcal/mole, includes 1.48 kcal/mole to
account for the term $\int C_p^-(e^-)dT$ (which is calculated assuming that
the electron is an ideal gas with a heat capacity of 5/2R). If
this value is used in conjunction with ionic enthalpies of formation
cited from threshold determinations[2] (for which the electron is
assumed to be a sub-atomic particle with a heat capacity of zero
at all temperatures), 1.48 kcal/mole should be subtracted from
$\Delta H_f(H^+)$ to give 365.7 kcal/mole or 1.48 kcal/mole should be added
to $\Delta H_f(M^+)$. A more correct treatment would be to recognize the
electron as a Fermi-Dirac gas with C_p equal to 1.023 R[3]. Some of
the scientific and semantic problems associated with the adaptation
of the concepts and terminology derived for macroscopic thermodynamic
measurements to microscopically-determined quantities are discussed
in an Appendix to this chapter written by Henry M. Rosenstock.

Recent work in various laboratories[4] has generated precise
and accurate data on the Gibbs free energy changes, $\Delta G(T)$, associated
with numerous bimolecular ion-molecule reactions occurring in the
gas phase:

$$A^+ + B \rightleftharpoons C^+ + D \tag{4}$$

through measurements of equilibrium constants, where:

$$K_{eq} = \frac{[C^+]}{[A^+]} \frac{[D]}{[B]} \tag{5}$$

and

$$-RT\ln K_{eq} = \Delta G \tag{6}$$

In such experiments, an equilibrium is established in a high pressure mass spectrometer, flowing afterglow apparatus, or ion cyclotron resonance spectrometer, and the equilibrium constant is
determined by observing the relative abundances of the two ions,
A^+ and C^+, at equilibrium. The neutral reactants, B and D, are
present in great abundance compared to the ionic reactants, and
therefore, the ratio [D]/[B] does not change as equilibrium is
established.

   Since the Gibbs free energy change is related to the enthalpy change associated with the reaction through the relationship:

$$\Delta G = \Delta H - T\Delta S \qquad (7)$$

it is possible to obtain from equilibrium constant determinations absolute measurements of the enthalpy change associated with a particular ion-molecule reaction provided that the term $T\Delta S$ in equation 7 can be evaluated. Thus, when the reaction is charge transfer:

$$A^+ + B \underset{\leftarrow}{\rightarrow} B^+ + A \qquad (8)$$

a determination of ΔG and ΔS can lead to the difference in the enthalpy of ionization at temperature T:

$$\Delta H(Rn\ 8) = [\Delta H_f(B^+) - \Delta H_f(B)] + [\Delta H_f(A) - \Delta H_f(A^+)] \qquad (9)$$

$$= IP_T(B) - IP_T(A)$$

When the reaction is proton transfer:

$$AH^+ + B \underset{\leftarrow}{\rightarrow} BH^+ + A \qquad (10)$$

equilibrium constant determinations can give the difference in the values of the enthalpy of deprotonation, commonly called the "proton affinity".

$$\Delta H(Rn\ 10) = [\Delta H_f(BH^+) - \Delta H_f(B)] + [\Delta H_f(A) - \Delta H_f(AH^+)] \qquad (11)$$

$$= PA(A) - PA(B)$$

and

$$PA(M) = \Delta H_f(H^+) + \Delta H_f(M) - \Delta H_f(MH^+) \qquad (12)$$

Similarly, a determination of values of ΔH for hydride transfer, halide transfer, or any other reaction type, can be used to define thermochemical quantities (halide affinities, etc.) which are functions of the heats of formation of an ion and its corresponding neutral species in the particular reaction.

   Further, if the enthalpies of formation of one of the ions and both neutral species in reaction 4 are known, equilibrium experiments can be used to derive absolute values for enthalpies of formation of ions, including species such as protonated molecules which are sometimes difficult or impossible to generate in threshold experiments.

At the present time, extensive interlocking ladders of free energy changes for ion-molecule reactions (particularly proton transfer, hydride transfer, and charge transfer) are being generated in a number of different laboratories[4]. These data are usually converted to scales of relative values of ΔH for the particular reaction type, evaluating the term TΔS in equation 7 through experimental results or through thermodynamic or statistical mechanical considerations. Finally, these relative scales of ΔH are often related to an absolute scale using energies or enthalpies of formation of ions determined in threshold experiments as standards of reference. In the following sections of this chapter, we shall consider (1) the evaluation of ΔS for ion-molecule reactions, and (2) the latest information about the absolute values of the enthalpies of formation of several reference standard ions often cited in the current literature, and how well or poorly these threshold ΔH_f values interrelate through the available thermodynamic ladders derived from equilibrium constant determinations.

EVALUATION OF ENTROPY CHANGES

For ion-molecule reactions involving only the transfer of an electron or atomic species (H, H^+, H^-, Cl^-, etc.) entropy changes associated with the reaction are usually (although not always) small, $<\sim 5$ cal/deg-mole. Thus, if it is sufficient for the purposes of a particular experiment to know the enthalpy change of a reaction to within 2 or 3 kcal/mole, the assumption which is sometimes made, that $\Delta H \sim \Delta G$, will be adequate[4]. However, it is possible using the experimental techniques currently available to generate scales of relative values of ΔG for ion-molecule reactions which have an internal consistency of better than ± 0.1 kcal/mole. Therefore, it appears worthwhile to examine the evaluation of the entropy changes for these reactions with the purpose of ascertaining with what degree of accuracy the enthalpy changes associated with ion-molecule reactions can be obtained.

There are three approaches to evaluating the value of ΔS for a chemical reaction. The experimental determination of ΔS (the so-called Second Law Method) consists of measuring ΔG as a function of temperature. It can be seen from equation 7 or the equivalent "Van't Hoff equation":

$$-\ln K_{eq} = \frac{\Delta H}{RT} - \frac{\Delta S}{R} \tag{13}$$

that if such a determination is made over a temperature range in which ΔH and ΔS are constant, a plot of ΔG versus T, or more commonly, $-\ln K_{eq}$ versus 1/T, will give a straight line whose slope and intercept are, respectively, $-\Delta S_T$ and ΔH_T (equation 7) or $-\Delta H_T/R$ and

$\Delta S_T/R$ (equation 13) (where the subscript T is an indication that the values obtained correspond to the average temperature T of the determinations, not to zero degrees Kelvin.

The Third Law Method for the determination of enthalpy changes is based on a knowledge of the absolute entropies of the reactants and products. Information about absolute entropies of ions is generally lacking, but in several studies[5], the assumption has been made that $S(M^+)$ is equal to the standard entropy of a species isoelectronic to M^+.

For most ion-molecule equilibria which have been reported to date, ΔS has been calculated from statistical mechanical considerations, that is, from ideal partition functions:

$$\Delta S = R\ln \frac{(Q_C+)(Q_D)}{(Q_A+)(Q_B)} \tag{14}$$

where

$$Q_X = (q_{tr}q_{rot}q_{vib}q_{elec})_X \tag{15}$$

From equations 14 and 15, it can be seen that the entropy change can be factored:

$$\Delta S_{Ideal} = \Delta S_{tr} + \Delta S_{rot} + \Delta S_{vib} + \Delta S_{elec} \tag{16}$$

Table 1 shows the expressions which are usually taken for the partition functions, as well as the corresponding expressions for the various components of ΔS. In general, the contribution of ΔS_{tr} is ignored, since the expression listed in Table 1 will lead to a negligible value unless there is a very large change in reduced mass between reactants and products. In most, though not all, ion-molecule reactions the degeneracies of the products and those of the reactants are the same, so ΔS_{elec} will be zero. Because of the general lack of data on the vibrational frequencies of ions, it is usually necessary to assume that the contribution of ΔS_{vib} is small. In the few cases for which ΔS_{vib} has been evaluated[6], this assumption has proved to be correct. Thus, in many evaluations of ΔS for ion-molecule equilibria, the simplifying assumption is made that:

$$\Delta S_{total} \sim \Delta S_{rot} \tag{17}$$

A further simplification is realized when the assumption is made that the ratios of moments of inertia in the expression for ΔS_{rot}

TABLE 1. Ideal Partition Functions and Corresponding Entropy
Changes for the Equilibrium $(A + B \rightleftharpoons C + D)$.

q(Ideal)   ΔS(Ideal)

Translation

$$\frac{(2\pi mkT)^{3/2}V}{h^3}$$ $$R\ln\left(\frac{m_C m_D}{m_A m_B}\right)^{3/2}$$

Rotation

(Non-Linear Molecule)

$$\frac{8\pi^2(8\pi^3 ABC)^{1/2}(kT)^{3/2}}{h^3\sigma}$$ $$R\ln\frac{(ABC)_C^{1/2}(ABC)_D^{1/2}\ \sigma_A\sigma_B}{(ABC)_A^{1/2}(ABC)_B^{1/2}\ \sigma_C\sigma_D}$$

$$\sim R\ln\frac{\sigma_A\sigma_B}{\sigma_C\sigma_D}$$

Vibration

$$\prod_{i=1}^{f}(1-e^{-h\nu_i/kT})^{-1}$$ $$R\ln\frac{\prod(1-e^{-h\nu_i/kT})_A\prod(1-e^{-h\nu_i/kT})_B}{\prod(1-e^{-h\nu_i/kT})_C\prod(1-e^{-h\nu_i/kT})_D}$$

Electronic

$$\sum_i g_e e^{-e_i/kT}$$ $$R\ln\frac{g_C g_D}{g_A g_B}$$

---

m = mass of reactant in grams, k = Boltzmann's Constant, T = degrees
Kelvin, h = Planck's Constant, V = volume, σ = symmetry number or
number of equivalent orientations in space, A, B, and C = the three
principle moments of inertia of a non-linear molecule, ν_i = fundamental vibrational frequencies, g_e = degeneracies of electronic
states of energy e_i.

are approximately equal to unity so that:

$$\Delta S_{rot} \sim R\ln\frac{(\sigma_{A+})(\sigma_B)}{(\sigma_{C+})(\sigma_D)} \tag{18}$$

The assumptions given in equations 17 and 18 are the basis for several recently published thermodynamic scales of values of ΔH for
proton transfer reactions (proton affinity scales) [4b,7].

It has been suggested[4c,8] that in the case of ion-molecule
reactions the correct evaluation of ΔS requires consideration of an
additional term to take into account the non-ideal nature of the
ion-molecule collision. This suggestion originally arose as a
logical extension of the well-known facts that exothermic ion-
molecule reactions do not exhibit activation energies and the rate
constants in many cases are essentially equal to the rate constants
predicted for the corresponding ion-molecule collisions. Hence:

$$k_f = Z_f P_f \tag{19}$$

(where the subscript f refers to the exothermic or "forward" direction of a reaction, Z is the ion-molecule collision rate constant,
and P is the probability that a collision will result in reaction
at temperature T). From this it was inferred that the rate constants of the corresponding endothermic or "reverse" reactions
should be equal to:

$$k_r = Z_r P_r e^{-\Delta H_r/RT} \tag{20}$$

(where the subscript r refers to the reverse reaction, and ΔH_r is
the energy barrier due to the endothermicity of reaction). Indeed,
an examination of rate constant data indicates that within the
limits of uncertainty of the prediction of the values of Z, the
rate constants of a number of ion-molecule reactions are adequately
predicted by equations 19 and 20. This is demonstrated by the
results presented in Tables 2 and 3, as well as by some data on
the rate constants of charge transfer reactions which appeared
recently[7]. The results given in Tables 2 and 3 show that for many
proton transfer reactions (observed at temperatures which are sufficiently low that the temperature dependence[9] of the collision
complex lifetime does not influence the observed rate constant) the
factor P can sometimes be approximately predicted from the expressions:

$$P = e^{\Delta S_{Ideal}/R} \text{ where } \Delta S_{Ideal} \leq 0 \tag{21a}$$

$$P = 1 \text{ where } \Delta S_{Ideal} \geq 0 \tag{21b}$$

(where ΔS_{Ideal} has been approximated by expression 18).

TABLE 2. Experimental Rate Constants of Exothermic Proton Transfer Reactions Compared with the Prediction: $k = Ze^{\Delta S/R}$*

$-cm^3/molecule-s \times 10^{10}-$

Reaction	$k_f{}^a$	ADO[b] Z_f	Var[c] Z_f	B-R[d] Z_f	$e^{\Delta S/R}$*	ΔH kcal/mole
$H_3S^+ + CH_3OH \rightarrow$	18.9	17.7	24.4	34.1	1.0	− 7.7
$CH_3SH_2{}^+ + (CH_3)_2CO \rightarrow$	27.6	23.4	32.2	44.6	1.0	− 8.0
$H_3S^+ + CH_3SH \rightarrow$	18.8	16.3	20.3	25.3	1.0	−11.4
$(CH_3)_2SH^+ + CH_3COOC_2H_5 \rightarrow$	7.0[e]	15.5	20.1	27.2	0.5	− 0.47
$(C_2H_5)_2OH^+ + CH_3COOC_2H_5 \rightarrow$	6.9	14.7	19.1	25.9	0.5	− 0.72
$CH_3CNH^+ + PH_3 \rightarrow$	3.4	12.4	14.2	16.7	0.25	− 0.90
$CH_3SH_2{}^+ + PH_3 \rightarrow$	3.3	12.0	13.7	16.2	0.25	−1.80
$CH_3CHOH^+ + PH_3 \rightarrow$	2.7	12.0	14.0	16.5	0.25	−2.70

*$\Delta S = \Delta S_{Ideal}$ assuming equations 17 and 18; for positive values of ΔS_{Ideal}, the term $e^{\Delta S/R}$ is taken to be unity. T = 315 K.

a) Reference 10 except where otherwise indicated. b) Reference 11. c) Variation formulation of Reference 12; estimated using Figure 2 in that chapter. d) Barker-Ridge formulation, Reference 13. e) Reference 7a.

TABLE 3. Experimental Rate Constants of Endothermic Proton Transfer Reactions. Comparison with $k_r = Z_r e^{\Delta S_{rot}/R}e^{-\Delta H_r/RT}$*

$cm^3/molecule-s \times 10^{10}$

Reaction	ΔH^a kcal/mole	$k_r{}^a$	ADO[b] $Ze^{\Delta S/R}e^{-\Delta H/RT}$*	Var[c] $Ze^{\Delta S/R}e^{-\Delta H/RT}$*
$PH_4{}^+ + CH_3CN \rightarrow$	+1.1	6.8	5.7(0.56)[d]	8.5(0.63)
$(HCOOCH_3)H^+ + CH_3CN \rightarrow$	+1.1	5.8	5.3(3.1)	7.9(4.2)
$CH_3SH_2{}^+ + CH_3CHO \rightarrow$	+1.2	4.5	3.4(2.3)	4.8(2.9)

*$\Delta S = \Delta S_{Ideal}$ assuming equations 17 and 18; for positive values of ΔS_{Ideal}, the term $e^{\Delta S/R}$ is taken to be unity. T = 315 K.

a) Experimental values taken from Reference 10. b) Reference 11. c) Variation formulation of Reference 12; estimated using Figure 2. d) Values in parentheses are rate constants predicted by equation 27.

If equations 19 and 20 are substituted into the equilibrium constant expression, one obtains:

$$K_{eq} = \frac{k_f}{k_r} = \frac{Z_f P_f}{Z_r P_r} e^{-\Delta H_f/RT} \tag{22}$$

Then since:

$$K_{eq} = e^{\Delta S/R} e^{-\Delta H/RT} \tag{23}$$

it follows that[4c,6,8]:

$$\Delta S = R\ln Z_f/Z_r + R\ln P_f/P_r \tag{24a}$$

$$\Delta S = R\ln Z_f/Z_r + \Delta S_{Ideal} \tag{24b}$$

Experimental determinations of ΔS for several ion-molecule reactions have provided evidence for the existence of a contribution to ΔS in addition to ΔS_{Ideal} as predicted by equation 24[6,14]. Some additional results are summarized in Table 4. Although the exact magnitude which is predicted for a term $R\ln Z_f/Z_r$ depends somewhat on the formulation used to predict the collision rate constants[11,12,13,15] it can be seen that the measured entropy changes are in reasonably good agreement with the values calculated using equation 24 and the various collision rate formulations available at this writing. Furthermore, as will be shown later, values of ΔH_{Rn} obtained for charge transfer reactions involving aromatic species[6] using equation 24 to evaluate ΔS are in excellent agreement with values of ΔH derived from independent data.

Nevertheless, at this writing, the concept of including a non-ideal term in the entropy change for ion-molecule reactions (equation 24) has not been generally accepted by ion-molecule kineticists, and the interpretation of the results given in Tables 2, 3, and 4 may be considered controversial. One dissenting discussion, presented by Timothy Su at the Panel on Potential Surfaces, is included as Appendix II at the end of this paper. There are essentially three major objections which have been voiced against equation 24, which will be considered below:

I.   "If one includes ratios in the equilibrium constant which are expressions of the attractive potentials which bring the ion-molecule pair together, then one must also include the ratios of the close-range physical forces which determine the lifetimes of the complexes, $A^+ \cdot B$ and $C^+ \cdot D$."

TABLE 4. Experimentally-Determined Values of ΔH and ΔS for Proton Transfer Reactions. Comparison of $\Delta S_{Experimental}$ With Values Calculated from Equation 24.

Equilibrium	Experimental[a] ΔH kcal/mole	Experimental[a] ΔS cal/mole-deg	$\dfrac{RlnZ_f/Z_r}{}$ ADO[b]	Var[c]	B-R[d]	ΔS_{Ideal}
$CH_3CNH^+ + HCOOCH_3 \rightleftharpoons$	-1.1	-1.1	-1.0	-1.3	-1.4	0
$CH_3SH_2^+ + PH_3 \rightleftharpoons$	-1.6	-3.4	-0.7	-0.6	-1.4	-2.8
$CH_3CHOH^+ + CH_3SH \rightleftharpoons$	-1.2	-0.9	-0.8	-1.0	-1.1	0
$CH_3CHOH^+ + PH_3 \rightleftharpoons$	-2.7	-4.2	-1.4	-1.8	-1.2	-2.8
$(HCOOCH_3)H^+ + PH_3 \rightleftharpoons$	-0.3	-4.5	-0.9	-1.1	-1.6	-2.8

a) Reference 10. b) Reference 11. c) Reference 12. Values of Z_f and Z_r estimated from Figure 2. d) Reference 13.

$$A^+ + B \underset{D_f}{\overset{Z_f}{\rightleftharpoons}} (A^+ \cdot B) \rightleftharpoons T^+ \rightleftharpoons (C^+ \cdot D) \underset{Z_r}{\overset{D_f}{\rightleftharpoons}} C^+ + D \tag{25}$$

The usual treatment of ion-molecule equilibrium constant data carries the implicit assumption, recently stated specifically[16], that the ratio, Z_f/Z_r, and that of the rate constants for the unimolecular decomposition of the complexes, D_r/D_f, will exactly cancel, so that neither needs to be considered in the evaluation of thermochemical data from the equilibrium constant. The justification for this assumption is that it is "required by microscopic reversibility". Microscopic reversiblity is grounded in the invariability of Newton's laws of motion under the transformation t = -t, but it does not imply that the detailed mechanism of the reverse reaction is identical to the mechanism one would observe if the forward reaction occurred with the clock running backward. Microscopic reversibility, when correctly interpreted[3,17], is merely a statement of the equilibrium condition, namely that the overall rate of the forward reaction is equal to the overall rate of the reverse reaction, and that the most probable path in one direction is also the most probable path in the other. The originator of the term "microscopic reversbility" has written[3], "The phrase is not very appropriately descriptive. It might be better to replace it by the phrase 'principle of equal frequency for reverse molecular processes at equilibrium'. The principle should in any case be distinguished from any statement as to equal frequencies for inverse molecular processes since these, in general, do not exist."

As for the inclusion in the evaluation of ΔS of the physical forces operating in the complex, it may be that future fine-tuning of experimental results and theoretical models of ion-molecule equilibria and collision rate constants will show that the operation of close range potentials must indeed infuence the observed equilibrium constants. Indeed, if one considers reactions occurring in a temperature range where a negative temperature dependence on the rate constants in the exothermic direction exists[9] (presumably because of a temperature-dependent shortening of the lifetime of the complex), it would be expected that D_f and D_r (equation 25) would appear as temperature dependent components of P_f and P_r (equations 19-24) and therefore would necessarily contribute to the entropy change observed for an equilibrium. However, simple considerations show that when this is the case, the effect must be to amplify the non-ideal term in the ΔS, not cancel it out. That is, an examination of thermochemical data determined for association reactions:

$$A^+ + B \underset{\leftarrow}{\rightarrow} A^+ \cdot B \qquad (26)$$

(for which there are no other exit channels from the complex than the return to the original reactants) allows us to isolate the first step of equation 25, and determine exactly how the forward collision rate and the unimolecular rate constant for dissociation of the complex are related, if they are related. The universal exact cancellation of the ratio Z_f/Z_r in equation 22 by the ratio D_r/D_f would require that a general relationship exist such that a fast collision rate would be balanced on a relative scale by a fast rate of dissociation to re-form the same collision partners, and vice versa. An examination of values of ΔH of association reactions as a function of the corresponding forward collision rates is presented in Figure 1, where it is seen that the factors which tend to increase the collision rate for a pair also tend to increase the well-depth of the complex. According to the arguments developed by Waage and Rabinovitch[18], this should lead to an inverse relationship between Z and D, contrary to the assumption implied by the usual treatment of ion-molecule equilibrium data.

It can also be noted that the assumption of cancellation of the term Z_f/Z_r in equation 22 implies that when the exothermic reaction occurs at the collision rate, equation 22 can only be satisfied if:

$$k_r = Z_r(Z_f/Z_r)P_r e^{-\Delta H_r/RT} \qquad (27)$$

That is, the rate of the endothermic reaction would somehow depend on the rate constant for collision of the partners in the exothermic reaction. Numbers in parentheses in Table 3 demonstrate that equation 27 does not give a good approximation to k_r.

II. "The inclusion of the ratios of the forward and reverse rate constants in the expression for the equilibrium constant implies that there is a different transition state for the forward and reverse reactions."

The confusion on this point perhaps results from earlier statements by the present author[4c],[6] pointing out that in the 1936 treatment of an ion-molecule reaction by Eyring, Hirschfelder, and Taylor[21] account is specifically taken of the modification of the rotational energy of the molecule because of the approach of the ion. These authors write an apparent rotational partition function for the transition state which includes the interaction potential plus the centrifugal potential. The resulting expression for the rate constant simplifies to give Z, the ion-molecule collision rate; the use of such a treatment for the forward and reverse rate constants in an equilibrium would lead to the inclusion of the non-ideal term in the entropy change. The problem arises from the fact that Eyring

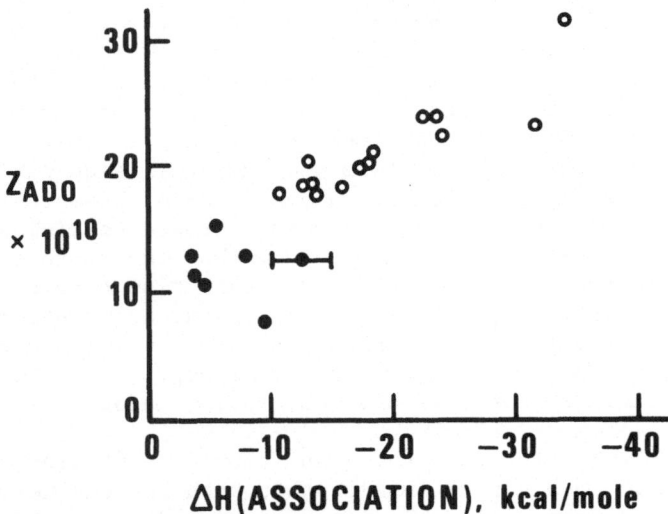

FIGURE 1. Rate constants for collision (ref. 11) of reactants in association reactions involving non-polar (•) and polar (o) molecules as functions of experimentally-determined values of ΔH (Reference 19, (•) and Reference 20 (o)).

et al[21] defined the rotational barrier to <u>collision</u> as the transition state for <u>reaction</u>, a treatment which is equivalent to the simple assumption that every collision leads to reaction. In such a model there would indeed be a different transition state for the forward and reverse reactions (although such a model makes little sense when applied to the endothermic case). However, there is abundant experimental evidence that for most thermal ion-molecule reactions, complex formation precedes reaction in both exothermic and endothermic reactions[22]. If the position of the transition state is recognized as being in the potential well corresponding to the complex(es) but the requirement is retained that reactants must cross the centrifugal barriers as a prerequisite to reaction, the model is in agreement with experiment. It is the description of the top of the rotational barrier as a "transition state for reaction" which is in error when the Eyring treatment is generalized to other reactions, not the inclusion of the intermolecular ion-molecule potentials in the partition functions, which is correct.

III.   "There can not be a non-ideal contribution to the entropy change for ion-molecule reactions when the equilibrium constants are measured under experimental conditions such that the ideal gas laws are obeyed. If one calculates the deviations from ideal behavior using the second virial coefficient, it is seen that the predicted deviations from ideality are too small by several orders of magnitude to account for a term in the entropy change as large as that predicted for many systems by $R \ln Z_f/Z_r$. Furthermore, according to the usual thermodynamic formulations, non-ideal contributions to the entropy change are pressure-dependent; the expression $R \ln Z_f/Z_r$ is not pressure-dependent."

This objection results from the mistaken conception that ideality is defined only in terms of the macroscopic PVT relationships of a gas. Actually, an ideal gas by definition is a gas in which there are no intermolecular potentials operating. In an ideal gas the probabilities of encounter of reactant species are simply dependent on the number density of the molecules and the reduced mass of the colliding pair. In a real gas, intermolecular potentials do exist and modify the rates of encounter between reactant species to a greater or lesser extent depending on the magnitude of the potentials involved. The operation of these potentials also modifies the observable PVT relationships of a gas or gas mixture at high pressures or low temperatures. In classical thermodynamics one corrects for deviations from the ideal gas laws using concepts such as fugacity or virial coefficients, which can be empirically determined from observed PVT relationships or derived from statistical mechanical considerations of the intermolecular potentials. When these corrections are made, it is generally assumed that the actual rates of encounter of the molecules in the system are not very much different from those which would prevail under ideal gas conditions. The effects of the intermolecular potentials on the collision rates are ignored, and indeed for most Van der Waals gases, the error introduced by this approximation is negligible.

In the case of ion-molecule equilibria, the relative concentrations of the two ions are measured directly in mass spectrometric experiments, and the pressures of the two neutral gases are measured directly at pressures low enough that it is reasonably correct to assume that the number of molecules per unit volume can be correctly estimated using the ideal gas law. However, although the relative molecular concentrations of all the reactant species are exactly represented in equation 5, these concentration ratios do not represent the relative rates of encounter of C^+ and D, or A^+ and B. Here the depths of the potential wells of the intermolecular potentials are one or more orders of magnitude greater than those generally encountered in neutral-neutral collisions. It is well established that ion-molecule collision rates are much greater than the corresponding rates predicted for ideal systems[11,12,13,15]. In equation 5, the microscopic non-ideality resulting from the existence of significant intermolecular potentials appears as a component of the equilibrium constant, which can no longer be described using only ideal partition functions.

If one were to correct the ratios on the right side of equation 5 so that they correspond to actual ratios of the probabilities of encounter:

$$K'_{eq} = \frac{k_f}{k_r} \frac{Z_r}{Z_f} = \frac{[C^+]}{[A^+]} \frac{[D]}{[B]} \frac{Z_r}{Z_f} \qquad (28)$$

then the quantities on the left side of the equation could be described entirely in terms of ideal partition functions, since the non-ideal component has been factored out. Since the equilibrium constant for ion-molecule systems has always been defined in terms of actual molecular concentrations (or of non-ideal rate constant ratios) rather than in terms of ratios of encounter rates, there results a term in the entropy change equal to $R \ln Z_f/Z_r$, which is an expression of the intermolecular ion-molecule potentials. If the ion-molecule equilibrium constant were defined as in equation 28, in terms of ideal partition functions, then the term Z_r/Z_f appears on the right side of the equation as a correction term. This, of course, is purely a matter of semantics, since the mathematical treatment of the data is identical in either case, as are the experimental observations.

ENTHALPIES OF FORMATION OF STANDARD REFERENCE IONS

Because equilibrium constant measurements lead to values of ΔG and possibly ΔH of ion-molecule reactions, the use of these data requires a value for an absolute heat of formation of at least one ion in order to generate absolute heats of formation of ions or to put thermodynamic ladders on an absolute scale. In the discussion which follows, all values cited for enthalpies of formation of ions will be based on the usual mass spectrometric convention which assigns a heat capacity of zero to the electron at all temperatures.

Charge Transfer Equilibria

In the case of charge transfer reactions (reaction 8), the problem of putting equilibrium data (equation 9) on an absolute basis is simplified by the existence of a large body of data on adiabatic ionization potentials of molecules. As shown in equation 2, the use of adiabatic ionization potential data in this way requires a correction of the zero degree ionization potential to the "enthalpy of ionization" at temperature T taking into account the integrated heat capacities of the ion and the molecule

$$IP_T = \Delta H_f(M^+)_T - \Delta H_f(M)_T \tag{29}$$

It is useful to examine the validity of the assumption usually made (equation 3) that $IP_0 = IP_T$, that is, that the integrated heat capacity of the ion and the corresponding molecule are equal. Except for the case in which there is a change from a linear to a non-linear species (or vice versa) in the course of ionization, $C_p(tr + rot)$ will be the same for M and M^+. For a given mode of vibration of a molecule at temperature T, there is a contribution to C_p of:

$$C_p(vib) = R\frac{x_i^2 e^{x_i}}{(e^x-1)^2} \tag{30}$$

where $x = h\nu_i c/kT$. Thus, for every vibrational frequency which changes when the molecule is ionized, there will be a difference between the heat of formation of the ion and molecule at high temperature which differs from that at zero degrees. This temperature effect on IP_T can be estimated by calculating $C_p(vib)$ for the ion and the neutral species for all of the vibrational frequencies which are different in the ion and the molecule.

$$\Delta(IP_T)_{0-T} = R \int_0^T C_p(vib)(M^+)dT - R \int_0^T C_p(vib)(M)dT \tag{31}$$

Generalizations about the effect of ionization on molecular vibrations are that (1) the bond or bonds whose vibrations are most affected are those from which the electron was removed, and (2) removal of a bonding electron reduces the vibrational frequency of the bond, removal of an anti-bonding electron increases the vibrational frequency, and removal of a non-bonding electron has little effect.

Most ions have a ground electronic state which is a doublet. If there is any splitting of the energies of the degenerate states, there will be a contribution to the heat capacity of the ion given by:

$$C_p(elec) = R\left(\frac{\Delta\epsilon}{kT}\right)^2 \frac{e^{\Delta\epsilon/kT}}{(1 + e^{\Delta\epsilon/kT})^2} \tag{32}$$

(where $\Delta\epsilon$ is the magnitude of the multiplet splitting) which will not be matched by a corresponding term for the neutral molecule, which usually has a singlet ground electronic state. In the temperature range from 0 to 400 K, only multiplet splittings of 0.05 to 3 kcal can contribute significantly to the heat of formation of an entity; it can be estimated[6] that the maximum contribution is about 0.2 kcal/mole.

Table 5 shows some values for enthalpies of ionization which have been calculated using expressions 30, 31, and 32 for several molecules at 350 K. The corresponding adiabatic ionization potentials are also shown. The significant differences (0.1–0.2 kcal/mole) observed between IP_0 and IP_{350} for NO and benzene come about as a result of multiplet splitting of degenerate electronic states in the NO molecule and the benzene ion (equation 32)[6]. The table also shows that for most aromatic species, in which ionization involves removal of a delocalized π-electron from the ring, the difference between IP_0 and IP_{350} is negligible. Results are also given for ethylene at 300, 350, and 400 K. The lowest ionization potential in olefins results from removal of an electron from the C=C π-bond; in ethylene this leads to a lowering of the frequency of the symmetric C=C stretch from 1623 cm^{-1} to 1230 cm^{-1}, and a lowering of the frequency of the twisting around the C=C bond from 1027 cm^{-1} to 430 cm^{-1}. These calculated values are included to demonstrate the pronounced variation of IP_T with temperature for systems in which there are substantial differences between the vibrational frequencies of the molecule and the corresponding ion.

In a recent study[6] an extensive interlocking ladder of values of ΔG for charge transfer equilibria involving mainly aromatic compounds was generated. Using the values of IP_{350} listed in Table 5 as reference standards, the 350 K enthalpies of ionization of a number of compounds were determined. Table 6 shows results for a few compounds included in the study for which reliable spectroscopic values of the adiabatic ionization potentials were available[2] and for which it is expected that the adiabatic ionization potential should be essentially identical to IP_{350}. The results show that there is excellent agreement between results derived from the thermodynamic ladder and those obtained from threshold measurements. Also, the experimentally-determined[6] entropy changes for the charge transfer equilibria are in agreement with those predicted on the assumption that the non-ideal contribution (equation 24) must be included.

TABLE 5.   Adiabatic Ionization Potentials (IP_0) and 350 K Enthalpies of Ionization (IP_{350}) Calculated from Differences in Vibrational Frequencies and Electronic Multiplet Splittings.

	IP_0	kcal/mole IP_{350}
$NO^+ - NO$	213.63 ± 0.01	213.50 ± 0.1
$C_6H_6^+ - C_6H_6$	213.24 ± 0.05	213.43 ± 0.1
$c\text{-}C_4H_4O^+ - c\text{-}C_4H_4O$	204.84 ± 0.02	204.84 ± 0.06
$C_6H_5F^+ - C_6H_5F$	212.15 ± 0.11	212.15 ± 0.11
$CH_3C_6H_5^+ + CH_3C_6H_5$	203.4 ± 0.2	203.4 ± 0.2
$C_2H_4^+ - C_2H_4$	242.3 ± 0.1	300 K:242.46 350 K:242.52 400 K:242.58

TABLE 6.   Values of IP_{350} of Organic Compounds Determined through Equilibrium Constant Measurements.   (Reference 6)

	IP_0	kcal/mole IP_{350}
Ethylbenzene	202.2 ± 0.2	202.2 ± 0.2
α,α,α-Trifluorotoluene	223.3 ± 0.1	223.3 ± 0.15
Methyl iodide	219.9 ± 0.07	219.9 ± 0.15
Ethyl iodide	215.5 ± 0.1	215.5 ± 0.15

Hydride and Proton Transfer Equilibria

A substantial body of data on hydride transfer and proton transfer equilibria has been placed on an absolute basis using the heat of formation of the t-butyl ion as a reference standard[a,b,7,23]. The heat of formation of this ion at 300 K has been obtained from measurements of the ionization potential of the t-butyl radical, assuming that the integrated heat capacities of the ion and the radical are identical (equation 2). This is a reasonable assumption; the loss of a non-bonding electron from the alkyl radical would not be expected to cause any significant changes in vibrational frequencies. A summary of values of $\Delta H_f(t-C_4H_9\cdot)$ and ionization potential measurements reported since 1959 is given in Table 7, along with the resulting values derived for the enthalpy of formation of $t-C_4H_9^+$. The Table also shows absolute values of $\Delta H_f(t-C_4H_9^+)$ derived from measurements of the appearance potential of the reaction:

$$neo-C_5H_{12} \rightarrow t-C_4H_9^+ + CH_3 + e^- \tag{33}$$

As described in Appendix I, appearance potential measurements, like ionization potential measurements, correspond to energies of reaction at zero degrees Kelvin, and therefore, the integrated heat capacities of all the species in reaction 33 must be known in order to derive a value of $\Delta H_f(t-C_4H_9^+)_{300}$ from the appearance potential measurements. However, since:

$$S_T = \int_0^T \frac{C_p}{T} dT \tag{34}$$

it is possible to estimate the probable contribution to the 300 K heat of formation of the ion from the integrated heat capacity terms. Consideration of the absolute entropies of the species on both sides of equation 33[1,24] indicates that the probable error of ignoring the correction from 0 K to 300 K is about 0.3 kcal/mole. It is seen that the most recent value for $\Delta H_f(t-C_4H_9^+)$ derived from ionization potential determinations is in excellent agreement with the maximum value (see Appendix I) obtained from the most recent determination of the appearance potential of reaction 33.

Houle and Beauchamp[25a] have recently redetermined the ionization potential of the benzyl radical to be 7.20 ± 0.02 eV (166.0 ± 0.5 kcal/mole). Accepting a value of 45 ± 1 for the heat of formation of the benzyl radical[25b], and making the usual assumption that the integrated heat capacities of the ion and the radical are the same, this leads to a value of $\Delta H_f(C_6H_5CH_2^+)$ of 211 ± 1 kcal/mole. The equilibrium constant of the reaction:

$$C_6H_5CH_2^+ + (CH_3)_3CCl \rightleftharpoons (CH_3)_3C^+ + C_6H_5CH_2Cl \qquad (35)$$

has been determined at 303 and 408 K by Abboud, Hehre and Taft[27a] and at 320 K by Jackson et al[27b]. The results lead to an estimate for the enthalpy change of reaction 35 of $\sim$ -0.06 kcal/mole at 300 to 400 K. Taking the value of $\Delta H_f(C_6H_5CH_2^+)$ of 211 $\pm$ 1 kcal/mole, this result leads to an estimate for $\Delta H_f(t-C_4H_9^+)$ of 162.7 $\pm$ 1.5 kcal/mole, in agreement with the revised heat of formation listed in Table 7.

TABLE 7. Values of $\Delta H_f(t-C_4H_9^+)$ Derived from Ionization Potential and Appearance Potential Measurements.

$$t-C_4H_9 \cdot \rightarrow t-C_4H_9^+ + e^-$$

$\Delta H_f(t-C_4H_9)$ kcal/mole	Ref	IP$(t-C_4H_9)$ eV	kcal/mole	Ref	$\Delta H_f(t-C_4H_9^+)$ kcal/mole
4.5	a	7.42 $\pm$ 0.07	171.1 $\pm$ 1.6	b	176
6.8 $\pm$ 1	c				177.9 $\pm$ 2.6
		6.93 $\pm$ 0.05	159.8 $\pm$ 1.1	d	166.6 $\pm$ 2.1
9.3 $\pm$ 1	e				169.1 $\pm$ 2.1
8.4	f				168.2
11.9 $\pm$ 1	g	6.58 $\pm$ 0.01	151.7 $\pm$ 0.2	h	163.6 $\pm$ 1.2

$$neo-C_5H_{12} \rightarrow t-C_4H_9^+ + CH_3 + e^-$$

$\Delta H_f(neo-C_5H_{12})$ kcal/mole	Ref	$\Delta H_f(CH_3)$ kcal/mole	Ref	AP$(t-C_4H_9^+)$ kcal/mole	Ref	$\Delta H_f(t-C_4H_9^+)$ kcal/mole
-39.67	i	33.2	j	243.3 $\pm$ 0.1	k	$\leq$ 170.4
		34.8	l			$\leq$ 168.8
				238.7	m	$\leq$ 164.2

Reference 26: a) Swarc, 1951 b) Lossing and DeSousa, 1959 c) Teranishi and Benson, 1963 d) Lossing and Semeluk, 1970 e) Tsang, 1972, f) Benson, 1976 g) Tsang, 1978 h) Dyke et al, 1978 i) Rossini et al, 1953 j) Wagman et al, 1968 k) Steiner, Geise, and Inghram, 1961 l) McCulloh and Dibeler, 1976 m) Chesnavich, Su and Bowers, 1978.

Experimental entropy changes[24b] determined for several proton
transfer reactions involving $t\text{-}C_4H_9^+$ lead to the conclusion that
the overall ideal entropy change associated with the change ($t\text{-}C_4H_9^+ \rightarrow i\text{-}C_4H_8$) is -1.1 to -1.5 cal/deg-mole. Consideration of external symmetry numbers only (equation 18) leads to a prediction of
a positive $\Delta S°$ associated with this change. The interpretation of
the experimental results is that the positive entropy change associated with the change in rotational symmetry numbers is over-balanced
by a negative entropy change due to the loss of an internal rotation
when a C-C bond in $t\text{-}C_4H_9^+$ is replaced by a double bond in $i\text{-}C_4H_8$.
The magnitude of this negative entropy change (-3.3 to -3.7 cal/
deg-mole) is the same as that observed for the loss of a methyl
rotation in going from CH_3CHCH_2 to $c\text{-}C_3H_6$ (-3.4 cal/deg-mole)[24a,26f].

In some studies of proton transfer equilibria, heats of formation of ions or proton affinity scales have been referred to the
heats of formation of protonated aldehydes as primary standards.
These have been determined from appearance potential measurements
in alcohols[28]. For example, the heat of formation of protonated
acetaldehyde, CH_3CHOH^+, has been obtained from appearance potential
measurements in CH_3CH_2OH, $(CH_3)_2CHOH$, and $CH_3CH(OH)C_2H_5$. The conclusion is that $\Delta H_f(CH_3CHOH^+)$ is $\leq$ 137.2 kcal/mole, assuming that
$AP_0 = AP_{300}$. (Consideration of the known heat capacities of these
alcohols and the corresponding neutral fragments[1a,24a] indicates
that a consistent value for the integrated heat capacity of CH_3CHOH^+
can be derived from the assumption that the difference between
AP_0 and AP_{300} is 0.0 ± 0.2 kcal/mole.) Taking the lowest value
determined in this way for $\Delta H_f(CH_3CHOH^+)$ as the preferred value[2],
one can now check this primary standard against $\Delta H_f(t\text{-}C_4H_9^+)$ since
they have been related through thermodynamic ladders from proton
transfer equilibrium measurements. The ΔG of the reaction:

$$CH_3CHOH^+ + i\text{-}C_4H_8 \rightleftarrows CH_3CHO + t\text{-}C_4H_9^+ \tag{36}$$

has been reported to be -7.9 kcal/mole[7a] or -7.3 kcal/mole[4b] at
300 K and -7.6 kcal/mole at 600 K[7b]. These values correspond to an
entropy change of 0 ± 1 cal/deg-mole. (Equation 24 predicts +0.1 to
+0.8 cal/deg-mole depending on which formulation is used to calculate Z_f and Z_r; equation 18, considering only the changes in external rotational symmetry numbers would predict -2.3 cal/deg-mole).
Thus, the enthalpy change of reaction 36 can be taken to be $-7.6 \pm$
0.3 kcal/mole. This leads to a value of the heat of formation of
CH_3CHOH^+ based on $\Delta H_f(t\text{-}C_4H_9^+)$ as standard, of 136.2 ± 1.5 kcal/mole,
in good agreement with the maximum value established by the appear-
ance potential measurements[28a].

Another primary standard which has been used to put proton affinity ladders on an absolute basis is the heat of formation of protonated formaldehyde, CH_2OH^+. Like protonated acetaldehyde discussed above, the heat of formation of this ion has been derived from appearance potential measurements in alcohols[28],[29], which give a value of 170.3 $\pm$ 1.1 kcal/mole for $\Delta H_f(H_2COH^+)$. Formaldehyde has been related to the primary standards discussed above through the thermodynamic ladder published by Taft et al[7a]. These experiments lead to a value for $\Delta H_f(H_2COH^+)$ of 160.4 $\pm$ 0.1 kcal/mole. Accepting this result (ΔG for proton transfer from H_3O^+ and H_3S^+ to H_2CO reported in that study is in agreement with values reported by Tanaka, Mackay, and Bohme[30], which tends to confirm the result), one must conclude that the appearance potential measurements lead to values of $\Delta H_f(H_2COH^+)$ which are too high by as much as 10 kcal/mole (0.4 eV).

Recently attention has been given to the determination of the proton affinity of ketene[31]. Because of discrepancies between thermodynamic ladders involving ketene in the literature[31a,b], the equilibrium constants for the reactions:

$$(CH_3)_2COH^+ + CH_2CO \rightleftarrows CH_3CO^+ + CH_3COCH_3 \tag{37}$$

and

$$CH_3CO^+ + CH_3COOCH_3 \rightleftarrows (CH_3COOCH_3)H^+ + CH_2CO \tag{38}$$

were re-determined in the NBS pulsed ion cyclotron resonance spectrometer at temperatures in the range 300–370 K. The values of ΔG resulting from these measurements and those reported by Davidson, Lau and Kebarle[31d] at 600 K are shown in Figure 2. Figure 3 shows the thermodynamic ladder of values of ΔH and the corresponding ladder of experimentally-determined values of ΔS derived from these experiments and those reported earlier for the proton transfer equilibrium linking acetone and isobutene[24b]. The proton affinity of ketene, taken relative to PA(i-C_4H_8) as a primary standard is 200.9 $\pm$ 1.2 kcal/mole, which corresponds to a value of 153.4 $\pm$ 1.6 kcal/mole for the enthalpy of formation of the CH_3CO^+ ion, based on a value of –11.4 $\pm$ 0.4 kcal/mole for the enthalpy of formation of CH_2CO[32d]. (An earlier recommended value[32e] for this heat of formation leads to a value of 150.6 $\pm$ 1.9 for $\Delta H_f(CH_3CO^+)$). The appearance potential of the process:

$$CH_3COCH_3 \rightarrow CH_3CO^+ + CH_3 + e^- \tag{39}$$

has been determined to be 10.37 eV[32a] or 10.36 eV[32b], corresponding to maximum values for the heat of formation of CH_3CO^+ of 152.4 or 152.2 kcal/mole. However, it has been reported that the fragmenta-

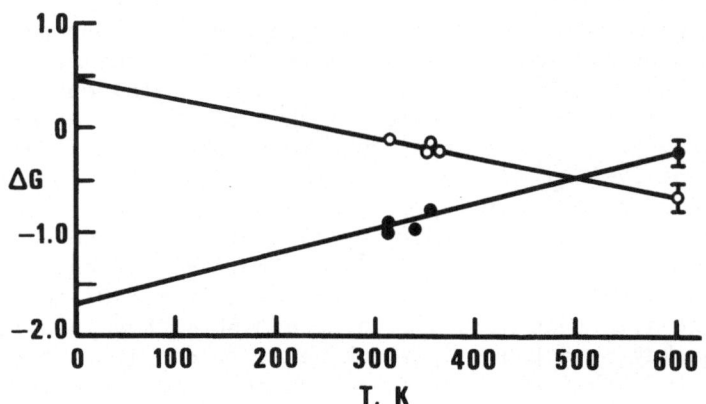

FIGURE 2.   ΔG as a function of temperature for the reactions:
$(CH_3COCH_3)H^+ + CH_2CO \rightleftarrows CH_3CO^+ + CH_3COCH_3$, •.
$CH_3CO^+ + CH_3COOCH_3 \rightleftarrows (CH_3COOCH_3)H^+ + CH_2CO$, ○.

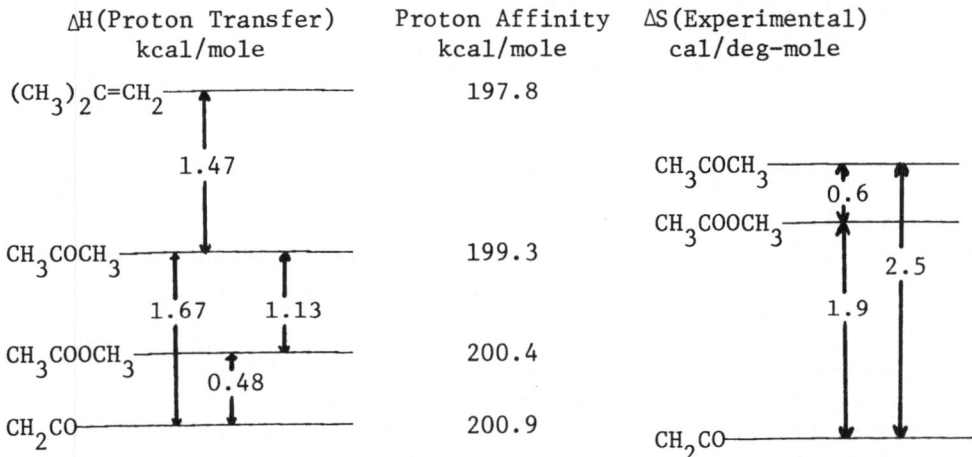

FIGURE 3.   Enthalpy changes and entropy changes for proton transfer
reactions from Figure 2 and Reference 19a, and the corresponding scale of proton affinity values.

tion process 39 has about 0.11 eV average translational energy of
decomposition at threshold[32c]; accepting this, the enthalpy of
formation of CH_3CO^+ must be taken to be 149.8 kcal/mole. The discrepancy between this value and the value of 153.4 kcal/mole
derived from the proton transfer equilibrium constant determinations is considerably greater than the error limits ascribed to any
of the measurements. The explanation may be (a) that the proton
affinity taken for the comparison standard, $i\text{-}C_4H_8$, is still somewhat too high, (b) the 2.5 kcal/mole average translational energy

of decomposition is over-estimated, or (c) the enthalpy of formation of neutral ketene is not well known.

Many relative scales of proton affinities in the literature have been related to the proton affinity of ammonia as a secondary standard. The ΔG of the proton transfer reactions from $t-C_4H_9^+$ and CH_3CHOH^+ to ammonia have been reported in thermodynamic ladders measured at 300 and at 600 K[7], so that "experimental" estimates of the entropy changes can be obtained. On this basis, a value of 207.6 ± 1.2 kcal/mole is obtained for the proton affinity of NH_3. After the preparation of this manuscript was completed, we became aware of a redetermination of the ionization potential of the t-butyl radical by Houle and Beauchamp (J. Am. Chem. Soc., submitted for publication). These authors recommend a value of 162.9 ± 1.2 kcal/mole for $\Delta H_f(t-C_4H_9^+)$ (or $PA(i-C_4H_8) = 198.5$ kcal/mole). Following reasoning identical to that followed here, they arrive at a value of 208.4 ± 2 kcal/mole for the proton affinity of NH_3.

CONCLUSIONS

At this writing, extensive scales of values of ΔG for ion-molecule reactions based on equilibrium constant measurements, are becoming available. Although in the past, entropy changes associated with such equilibria have often been evaluated only approximately[+] experimental determinations of ΔS will increasingly make available accurate values of ΔH for ion-molecule reactions. Thus, relative scales of finite temperature thermodynmic functions such as hydride affinities, proton affinities, or enthalpies of ionization can be determined with high internal accuracy and precision. The measurement and interpretation of ionization thresholds to obtain absolute values for enthalpies of formation of ions (to which the relative thermodynamic scales can be related) is intrinsically more difficult. Except for spectroscopic ionization potentials which relate to the recently reported[6] scale of ΔH determinations for charge transfer equilibria with a high degree of precision, the error limits associated with most <u>absolute</u> values of heats of formation (mainly obtained from appearance potential measurements) of ions are wider by an order of magnitude or more than the error limits of the <u>relative</u> values for the heats of formation obtained from equilibrium measurements, as shown by the few examples discussed above of standards for a scale of proton affinities. These examples also illustrate the fact that in many cases the major obstacle to deriving heats of formation of ions is the uncertainty associated with the heat of formation of the corresponding neutral species.

Acknowledgment: I would like to thank Dr. P. Ausloos for his continuing encouragement, and for the in-depth discussions and criticisms which went into the preparation of this manuscript and the lecture which was delivered at La Baule.

APPENDIX: STANDARD STATES IN GAS PHASE ION THERMOCHEMISTRY

Henry M. Rosenstock, National Bureau of Standards
Washington, DC, 20234, U.S.A.

The principal tools which give numerical information on energy requirements for ionizing processes are threshold and spectroscopic measurements on isolated molecules. Those relating the energy, entropy and free energy changes between gas phase ions are the steady state and equilibrium measurements of high pressure mass spectrometry, ion cyclotron resonance spectrometry, and flowing afterglow experiments. It is useful to connect these in a compatible manner to the corpus of neutral thermochemistry, most of which is based on macroscopic measurements.

In recent years, workers in ion chemistry have relied on several thermochemical data compilations for auxiliary information on both neutral and ion species. These compilations include the National Bureau of Standards Technical Note 270 Series (TN270)[1a], the JANAF Tables[1b], and the recent compilation of Gaseous Ion Energetics (GIE) as well as its precursor, Ionization Potentials, Appearance Potentials, and Heats of Formation of Gaseous Positive Ions[2]. We would like to point out that there is a difference in convention regarding ion heats of formation at finite temperature between TN270 and JANAF on the one hand, and GIE on the other. The standard enthalpy of formation of the ion M^+ in kcal/mole at 298.15 K is related among the three compilations as follows:

$$\Delta H_{f_{298}} (M^+)(JANAF, TN270) = \Delta H_{f_{298}} (M^+)(GIE) + 1.481$$

whereas at absolute zero degrees, the three quantities agree. The differences arise solely and exclusively from the arbitrary conventions adopted for the heat capacity of the electron in the three compilations.

The TN270 and JANAF compilations both treat the electron as a conventional chemical element with a standard enthalpy of formation of zero at all temperatures. Further, they assume that the electron is a classical ideal gas, rather than a Fermi gas[3], at all temperatures and calculate enthalpy functions:

$$H° - H°_{298} = \int_0^T C_p(T)\, dT$$

accordingly, assuming that the heat capacity of the electron is that of an ideal gas, 5/2R. The Gaseous Ion Energetics compilations, (GIE and NSRDS-NBS-26) assume that the electron is at rest at all temperatures and, correspondingly, has no integrated heat capacity correction at any temperature. Unfortunately, this assumption was not properly phrased in the prefaces of the compilations.

The TN270-JANAF convention has the virtue of defining thermo-dynamic functions for the electron in a manner consistent with that for any chemical element, so that one may immediately write down numerical values for calculating ionization equilibria of the form:

$$M(g) \; \overset{\rightarrow}{\leftarrow} \; M^+(g) + e^-(g)$$

However, these equilibria are difficult to realize in the laboratory and the assumption of classical electron behavior and the approxi-mation of ideal gas behavior of the ionized products appears to be very approximate. With this convention, the heat (enthalpy) of formation of an ion at room temperature is assumed to differ from that of the corresponding molecule by an amount equal to:

$$\Delta H_{f_{298}}(M^+) - \Delta H_{f_{298}}(M) = I_z + 5/2RT = I_z + 1.481 \text{ kcal/mole}$$

To our knowledge, no publication giving threshold measurements has every considered the heat capacity of the electron in deriving ion heats of formation.

The GIE convention ignores the electron and its heat capacity; it assumes the electron at infinite separation from the ion and at rest at all temperatures. This, of course, leaves the question of ion-electron-neutral equilibria incomplete, as well it should be. However, it leaves intact the approximate relation:

$$\Delta H_{f_{298}}(M^+) - \Delta H_{f_{298}}(M) \sim I_z$$

Finite Temperature Thermochemistry of Threshold Processes

At absolute zero it is rigorously true that for

$$M \rightarrow M^+ + e^- \qquad \Delta E_0 = \Delta H_0 = \Delta H_{f_0}(M^+) + \Delta H_{f_0}(e^-) - \Delta H_{f_0}(M) \qquad \text{(I)}$$

$$AB \rightarrow A^+ + B + e^- \qquad \Delta E_0 = \Delta H_0 = \Delta H_{f_0}(A^+) + \Delta H_{f_0}(B) + \Delta H_{f_0}(e^-)$$
$$-\Delta H_{f_0}(AB) \qquad \text{(II)}$$

$$AH^+ \rightarrow A + H^+ \qquad \Delta E_0 = \Delta H_0 = \Delta H_{f_0}(A) + \Delta H_{f_0}(H^+) - \Delta H_{f_0}(AH^+) \qquad \text{(III)}$$

For a general gas phase process at temperature T:

$$\Delta E_T = \Delta E_0 + \int_0^T (C_{v_{products}} - C_{v_{reactants}})dT \qquad (IV)$$

$$\Delta H_T = \Delta H_0 + \int_0^T (C_{p_{products}} - C_{p_{reactants}})dT \qquad (V)$$

$$= \Delta E_T + \Delta nRT$$

where C_v and C_p are heat capacities, and the final equality arises from the PV work term in the enthalpy change, and the assumption that the system behaves as an ideal gas. Δn is the change in the mole number in going from reactants to products.

From equation I, it is seen that the zero degree heat of formation of the ion depends on the chemistry of the ionization process, the standard state of the neutral species, and the defined heat or energy of formation of the electron. (One can call it zero, or, just for example, one-half of the electron-positron pair creation energy). At finite temperature, ΔH_T differs from ΔE_T by an amount that depends on what we assume about the work performed in forming an electron-ion pair. In the GIE convention, it is ignored, while in the TN270-JANAF convention, it becomes an artifice. In reality, it is a difficult question pertinent to plasma physics. In addition, at finite temperatures the quantities differ from the absolute zero values by the difference in integrated heat capacity of the ion and neutral. As discussed in the body of this chapter, this is generally small unless there are one or more low frequency ($\leq$ kT) vibrations or torsional oscillations in one species which are not present in the other. Also, it may happen that one species is linear and its charged (uncharged) counterpart bent, in which case a contribution of 1/2 kT to the rotational energy must be added, and a contribution to the vibrational energy subtracted. Lastly, there is the possibility of minor contributions from electronic levels (most ions are doublets). We note in passing that no threshold measurement at finite temperature can directly give the finite temperature difference between the ion and neutral enthalpies of formation.

In equation II, the value of the heat of formation of the ion at zero degrees depends on the heats of formation of the neutral species, the standard state of the electron, and the chemistry of the fragmentation process. Of course, equation II is correct only if one succeeds in measuring the energy required to form the fragments in the ground rotational, vibrational, and electronic states, and with infinitesimally slow velocities of separation. In correcting from 0 K to a finite temperature (equations IV and V) we see that ΔE changes primarily due to rotational and translational heat capacity terms, while ΔH, in addition, requires PV work whose magnitude depends on whether or not we put the electron to work. There is no shortcut to using threshold techniques to deduce 0 K values for enthalpies or energies of formation and then carrying out the appropriate heat capacity calculations[2].

Equation III defines the "enthalpy of deprotonation" commonly called the "proton affinity"; although defined as a threshold process, information about proton affinities is rarely obtained from direct observations of reaction III. In this equation positive ions appear on both sides of the equation so the electron standard state cancels out. The energy or enthalpy of protonation is independent of the electron standard state (provided the same standard state is used to define the ionic heats of formation on both sides of the equation!), i.e. either the TN270-JANAF or the GIE conventions. The energy and enthalpy of reaction change approximately as 3/2RT and 5/2RT, respectively, due to translational heat capacities and PV work. As before, changes in low frequency vibrations and internal rotations may also effect minor changes. Thus, the energy or enthalpy or protonation (or deprotonation) exhibit a temperature dependence which is almost exclusively due to factors other than molecular bonding.

In ion thermochemistry, there is need for a more carefully defined nomenclature for both molecular or threshold measurements and derived or measured thermochemical quantities. For example, current usage of "electron affinity" implies a threshold measurement, whereas "proton affinity" implies an enthalpy change at some temperature (including absolute zero). This is unfortunate, and suggests that the term "proton affinity" be reserved for the quantity at absolute zero, while "enthalpy of protonation" should be adopted for the finite temperature quantity. Also the differences in electron standard states discussed above suggests the term "ion-electron pair heat of formation" for the JANAF-TN270 convention, whereas "ion heat of formation" may reasonably describe the GIE convention. In that framework, the enthalpy difference between an ion and its neutral might be termed the enthalpy of ion-electron pair formation or of ionization respectively.

Acknowledgment: We would like to acknowledge helpful discussions with Drs. M. Chase, A. Syverud, and D. Wagman.

APPENDIX: ION-MOLECULE EQUILIBRIA AND ENTROPY CHANGE

Timothy Su, Southeastern Massachusetts University,
North Dartmouth, MA, 02747, U.S.A.

Lias and Ausloos have recently measured the entropy change in
proton transfer[24b] and charge transfer[6],[8] ion-molecule equilibria
by observing the equilibrium constant as a function of temperature
in an ICR spectrometer. From the comparison of their experimental
results with theoretical calculations of entropy changes, i.e. the
ratio of the partition function of the products to the reactants,
they have concluded[4c],[6],[8],[24] that a term must be included to
accurately estimate the theoretical entropy change. This term has
been called the intermolecular entropy term which is derived based
on collision theory formulation:

$$\Delta S_{\text{intermolecular}} = R \ln \frac{Z_f}{Z_r}$$

where Z_f and Z_r are collision frequencies of the forward and reverse
processes respectively. The above conclusion is questionable since
ΔS is independent of reaction path. Also, an intermolecular contribution to the entropy change seems unlikely for the pressure range
used in the experiments (10^{-6}-10^{-5} torr). It can be shown from
the statistical mechanics of non-ideal gases[33] that the intermolecular contribution is negligible. Lias[4c] has pointed out that in
the theoretical treatment of Eyring, Hirshfelder and Taylor[21] of
ion-molecule reactions, the rotational partition functions of the
molecules are modified due to the approach of the ion. The modification is not on the rotational partition functions of the reactants
or the products but is on the partition function of the orbiting
transition state (the transition state of the complex formation).
The partition function of the reactants used are the ideal gas partition functions.

In fact, the derivations of both the Langevin and ADO collision
frequencies assume ideal gas reactants. Using these theoretical
collision frequencies to represent non-ideal behavior of the reactants or products in the first equation is not consistent.
The above argument shows that the term $R \ln Z_f/Z_r$ does not represent
$\Delta S_{\text{intermolecular}}$. The agreement of the experimental results of
Lias and Ausloos with the first equation is due to some other reason. The following theoretical argument gives a possible explanation. Because of the strong attractive forces between the ion-
molecular partners, the intermediate complex has a rather high
angular momentum. Since the attractive forces of the forward and
reverse processes may not be the same, the intermediate complexes
formed from the opposite directions have different rotational
energy (angular momentum) distributions. That is, they pass through

intermediate complexes (paths). Furthermore, when the intermediate dissociates, the translational energy release is not thermal[34]. This violates the principle of microscopic reversiblity for equilibrium. However, based on the law of conservation of angular momentum, this situation will occur. Thus, the above argument suggests that in gas phase equilibrium measurements, equilibrium may never be established. Instead, a steady state cyclic reaction scheme occurs:

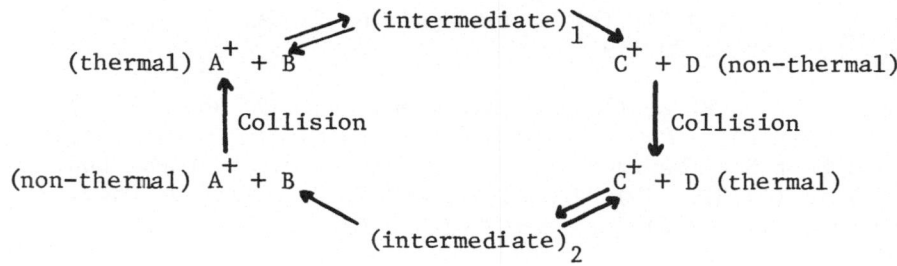

If this cyclic reaction does happen, which is very probable, the measured equilibrium constants and entropy changes are apparent values and not the true values. Thus a correlation should be made, which may be approximated by the Z_f/Z_r ratio. The above suggestion deserves further investigation both theoretically and experimentally.

REFERENCES

1. (a) D. D. Wagman, W. H. Evans, V. B. Parker, I. Halow, S. M. Bailey and R. H. Schumm, Selected Values of Chemical Thermodynamic Properties, NBS Technical Note 270-3 (1968); (b) JANAF Thermochemical Tables, Nat. Stand. Ref. Data Ser., Nat. Bur. Stand. (U.S.) 37 (1971).

2. (a) H. M. Rosenstock, K. Draxl, B. W. Steiner, and J. T. Herron, J. Phys. and Chem. Ref. Data 6, Supplement 1 (1977); (b) J. L. Franklin, J. G. Dillard, H. M. Rosenstock, J. T. Herron, K. Draxl, and F. H. Field, Nat. Stand. Ref. Data Series, Nat. Bur. Stand. (U.S.) 26 (1969).

3. Richard C. Tolman, "The Principles of Statistical Mechanics," The Clarendon Press, Oxford (1938) pp. 159-165; pp. 388-394.

4. (a) P. Kebarle, Ann. Rev. Phys. Chem. 28, 445 (1977); (b) D. H. Aue and M. T. Bowers in "Gas-Phase Ion Chemistry," (M. T. Bowers, Ed.), Academic Press, New York, NY (1979); (c) S. G. Lias in "Lecture Notes in Chemistry; Ion Cyclotron Resonance Spectrometry," (H. Hartmann and K.-P. Wanczek, Eds.), Springer-Verlag, Berlin, Heidelberg, New York (1978).

5. See for example: (a) A. E. Roche, M. M. Sutton, D. K. Bohme, and H. I. Schiff, J. Chem. Phys. 55, 5480 (1971); (b) J. D. Payzant, H. I. Schiff, and D. K. Bohme, J. Chem. Phys. 63, 149 (1975).

6. S. G. Lias and P. Ausloos, J. Am. Chem. Soc. 100, 6027 (1978).

7. (a) J. F. Wolf, R. H. Staley, I. Koppel, M. Taagepera, R. J. McIver, Jr., J. L. Beauchamp, and R. W. Taft, J. Am. Chem. Soc. 99, 5417 (1977); (b) R. Yamdagni and P. Kebarle, J. Am. Chem. Soc. 98, 1320 (1976).

8. S. G. Lias and P. Ausloos, J. Am. Chem. Soc. 99, 4831 (1977).

9. (a) M. Meot-Ner (Mautner) and F. H. Field, J. Am. Chem. Soc. 100, 1356 (1978). (b) M. Meot-Ner and F. H. Field, J. Chem. Phys. 64, 277 (1976).

10. S. G. Lias and P. Ausloos, to be published.

11. (a) T. Su and M. T. Bowers, Int. J. Mass Spectrom. Ion Phys. 12, 347 (1973); (b) T. Su and M. T. Bowers, J. Am. Chem. Soc. 95, 1370 (1973); M. T. Bowers and T. Su in "Interactions Between Ions and Molecules," (P. Ausloos, Ed.) Plenum Press, New York (1975).

12. W. J. Chesnavich, T. Su, and M. T. Bowers, this volume.

13. R. A. Barker and D. P. Ridge, J. Chem. Phys. 64, 4411 (1976).

14. K. Hartman and S. G. Lias, Int. J. Mass Spectrom. Ion Phys. in press.

15. D. P. Ridge, this volume.

16. J. M. Jasinski, R. N. Rosenfeld, D. M. Golden and J. I. Brauman, private communication.

17. See for example: (a) E. H. Kennard, "The Kinetic Theory of Gases", McGraw-Hill, New York and London (1938), pp. 55-58; (b) K. J. Laidler, "Chemical Kinetics", McGraw-Hill, New York and London (1965), pp. 110-112.

18. E. V. Waage and B. S. Rabinovitch, Chem. Rev. 70, 377 (1970).

19. (a) F. H. Field, P. Hamlet, and W. F. Libby, J. Am. Chem. Soc. 91, 2839 (1969); (b) S. L. Bennett and F. H. Field, J. Am. Chem. Soc. 94, 8669 (1972); (c) S. L. Bennett and F. H. Field, J. Am. Chem. Soc. 94, 6305 (1972); (d) S. L. Bennett

and F. H. Field, J. Am. Chem. Soc. 94, 5188 (1972); (e) F. H. Field and D. P. Beggs, J. Am. Chem. Soc. 93, 1585 (1971); (f) J. D. Payzant, A. J. Cunningham and P. Kebarle, J. Chem. Phys. 59, 5615 (1973); (g) J. D. Payzant and P. Kebarle, J. Chem. Phys. 53, 4723 (1970).

20. (a) E. P. Grimsrud and P. Kebarle, J. Am. Chem. Soc. 95, 7939 (1973); (b) I. Dzidic and P. Kebarle, J. Phys. Chem. 74, 1466 (1970); (c) M. Arshadi, R. Yamdagni, and P. Kebarle, J. Phys. Chem. 74, 1475 (1970); (d) M. Arshadi and P. Kebarle, J. Phys. Chem. 74, 1483 (1970); (e) D. P. Beggs and F. H. Field, J. Am. Chem. Soc. 93, 1576 (1971); (f) F. H. Field, J. Am. Chem. Soc. 91, 2827 (1969).

21. H. Eyring, J. O. Hirschfelder, and H. S. Taylor, J. Chem. Phys. 4, 479 (1936).

22. P. Ausloos and S. G. Lias, to be published.

23. (a) A. Goren and B. Munson, J. Phys. Chem. 80, 2848 (1976); (b) J. J. Solomon and F. H. Field, J. Am. Chem. Soc. 97, 2625 (1975).

24. (a) S. W. Benson and H. E. O'Neal, Nat. Stand. Ref. Data Series, Nat. Bur. Stand. (U.S.) 21 (1970); (b) P. Ausloos and S. G. Lias, J. Am. Chem. Soc. 100, 1953 (1978).

25. (a) F. A. Houle and J. L. Beauchamp, J. Am. Chem. Soc. 100, 3290 (1978); (b) J. A. Kerr, Chem. Rev. 66, 465 (1966).

26. (a) M. Swarc, Disc. Faraday Soc. 10, 336 (1951); (b) F. P. Lossing and J. B. DeSousa, J. Am. Chem. Soc. 81, 281 (1959); (c) H. Teranishi and S. W. Benson, J. Am. Chem. Soc. 85, 2887 (1963); (d) F. P. Lossing and G. P. Semeluk, Can. J. Chem. 48, 955 (1970); (e) W. Tsang, J. Phys. Chem. 76, 143 (1972); (f) S. W. Benson, "Thermochemical Kinetics, 2nd Edition", John Wiley & Sons, Inc., New York (1976); (g) W. Tsang, Int. J. Chem. Kinetics 10, 821 (1978); (h) J. Dyke, N. Jonathan, E. Lee, A. Morris and M. Winter, submitted for publication and referred to in Reference 25a; (i) F. D. Rossini, K. S. Pitzer, R. L. Arnett, R. M. Braun, and G. C. Pimental, API Research Project 44, Pittsburgh (1953); (j) Reference 1a; (k) B. Steiner, C. F. Giese, and M. G. Inghram, J. Chem. Phys. 34, 189 (1961); (l) K. E. McCulloh and V. H. Dibeler, J. Chem. Phys. 64, 4445 (1976); (m) W. J. Chesnavich, T. Su and M. T. Bowers, J. Am. Chem. Soc. 100, 4362 (1978).

27. (a) J.-L. M. Abboud, W. J. Hehre, and R. W. Taft, J. Am. Chem.
     Soc. 98, 6072 (1976); (b) J.-A. A. Jackson, S. G. Lias, and P.
     Ausloos, J. Am. Chem. Soc. 99, 7515 (1977).

28. (a) F. P. Lossing, J. Am. Chem. Soc. 99, 7526 (1977); (b) K.
     M. A. Rafaey and W. A. Chupka, J. Chem. Phys. 48, 5205 (1968).

29. F. M. Benoit and A. G. Harrison, J. Am. Chem. Soc. 99, 3980
     (1977).

30. K. Tanaka, G. I. Mackay, and D. K. Bohme, Can. J. Chem. 56,
     193 (1978).

31. (a) P. Ausloos and S. G. Lias, Chem. Phys. Lett. 51, 53 (1977);
     (b) J. Vogt, A. D. Williamson, and J. L. Beauchamp, J. Am.
     Chem. Soc. 100, 3478 (1978); (c) D. B. Debrou, J. E. Fulford,
     E. G. Lewars, and R. E. March, Int. J. Mass Spectrom. Ion
     Phys. 26, 345 (1978); (d) W. R. Davidson, Y. K. Lau and P.
     Kebarle, Can. J. Chem. 56, 1016 (1978).

32. (a) E. Murad and M. G. Inghram, J. Chem. Phys. 41, 404 (1964);
     (b) R. H. Staley, R. D. Wieting, and J. L. Beauchamp, J. Am.
     Chem. Soc. 99, 5964 (1977); (c) M. A. Haney and J. L. Franklin,
     Trans. Faraday Soc. 65, 1794 (1969); (d) R. L. Nuttall, A. H.
     Laufer, and M. V. Kilday, J. Chem. Thermodynamics 3, 167
     (1971); (e) J. D. Cox and G. Pilcher, "Thermochemistry of
     Organic and Organometallic Compounds", Academic Press, London
     and New York (1970).

33. See for example: "Chemical Thermodynamics", F. T. Wall, 3rd
     Edition, W. H. Freeman, San Francisco, 1974, pp. 292-295.

34. (a) W. J. Chesnavich and M. T. Bowers, J. Am. Chem. Soc. 98,
     8301 (1976); (b) W. J. Chesnavish and M. T. Bowers, J. Am.
     Chem. Soc. 99, 1705 (1977).

EQUILIBRIUM STUDIES OF NEGATIVE ION-MOLECULE REACTIONS

Robert T. McIver, Jr.

Department of Chemistry, University of California

Irvine, California   92717

There exists in the literature an extensive body of thermochemical data for positive ions in the gas phase.[1] These data have
been derived primarily from ionization potentials and appearance
potentials measured by electron impact and photoionization mass
spectrometry. In contrast to this, very little is known about the
thermochemistry of negative ions. Negative ions can be detected
by electron impact and photoionization mass spectrometry, but the
cross section for their production is typically more than a hundred
times less than for production of positive ions. Furthermore,
negative ions produced by electron impact often have excess translational energy which hinders their collection in sector-type
instruments. For these reasons, measurements of thresholds for
production of negative ions has not been nearly as fruitful as for
positive ions.

During the last four years, equilibrium studies of gaseous
ion-molecule reactions has emerged as a powerful method for
acquiring thermochemical data for isolated positive and negative
ions. Equilibrium proton transfer reactions of the type

$$B_1H^+ + B_2 = B_2H^+ + B_1 \tag{1}$$

have provided extensive data for the basicity of molecules in the
gas phase.[2] In addition, valuable insight into the nature of
solvation effects has been gained by comparing solution phase and
gas phase basicities (see the chapter by R. W. Taft). The amount
and quality of the thermochemical data for gaseous negative ions
has been greatly increased by studying equilibrium ion-molecule
reactions such as

$$A^- + BH = B^- + AH \tag{2}$$

where BH and AH are molecules which can function as Brönsted acids. The standard Gibb's free energy change, $\delta\Delta G^o_{acid} = -RT \ln K$, for reaction 2 is a measure of the relative acidity of AH and BH in the gas phase. A series of acids can be studied to establish a scale of relative acidities in the same manner as pK_a values are determined in solution. In addition, an absolute scale corresponding to the process

$$AH = A^- + H^+ \tag{3}$$

can be established by incorporating into the relative scale certain standards such as $H_2 = H^- + H^+$ and $HF = F^- + H^+$ for which ΔH^o_{acid} and ΔG^o_{acid} can be calculated from available data. Measurement of gas-phase acidities in this way provides through reaction 3 a powerful and general route to heats of formation of anions.

This chapter focuses on the thermochemical data derived from equilibrium negative ion-molecule reactions. In particular, data obtained by high pressure mass spectrometry, flowing afterglow, and trapped ion cyclotron resonance spectrometry are combined to yield a gas-phase acidity scale containing 171 compounds. Comparisons of equilibrium constant data from various laboratories and with acidities calculated from available thermochemical data indicate that the accuracy of the ion-molecule methods are quite good. In addition, this chapter contains a description and comparison of pulsed high pressure mass spectrometry and pulsed ion cyclotron resonance spectrometry, the two main techniques used in studying equilibrium negative ion-molecule reactions.

EXPERIMENTAL METHODS

Most of the equilibrium measurements of relative acidities in the gas phase have been done by Kebarle and coworkers at the University of Alberta using pulsed high pressure mass spectrometry (HPMS) and by McIver and coworkers at the University of California, Irvine, using pulsed ion cyclotron resonance spectroscopy (ICR). The two techniques use the same basic chemical reactions, as presented in Scheme I:

Scheme I:

$$e^- + XY \longrightarrow X^- + Y \tag{4}$$

$$X^- + AH \longrightarrow A^- + HX \tag{5}$$

$$X^- + BH \longrightarrow B^- + HX$$

$$A^- + BH \underset{k_{-1}}{\overset{k_1}{\rightleftharpoons}} B^- + AH \qquad (6)$$

$$A^- \xrightarrow{k_{LOSS}} \text{ion loss} \qquad (7)$$

$$B^- \xrightarrow{k_{LOSS}} \text{ion loss} \qquad (8)$$

An experiment is initiated by a short pulse of electrons. Compound XY captures the electrons and undergoes dissociative electron capture, reaction 4 , to produce the negative ion base X^-. Methyl nitrite (CH_3ONO), as a source of CH_3O^-, is most often used in our pulsed ICR experiments at the University of California, Irvine, but there are many other useful compounds such as N_2O, NF_3, SO_2F_2, NH_3, and H_2O which have reasonably high cross sections for formation of negative ions.   The next step in the reaction sequence, reactions 5 , involve rapid, irreversible deprotonation of the acids of interest, AH and BH. After a sufficient number of collisions, equilibrium according to reaction 6 is established and the ratio of the signals due to A^- and B^- is determined. When this is combined with the ratio of AH and BH introduced into the reaction chamber, the equilibrium constant and the standard Gibb's free energy change can be calculated.

$$K = \frac{[B^-][AH]}{[A^-][BH]} \qquad (9)$$

$$\delta\Delta G_{acid} = -RT \ln K \qquad (10)$$

Reliable equilibrium data can only be obtained when the time constant for attainment of equilibrium is much shorter than the lifetime of A^- and B^- in the apparatus. Usually the lifetime of ions is limited by diffusion to the walls of the reaction chamber. However, other loss mechanisms such as reactions with acidic impurities in the sample or additional ion–molecule reactions, such as clustering, which compete with proton abstraction must also be considered sometimes.

   In HPMS ions are generated in a field-free chamber containing ca. 10^{-2} Torr of neutral acids and several Torr of methane to serve as a thermalizing collision gas.[3] After an initial ionizing electron pulse, a small fraction of the ions continuously escape from the chamber through an exit port into a low pressure region where they are accelerated and mass analyzed in a magnetic sector. The ions are observed for up to 3 msec and suffer several hundred collisions with the neutral acids before they are neutralized on the walls or diffuse out of the collision chamber. Equilibrium is assumed to be attained if the ratio of ion abundances $[A^-]/[B^-]$ is invariable with time. At the relatively high pressures used in the source region of the mass spectrometer, formation of "cluster"

ions such as AHA^- and $A^-(HA)_2$ occurs rapidly. While the chemistry
of such species as a stepwise model of solvation is of considerable
interest, their formation tends to interfere with HPMS measurements
of relative acidities. Thus, source temperatures of 600K are
typically used to suppress the formation of the cluster ions. In
order to relate equilibria at 600K to thermochemical data at 298K,
the experimental $\delta\Delta G^o_{acid}$ values are adjusted using appropriate
entropy corrections, as is discussed below.

In pulsed ICR studies of ion-molecule equilibrium reactions,
much lower pressures (10^{-6} - 10^{-5} Torr) and lower temperatures
(320K) are used.[4] In order to achieve sufficient extents of reaction, the ions are stored for a considerably longer time (up to
several seconds) in a static magnetic ion trap, shown in Fig. 1.
While the ions are reacting, they are trapped by static magnetic
and electric fields, and each ion undergoes several hundred collisions before it is lost from the trap. After a certain reaction
period, ions of a chosen mass-to-charge ratio are detected by a
capacitance bridge detector connected to the upper and lower plates
of the ICR cell.[5] This detector works by applying an RF pulse at
frequency ω_1 to the upper plate of the cell. When ω_1 is the same
as the cyclotron frequency of an ion, $\omega = qB/m$, the ion is accelerated to higher translational energy. Coherent motion of the

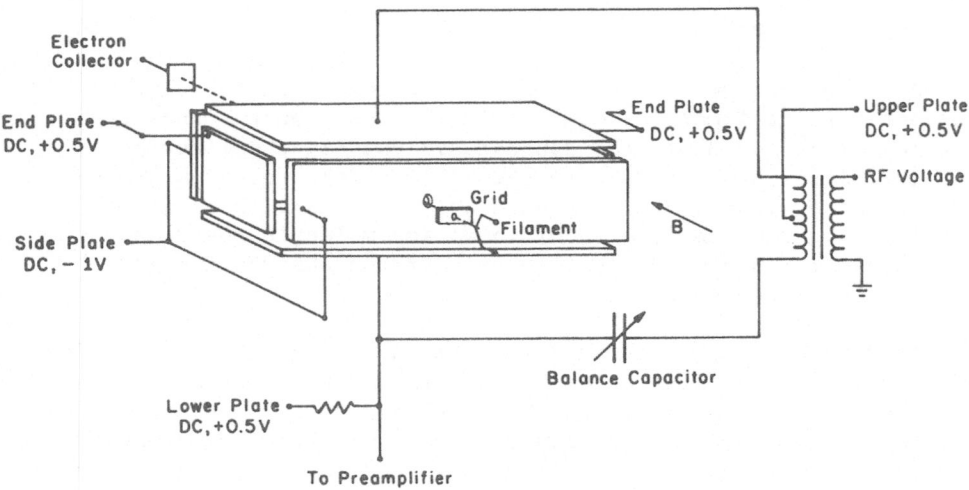

Figure 1. Schematic drawing of a one-region ICR cell shown with
DC voltages appropriate for trapping negative ions.

accelerated ions induces image currents at frequency ω in the lower plate of the ICR cell which can be amplified and detected. The mass resolution and sensitivity of this detector are excellent, as is shown by the high resolution ICR signal in Fig. 2. In addition, the capacitance bridge detector is more versatile than marginal oscillator detectors used previously in ICR experiments because ions of different m/e can be detected by holding the magnetic field strength B constant and scanning the frequency ω_1 of the excitation signal. This capability has led the development of the rapid scan ICR where an entire mass spectrum is recorded in less than one second.[5]

In both the HPMS and ICR experiments, gaseous ions are stored for an extended period of time and caused to suffer hundreds of stabilizing ion-neutral collisions. This is essential to insure thermalization of excited ions produced by electron impact or exothermic ion-molecule reactions. Mechanisms for deactivation of excited ions $(A^-)^*$ are shown below.

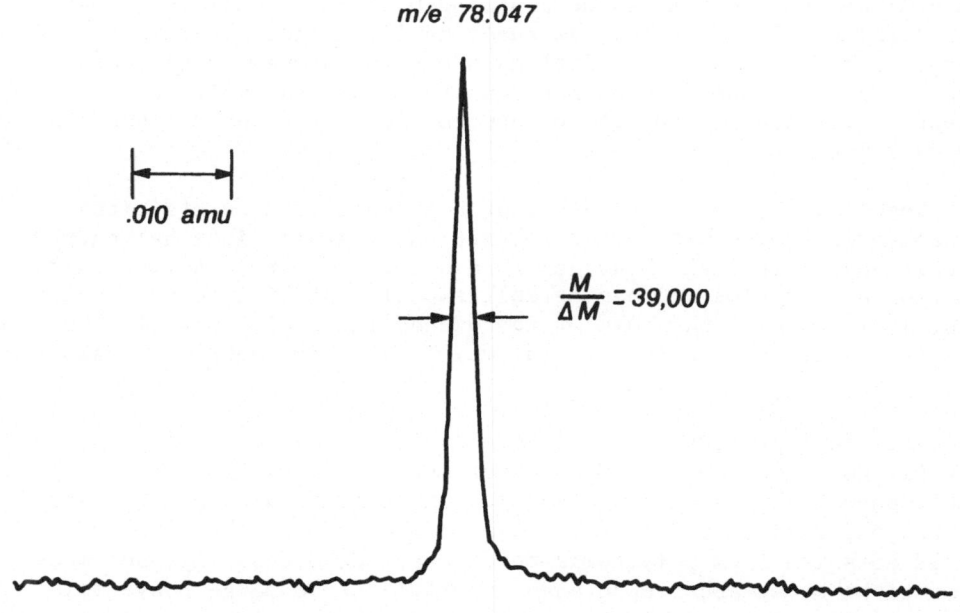

Figure 2. High resolution ICR signal for m/e 78 ions in benzene at 1.8×10^{-7} Torr and a magnetic field strength of 1.28 T.

Scheme II:

Collisional Deactivation

$$(A^-)^* + M \longrightarrow A^- + M^*$$ (11)

Proton Exchange

$$(A^-)^* + AH \longrightarrow [A\cdots H^- \cdots A]^* \longrightarrow A^- + AH^*$$ (12)

Spontaneous Emission

$$(A^-)^* \longrightarrow A^- + h\nu$$ (13)

Relaxation of excess translational momentum in the ion can occur by a nonspecific deactivation mechanism such as reaction 11. The efficiency of this process depends on the relative mass of A^- and molecule M, but generally about 10 collisions are sufficient. In the ICR cell, ions are formed initially with excess translational energy due to the electrostatic trapping well parallel to the magnetic field.[6] However, ions quickly relax to the center of the ICR cell and the bottom of the trapping well as a result of momentum relaxing collisions such as reaction 11. Since experiments are always carried out under conditions where the number density of ions is far less than the number density of neutral molecules, excess energy transferred to the neutral molecules never finds its way back into the ions.

Reaction 12, near resonant proton transfer, is an efficient mechanism in these systems for relaxation of excess internal energy in the ions. Since the hydrogen bonded intermediates $(A\ldots H\ldots A)^-$ live for a long time and can be collisionally stabilized as cluster ions, it is likely that some of the excess internal energy in the ions is transferred to the neutral molecule as the complex breaks up.

The final process shown in Scheme II, spontaneous emission, is probably not important for HPMS experiments. However, it is a likely possibility for the pulsed ICR experiments because the ions are trapped typically for 1 s, a time comparable to the lifetime for IR emission from polyatomic molecules. Unfortunately, not much seems to be known about spontaneous emission from large polyatomic molecules, but there is indirect evidence like bimolecular formation of cluster ions and generation of stable negative ion radicals from direct capture of electrons.[7]

ESTIMATION OF ENTROPIES

In order to place acidity measurements done at various temperatures on a common ΔH^o_{acid} scale, it is necessary to determine the entropy change ΔS^o_{acid} for reaction 3 where

$$\Delta S^o_{acid} = S^o(H^+) + S^o(A^-) - S^o(AH) \qquad (14)$$

The absolute entropy of the proton is well known, 26.012 eu, and the entropies of the neutral acids can be obtained from the literature or estimated by group additivity methods.[8] The absolute entropies of the anions are generally not available in the literature and must be estimated in various ways. Cumming and Kebarle have discussed a statistical mechanical approach which considers the translational, electronic, vibrational, and rotational contributions to the quantity $S^o(A^-) - S^o(AH)$.[3] For acids of molecular weight greater than 30 amu, the translational entropies of AH and A^- are less than 0.1 eu different, so essentially cancel and can be neglected. The difference in electronic entropies of AH and A^- is also expected to be small since both the acid and its anion are isoelectronic, closed shell species. The structures of AH and A^- are usually similar enough that those vibrationals less than 1000 cm^{-1} which contributes to S^o_{VIB} at 298K should nearly cancel. Therefore, it is rotational symmetry, primarily from loss of internal rotations and symmetry changes upon proton removal, that contributes primarily to $S^o(A^-) - S^o(AH)$.[8]

An alternate way of approximating the entropies is to assume the entropy of the anion is the same as the experimentally measured entropy of the appropriate isoelectronic neutral species; i.e., $S^o(CH_3O^-) \cong S^o(CH_3F)$. Generally, the isoelectronic approximations are within 1 eu of the entropies estimated by the statistical mechanical method.[9]

SUMMARY OF THERMOCHEMICAL DATA

Figure 3 on the next three pages summarizes all the currently available data for the enthalpy change, ΔH^o_{acid}, for the process AH = A^- + H$^+$ in the gas phase.

Relative acidities measured by pulsed ICR cover the range from MeOH (379.1 kcal/mol) to phenol (350.4 kcal/mol). Experimental equilibrium constants for an extensive series of proton transfer reactions were converted to enthalpy changes using the equation

$$\delta \Delta H_{acid} = -RT \ln K + T \delta \Delta S^o_{acid} \qquad (15)$$

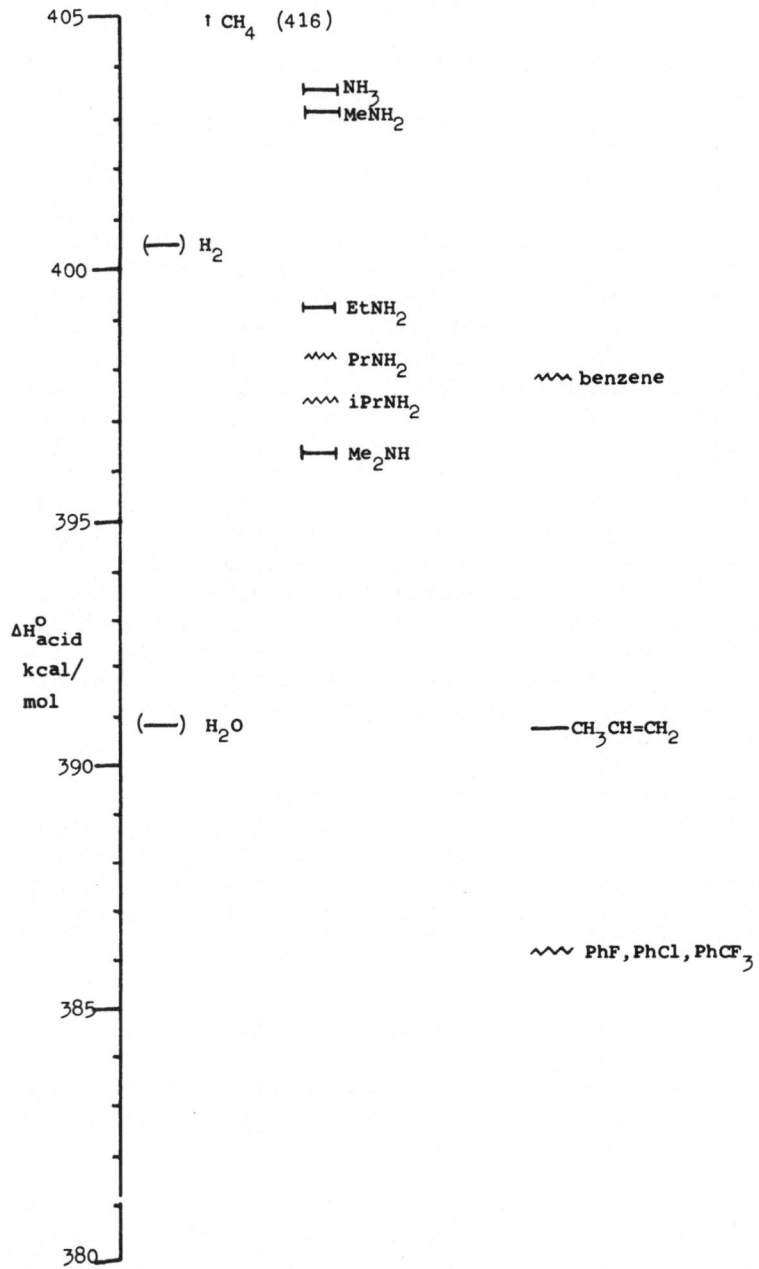

Figure 3. The enthalpy change ΔH^o_{ACID} for AH = H$^+$ + A$^-$ in the gas phase at 298K. Symbols: —— indicates ICR data; ---- indicates HPMS data; (——) indicates calculated according to Scheme III; ⊢——⊣ indicates flowing afterglow data in reference 10; and ⋀⋀⋀ indicates qualitative data from bracketing studies.   ⟶

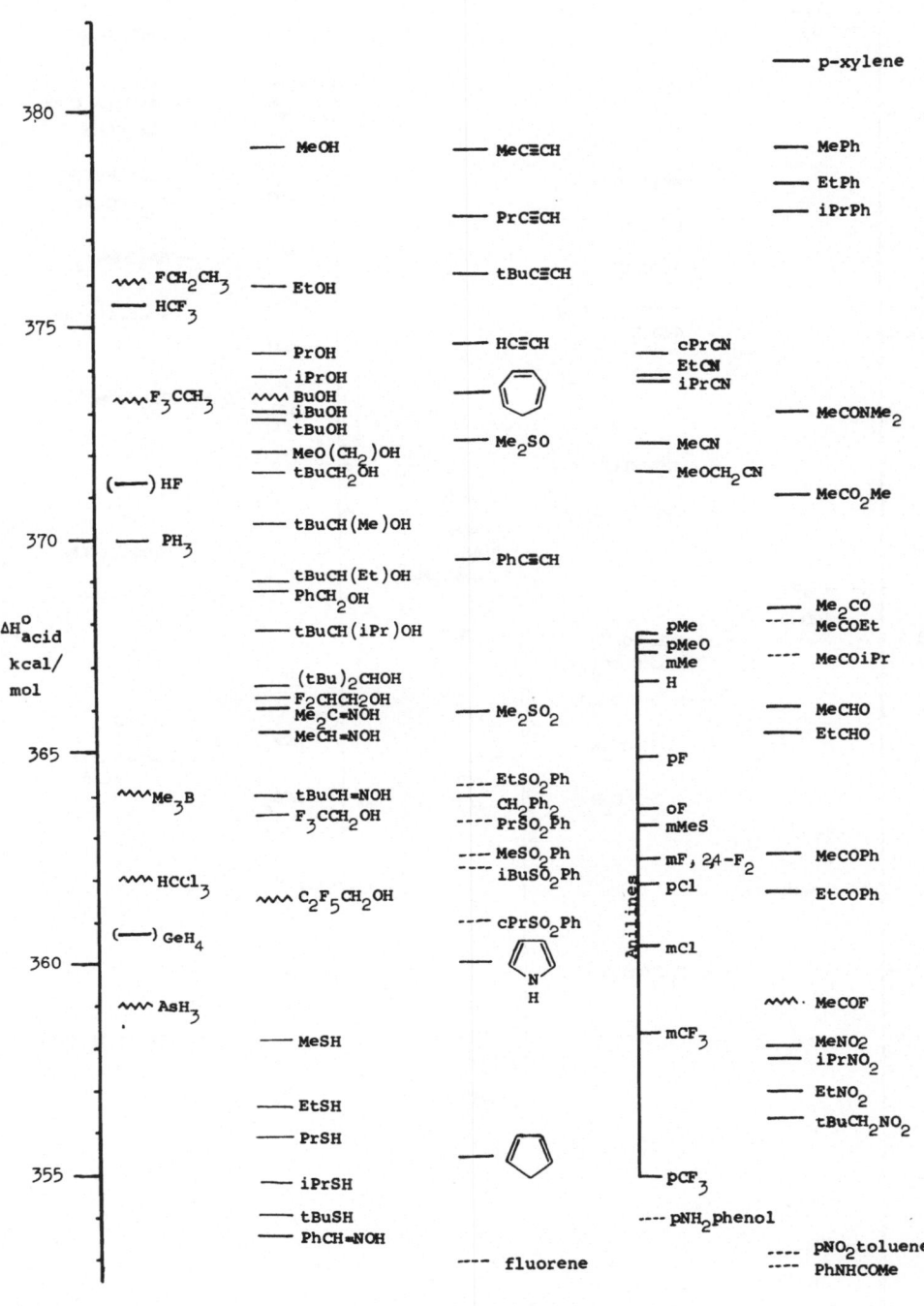

Fig. 3 (continued) →

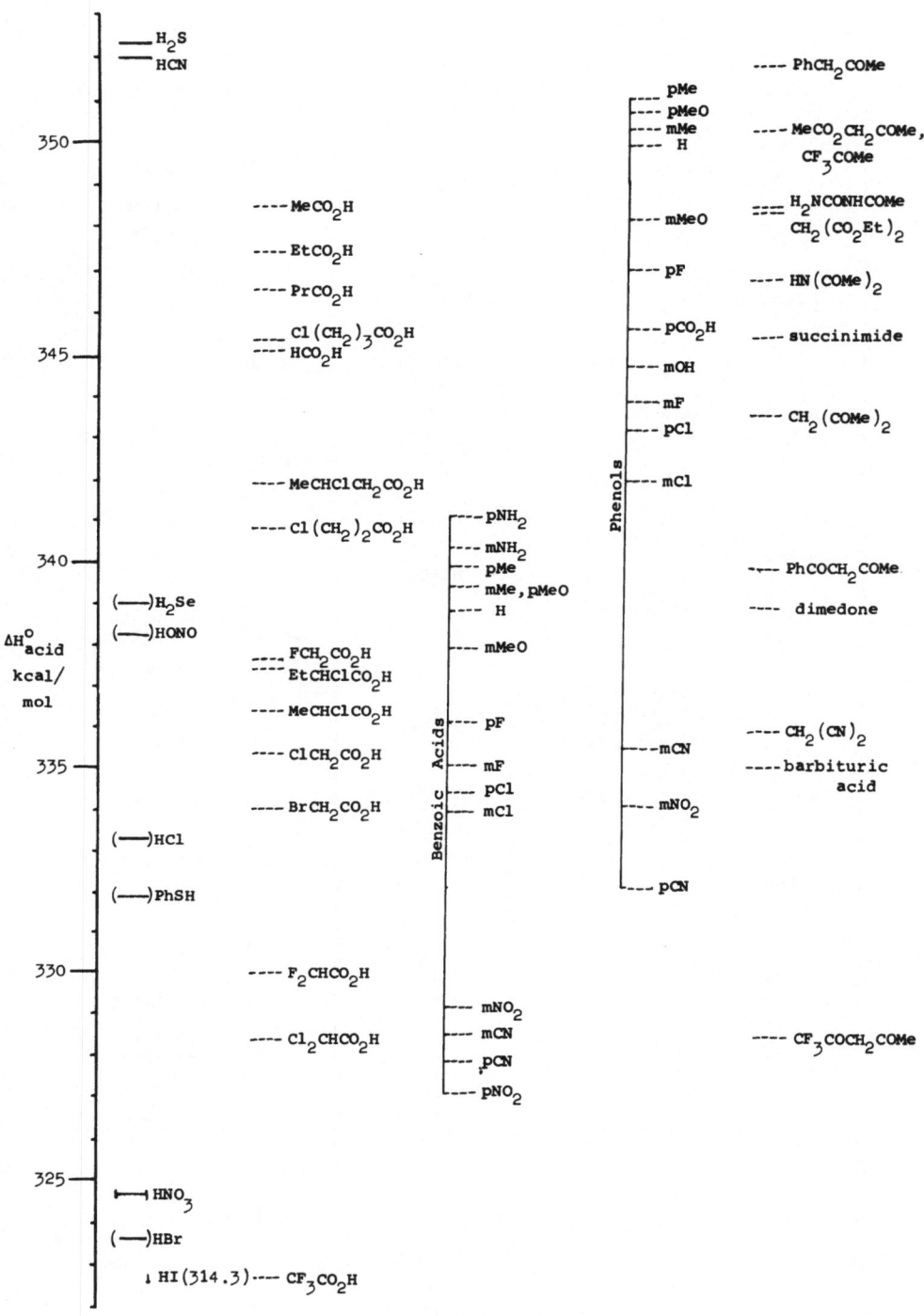

Fig. 3 (continued)

where $\delta\Delta H^{\circ}_{acid}$ and $\delta\Delta S^{\circ}_{acid}$ refer to the enthalpy change and estimated entropy change for reaction 2. There is some uncertainty in
the temperature T of the system due to temperature gradients in the
ICR cell. By connecting thermocouples to various parts of the ICR
cell, we have observed increases in the temperature of the cell as
the current to the electron beam filament is increased. The
temperature increases approximately as the square of the current
supplied to the filament. The plates of the ICR cell are about
$20C^{\circ}$ warmer than the walls of the vacuum system which are at
ambient 300 K. Temperature gradients of this magnitude are hardly
noticeable in calculating $\delta\Delta G^{\circ}_{acid} = -RT \ln K$ when K is in the
directly measureable range of 1 to 200. However, the effect is
significant (about 7%) over the 30 kcal/mol range from methanol to
phenol. For this reason, the temperature of the neutral molecules
was taken to be the average of the measured temperatures for the
two side plates of the ICR cell. Generally, T was about 320 ± 5 K.
The measured relative acidities are anchored to HF at 371.3 kcal/mol.
The choice of HF to anchor the scale is due to the low uncertainty
in ΔH°_{acid} and ΔS°_{acid}, and to its central position in the scale.

   Kebarle and coworkers have constructed a ΔH°_{acid} scale from
dimethylsulfoxide to CF_3CO_2H in a similar way using data from HPMS at
500 K and 600 K.[3] The anchor point for their scale is HCl at
333.6 kcal/mol. Appropriate ΔC°_p corrections were used to convert
ΔH°_{acid} (600 K) to ΔH°_{acid} (300 K).

   Also included in Fig. 3 are values of ΔH°_{acid} calculated using
the thermochemical cycle shown below.

Scheme III:

$$AH \quad = A \; + H \; , \quad \Delta H^{\circ}_{298} = DH^{\circ}(A-H) \tag{16}$$

$$\underline{A + H = A^- + H^+}, \quad \Delta H^{\circ}_{298} \cong \Delta H^{\circ}_0 = IP(H) - EA(A) \tag{17}$$

$$AH \quad = A^- + H^+, \quad \Delta H^{\circ}_{298} = \Delta H^{\circ}_{acid} \tag{18}$$

$$\Delta H^{\circ}_{acid} = DH^{\circ}(A-H) + IP(H) - EA(A) \tag{19}$$

Since the ionization potential of hydrogen, IP(H), is 313.5 kcal/mol[1]
and a constant for all acids, the enthalpy change ΔH°_{acid} is determined by the difference between the bond strength, $DH^{\circ}(A-H)$, and the
electron affinity of the radical, EA(A). A weak A-H bond and/or a
high electron affinity of the radical A causes molecules to be
strong acids. A major limitation of this approach in the past has
been the paucity of reliable EA values, but great progress has been
made in the last five years due largely to the development of new
electron photodetachment and negative ion photoelectron spectroscopy

experiments.[11]   Another problem encountered with calculations of
this type is that bond dissociation energies, $DH^{\circ}(A-H)$, are usually
reported at 298K, whereas electron affinities and ionization
potentials obtained from spectroscopic measurements refer to
enthalpies at 0K.   We find, however, that ΔC_p for the electron
transfer reaction 17 is generally quite small due to the structural
similarity of the reactants and products.   Thus, for the reactions

$$F\cdot \ + H\cdot \ = H^+ + F^- \tag{20}$$

$$OH\cdot + H\cdot \ = H^+ + OH^- \tag{21}$$

available data indicate $\Delta H^{\circ}_0 - \Delta H^{\circ}_{298}$ is 0.12 kcal/mol and 0.16
kcal/mol, respectively.[12]

There is a region of overlap from ΔH°_{acid} 374 kcal/mol to
349 kcal/mol in the HPMS data and the ICR data.   Comparison of
data from the two methods is shown in Table I.   Generally the
agreement is excellent considering the vastly different pressures,
temperatures, and reaction times for the two techniques.   The data
deviate most for weak carbon acids like acetone and dimethyl
sulfoxide.   These compounds are difficult to work with because
their proton transfer rates are very slow.[13]   In the ICR studies,
carbon acids were always measured relative to heteroatom acids
like anilines to insure that the proton transfer rates were fast
enough for a reliable determination.

Table II shows a comparison between experimental data and
ΔH°_{acid} calculated from Scheme III.   The bond strengths and electron
affinities were taken from the literature, and the experimental
data is from ICR measurements.   Once again the agreement over a
29 kcal/mol range is excellent within the error limits of the
measurements.

These comparisons are convincing evidence that reliable
thermochemical data for negative ions is available HPMS and ICR
measurements of proton transfer equilibria.   Once ΔH°_{acid} is
available for the process $AH = H^+ + A^-$, the data can be used to
calculate heats of formation of gaseous negative ions, electron
affinities for radicals when the A-H bond strength is known, or
conversely, bond strengths when the electron affinity of the
radical is known.[14]   Heats of formation of negative ions can also
be used in calculating hydride affinities of unsaturated compounds,
as in

$$CH_3O^- = CH_2{=}O + H^- \tag{22}$$

and methyl cation affinities, as in

Table I. Comparison of ΔH°_{acid} Values from HPMS and ICR

Acid	ΔH°_{300}(HPMS)[a]	ΔH°_{300}(ICR)[b]	$\delta \Delta H^{\circ}$[c]
CH_3SOCH_3	374.6	372.3	2.3
CH_3CN	373.5	372.2	1.3
CH_3COCH_3	370.0	368.0	2.0
CH_3CHO	367.0	365.9	1.1
$MeCH_2CHO$	366.4	365.3	1.1
$CH_3SO_2CH_3$	366.4	365.9	0.5
Ph_2CH_2	362.7	363.9	-1.2
CH_3COPh	362.8	362.5	0.3
$MeCH_2COPh$	361.2	361.6	-0.4
pyrrole	359.2	360.0	-0.8
CH_3NO_2	357.6	358.0	-0.4
$MeCH_2NO_2$	357.3	356.9	0.4
Me_2CHNO_2	356.6.	357.7	-1.1
cyclopentadiene	355.5	355.3	0.2
H_2S	352.0	352.3	-0.3
$PhCH_2COMe$	351.8	351.8	0.0
$PhCH_2CN$	351.9	352.1	-0.2
$pMeC_6H_4OH$	351.7	351.5	0.2
PhOH	349.8	350.4	-0.6

a. reference 3

b. reference 9

c. $\delta \Delta H^{\circ} = \Delta H^{\circ}_{acid}$ (HPMS) $- \Delta H^{\circ}_{acid}$ (ICR)

Table II. Comparison of Calculated and Equilibrium

Values of ΔH°_{298} for HA $= A^- + H^+$

Acid	DH°(A-H)	E.A. (A·)	ΔH°_{acid}	ΔH°_{exp}	$\delta\Delta H^\circ$
MeOH	104 ± 1	36.2 ± 0.7	381.4	379.1	2.3
PhCH$_3$	85 ± 1	20.4 ± 1.5	378.2	378.9	-0.7
EtOH	104 ± 1	39.8 ± 0.7	377.8	375.9	1.9
n-PrOH	104 ± 1	41.2 ± 0.7	376.4	374.3	2.1
t-BuOH	105.1 ± 1	43.8 ± 0.7	374.9	372.9	2.0
CH$_3$CN	92.9 ± 2.5	34.8 ± 0.4	371.7	372.2	-0.5
HF	136.1 ± 0.6	78.4 ± 0.05	371.3	(371.3)	(0.0)
PH$_3$	83.9 ± 3	28.8 ± 0.7	368.6	370.0	-1.4
CH$_3\overset{\text{O}}{\overset{\|}{C}}CH_3$	98.0 ± 2.6	45.3 ± 1.0	366.3	368.0	-1.7
MeSH	87 ± 2	44.3 ± 0.7	356.3	358.2	-1.9
H$_2$S	91.6 ± 4	53.5 ± 0.2	351.7	352.3	-0.6
HCN	123.8 ± 2	88.1 ± 0.4	349.3	352.0	-2.7
PhOH	86.5 ± 2	54.4 ± 1.4	345.7	350.4	-4.7

$\Delta H^\circ_{acid} = DH - E.A. + 313.6$

$$CH_3OH = CH_3^+ + OH^- \tag{23}$$

The data can also be used as shown in Table III to evaluate the relative solvation enthalpies of ions from the gas phase to solution.[15]

Table III. Solvation Enthalpies for

Alkoxide Ions in Dimethylsulfoxide

GAS PHASE $\qquad$ CH_3O^- + ROH = RO^- + CH_3OH $\quad$ ΔH°_g

$\downarrow$ $\qquad$ $\downarrow$ $\qquad$ $\downarrow$ $\qquad$ $\downarrow$

DMSO SOLUTION $\quad$ $CH_3O^-_{(s)}$ + $ROH_{(s)}$ = $RO^-_{(s)}$ + $CH_3OH_{(s)}$ $\quad$ ΔH°_{DMSO}

RO^-	ΔH°_g	ΔH°_{DMSO}	ΔH_s
CH_3O^-	(0.0)	(0.0)	(0.0)
$C_2H_5O^-$	-3.2	+4.9	+7.5
$i-C_3H_7O^-$	-5.3	+6.5	+12.1
$(CH_3)_3CO^-$	-6.2	+6.8	+13.1
$(CH_3)_3CCH_2O^-$	-7.5	+5.4	+11.7
$((CH_3)_3C)_2CHO^-$	-12.6	+8.5	+18.2
OH^-	+11.7	+3.6	-10.6

ALL VALUES kcal/mole

ACKNOWLEDGEMENTS

The author would like to acknowledge the important contributions of Prof. John E. Bartmess, now at the University of Indiana, and Prof. Judith A. Scott, now at Metropolitan State College in Denver, Colorado, in performing the ICR measurements. The keen chemical insight and support of Prof. R. W. Taft at U. C. Irvine has contributed invaluably to the development of this project. Funds for construction of the ICR spectrometers were provided by the National Science Foundation.

BIBLIOGRAPHY

1. H. M. Rosenstock, K. Draxl, B. W. Steiner, and J. T. Herron, J. Phys. and Chem. Ref. Data, 6, Suppl. 1, 1977.

2. (a) J. F. Wolf, R. H. Staley, I. Koppel, M. Taagepera, R. T. McIver, Jr., J. L. Beauchamp, and R. W. Taft, J. Am. Chem. Soc., 99, 5417 (1977); (b) R. Yamdagni and P. Kebarle, J. Am. Chem. Soc., 98, 1320 (1976); and (c) R. W. Taft in "Proton Transfer Reactions," E. F. Caldin and V. Gold, Ed., Chapman and Hall, London, 1975, p. 31.

3. J. B. Cumming and P. Kebarle, Can. J. Chem., 56, 1 (1978).

4. R. T. McIver, Jr., Rev. Sci. Instrum., 49, 111 (1978) and references therein.

5. R. L. Hunter and R. T. McIver, Jr., Chem. Phys. Lett., 49, 577 (1977).

6. T. E. Sharp, J. R. Eyler, and E. Li, Int. J. Mass Spectrom. Ion Phys., 9, 421 (1972).

7. M. S. Foster and J. L. Beauchamp, Chem. Phys. Lett., 31, 482 (1975).

8. (a) S. W. Benson, "Thermochemical Kinetics," 2nd Ed., Wiley-Interscience, New York, 1976, and (b) W. F. Bailey and A. S. Monahan, J. Chem. Ed., 55, 489 (1978).

9. J. E. Bartmess and R. T. McIver, Jr. in "Gas-Phase Ion Chemistry," M. T. Bowers, Ed., Academic Press, 1979.

10. G. I. Mackay, R. S. Hemsworth, and D. K. Bohme, Can. J. Chem., 54, 1624 (1976).

11. (a) W. C. Lineberger in "Chemical and Biochemical Applications of Lasers, Vol. I," C. B. Moore, Ed., Academic Press, New York, 1974; (b) A. H. Zimmerman, K. J. Reed, and J. I. Brauman, J. Am. Chem. Soc., 99, 7203 (1977).

12. "JANAF Thermochemical Tables," D. R. Stull and H. Prophet, Eds., NSRDS-NBS 37, U. S. Government Printing Office, Washington, D. C., 1971.

13. W. E. Farneth and J. I. Brauman, J. Am. Chem. Soc., 98, 7891 (1976).

14. K. J. Reed and J. I. Brauman, J. Am. Chem. Soc., 97, 1625 (1975).

15. E. M. Arnett, L. E. Small, R. T. McIver, Jr., and J. S. Miller, J. Am. Chem. Soc., 96, 5638 (1974).

PROTON TRANSFER EQUILIBRIA IN THE GAS AND SOLUTION PHASES

Robert W. Taft

Department of Chemistry, University of California, Irvine

Irvine, CA, 92717 U.S.A.

Structural theories of organic chemistry have been heavily based for support upon measurements of proton transfer equilibria in solution--largely for -NH and -OH acids and bases in aqueous solution.[1] The first quantitative measurements of proton transfer equilibria in the gas phase by either ion cyclotron resonance, high pressure mass spectrometry, or flowing afterglow methods[2] were excitedly greeted as providing the needed means to separate intrinsic structural effects from solvation effects. However, most of the initial observations were so drastically unrelated to solution results that there developed among many chemists the notion that the results were only isolated curiosities of the gas phase--largely irrelevant to the practice of acid-base chemistry.

In this lecture, two themes are developed. First, it is shown that a number of intrinsic effects of molecular structure were correctly identified by the earlier solution measurements, since solvation effects only lead to quantitative rather than qualitative modifications in relative basicities or acidities. Further, it will be shown that many dramatic examples of reversals in acidity or basicity order in the gas phase and in solution can be readily understood (and predicted) in simplified first order terms as the consequence of specific binding of solvent molecules to cations and anions. Second, the notion will be developed that gas phase acidities and basicities can be used with great utility as standards for the understanding of, and in particular, in anticipating major new advances in the practice of acid-base reactions in solution.

INTRINSIC EFFECTS OF MOLECULAR STRUCTURE

Site of Protonation

The proton affinities of the hydrides of Groups IV, V, VI, and VII show marked effects of the atomic site of protonation on base strength[3]. The meager available solution basicities indicate that the the same basicity order holds in solution as in the gas phase for the hydrides of a given period of the periodic table. A similar relationship is found for the gas phase and solution acidities of these hydrides[4].

As the valence state electronegativity exerted upon the lone pair of electrons increases, basicity decreases and acidity increases. In addition, X-H bond energy considerations are also of importance. The markedly decreasing proton affinities[3] of CH_3OH, $H_2C=O$ and $C\equiv O$, or of CH_3NH_2, $CH_2=NH$, and $H\equiv CN$ provide another type of illustration of decreasing basicity with the increasing valence state electronegativities of O and N, respectively. Here again there are indirect indications that the same basicity orders apply also in solution.

Dramatic reversals in acid-base behavior between the gas phase and solution are associated either with comparisons between elements having both different group and period numbers or with substituent effects on acid or base strength at the same kind of protonation site. This paper is largely concerned with structural and solvent effects of the latter type.

Substituent Effects

This lecture is limited to discussion of our attempts to isolate and study three kinds of intrinsic substituent effects which may contribute to the observed effects of molecular structure on gas phase basicities (or acidities). Additional kinds of substituent effects which have been identified are discussed in the following literature citations: hybridization change in the substituent groups;[2b,5] hydrogen-bond chelation between substituent and conjugate acid[6] or base site;[4,7] steric effects;[2b,8] entropy effects[3,4,9]. A schematic representation of the effects on gas phase basicity for a series of substituents, X, involving a given kind of molecular cavity and protonation site, B, is given by the generalized proton transfer equilibrium:

$$X-BH^+ + :BH \rightleftarrows H-BH^+ + X-B \qquad (1)$$

If the above mentioned substituent effects are small or negligible, the observed standard free energy change for any such reaction (1) may be regarded as a potential composite of an inductive-field effect, I; a polarizability effect, P; a resonance or pi electron delocalization effect, R, i.e.,

$$-2.303RT\ln K_{(1)} = \delta\Delta G° \underset{\sim}{} I + P + R \tag{2}$$

The inductive-field effect (I) is visualized (in oversimplified terms, of course) as arising from the differential (between $X-BH^+$ and $H-BH^+$) in electrostatic charge-dipole stabilization or destabilization. In the point charge-dipole approximation this energy is given as:

$$E = \frac{q\mu\cos\theta}{Dr^2} \tag{3}$$

where q is the charge, μ is the dipole moment, r is the distance of separation, θ is the orientation angle, and D is the dielectric constant. Thus, for example, a large dipole moment localized in X with orientation of its positive end toward the cationic center, BH^+, will destabilize $X-BH^+$ and contribute a favorable negative term to $\delta\Delta G°$ (i.e., I = -). Dipole orientation and sign of the charge (negative, if acidities of neutral compounds are being considered) determine the sign of the I effect.

The polarizability effect (P) is similarly visualized as arising from the differential in stabilizing charge-induced dipole interaction[10]. In point charge-polarizability approximation this energy is given as:

$$E = \frac{\alpha q^2}{2Dr^4} \tag{4}$$

where q is the charge, α is the polarizability, r and D as above. A highly polarizable substituent, X, will preferentially stabilize $X-BH^+$ compared to $H-BH^+$, contributing an unfavorable positive term to $\delta\Delta G°$ (i.e., P = +). A proton transfer equilibrium will be shifted by the P effect in the direction which places the most polarizable substituent X with the charge, irrespective of the charge type (cation or anion).

The resonance effect[1](R) is visualized as a quantum mechanical binding energy contribution which is associated with molecular stabilization through pi electron delocalization between the substituent X and either or both the B: or BH^+ centers. A proton transfer equilibrium will be shifted preferentially by the R effect in the direction of the molecule or ion having the greatest stabilization by pi electron delocalization (for example, if this is $X-BH^+$, there will be an unfavorable positive R term contribution to $\delta\Delta G°$).

The R effect is made small or eliminated by considering structures in which there are extended chains of "saturated" carbon atoms intervening between X and B:/BH$^+$. It is this type of structure which will be considered first, to demonstrate I and P effects and their separability.

Polarizability Effects

The polarizability effect on proton transfere equilibria was not identified until the measurement of gas phase equilibria[10a]. The other substituent effects listed above were all first identified by acid-base measurements in solution,[1] although gas phase results offer clearer more conclusive evidence[2b,c].

The gas phase basicities of n-bases with the two straight chain substituents, Et and n-Pr, relative to that for their corresponding dimethyl derivatives, provide an excellent illustration of the dominant polarizability effects associated with alkyl groups (cf. Table 1). There are marked increases in basicity in the gas phase due to these simple alkyl groups, which are in stark contrast with the very small effects observed in aqueous solutions (also given if available in Table 1). Investigators have shown that the effects of both alkyl and isolated unsaturated substituents on gas phase basicities and acidities depend predominantly upon the polarizability effects associated with replacing hydrogen by carbon atoms at positions close to the charge centers[2b,11]. The results in Table 1 show that the polarizability effects are largest in the ether series since oxygen is least able to accommodate the positive charge, and are the smallest in the methyl phosphine series where phosphorus is best able to accommodate the charge. It may be inferred from these results that some charge is relayed to the alkyl groups and that the magnitude of this transfer of charge influences the observed polarizability effects. This same conclusion is also indicated by the non-additive effects of corresponding alkyl substituents in the basicities of primary, secondary, and tertiary amines[2b].

Inductive Field Effects

The I effect is made predominant by taking structures which minimize both P and R effects. The inductive effects of distant -CF$_3$ and HC≡C- substituents are large and very predominant in the gas phase proton transfer equilibrium results given in Table 2. Isolation of I from P effects depends here upon the use of standards of comparison which have the same number of carbon atoms in approximately equivalent structures.

Table 1. Effects of Lengthening the Alkyl Chain on Basicities of R_2Y Relative to Me_2Y n-Bases

R Y	>O (g)	>O (aq)	>NH (g)	>NH (aq)	>S (g)	>S (aq)	>PMe (g)	>PMe (aq)
Me	(0.0)	(0.0)	(0.0)	(0.0)	(0.0)	(0.0)	(0.0)	----
Et	7.3	0.1	4.8	0.3	5.0	0.2[d]	3.6[c]	----
n-Pr	9.2[b]	---	6.8	0.3	6.4[b]	---	5.4[c]	----

a) results are $\delta\Delta G°$ in kcal/mole. Positive value denotes greater base strength. The results unless otherwise indicated are from references 2b and 3.
b) J. L. M. Abboud and R. W. Taft, unpublished results.
c) P. Gately, K. Marsi, and R. W. Taft, unpublished results.
d) G. Scorrano, private communication.

Table 2. Predominant Inductive Effects of CF_3 and $HC\equiv C-$ Substituents[a]

$$C_2H_5NMe_2 + CF_3CH_2N(Me)_2H^+ \rightleftarrows C_2H_5N(Me)_2H^+ + CF_3CH_2NMe_2 \qquad K=10^{8.8b}$$

$$C_2H_5OEt + CF_3CH_2O(Et)H^+ \rightleftarrows C_2H_5O(Et)H^+ + CF_3CH_2OEt \qquad K=10^{9.2c}$$

$$C_2H_5SEt + CF_3CH_2S(Et)H^+ \rightleftarrows C_2H_5S(Et)H^+ + CF_3CH_2SEt \qquad K=10^{8.7c}$$

$$n-C_3H_7NH_2 + HC\equiv CCH_2NH_3^+ \rightleftarrows n-C_3H_7NH_3^+ + HC\equiv CCH_2NH_2 \qquad K=10^{4.8d}$$

a) Results are given as equilibrium constants at room temperature. P effects are minimized by using a reference base in each equilibrium which has the same number of carbon atoms in roughly equivalent structures.
b) Reference 12
c) J. L. M. Abboud and R. W. Taft, unpublished results.
d) Reference 2b.

Using this same approach with an extended series of N,N-dimethyl substituted methyl amines, the difference in the gas phase basicity between $X-CH_2NMe_2$ and $X'-CH_2NMe_2$ plots linearly[13] vs $\Delta\sigma_I = \sigma_{I(X)} - \sigma_{I(X')}$, $(\delta\Delta G°_{(X)} - \delta\Delta G°_{(X')} = -26.7 \Delta\sigma_I)$. The empirical inductive substitutent constants $\sigma_{I(X)}$ are from various solution

measurements[14]. The substitutent X' is selected to give about the
same P effect as X at the distance of separation involved.

Separation of P and I Effects

Combining appropriate gas phase equilibria provides an alternate
but powerful tool for eliminating certain effects in order to isolate others. For the series of dimethylamines, $RNMe_2$, the gas phase
basicities increase markedly in the sequence R = CF_3 < CF_3CH_2 <
H < CH_3 < C_2H_5, covering a range of 33 kcal/mole[12]. These substituent effects are largely a combination of I and P effects. By use
of the adiabatic ionization ptentials of these amines, the following
H-atom transfer equilibria have been determined[12]:

$$\overset{.+}{R-NMe_2} + H-\overset{\overset{H}{|}}{N}Me_2^+ \rightleftarrows R-\overset{\overset{H}{|}}{N}Me_2^+ + \overset{.+}{H-NMe_2} \qquad (5)$$

Since this equilibrium involves four ions of crudely similar structure, charge-dipole interactions should cancel, approximately. This
is confirmed by the results, which show only relatively small substi-
tuent effects ($\delta\Delta G^\circ_{(5)}$ = 3-5 kcal/mole for R = CF_3, CF_3CH_2, CH_3, and
C_2H_5, indicating slightly greater stabilization of the cation radical
than the corresponding ammonium ions by P and R effects).

The gas phase basicities (eq. 6) and acidities (eq. 7) of alcohols, ROH, derived by substitution of the H's of the CH_3 group of
methanol, have been combined to provide a separation of I and P
effects[15].

$$ROH_2^+ + CH_3OH \rightleftarrows CH_3OH_2^+ + ROH; \quad \delta\Delta G^\circ_{(6)} \underset{\sim}{\sim} I + P \qquad (6)$$

$$RO^- + CH_3OH \rightleftarrows CH_3O^- + ROH; \quad \delta\Delta G^\circ_{(7)} \underset{\sim}{\sim} -I + P \qquad (7)$$

Equation (4) indicates that charge-induced dipole interaction will
stabilize both ROH^+ relative to $CH_3OH_2^+$ as well as RO^- relative to
CH_3O^- (the P effect)[10], whereas eq. (3) indicates, for example, that
a charge-dipole interaction which stabilizes RO^- relative to CH_3O^-
(-I effect) will destabilize ROH_2^+ relative to $CH_3OH_2^+$[12, 16]. In
view of the nature of reactions (6) and (7), the simplifying assump-
tion was made (as expressed in eqns. (6) and (7) that $P_{(6)} \underset{\sim}{\sim} P_{(7)}$ and
$I_{(6)} \underset{\sim}{\sim} I_{(7)}$. This assumption is adequate for present pruposes
although proportionality rather than precise equality likely describes
these relationships more accurately.

Subtracting (7) from (6) gives a (hypothetical) double proton
transfer equilibrium between four ions, eq. (8), for which the P
effect has been minimized or eliminated, i.e.,

$$ROH_2^+ + CH_3O^- \overset{\rightarrow}{\leftarrow} RO^- + CH_3OH_2^+; \quad \Delta G^\circ_{(8)} \simeq 2I \tag{8}$$

From eqns. (6) and (8) one obtains:

$$I = \delta\Delta G^\circ_{(8)}/2 \text{ and } P = \delta\Delta G^\circ_{(6)} - I \tag{9}$$

Values of $\delta\Delta G^\circ_{(6)}$ and $\delta\Delta G^\circ_{(7)}$ from our combined studies with Profs.
McIver and Hehre are given in Table 3, together with the I and P
values obtained from eqn. (9). The I values obtained are in the
"classical" inductive order and correlate very well with the empirical inductive-field substituent parameters, σ_I, i.e., $I = \rho_I\sigma_I$,
where $\rho_I = -70$ kcal/σ_I.

   For alkyl substitutents, the "electron-releasing" inductive
effects of R relative to CH_3 are substantial, i.e., up to 2.3 kcal/
mole for t-Bu, but nonetheless, these are 3 to 7 times smaller than
for the corresponding predominant polarizability (P) effects.
I values are nearly additive in the two series: Me, Et, i-Pr, t-Bu
and in Et, CH_2CHF_2, CH_2CF_3, whereas P values show that saturation
occurs in the former series (increments of 3.8, 2.7 and 2.0 kcal)
and in the latter series P is approximately constant. The introduction of CH_3 substituents for H on the β-carbon gives I values which
are the same within the combined experimental errors, whereas P
values increase substantially. Evidently, the conformations with
β-alkyl substituents "bent around" are favored to obtain optimal
stabilization by the polarizability effect (giving, for example,
$P_{t-Bu} \simeq P_{neopent}$). P values are well within the order of magnitude
calculations of Aue, Webb, and Bowers[17], which were based upon the
relative polarizabilities of CH_3 and H substituents and assumed distance relationships. Equation (8), $\delta\Delta G^\circ_{(8)} \simeq 2I$, is not applicable
to the H-substituent on oxygen, i.e., for H_2O, since there is hyperconjugative stabilization of alkoxides compared to hydroxide[18]. Small
differential stabilizations of the alkoxide ions due to differing
C-H and C-C hyperconjugative interactions probably contribute to
$\delta\Delta G^\circ_{(7)}$ values, but these and P contributions have evidently nearly
cancelled in the $\delta\Delta G^\circ_{(8)}$ values.

   A theoretical treatment of inductive-field effects at the level
of approximation of STO-3G ab initio molecular orbital calculations
has been carried out[19]. Hyperconjugative and polarizability effects
on proton affinities were minimized and chelation effects excluded
through the use of distant substituents in β-substituted ethyl amines.
The calculations were carried out with the amines and their conjugate
acids held in a fully extended all trans conformation to give pre-
dominantly the I effects, i.e., it was found that $\delta\Delta E^\circ$(calc) $\simeq I \simeq$
$\rho_I\sigma_I$.

Table 3.   Evaluation of Inductive and Polarizability Effect
Contributions to $\delta\Delta G°$ Values for Reactions (6) and
(7)

R	$\delta\Delta G°_{(6)}$ [a]	$\delta\Delta G°_{(7)}$ [a]	I [b]	P [c]
t-Bu	10.8	6.2	2.3	8.5
i-Pr	8.0	5.1	1.5	6.5
neo-Pent	9.1	7.3	0.9	8.2
i-Bu	8.0	6.1	1.0	7.0
n-Pr	6.5	4.8	0.9	5.6
Et	4.5	3.1	0.7	3.8
Me	(0.0)	(0.0)	(0.0)	(0.0)
$C_6H_5CH_2$	4.8	10.3	-2.8	7.6
$MeO(CH_2)_2$	1.0	7.0	-3.0	4.0
F_2CHCH_2	-4.3	13.4	-8.9	4.6
CF_3CH_2	-10.0	16.6	-13.3	3.3

a) From reference 15.   In kcal-mole^{-1}, precision $\pm$ 0.2 kcal-mole^{-1}.
b) $I = (1/2)(\delta\Delta G°_{(8)})$, kcal-mole^{-1}
c) $P = \delta\Delta G°_{(6)} - I$, kcal-mole^{-1}.

Resonance Effects

In solution chemistry, resonance stabilization is commonly the
predominant structural driving force for a reaction and this concept
is widely used by organic chemists as an operating basis. Gas phase
proton transfer equilibria offer a striking contrast in which the R-
effects are frequently found to be secondary to a predominant combination of I- and P-effects. Accordingly, it is necessary with gas
phase basicities or acidities of conjugated compounds to use standards of comparison which have the same number of carbon atoms, as
well as similar substituents and structures, to obtain measures of
predominant R-effects. Reactions 10-12 are three such examples of
proton transfer equilibria results[2b,c,4,20]:

$$\text{(structures)} \quad ; \quad K = 10^{11.8} \quad (10)$$

$$\text{(structures)} \quad ; \quad (11)$$

$$K = 10^{9.5}$$

$$\text{(structures)} \quad (12)$$

$$K = 10^{3.2}$$

In reaction (1) the N-methyl imidazolium ion is strongly resonance stabilized as represented by the resonance structures:

$$\text{(resonance structures)}$$

Similar resonance structures for N-methyl imidazole involve a less important charge separated structure,

$$\text{(structure)}$$

so that the greater resonance stabilization of the former is the predominant contribution determining the very large room temperature equilibrium constant. Similarly, in reaction (11) there is preferred resonance stabilization of the coplanar dibenzocyclopentadienide carbanion (fluorenide ion) which provides the predominant driving force for the indicated equilibrium constant. Reaction (12) provides an example of a reaction driven by preferential resonance stabilization of the neutral N-phenyl piperidine[2b]. The equilibrium constant obtained provides a good estimate of the extra resonance stabilization energy (4.4 kcal/mole) of this base since the two ions are of closely similar energy and resonance interaction of the lone pair of electrons in the benzoquinuclidine is totally inhibited sterically[21].

Isolation of R from P-effects provides a greater challenge than does the separation of I and P effects. This separation also provides

uncertain conceputal difficulties, but the solvent effects on certain proton transfer equilibria appear to provide a tool permitting
such a separation (cf. subsequent discussion).

Solvent Effects on Proton Transfer Equilibria

A complete thermodynamic analysis of solvent effects has been
carried out for many bases in water and fluorosulfuric acid media[22].
Although these results reveal many of the major features of struc-
ture-solvation effects, the data are relatively meager with respect
to a wide variation in structural types. Consequently, the presently
most generally useful first order analysis of medium effects on
proton transfer equilibria involves the evaluation obtained[23] from
the ratio of the equilibrium constant for the formal reaction in
solution, K_s, to that for the corresponding reaction in the gas
phase, i.e., $K(s)/K(g)$. The solvent effect on the standard free
energy change is given as:

$$\delta_s \Delta G° = -RT\ln(K(s)/K(G)). \tag{13}$$

Before proceeding to these results, it proves useful to their
understanding to classify solvation according to two types (a) nonspecific or general solvation; (b) specific solvation, involving the
binding of solvent molecules to the solute. Examples of solutes
falling into category (a) are unsaturated hydrocarbon acids and
bases (neutral molecules) and cations or anions in which the charge
is not strongly concentrated in solvent accessible space. Examples
of type (b) solutes include neutral molecules bound (stabilized)
by specific hydrogen bonds to the solvent (e.g., -OH---:B) or cations
or anions with strongly localized charges which are stabilized by
specific binding of solvent molecules through interactions pictured
as ion-dipole or hydrogen bonding in nature.

In the absence of specific solvation, coulombic ion solvation,
van der Waals dispersion force interactions and solvent structure
making and breaking interactions will tend to cancel between the
right and lefthand sides of a proton transfer equilibrium[22,24].
Thus, $K(s)/K(g) \rightarrow 1$ or $\delta_s \Delta G° \rightarrow 0$ for any two non-specifically solvated acid-base pairs ($B-BH^+$ or $HA-A^-$). On the other hand, if one
uses an acid-base pair of the nonspecifically solvated type as a
standard toward another acid-base pair which are specifically solvated, then (because specific ion solvation is generally much greater
than that for neutral molecules) $K(s)/K(g)$ will be a large number.
The measured $\delta_s \Delta G°$ gives then the standard free energy of solvation
of the specifically solvated ion relative to that for its free acid
or base form.

Solvent Effects on Basicities

Figure 1 illustrates both of the above types of behavior. In this plot, gas phase basicity is plotted (ordinate) vs corresponding aqueous solution basicity (abscissa), using hexamethylbenzene as the reference base (the generalized proton transfer reaction is indicated in Fig. 1). The closed circle points in Fig. 1 are those for a series of nonspecifically solvated unsaturated hydrocarbon-charge delocalized carbocation pairs[25]. A range of ~ 25 kcal-mole^{-1} in basicity is covered with near unit slope, i.e., there is little or no solvent effect of water on these proton transfer equilibria even though there is appreciable variation in the size of, the shape of, and the number of carbon atoms (8 to 17) in the carbocations.

The open circle points in Fig. 1 are those for BH^+ ions which are known to specifically bind water molecules in the gas phase[2c] or may be inferred to do so. There is a milky-way of open circle points which accumulate in Fig. 1 to the right of the line defined by the nonspecifically solvated standards (much of the available data is not shown in Fig. 1 in order to prevent overcrowding). The horizontal deviations of the open circle points from the line in Fig. 1 provide a wealth of data on specific solvation effects. The magnitude of the horizontal deviation increases (i.e., apparent basicity in water increases) as the solvent accessible protons in BH^+ increase in number and in their localization of the positive charge. Protonic centers bound to strongly electronegative atoms tend to exert stronger electric field strengths and consequently bind more strongly lone-pair electrons of solvent molecules, as represented formally by a H-bond structure, e.g., $BH^+\dots(OH_2)_n$. Table 4 lists a number of typical numerical values of specific solvent effects in water (obtained as the horizontal deviation in Fig. 1) which illustrate this relationship. The solvent effect for the most highly specifically solvated ion, H_3O^+, indicates that the inherent acidity is reduced by $\sim 10^{35}$ due to the hydration. The solvent effects of Table 4 parallel the available data for the gas phase binding energies for the attachment of one water molecule to each protonic site of the cation[2c], $\Delta H_n^{\circ}(g)$, as shown by the figures given (n is the total number of water molecules discociated from the cluster ion).

There are several reasons for believing that the specific solvent effects on proton transfer equilibria are associated with the attachment of a relatively small number of solvent molecules. First, the attachments compete with self-association of the solvent. Mass law action (the high solvent concentration) favors the latter ($nHS \rightleftarrows (HS)_n$) compared to equilibria, e.g., $BH^+ + HS \rightleftarrows BH^+\dots SH$ or $BH^+ + (HS)_n \rightleftarrows BH^+\dots(\overline{SH})_{n-1} + HS$. Second, ionic solvation, in particular, is accompanied by substantial loss of entropy. Third, gas phase attachement energies are highly specific to the cation structure

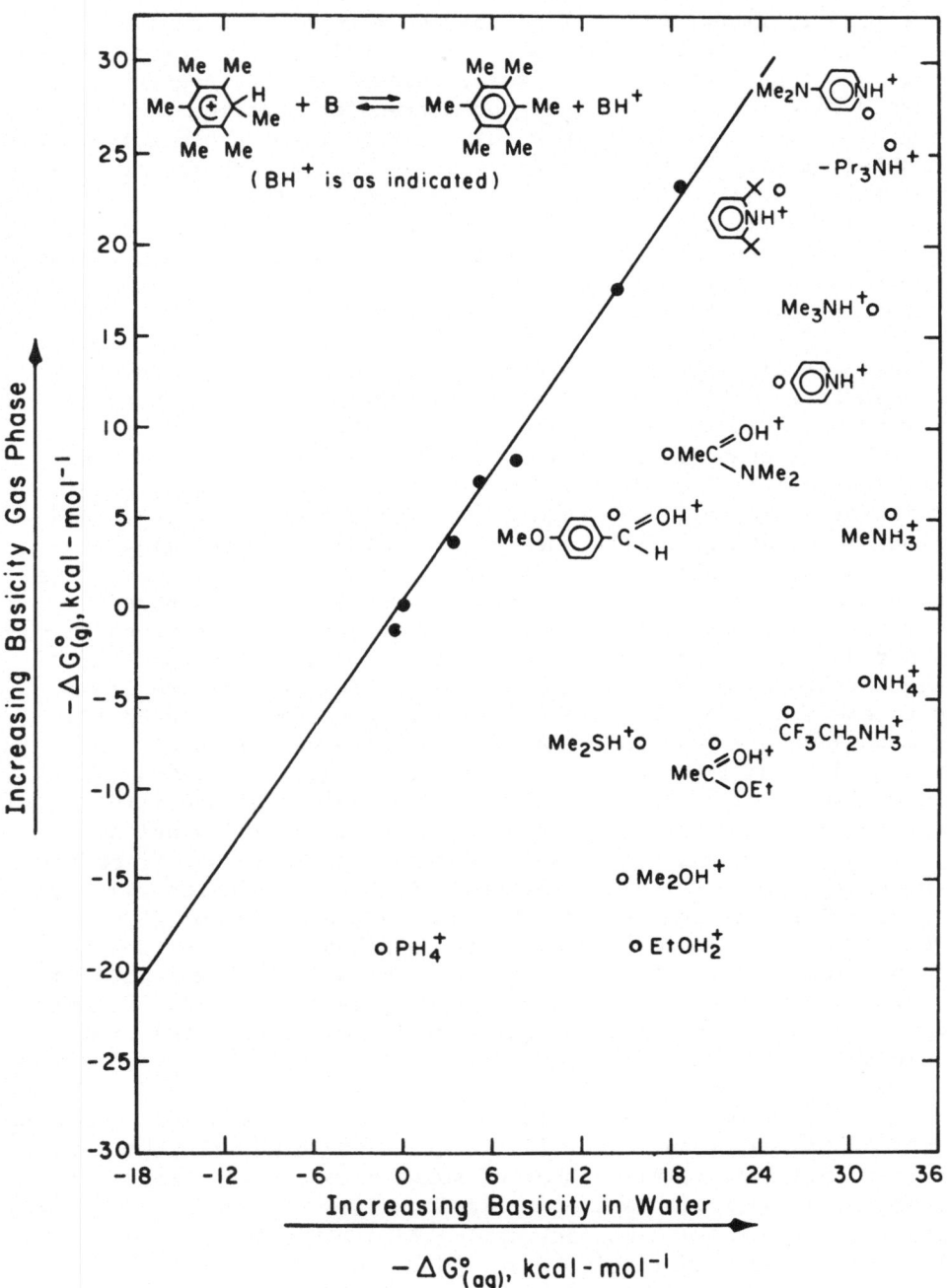

FIGURE 1.   Gas phase basicity versus basicity in water.

Table 4. Specific Solvent Effects in Aqueous Solution. The Effect of Increasing Protonic Charge Localization on Increasing Apparent Solution Basicity.

BH^+	$\delta_{aq}\Delta G°$, kcal-mole^{-1}	$\Delta H°_{n(g)}$ [a]	n
OH_3^+	46.1	70.	3
NH_4^+	33.9	57.	4
$EtOH_2^+$	31.2	43.	2
Me_2OH^+	27.0	17.	1
$C_6H_5C{\overset{OH^+}{\underset{H}{}}}$	18.7	---	---
Me_3NH^+	17.7	14. [b]	1
(pyridine) NH^+	16.1	15. [c]	1
$MeO{-}C_6H_4{-}C{\overset{OH^+}{\underset{H}{}}}$	10.5	---	---
$Me_2N{-}C_6H_4{-}NH^+$	7.2	12.3 [c]	1
(hindered pyridine) NH^+	5.7	----	---

a) Values from reference 2c unless otherwise cited
b) Cited in reference 22.
c) W. R. Davidson, J. Sunner and P. Kebarle, private communication.

for the binding of a single water molecule (for example, H^+, 170 kcal; CH_3^+, 66 kcal; Li^+, 34 kcal; H_3O^+, 32 kcal; OH^-, 25 kcal; F^-, 23 kcal; Pb^+, 22 kcal; NH_4^+, 17 kcal; Cs^+, 14 kcal), but the binding energy of a fifth water molecule to the tetrahydrates of these ions is small and very nearly nonspecific (ave value = 12 $\pm$ 2 kcal)[2c]. These considerations suggest that unless the binding energy of a solvent molecule to an ion is greater than 10-15 kcal/mole, there will be very little contribution from such binding to the specific medium effect on a proton transfer equilibrium. A somewhat lower limiting value of 3-5 kcal-mole^{-1} probably applies for specific binding of solvent to neutral solutes.

The specific binding of solvent molecules (or counterions) to BH^+ ions in solution can lead to marked reversals in basicity compared to the gas phase. A frequent cause of this reversal is the loss of stabilization of the specifically solvated ion by the P-effect. For example, by conventional arguments, benzylamine is thought to be a weaker base than methyl amine because of the "electron-withdrawing" I effect of the C_6H_5 substituent:

$$\text{\raisebox{0pt}{\Large $\bigcirc$}}\!-\!CH_2NH_3[(OH_2)_n]_3^+{}_{(aq)} + CH_3NH_2{}_{(aq)} \; \rightleftarrows \qquad\qquad (14)$$

$$CH_3NH_3[(OH_2)_n]_3^+ + \text{\raisebox{0pt}{\Large $\bigcirc$}}\!-\!CH_2NH_2{}_{(aq)}$$

In aqueous solution at $25°$, $K = 10^{+1.3}$. For the same formal reaction in CH_3CN solution[26] at $25°$, $K = 10^{+1.6}$ (CH_3CN molecules replace H_2O molecules as H-bond acceptors). However, for the simple proton transfer reaction in the gas phase[2b] at $25°$, $K = 10^{-2.3}$. This result is expected, as discussed earlier, by the preferential stabilization of the benzylammonium ion by charge-induced dipole stabilization. This behavior may be understood in simplifed terms by equations (3) and (4). Specific binding of solvent molecules is accompanied by dispersal of charge to the H-bonded solvent, in effect increasing the distance r in these equations. There is a much greater decrease in the P-effect than the I-effect since the former falls-off with r^4, the latter with only r^2. The predominant P-effect in the gas phase proton transfer equilibrium, and the reverse predominant I-effect in the corresponding solution equilibria are thus accounted for. This conclusion is strongly supported by correlations of substituent effects on reaction (15) with the I and P values of Table 3.

$$RNH_3^+ + CH_3NH_2 \; \underset{\longleftarrow}{\longrightarrow} \; RNH_2 + CH_3NH_3^+ \qquad\qquad (15)$$

$$\delta\Delta G°_{(g)} = (.92)I + (.67)P \quad R = .998, \; n = 10$$

$$\delta\Delta G°_{(aq)} = (.56)I - .4 \qquad R = .975, \; n = 10$$

$$\delta\Delta H°_{(aq)} = (.37)I \qquad\qquad R = .980, \; n = 10$$

In solution, the dependence on I values is reduced by a factor of 2-3 compared to the gas phase. The dependence on the P values, which is the predominant term for alkyl substituents in the gas phase[10], is reduced to such an extent in solution as to be negligible[15].

It is important to recognize that inversions in basicity order between the gas phase and solution due to a pronounced attenuation in the P-effect in solution only occur when there is charge dispersal from the ion to H-bonded solvent molecules (specific solvation). In Fig. 1, the unsaturated hydrocarbon-carbocation standards (closed circle points) must involve important P- (as well as R- and I-) effects on the basicities since the number of carbon atoms varies from 8 to 17. The lack of any important effect of water solvent on these basicities is explained in terms of the non-specific solvation ($C-H \overset{\delta+}{\ldots} (OH_2)_n$ binding being of little or no importance) so that stabilization of the carbocations by charge-induced dipole interactions is nearly the same in water as in the gas phase[25].

Solvent Effects on Acidities

Figure 2 illustrates that similar phenomena are apparently involved for the acidities of neutral molecules. The gas phase acidities from the work of Kebarle[3c] and McIver[4] are plotted (ordinate) vs the solution acidities (abscissa) in dimethyl sulfoxide[27], DMSO (closed circle points) and in water[27a,28] (open circle points), using toluene as the reference acid. The plot serves as an updating of that presented earlier by Bordwell[29], and a somewhat modified interpretation of it will be made here.

In Fig. 2, the available data for three unsaturated hydrocarbons which give carbanions with highly delocalized charge give a line of unit slope: $C_6H_5CH_3$ ($pK_A = 42$), $(C_6H_5)_2CH_2$ ($pK_A = 32.2$) and fluorene ($pK_A = 22.6$), where pK_A's are values measured in DMSO and also estimated to apply in water. Four additional points also fall essentially on this line of unit slope:

$$C_6H_5SO_2CH_3 \qquad\qquad pK_A = 29.0$$

$$C_6H_5CH_2CN \qquad\qquad pK_A = 21.9$$

$$C_6H_5\underset{O}{\overset{\|}{C}}CH_2\underset{O}{\overset{\|}{C}}CH_3 \qquad\qquad pK_A = 12.7$$

and $\quad CH_2(CN)_2 \qquad\qquad pK_A = 11.1$

All of the points may be characterized by the structural feature that there is relatively little localization of the anion charge in lone-pair electrons of a strongly electronegative atom[27a]. The expected consequence is little or no specific binding of solvent molecules. Streitwieser has found essentially the same relative acidities for acids of this kind in cyclohexylamine solution with Cs^+ ion-pairs[30].

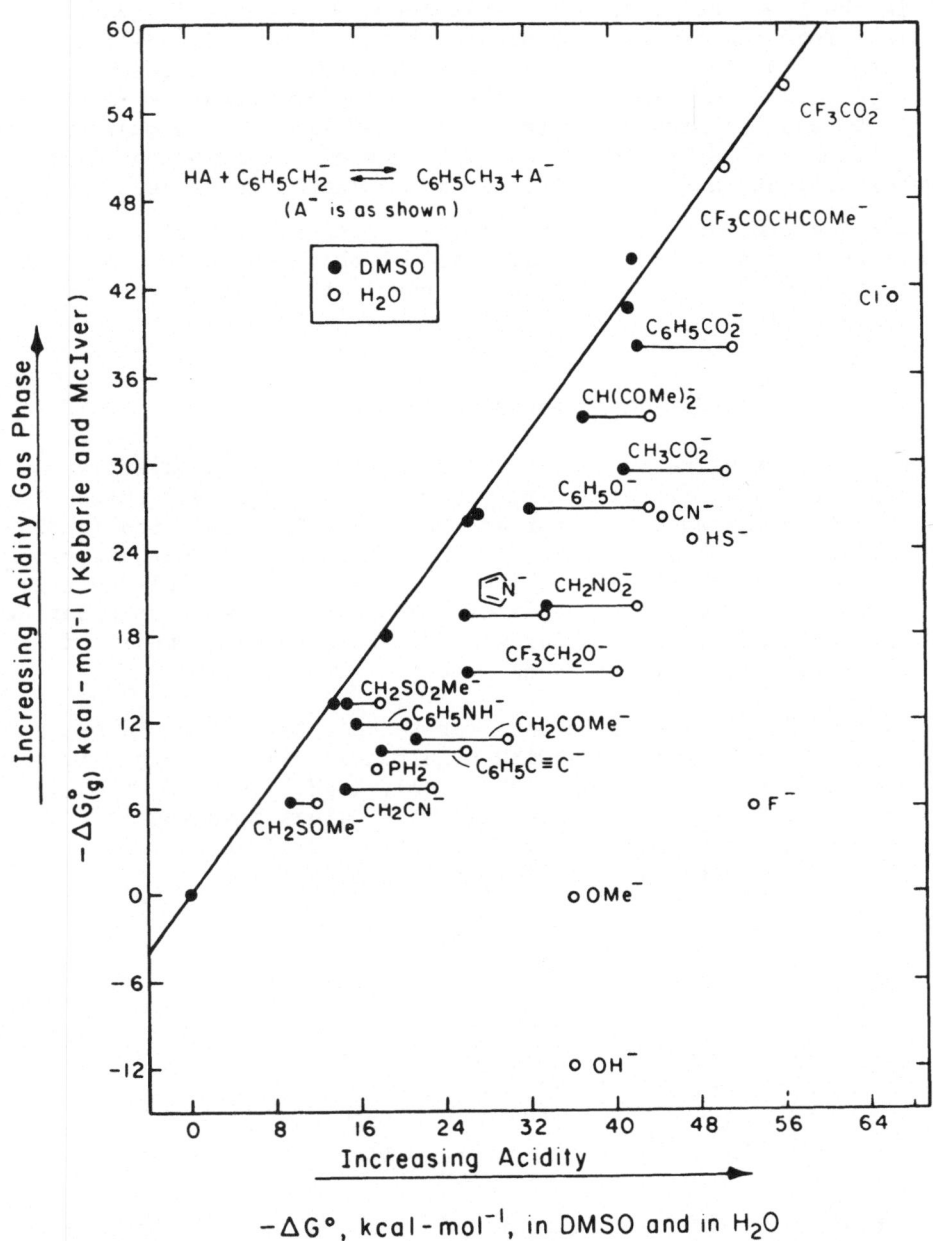

FIGURE 2.   Gas phase acidity versus acidity in water.

A milky-way of points is again found accumulating to the
right of the line of unit slope. The horizontal deviations from
this line (the apparent increases in acidity due to specific solvation)
may be taken to measure the standard free energy of specific binding
of solvent molecules to the anions which have appreciable localization of negative charge in a relatively small volume (as generally
typified by the association of charge with a strongly electronegative
atom), relative to that for their neutral acid forms (which in cases
like CF_3CO_2H are relatively large). The horizontal deviations
are at least twice as large in water as in DMSO, as expected of the
better anion solvating properties of water (ion-dipole interaction
with DMSO is somewhat sterically hindered)[27a]. Since many of the
acidities in water have been extrapolated from other solvent systems,
the specific solvent effects for water are estimates only. Nonetheless, the results obtained parallel the available gas phase binding energies for the attachment of four water molecules[2a], as well
as relative anion hydration ratios[31]. The deviations for F^- and OH^-
indicate that their basicity is reduced from their intrinsic value
by specific aqueous hydration by $\sim 10^{35}$. Discussion of other structural effects on the solvent effects on acidity is given elsewhere[32].

Specific binding of solvent molecules (or counterions) to A-ions
in solution leads also to certian marked reversals in acidity compared to the gas phase. The loss of charged-induced dipole stabilization by polarizable substituents in specifically solvated oxyanions
is a major contributor. By conventional arguments, for example,
benzoic acid has been accepted to be a weaker acid than formic acid
due to the extra resonance stabilization energy of the free benzoic
acid molecule:

$$\tag{16}$$

In aqueous solution at $25°$, $K = 10^{+0.4}$. However, for the simple
proton transfer reaction in the gas phase[2c] at $25°$, $K = 10^{-4.8}$. This
result is atributalbe to a predominant P-effect (C_6H_5 much more
polarizable than H). This P-effect must be attenuated away by
specific hydration before the predominant (but presumably reduced)
R effect may be seen in solution. Again there is support for this
conclusion from the comparison of the correlation of the acidities
(eqn 7) of the alcohols of Table 3 in the gas and in the aqueous
phase[33] with I and P parameters:

$$\delta\Delta G^\circ_{(g)} = -I + P \qquad R = (.999) \quad n = 11$$

$$\delta\Delta G^\circ_{(aq)} = -(.33)I - .03 \qquad R = .990 \quad n = 4$$

The I-effect is reduced by a facotr of 3 by specific solvation, but the P effect is attenuated to the point of insigifnicance because of the dispersal of the anionic charges into the H-bonded solvent. The reverse order of acidities of aliphatic alcohols in the gas phase and in hydroxylic solvents is explained in those terms. The P effects determine the gas phase acidity order,[10] but the I effects of the opposite order (reduced but not eliminated by solvation) determine the solution acidity order[15].

Significance to Acid-Base Chemistry in Solution

The present discussion leads to the prediction that acidity and basicity in a condensed phase medium can be made to approach the corresponding gas phase chemistry, if appropriate means are found to eliminate or minimize specific binding of <u>both</u> solvent molecules and counterions to BH^+ and A^- ions. For example, the potential, as indicated earlier, is to increase the acidity of H_3O^+ and the basicity of OH^- <u>both</u> by $\sim 10^{35}$ compared to their aqueous behavior. If potentials such as this can be realized, many revolutionary new practices of acid-base chemistry in solution are within grasp, using the gas phase proton transfer equilibrium as a basis.

Aprotic solvents which poorly bind to BH^+ or A^- with strongly localized charges are generally ineffective at preventing specific binding by the counterion. Condensed phase conditions are required which prevent the binding of both solvent and counterion. Preliminary experiments indicate that there is much promise in applying and improving existing technology to effectively accomplish this task.

For example, reaction (17) is very favorable in the gas phase[4]:

$$CH_3O^-_{(g)} + (C_6H_5)_2CH_{2(g)} \rightleftharpoons (C_6H_5)_2CH^-_{(g)} + CH_3OH_{(g)}; \quad K = 10^{10.3} \qquad (17)$$

However, specific binding of hydroxylic solvents by CH_3O^- causes a specific solvent effect of ~ 34 kcal forcing the reaction to the left, <u>e.g.</u>,

$$CH_3O^- \cdots (H_2O)_{n(aq)} + (C_6H_5)_2CH_{2(aq)} \rightleftharpoons (C_6H_5)CH^-_{(aq)} + MeOH_{(aq)};$$

$$K_{(aq)} \simeq 10^{-17}$$

Dr. Chawla has observed[34] that with the use of the solvent hexamethyl phosphoramide (an excellent cation trap, but poor solvator of anions) and potassium methoxide--18 crown-6-polyether, proton transfer in the condensed phase is spontaneous:

$$CH_3O^- \cdots \text{K}^+ + (C_6H_5)_2CH_2 \rightleftharpoons (C_6H_5)_2CH^- + \text{K}^+ + MeOH; \quad K \sim 1$$

This result, together with kinetic evidence[35] and synthetic utility[36] found for enhanced reactivity of charge localized ions under conditions of poor interaction with solvent and counterion, are impressive, but further improvements seem possible.

For cations with strong localization of the charge in solvent accessible protonic sites, not oxygen containing solvent will likely be adequate to prevent specific binding (since even super acid media is too good as an H-bond acceptor[37]). The use of the poor H-bond acceptor solvent dichloroethane, in combination with the formation of the homoconjugate of the counterion, has given the promising initial results[38] given in Table 5. In these enthalpy titration experiments, the bases were injected (giving $\sim$ 0.04 M concentrations) into $\sim$0.6 M CF_3CO_2H in CH_2Cl_2 solutions. Presumably, BH^+ and $(CF_3CO_2 \cdots HO_2CCF_3)^-$ are completely formed. Guaiazulene is intrinsically more basic than all but the 4-NMe$_2$ pyridine (as shown by the $-\Delta G^\circ_{(g)}$ values of Table 5). In water, all of the nitrogen bases are significantly stronger than guaiazulene (as shown by $-\Delta G^\circ_{(aq)}$ values), because of their specific hydration effects. The reduced

Table 5. Medium Effects on Proton Transfer Equilibria

BH^+	$-\Delta G^\circ_{(g)}$ kcal-mole^{-1}	$-\Delta G^\circ_{(aq)}$ kcal-mole^{-1}	$-\Delta H^\circ (CF_3CO_2H, CH_2Cl_2)$, kcal-mole^{-1} $BH^+(CF_3CO_2 \cdots HO_2CCF_3)^-$
$C_6H_5NH_3^+$	−20.3	4.36	−9.0
⬡NH$^+$	−10.4	5.31	−6.4
Me$_2$N=⬡=NH$^+$	4.2	11.24	3.8

binding to these BH^+ offered by the combination of solvent and counterion in the $CH_2Cl_2-CF_3CO_2H$ system is apparent in the reaction heats, $-\Delta H°$. With anilinium ion, the aqueous medium effect of ~ 25 kcal $= -\Delta G°_{(aq)} + \Delta G°_{(g)}$ is reduced to 11 kcal $= \Delta H° + \Delta G°_{(g)}$; with pyridinium ion, the aqueous medium effect of ~ 16 kcal is reduced to 4 kcal, and with the more charge delocalized $4-NMe_2$ pyridinium ion, the aqueous medium effect of 7 kcal is reduced to zero-- i.e., inherent gas phase results then appear to be observed in solution. Further improvement is surely possible in minimizing binding by both solvent and counterion, so acid-base behavior in the condensed phase even more closely approaching the gas phase results may be anticipated.

Acknowledgments: I am greatly indebted to Professors J. L. Beauchamp and R. T. McIver, Jr., for their instruction and use of ICR spectrometry and for many helpful discussions. In particular, thanks are due to numerous coworkers, acknowledged in the references, for many hours of diligent work. This work has been supported in part by grants from the National Science Foundation and the Public Health Service.

REFERENCES

1. (a) R. P. Bell, "The Proton in Chemistry," Cornell University Press, Ithaca, NY, 1973; (b) L. P. Hammett, "Physical Organic Chemistry," McGraw-Hill Book Co., NY, 1970; (c) G. E. K. Branch and M. Calvin, "The Theory of Organic Chemistry," Prentice-Hall, NY, 1941; (d) C. K. Ingold, "Structure and Mechanism in Organic Chemistry," Cornell University Press, Ithaca, NY, 1953; (e) G. W. Wheland, "Resonance in Organic Chemistry," John Wiley and Sons, NY, 1955; (f) C. A. Coulson, "Valence", Oxford University Press, Oxford, 1956; (g) H. C. Brown, D. H. McDaniel, and O. Haflinger, "Determination of Organic Structures by Physical Methods," E. A. Braude and F. C. Nachod, Editors, Academic Press, NY, 1, 634-643 (1955).

2. For reviews with literature references cf. J. L. Beauchamp, Ann. Rev. Phys. Chem. 22, 527 (1971); R. W. Taft, in "Proton Transfer Reactions," ed. E. F. Caldin and V. Gold, Chapman and Hall, London, 1975, p. 31; (c) P. Kebarle, Am. Rev. Phys. Chem. 28, 445 (1977).

3. J. F. Wolf, R. G. Staley, I. Koppel, M. Taagepera, R. T. McIver, Jr., J. L. Beauchamp, and R. W. Taft, J. Am. Chem. Soc. 99, 5417 (1977).

4. J. E. Bartmess and R. T. McIver, Jr., in "Gas Phase Ion Chemistry," ed. M. T. Bowers, Academic Press, in press.

5. D. H. Aue, H. M. Webb, and M. T. Bowers, J. Am. Chem. Soc. 97, 4137 (1975).

6. (a) T. H. Morton and J. L. Beauchamp, ibid., 94, 3671 (1972); (b) D. H. Aue, H. M. Webb, and M. T. Bowers, ibid., 95, 2699 (1973); (c) R. Yamdagni and P. Kebarle, ibid., 95, 3504 (1973).

7. T. B. McMahon and P. Kebarle, ibid., 99, 2222 (1977).

8. D. H. Aue, H. M. Webb, and M. T. Bowers, ibid., 97, 4136 (1975).

9. P. Ausloos and S. G. Lias, ibid., 100, 1953 (1978).

10. (a) J. I. Brauman and L. K. Blair, ibid., 90, 5636, 6501 (1968); (b) J. I. Brauman, J. M. Riveros, and L. K. Blair, ibid., 93, 3914 (1971); (c) D. K. Bohme, E. Lee-Ruff, and L. B. Young, ibid., 93, 4608 (1971).

11. D. H. Aue, H. M. Webb, and M. T. Bowers, ibid., 98, 311 (1976).

12. R. H. Staley, M. Taagepera, W. G. Henderson, I. Koppel, J. L. Beauchamp, and R. W. Taft, ibid., 99, 326 (1977).

13. R. W. Taft, presented at the Structure-Energy Conference, Santa Barbara, CA, February, 1977.

14. (a) R. W. Taft, in "Steric Effects in Organic Chemistry," M. S. Newman, ed., John Wiley, NY, 1956, Chapt. 13; (b) S. Ehrenson, R. T. C. Brownlee, and R. W. Taft, Prog. Phys. Org. Chem. 10, 1 (1973).

15. (a) R. W. Taft, M. Taagepera, J. L. M. Abboud, J. F. Wolf, D. J. Defrees, W. J. Hehre, J. E. Bartmess, and R. T. McIver, Jr., J. Am. Chem. Soc. 100, 0000 (1978); (b) G. I. Mackay and D. K. Bohme, ibid., 100, 327 (1978).

16. (a) C. K. Ingold, J. Chem. Soc., 1032 (1930); (b) R. W. Taft, J. Am. Chem. Soc. 74, 3120 (1952); (c) R. W. Taft, in "Steric Effects in Organic Chemistry," M. S. Newman, Ed., John Wiley NY, 1956, Chapt. 13.

17. D. H. Aue, H. M. Webb, and M. T. Bowers, J. Am. Chem. Soc. 98, 311 (1976).

18. D. J. DeFrees, J. E. Bartmess, J. K. Kim, R. T. McIver, Jr., and W. J. Hehre, J. Am. Chem. Soc. 99, 6461 (1977).

19. M. Taagepera, W. J. Hehre, R. D. Topsom, and R. W. Taft, ibid., 98, 7438 (1976).

20. M. Taagepera, T. D. Singh, and R. W. Taft, unpublished results.

21. B. M. Wepster, "Progr. in Stereochemistry," Vol. 2, W. Klyne and P. B. de la Mare, Eds., Butterworths, London, 1958.

22. R. W. Taft, J. F. Wolf, J. L. Beauchamp, G. Scorrano, E. M. Arnett, J. Am. Chem. Soc. 100, 1240 (1978).

23. R. W. Taft, presented at the Symposium on Gaseous Ion Thermochemistry, Am. Chem. Soc. Regional Meeting, Anaheim, CA, March 15, 1978.

24. E. M. Arnett, B. Chawla, L. Bell, M. Taagepera, W. J. Hehre, and R. W. Taft, J. Am. Chem. Soc. 99, 5729 (1977).

25. J. F. Wolf, J. L. M. Abboud, and R. W. Taft, J. Org. Chem. 42, 3316 (1977).

26. J. F. Coetzee, Progr. Phys. Org. Chem. 4, 76 (1967).

27. (a) F. G. Bordwell, Pure and Appl. Chem. 49, 963 (1977); (b) F. G. Bordwell, D. Algrim and N. R., Vanier, J. Org. Chem. 42, 1819 (1977); (c) F. G. Bordwell, private communication.

28. (a) R. G. Pearson and R. G. Dillon, J. Am. Chem. Soc. 75, 2439 (1953); (b) A. Streitwieser, Jr., and L. L. Nebenzahl, ibid., 98, 2188 (1976); (c) D. D. Perrin, in "Dissociation Constants of Inorganic Acids and Bases in Aqueous Solution," Butterworths, London, 1969.

29. F. G. Bordwell, W. S. Mathews, G. E. Drucker, Z. Margolin, and J. E. Bartmess, J. Am. Chem. Soc. 97, 3226 (1975).

30. (a) A. Streitwieser, Jr., and L. L. Nebenzahl, J. Syn. Org. Japan, 33, 889 (1975); (b) A. Streitwieser, Jr., J. R. Murdock, G. Hafelinger, and C. J. Chang , J. Am. Chem. Soc. 95, 4248 (1973).

31. E. M. Arnett, B. Chawla, and N. J. Hornung, J. Soln. Chemistry 6, 781 (1977).

32. R. W. Taft, in preparation for Progr. Phys. Org. Chem.

33. P. Ballinger and F. A. Long, J. Am. Chem. Soc. 82, 795 (1960).

34. B. Chawla and R. W. Taft, unpublished results.

35. cf. D. Landini, A. Maia, and F. Montanari, J. Am. Chem. Soc. 100, 2796 (1978).

36. E. Buncel and B. Menon, J. Chem. Soc. Comm., 648 (1976).

37. E. M. Arnett and J. F. Wolf, J. Am. Chem. Soc. 97, 3262 (1975).

38. B. Yang and R. W. Taft, unpublished results.

STUDIES OF ION CLUSTERS: RELATIONSHIP TO UNDERSTANDING NUCLEATION AND SOLVATION PHENOMENA

A. W. Castleman, Jr.

Department of Chemistry and Chemical Physics Laboratory
CIRES
University of Colorado, Boulder, Colorado, 80309, U.S.A.

Introduction

Study of the interactions of ions and molecules provides detailed knowledge of reaction kinetics, especially for certain bimolecular channels, and it is clear that these investigations also yield invaluable information concerning mechanisms of association reactions and related processes of unimolecular decomposition. Now, it is becoming recognized that there is another important area in the field of chemical physics to which such studies also contribute, namely, to an understanding of phase transitions and properties of the condensed state.

In passing through a phase boundary between the gaseous and condensed state, molecules must first cluster together and subsequently grow to critical size before the gas phase is converted to either a solid or liquid. In principle, condensation is possible whenever the partial pressure of a vapor exceeds the vapor pressure of a condensed phase. Nevertheless, meta-stable states exist and, a certain minimum degree of supersaturation is invariably required before an actual phase transformation occurs. There are many cases in nature where such situations are observed: for instance, water can be super-cooled before it freezes, and a liquid can be super-heated before boiling takes place. Despite the fact that such phenomenon have long been recognized, progress in theoretically formulating the rates of phase transformation, and clarifying the molecular processes involved, has only just begun to be made.

The general requirement that a system becomes supersaturated before condensation occurs is related to the fact that cluster

growth must proceed to the formation of a critical size nucleus.
This must compete with the process of cluster decomposition, and
from a conceptual point of view nucleation can be formulated in
terms of theories of association reactions and those of unimolecular
decomposition. Thereby, it should be evident that a study of certain
classes of ion molecule reactions will, indeed, lead to a more modern
understanding of this important phenomenon.

Progress in understanding phase transformations has been
partly hampered by the lack of suitable experimental techniques for
studying the molecular aspects of the process. For example, much
of the existing data on vapor phase nucleation is based on light
scattering measurements made in cloud chambers. However, before
appreciable light scattering takes place, clusters must develop to
sizes which are close in magnitude to the wavelength of the light.
This generally requires clustering together of hundreds of molecules,
which far exceeds the usual regime of the critical size cluster.
Likewise, attempts at studying nucleation by mass spectrometrically
analyzing neutral cluster distributions produced in free-or super-
sonic-jet experiments have also been fraught with difficulties.
The major problem is associated with the fact that neutral clusters
may fragment upon electron impact ionization. Furthermore, ionization efficiencies of clusters are not well known which makes it exceedingly difficult to establish absolute ratios of cluster intensities. Yet, knowledge of cluster distributions are required in
examining theoretical expressions based on either kinetic or steady-
state formulations.

In the case of ion induced nucleation, these difficulties have
been overcome with recent advances in high pressure mass spectrometry.
Studies of molecules clustered about preformed, thermalized, ions
involve no uncertainty of cluster fragmentation. Therefore, it is
expected that investigations of the latter will provide the basis
for a more complete understanding of the molecular aspects of all
nucleation phenomenon.

Closely related to the process of ion induced nucleation is
the solvation of ions. In fact, when condensable ligand molecules
progressively attach to an ion in large enough number for a phase
transformation to occur, a small liquid droplet is formed. While
the droplet is not quite an electrolyte because ions of only one
sign are present, detailed studies of such phenomenon in the gas
phase lead to further elucidation of the molecular interactions
analogous to those of ions in the liquid phase. The interrelation-
ship of solvation and nucleation phenomenon is evident, and the purpose of this review is to compare these two processes, point out
similarities, differences, present unknowns, and directions for
future research. Herein, discussion is confined to the subject of
nucleation from the gaseous to the condensed state. Major emphasis

is focused on information concerning energetics and structure of
small clusters which can be derived from mass spectrometric measurements combined with ab initio semi-empirical quantum mechanical,
and electrostatic calculations.

Phase Transformation: Nucleation Phenomena

Nucleation phenomena may be broadly classified into three
categories. The first is so-called homogeneous nucleation which
involves the interaction of like vapor-phase molecules, subsequent
formation of prenucleation embryos, and ultimately phase transition.
At the other end of the scale is heterogeneous nucleation in which
vapor condenses onto a pre-existing surface. Clearly, the latter
involves additional interaction between the vapor phase molecules
and the already existing condensed phase. Midway between these
two is the phenomenon termed heteromolecular nucleation. This is
defined as the class influenced by the presence of an ion, atom or
foreign molecule of difference chemical composition from the bulk
nucleating phase. Compared to homogeneous nucleation, the general
heteromolecular case involves additional forces between the molecules participating in the formation of the prenucleation clusters.
Consequently, heteromolecular nucleation is observed when the
attractive forces between the foreign center and the bulk nucleating phase lower the supersaturation level beyond that necessary for
homogeneous transformation to a condensed phase. Clearly, advances
in understanding heteromolecular nucleation will also provide a
basis for understanding other classes of the phenomenon as well.

Conceptually, ion induced nucleation differs little from that
of homogeneous nucleation. It may be thought of as a sequence of
clustering reactions beginning with:

$$A^{\pm} + B + M \rightleftharpoons A^{\pm} \cdot B + M \tag{1}$$

Here $A^{\pm}$ designates the ion (of either sign) about which molecule B
is clustering; M is the third-body stabilizing ion cluster formation.
Whether or not the kinetics eventually become effectively second
order, involves a consideration of cluster lifetimes and times between collisions with M. In any event, the kinetics can always be
expressed in terms of effective second-order reactions of the following type:

$$A^{\pm}(B)_{n-1} + B \underset{k_{r,n}}{\overset{k_{f,n-1}}{\rightleftharpoons}} A^{\pm}(B)_{n} \tag{2}$$

$$A^{\pm}(B)_{n*-1} + B \underset{k_{r,n*}}{\overset{k_{f,n*-1}}{\rightleftharpoons}} A^{\pm}(B)_{n*}$$

Here n^* designates the cluster of critical size, and subscripts
f and r denote forward and reverse rate constants, respectively.
It is frequently assumed that, after the critical size cluster is
developed, its subsequent collision with another B molecule leads
directly to a transition from the gaseous to the condensed state.
Obviously, this is a mathematical convenience rather than a
physical reality. Under such assumptions, the rate of phase transformation may be expressed in terms of the rate constant for the
subsequent clustering of a molecule with the critical size cluster,
and a product of equilibrium constants as follows (Castleman,
Holland, Keesee, 1978):

$$R = k_{n^*} \left[B \right]^{n^*+1} \prod_{n=1}^{n^*} K_{n-1,n} \tag{3}$$

with

$$\prod_{n=1}^{n^*} K_{n-1,n} \left[B \right]^n = e^{-\Delta\phi/kT} \tag{4}$$

In deducing these expressions, equilibrium has been assumed. In
case of a finite rate of nucleation, the steady state concentrations
of the clusters will depart from their equilibrium values and introduction of the well known Zeldovich pre-exponential factor
(Zettlemoyer, 1969; Abraham, 1974) must be included in the rate expressions. The energy barrier defined by $\Delta\phi$ in the above expression is referenced to a standard state of one atmosphere. Account
must be taken of the vapor pressure of the condensing phase in order
to relate the numerical values to conventional supersaturation
ratios.

In principle, the energy barrier to both homogeneous and
heteromolecular nucleation can be readily calculated in terms of
the free energy of formation of small microsize clusters containing
n molecules of B clustered about the central ion (Reiss, 1977;
Zettlemoyer, 1969). The appropriate expression is given by

$$- \frac{\Delta G^o}{kT} = \frac{n\mu_1^o - \mu_n^o}{kT} = \ln\left[\frac{1}{n!} \left\{ \frac{Q_n/V}{\left[Q_1/\overline{V} \right]^n} \right\} \right] \tag{5}$$

The symbol ΔG^o is the free energy of formation of the nth size
cluster, and μ_1^o and μ_n^o correspond to the chemical potential of the
monomer and the n-mer, respectively. In the case of condensation
from vapor to solid, rather than liquid, n! must be replaced by
the symmetry number of the crystal. Q_n and Q_1 correspond to the

partition function of the n-mer and monomer respectively, and V
is volume. In principle, this expression can be evaluated; however major difficulty is associated with an appropriate calculation
of Q_n. Numerous theoretical attempts have been made at evaluating
ΔG^O, but it has often been necessary to make very simplyfing
approximations. Herein lies the difficulty and the reasons for the
great discrepancies in the predicted energy barriers using the
various theoretical treatments. Due to the extreme difficulty of
deducing values for the free energy of formation employing statistical mechanical procedures, a continuum liquid-drop expression
known as the Thomson equation (Thomson, 1888, 1928) has often
been used. This equation, and modifications thereto, is based on
the Kelvin equation and therefore assumes clusters to have a well-
defined surface tension. Furthermore, the Thomson equation also
includes an additional term to account for the change in field
energy due to the presence of a sphere of molecules, having a
dielectric constant, surrounding the ion.

In considering homogeneous nucleation, Frankel and Kuhrt
(Frankel, Kuhrt, 1952) were among the first to note that the
classical Kelvin expression would only be rigorously valid for the
case of a stationary droplet. Since that time, Pound, et al
(Lothe, Pound, 1969; Lothe, Pound, 1966) have done much to call
attention to the need for a correction to the conventional macroscopic liquid drop formulations. The need for such a factor,
generally termed "replacement free energy", has led to considerable discussion and some disagreement in the literature (Reiss,
1977; Zettlemoyer, 1969; Abraham, 1974). Although the need for
such a "correction" factor is now generally conceded, the magnitude of the replacement term is still not settled.

Reiss (Reiss, 1977) has recently discussed the effect in
considerable detail, and correctly demonstrated the origin of the
"replacement term" in homogeneous nucleation. It arises because
of the use of conventional thermodynamic relationships for droplets in deriving numerical values for their free energy of formation. Although less attention has been given to these factors in
the case of ion induced nucleation, the same general considerations
apply (Castleman, Tang, 1972; Kortzeborn, Abraham, 1973; Castleman,
Holland, Keesee, 1978).

For homogeneous nucleation, the standard free energy of
cluster formation includes a "replacement" partition function
accounting for internal degrees of freedom of condensing molecules
replaced by translation and rotation. Russell, (Russell, 1969)
and Kortzeborn and Abraham (Kortzeborn and Abraham, 1973), have
considered the case of ion induced nucleation including the replacement term of homogeneous nucleation as a correction to the
classical equation due to Thomson.

Castleman and coworkers (Castleman, Holland, Keesee, 1978) have considered the equivalence of the rate expressions derived from the kinetic and steady state formulations of ion induced nucleation. The energy barriers for various supersaturation ratios are sketched in Figure 1. These are seen to be infinite for unsaturated or just saturated systems, then finite, and eventually nonexistent at high values of supersaturation. For the case of a finite energy barrier, there must be a reversal in the relative concentrations of successive cluster sizes at two points, that of the first minimum, and at the point of the maximum in barrier height (Zettlemoyer, 1969; Castleman, Tang, 1972). In the approximation that there is a continuum of clusters in these regimes, this leads to

$$\left[A^+(B)_n \right] \Big/ \left[A^+(B)_{n-1} \right] = 1 = \frac{k_f \left[B \right]}{k_r} \tag{6}$$

The symbol $\left[B \right]$ represents the vapor phase concentration of the condensing molecule and is defined such that $\left[B \right]$ equals unity for the concentration corresponding to vapor-liquid equilibrium. The symbols k_f and k_r designate the respective, forward and reverse, rate constants for clustering. In terms of the kinetic approach, it is thereby predicted that there are crossing points of the respective forward-and reverse-rate curves. In accordance with

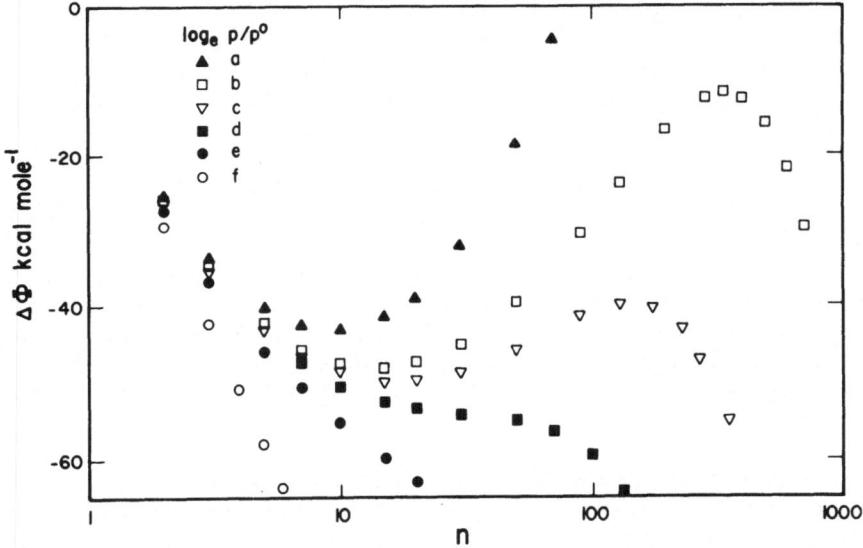

Fig. 1. Energy barriers to ion induced nucleation as a function of cluster size. The parameters are the natural logarithm of the supersaturation S ($S=p/p^0$; a:0.0; b:0.7; c:0.9; d:1.2; e:2.0; f:7.0) (Adapted from Castleman, Holland, Keesee, 1978)

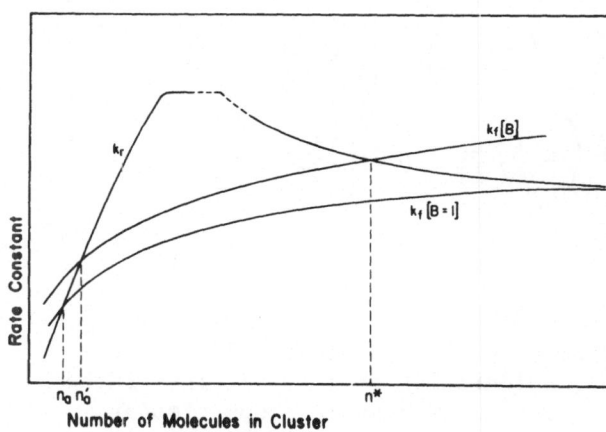

Fig. 2. Sketch of forward and reverse rates showing crossing points. $[B]$ >1 designates that the condensing neutral is at a supersaturation (p/p^0>1). (Castleman, Holland, Keesee, 1978)

equation 6, the crossing points would occur at values where the product of k_f and $[B]$ is equal to k_r. See Figure 2.

At the thermodynamic condition corresponding to vapor-liquid equilibrium, intersection of the curves is possible only at infinite cluster sizes, or at the very small cluster sizes representative of an equilibrium cluster distribution. However, at moderate supersaturations the forward curve crosses the reverse one, which is equivalent to having a finite energy barrier. At a high value of supersaturation, where the energy barrier of Figure 1 has ceased to exist, the forward curve (Figure 2) crosses the reverse curve at all cluster sizes. Under such conditions, nucleation becomes kinetically controlled by the forward rate of clustering and a steady state distribution of clusters is not obtained.

Evaluation of the Classical Equations for the Predicting Energy
Barrier to Nucleation - Results of Ion-Molecule Clustering Studies

An expression for the free energy change per molecule is readily derived by differentiating the Thomson equation with respect to radius (Castleman, Holland, Keesee, 1978)

$$\Delta G^0_{n-1,n} = nkT \ln p^0 + \left(\frac{32\pi NM^2}{3\rho^2 n}\right)^{1/3} \sigma - \frac{q^2}{8\pi}\left(1-\frac{1}{\epsilon}\right)\left(\frac{4\pi N}{3n}\right)^{4/3}\left(\frac{\rho}{M}\right)^{1/3} \quad (7)$$

Here p^0 is the vapor pressure of the condensing molecule at the temperature of the system, N is Avogadro's number, M the molecular weight, ρ the bulk density of the clustering molecules, σ the surface tension, and q the elementary charge. Subsequent differentiation with respect to temperature enables an evaluation of entropy; enthalpy is then deduced by substitution into well-known thermodynamic relationships.

$$-\Delta S^O_{n-1,n} = R \, \ell n p^O + RT \, \frac{d \ell n p^O}{dT} + \left(\frac{32 \pi N M^2}{3n}\right)^{1/3} \left(\frac{-2}{\rho}^{/3} \frac{d\sigma}{dT} - \frac{2}{3} \, \sigma \rho^{-5/3} \frac{d\rho}{dT}\right)$$

$$- \frac{q^2 N}{6n} \left(\frac{4 \pi N}{3 M n}\right)^{1/3} \left[\frac{1}{3} \bar{\rho}^{2/3} \left(1 - \frac{1}{\varepsilon}\right) \frac{d\rho}{dT} + \rho^{1/3} \varepsilon^{-2} \frac{d\varepsilon}{dT}\right] \qquad (8)$$

and

$$\Delta G^O = \Delta H^O - T \Delta S^O. \qquad (9)$$

Castleman, et al. (Castleman, Holland, Keesee, 1978) have recently undertaken detailed consideration of the validity of the Thomson equation in expressing the thermodynamic properties of small ion clusters. In making the evaluation, data for two iso-electronic ligands, water and ammonia clustered about positive ions of closed electronic configuration, and water about a series of negative ions, proved to be particularly revealing. Exact agreement between experiment and theory is not expected since the model assumes a point-charge. Nevertheless, it is instructive to consider the minimum sizes of clusters, if any, which can be suitably expressed by the Thomson equation.

The assessment showed the Thomson equation to be adequate for expressing the enthalpy changes of water clustered about positive ions comprised of the alkali metal ions and Pb^+ at cluster sizes greater than five. In the case of Sr^+, however, somewhat more than ten ligand molecules are apparently required. This is readily seen from the comparison of the theory and experimental data plotted in Figure 3a. In the case of ammonia, the exact cluster size where the enthalpy is adequately described by the Thomson equation is less certain. Nevertheless, the trend for the experiment enthalpies to merge with the values predicted by the Thomson equation is also clearly indicated as seen in Figure 3b. In the case of water clustered about negative ions, the size where experimental and theoretical values merge is even less certain. But this is largely due to the fact that clustering data are not generally available to as large a cluster size as for many of the positive ion systems.

An earlier examination of the Thomson equation and its usefulness in deducing entropy of clustering was made for the ions bismuth and lead clustered with water. In these systems it was found that the Thomson equation was a very good approximation at cluster sizes around 5 to 7. Subsequent examination of the Thomson equation and its ability to evaluate entropies of clustering were made for the alkali metal systems clustered with both water and ammonia, and for water clustered to negative ions. In the case of water, about positive ions, a reasonable correspondence between experimental and predicted values was found.

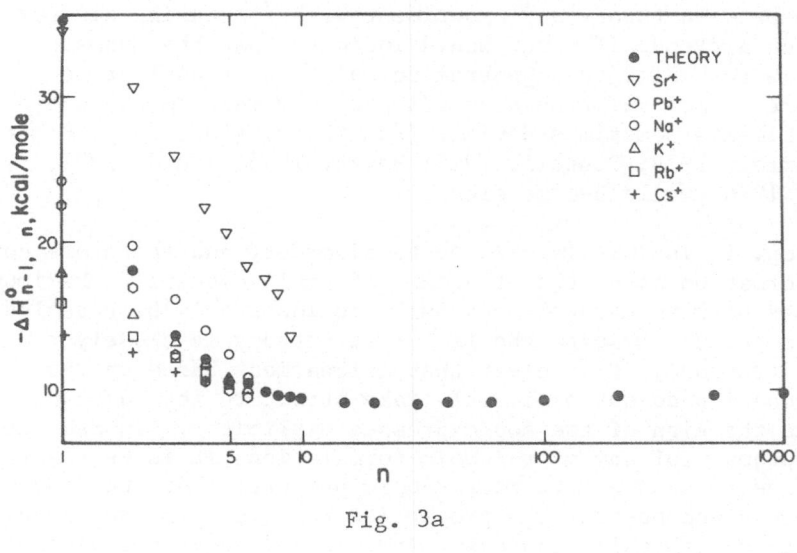

Fig. 3a

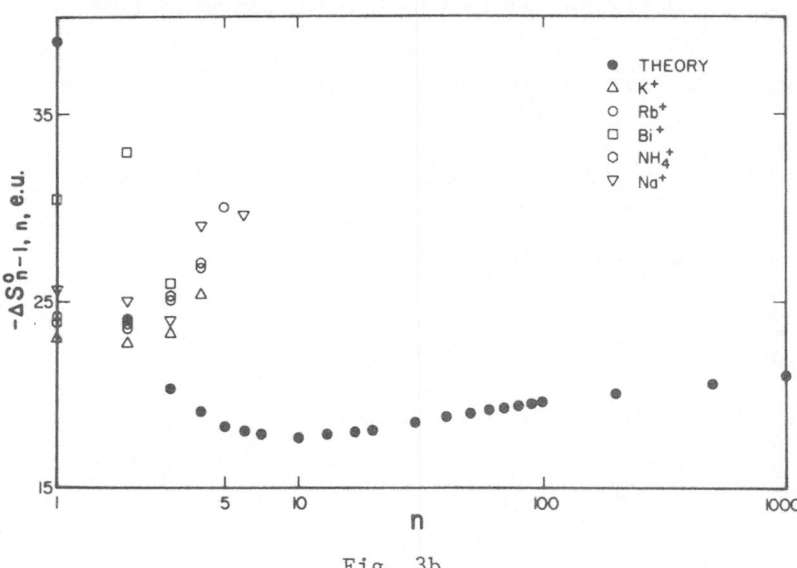

Fig. 3b

Fig. 3. Gas-phase enthalpies a) hydration, b) ammoniation. Theoretical calculations are based on the Thomson equation (Castleman, Holland, Keesee, 1978).

However, an examination of the entropy data in Figure 4 for water clustered about negative ions shows a trend for a greater departure between theory and experiment with increasing cluster size. Such a trend, if true, would indicate that the Thomson equation is not valid as a general formulation of nucleation. Furthermore, a molecular understanding of why macroscopic nucleation experiments sometimes deviate from theory (Loeb, Kip, Einarsson, 1938; Scharrer, 1939; Russell, 1969; Hertz, 1956; Pound, 1972; Heicklen, 1976) would become clear.

Entropy is intimately related to disorder, and therefore contains information about the structure of small clusters. Maximum differences between clusters and bulk liquids should be revealed in these aspects. Despite the fact that entropy is closely related to structure, it is clear that calcuations based on the Thomson equation do not explicitly take structure into account; neither is the sign of the ion expressed explicitly. In ascertaining the adequacy of any macroscopic formulation, it is necessary to determine to what extent macroscopic properties of the condensing phase can account for the properties related to microscopic configurations of small clusters. Ammonia and water are isoelectronic molecules and might be expected to display rather similar entropy values if the liquid drop-like assumptions of the Thomson equation are not valid, and cluster structure plays an important

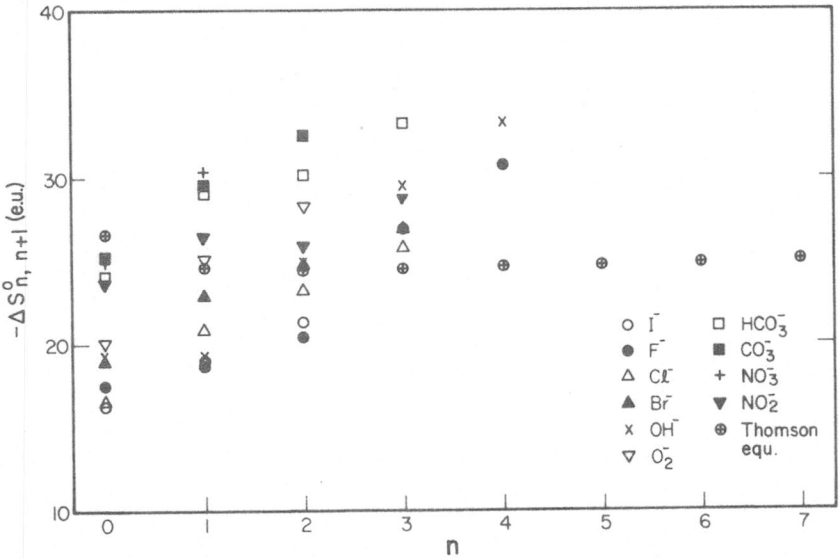

Fig. 4. Entropies of hydration for certain negative ions. All data from Kebarle, 1977 except: CO_3^-, HCO_3^- — Keesee, Lee, Castleman, 1978; NO_3^-, NO_2^- — Lee, Keesee, Castleman, 1978.

role. Calculations were made which showed that if the Thomson
equation were valid as a general equation, the entropies for
alkali metal ions clustered with ammonia should differ significantly from those with water. See Figure 5.

A comparison of the experimental and predicted values for
ammonia clustered about the alkali metal ions is given in Figure
6. The data reveal that there are significant departures between
theory and experiment of at least 8 entropy units or more at
cluster sizes exceeding five. Not only are the departures significant in magnitude, but they display a trend of becoming increasingly greater with cluster number. Furthermore, comparing the
experimental entropy values for water and ammonia about positive
ions shows fairly close agreement in contrast to the theoretical
predictions given in Figure 5.

In this regard, it is interesting to consider the relative
contributions to entropy of the terms in the Thomson equation
involving surface tension and dielectric constant. Table I lists
the contributions of the individual terms to the overall entropy
value. Bulk liquid phase values have been taken for both the
dielectric constants and the surface tensions of water and ammonia.
Calculations were subsequently made in which they were individually varied to assess their contribution to the entropy

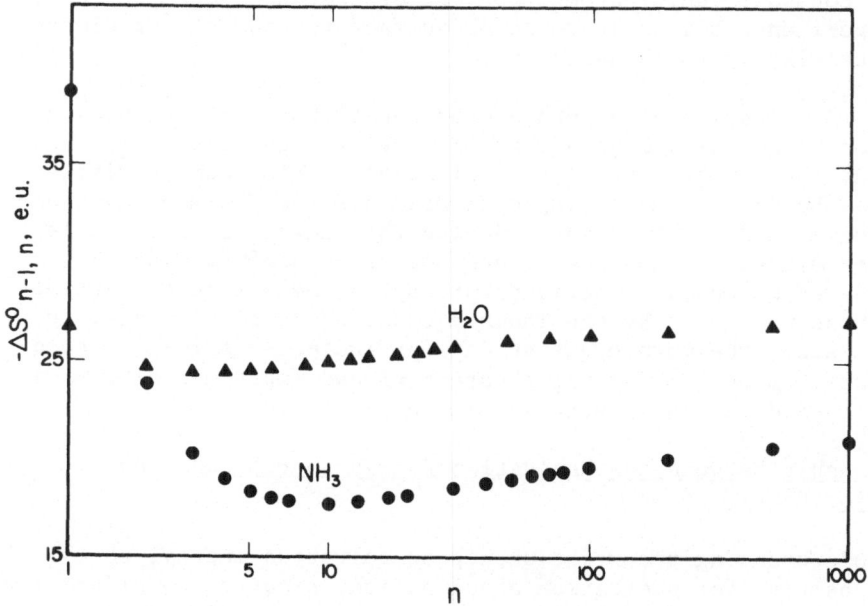

Fig. 5. Comparison of predicted entropies of hydration and
ammoniation on the basis of the Thomson equation (Castleman, 1978)

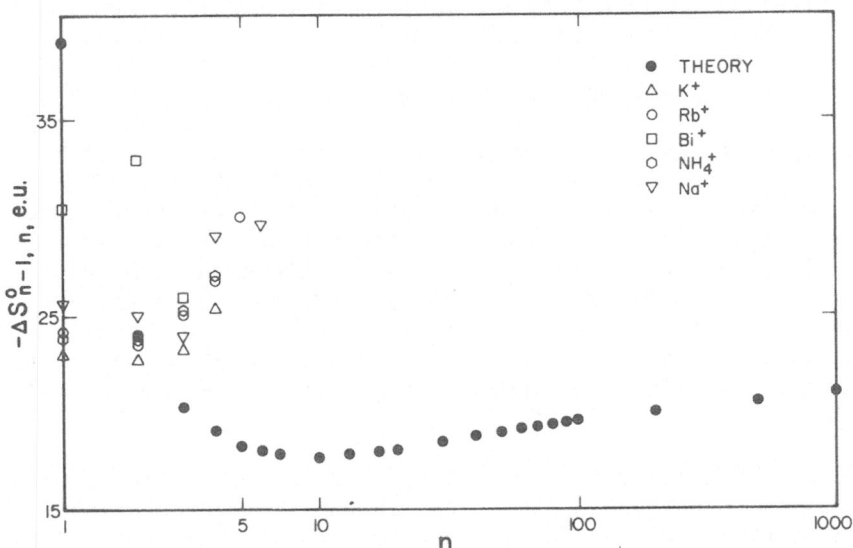

Fig. 6 Gas-phase ammoniation entropies. The theoretical points
are calculated from the Thomson equation (Castleman, Holland,
Keesee, 1978).

change predicted by the model. No reasonable adjustment of dielectric constant could account for the discrepancy between theory
and experiment, but an increase in surface tension of the clusters
substantially improved agreement.

It is clear from the above considerations that the Thomson
equation is not useful as a general model for quantitatively calculating the energy barrier to nucleation. This then clarifies
reasons why in some macroscopic experiments the Thomson equation
has been deemed as valid for deducing the energy barrier, while
in other systems it has been found not to be useful. The fact
that the experimentally derived entropy values are more negative
than those predicted by the Thomson equation is particularly revealing since the more negative entropy indicates a more ordered
structure. Clearly, further theoretical and experimental work is
needed to deduce ion cluster structures.

Relationship between the Nucleation Energy Barrier and the Solvation Energy of an Ion

Understanding the relationship of solvation energy, to the
energy barrier for nucleation about an ion, requires consideration
of the ion cluster distributions (i.e. relative concentrations)
existing and evolving during a nucleation experiment. In the
absence of a finite nucleation rate,cluster concentration can be

Table I. Calculated Entropies at 313°K Using Equation 8

System: Point Charge-Water

$-\Delta S^{O}_{n-1,n}$ e.u.

Cluster Size	Surface Contribution	Dielectric Contribution	Total
2	-5.2	2.1	24.7
4	-4.1	0.9	24.5
6	-3.6	0.5	24.7
8	-3.3	0.3	24.8
10	-3.0	0.2	25.0
15	-2.7	0.2	25.3
20	-2.4	0.1	25.5
50	-1.8	0.0	26.0
100	-1.4	0.0	26.4

System: Point Charge-Ammonia

$-\Delta S^{O}_{n-1,n}$ e.u.

Cluster Size	Surface Contribution	Dielectric Contribution	Total
2	-9.8	11.4	23.8
4	-7.8	4.5	18.9
6	-6.8	2.6	18.0
8	-6.2	1.8	17.8
10	-5.7	1.3	17.8
15	-5.0	0.8	17.9
20	-4.5	0.5	18.1
50	-3.3	0.2	19.0
100	-2.7	0.1	19.6

expressed by

$$C_n = C_t \exp(-\Delta\phi/kT), \qquad (10)$$

where C_n represents the concentration of the n-th cluster; the symbol $\Delta\phi$ was defined by Equation 4. As discussed earlier, this relationship must be modified by the Zeldovich factor when nucleation is proceeding at a finite rate. The total concentration of clusters, C_t, is given by

$$C_t = \sum_n C_n \qquad (11)$$

where a choice is required to specify either the presence of a finite number of ions in the system, or a fixed concentration of a particular cluster size. With the definition

$$K_{n-1,n} \equiv \exp \left(-\Delta G^O_{o,n-1,n}/kT\right) \tag{12}$$

and the approximation that, at low pressures, the concentration
is a direct measure of the chemical activity of the given species

$$K_{n-1,n} = \frac{C_n}{C_{n-1}P_B} \tag{13}$$

From the foregoing, it follows that

$$\exp \left(-\frac{\Delta\phi}{kT}\right) = \frac{\exp \left[(-\Delta G^O_{o,n}/kT) + n\ell np\right]}{\sum\limits_{n} \exp \left[(-\Delta G^O_{o,n}/kT) + n\ell np\right]} \tag{14}$$

In the typical ion nucleation experiment made in either a
cloud chamber or in a flowing system, a quasi-stable small cluster
spectrum is first established. An experiment is usually designed
to detect the appearance of the first droplet of the "new" condensed phase. In this situation (Castleman, Tang, 1972), equation
11 is normalized to the peak in the small cluster spectrum, designated C_{n_a}, rather than to C_t. A more rigorous procedure would involve normalization to the total small cluster population, but
a priori knowledge of this value is not generally available, and
the former is adopted for convenience.

On this basis,

$$C_n = C_{n_a} \exp \left(-\Delta\phi'/kT\right), \tag{15}$$

where

$$\Delta\phi'_n = -kT \sum\limits_{n_a+1}^{n} \left(\ell n\frac{C_n}{C_{n-i}}\right) \tag{16}$$

In order to calculate the distribution of clusters about the
critical size nucleus, a knowledge of $\Delta\phi'$ as a function of n is required. This can be obtained by appropriate summations of the free
energies of cluster formation as follows:

$$\Delta\phi'_n = \left[\sum\limits_{n_a+1}^{n} \Delta G^O_{n-1,n}\right] -(n-n_a)kT \, \ell n \, P_B \tag{17}$$

Here, the free energies of individual clustering reactions are taken with respect to the usual standard state of 1 atmosphere.

As a general rule, the free energy of successive cluster attachment becomes independent of the nature of the ion involved at some cluster size (designed n_c) prior to the critical nucleus. It follows that the energy barrier to nucleation for n greater than n_c is given by:

$$\Delta \phi'_n = \Delta G^O_{a,c} + \Delta G^O_{c,n} - (n-n_a)kT \ln p_B \qquad (18)$$

Clearly, the free energy $\Delta G^O_{a,c} + \Delta G^O_{c,n*}$, including the product of $(n*-n_a)$ with kT and $\ln p_B$, governs the absolute barrier height for nucleation. Differences between ions in their ability to promote nucleation are expected to be manifested principally in the term $\Delta G^O_{a,c}$, with a somewhat less effect arising from the magnitude n_a. The differences between ions occur because of the nature of the bonding of ligands to the ion clusters over the size range n_a to n_c.

In contrast, the free energy of solution, ΔG^O_s, is a direct measure of the affinity of the ion for the solvent, and is given by

$$\Delta G^O_s = \sum_1^\infty (\Delta G^O_{n-1,n} + \Delta G^O_v). \qquad (19)$$

The symbol ΔG^O_v is the free energy of vaporization of the system and is a function of temperature only. Therefore, in addition to the constants of the system represented by ΔG^O_v and the logarithm of the pressure, $\Delta \phi'$ and ΔG^O_s also differ by the term $\Delta G^O_{o,a}$. Consequently, in the case of a system for which $\Delta G^O_{o,a}$ makes an appreciable contribution to ΔG^O_s, the energy barrier to nucleation may not directly reflect the energy of solvation. Consideration of the bonding of the small clusters which play a role as prenucleation embryos, and their direct relationship to understanding solvation, is deferred to a later section of this chapter.

Studies of Large Clusters in the Gas Phase

It is interesting to consider the effect of vapor phase concentration on the distribution of the "quasi-stable" small clusters. Clearly, ion cluster concentrations should display a maximum where $\Delta \phi$ (or $\Delta \phi'$) exhibits a minimum value. Referring to Figure 1, it can be seen that a system at (or just below) saturation has one minimum and no finite maxima; therefore, only one maximum will exist in the cluster concentration. Increasing the partial pressure of the condensing vapor should lead to only a minor shift in the cluster size which displays the peak concentration. This effect is

always apparent in experiments where ions are sampled from high
pressure reaction cells. A typical example is given in Figure 7.

The exact position of the peak in the small cluster spectrum,
as well as the breadth of the distribution, will vary from system
to system. This reflects the magnitude of the forces responsible
for cluster stability.

Under conditions exceeding thermodynamic saturation, the
energy barrier becomes finite and there is the probability of a
phase transformation. Clearly, the cluster distributions of a
nucleating vapor should still show a peak at small sizes. There-
after, the distribution would be expected to progressively decrease through the critical size cluster.

Referring to curves a-f in Figure 1, the effect on the distribution of small size clusters can be predicted. The position
and shape at the first minimum in $\Delta\phi$ is influenced relatively
little by a moderate change in supersaturation. Therefore, the
small cluster distribution should be shifted only marginally to
higher sizes, and broadened to only a minor extent.

It is only now being realized that dynamic flow experiments
can be utilized to elucidate mechanisms of ion clustering and re-

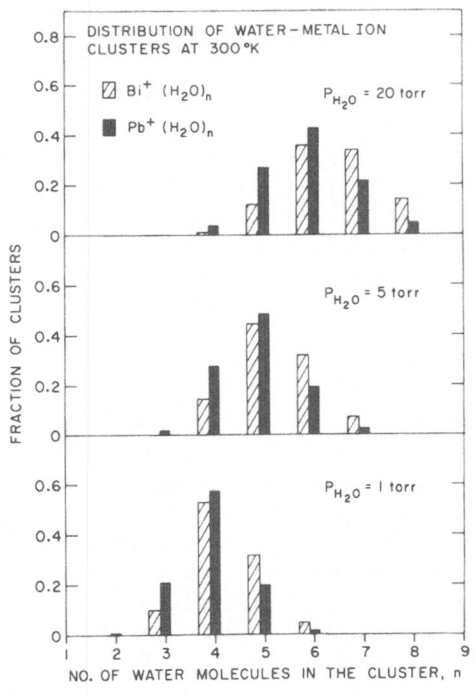

Fig. 7. Distribution of
bismuth-and lead-ion hy-
drates as a function of
water vapor partial pres-
sure at 300°K. (Tang,
Castleman, 1974)

lated phenomenon of ion induced nucleation. A limited number of relevant experiments have been reported, mainly by Searcy and Fenn (Searcy, Fenn, 1974, 1975), and Burke (Burke, 1972, 1978). Searcy and Fenn have observed the existence of numerous water molecules clustered with H^+ under free jet expansion conditions. A particularly interesting finding was the existence of an anomalously large concentration of a cluster containing 21 water molecules. In view of the considerations discussed above, the existence of an unexpectedly high concentration of 21 molecules with H^+ has generated considerable interest.

As pointed out by Searcy and Fenn, the clustering environment in their free jet expansion changes so rapidly that it is doubtful whether the clusters actually attain equilibrium with the ambient gas. Since the supersaturation existing during the experiments (Searcy, 1975) is not very large, it is doubtful that $k_f \boxed{B}$ has crossed the reverse curve sketched in Figure 2. Consequently, the distributions probably have not arisen as a result of clustering driven only by forward reactions. Another possible explanation is that the data give the appearance characteristic of equilibrium distributions only because of the fact that a limited number of clusters are available onto which neutral molecules can condense, and the distributions become "frozen-in" during the expansion process. An alternate explanation is that the data reflect a kinetic process influenced by the supply of condensable vapor.

More recently Burke (Burke, 1978) has also reported the observation of an unusually large concentration of 21 water molecules with H^+; he employed a completely different expansion technique in making the measurements. In addition, Lin (Lin, 1973) has observed the existence of this cluster originating from the electron impact ionization of neutral water clusters. Very recently, Kay, et al. (Kay, Hermann, Castleman, 1978) made similar observations during the course of experiments on neutral cluster-beams. Furthermore, both the ion cluster experiments of Burke, and the neutral ones of Kay, et al. reveal that a cluster of 28 water molecules with H^+ also exhibits a pronounced concentration anomaly.

Certain constraints are imposed by thermodynamics in order that a cluster of unusual stability can exist within a continuous distribution. This can be seen most easily in terms of equations 9 and 13 with the realization that $\Delta H^O_{n-1,n}$ and $\Delta S^O_{n-1,n}$ must be negative for systems under consideration.

$$\frac{C_n}{C_{n-1}} = P_B \exp \left\{ \frac{-\Delta H^O_{n-1,n}}{RT} + \frac{\Delta S^O_{n-1,n}}{R} \right\} \qquad (20)$$

If ΔS^O exhibits a smooth trend, an increase in the ratio C_n/C_{n-1} with n = 21 would require ΔH^O to suddenly become more negative in going from the less to the more stable size, with the reverse trend for the next addition. On the other hand, if ΔH^O were considered to display a smooth trend with cluster size, ΔS^O must become less negative in going from cluster size 20 to 21, with the reverse trend for cluster 22. The latter is opposite to expectation since the formation of a more ordered structure would be accompanied by a more negative entropy change. Consequently, the appearance of unusually stable structures should be discernible from data on the enthalpy of cluster formation. This is fortunate since with existing techniques, enthalpy data can be derived more accurately than entropies.

On the basis of the assumption of local thermodynamic equilibrium, gas-dynamic considerations in conjunction with the Thomson equation can be used to estimate the local temperature at which the cluster distributions were attained in the experiments of Searcy and Fenn (Searcy, Fenn, 1974) and those of Castleman (Castleman, 1978). Since the Thomson equation approximates the entropy of moderate sizes of positive ion clusters with water reasonably well, it has been used to make estimates of the values for the cluster range 9 to 26 molecules.

In this manner, approximate enthalpy values have been derived from the free jet expansion experiments of Castleman, and those of Searcy and Fenn. These values are plotted in Figure 8 along with data for small clusters due to Kebarle, et al. (Kebarle, 1972). For comparison, a solid line is plotted which indicates the ΔH^O values predicted by the Thomson equation. The less negative value for the enthalpy of formation of the 22nd, compared to the 21st cluster about H_3O^+, is evident. Furthermore, the ΔH^O for the 21st cluster is seen to be slightly more exothermic than for the 20th. The trend is in accord with expectation based on thermodynamic considerations, but the decrease is small and may merely reflect scattering of the experimental data.

It is interesting to speculate on the nature of the structure and reasons for the unusual stability of the 21st cluster. Burke (Burke, 1978) has suggested that condensation may proceed to an ice-like structure. This gives a plausible explanation for the diminishing rate observed in his experiments for the addition of the 22nd and 29th water molecule to the preceding clusters. However, this model does not seem to account for the anamously large ion clusters formed by ionizing neutral ones. Kassner and coworkers (Hagen, Kassner, 1974; Plummer, Hale, 1972) have speculated that the 21st cluster is one of unusual stability because it can form a dodecahedron structure with an H_3O^+ ion located at its center.

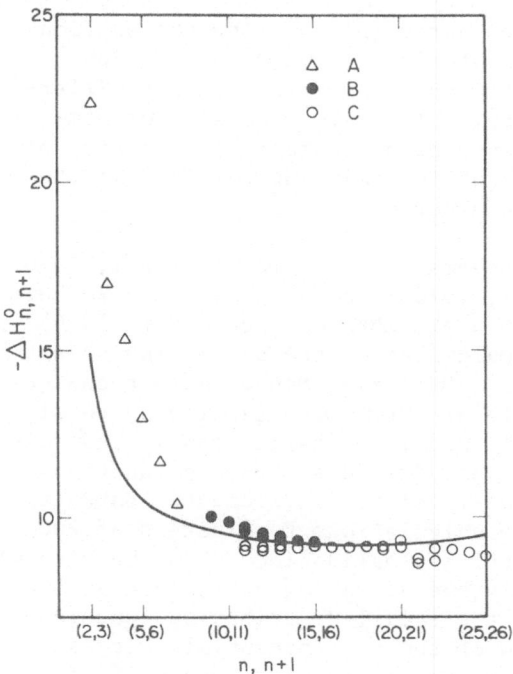

Fig. 8. Enthalpy of hydration versus cluster number, n. The solid curve is calculated from the Thomson equation. Points: A (Castleman, unpublished results), B (Kebarle, 1972), and C (Keesee, Lee, Castleman, 1978).

Earlier, before any experimental observations were made, Siksna predicted that a cluster of 21 water molecules, clustered with H^+, might be very stable. His model also presumes that the 21st cluster is a clathrate in the form of the dodecahedren, but, instead of the H_3O^+ ion being located at the center he assumes that the proton is located either on, or attached to, the clathrate surface. A crucial feature of this model is the possibility that migration of the excess proton through hydrogen bonds over the surface of the clathrate might stabilize the structure. Recently Newton (Newton, 1977) has made _ab initio_ calculations of the structure of H_3O^+ water clusters, and has shown that only three water molecules bond directly to the H_3O^+ ion; the addition of a fourth molecule was found to be repulsive. This suggests (Holland, Castleman, 1978) that the transient configuration in which a proton is localized on the surface, should have favorable bonding. Consequently, the delocalization of charge might have the effect of stablizing and reinforcing the clathrate structure by bonding the transient H_3O^+ ion. This is seen to have features in common with the Eigen structure (Eigen, 1958) for H_3O^+ in water. Despite the attractive features of this model, it apparently fails to account for the reported stability of a cluster involving 28 water molecules.

Relationship of Gas Phase Clustering to Ion Solvation Phenomenon

Several factors must be considered in assessing the application of data on gas phase ion clustering to understanding ion solvation. These include effects, and therefore possible differences, due to: 1) neighboring molecules (e.g., density or compressive effects), 2) the existence of a "surface" in the case of the ion cluster, and 3) the lack of ion pair interactions in the case of the usual ion clustering experiment.

Clementi and coworkers (Kistenmacher, Popkie, Clementi, 1976, 1973,1974; Clementi and Barsotti, 1978), and Abraham and coworkers (Abraham, Mruzik, Pound, 1976; Mruzik, Abraham, Schreiber, 1976) have undertaken theoretical calculations of the properties of small clusters about closed shell ions. Utilizing Monte Carlo techniques, Abraham has deduced the structure of clusters containing up to six water molecules about both alkali metal and halide ions. The calculations have been used to derive radial distribution functions for H and O about the central ions. Clementi (Clementi, Barsotti, 1978) has recently extended such calculations to clusters of 200 water molecules about alkali metal and halide ions. In the case of Abraham's calculations where only one solvation shell is possible, it is expected that the peak in the radial distribution function will be located near the minimum of the pair potential. It is interesting to note that the position of the first peaks (first solvation shell) in the radial distribution function for the larger clusters are located at virtually the same distance from the central ion. Yet, in the case of the large clusters, higher order solvation shells also exist.

Regarding the possible influence of a "surface", it is not evident that one actually exists for the very small clusters typically investigated in high pressure ion cluster experiments. However, even if surface effects do occur, their contribution can be seen to be relatively small even regarding their influence on entropies. (See Table I).

Kebarle has considered the application of gas phase ion clustering experiments as a tool in understanding the liquid phase. The reader is referred to recent reviews (Kebarle, 1972, 1975,1977) and references contained therein. Castleman and coworkers (Castleman, Holland, Lindsay, Peterson, 1978) have recently undertaken a study of ammonia clustered to alkali metal ions to investigate analogies between gas-phase clusters and complexes thought to exist in the liquid phase. In the case of the larger alkali metal ions, K^+ and Rb^+, the gas-phase derived bond energies were found to display a continuous decrease with increasing cluster size. However, in the case of Li^+ and Na^+, the data show a significant effect, where the bond energy between the fourth and fifth ammonia cluster diminishes abruptly. These data suggest

that a well-defined "coordination shell" may exist for ammonia about
the smaller alkali metal ions. Interestingly, the results of Raman
spectra studies with Li^+ in liquid ammonia (Gans, Gill, 1976)
also indicate that the coordination number of Li^+ is four. Taken
together, these findings strongly suggest that gas phase ion clustering studies do reflect the structure of ions solvated in liquids.

A comparison of the experimental bond energies for the attachment of the first cluster of water and ammonia to the alkali metal
ions shows that the bonding of ammonia is invariably greater. This
is in accord with the greater polarizability of ammonia compared to
water ($2.26\mathring{A}^3$ compared to $1.45\mathring{A}^3$, respectively (Weast, 1976)).
Since their respective dipole moments are 1.47 and 1.84 Debye
(Coles, Good, Bragg, Sharbough, 1951; Robinson, Stokes, 1959), the
findings indicate the important role of ligand polarization in the
bonding. This finding is consistent with recent extended Hückel
calculations (Castleman, 1978) which show that, even though there
is little transfer of charge in either ligand system, there is
slightly more in the case of ammonia than water; this fact explains
its enhanced bonding.

Generalizations which are valid for explaining the relative
bond strength for the first ligand attachment, often (Kebarle, 1975)
hold for attachment of the second as well, but crossover in relative
bond strengths are frequently noted (Kebarle, 1975; Castleman,
Holland, Lindsay, Peterson, 1978; Castleman, 1978) at higher cluster sizes.

The importance of these crossovers in governing relative
cluster stabilities can be readily seen by considering, as example,
the bonding of NH_3 and H_2O to the alkali metal ions. Referring to
Figure 9, it can be seen that in the case of each of the ions Li^+,
Na^+, K^+, and Rb^+, the bond energy for the attachment of NH_3 is
greater than H_2O for at least the mono and diligand clusters. But,
beyond the fourth cluster, water is invariably more strongly bonded.

A consideration of trends such as these leads to a ready
understanding of why ions, which have a greater affinity for solvation by a given molecule, do not always display the same tendency
in promoting nucleation. It is evident that if sufficient neutral
(ligand) molecules cluster about an ion, it eventually approaches
a state equivalent to that of a liquid droplet. Summation of the
various enthalpy changes for the successive clustering reactions
up to infinite cluster size, would clearly lead to a direct measure
of the solvation energy. In fact, however, the distance over which
the central charge has an appreciable effect is rather short. Therefore, after a relatively small number of ligands are clustered, the
various ion-ligand bond energies become nearly equal. Interest,
then, centers on the small clusters, and these are the ones most

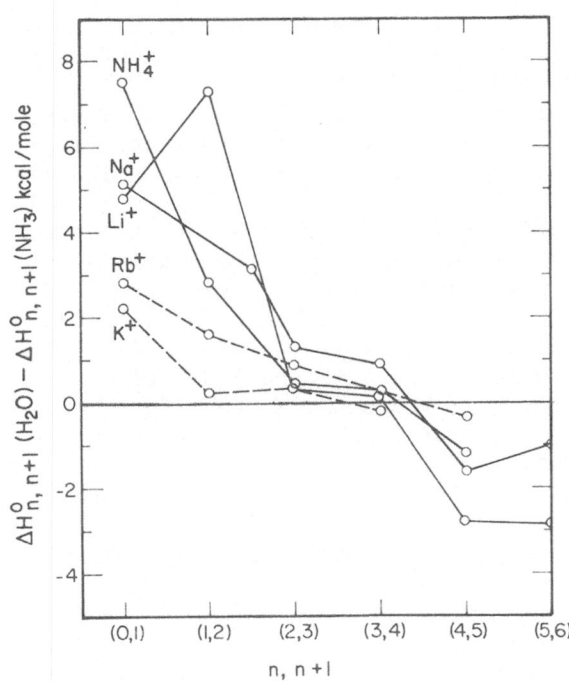

Fig. 9. Comparison of
the enthalpy change for
the hydration, compared
to the ammoniation, of
selected ions. Hydration
data from Kebarle, 1972;
alkali metal data of
ammoniation from Castle-
man, Holland, Lindsay,
Peterson, 1978; NH_4^+
data from Tang and Castleman, 1975.

accessible to direct experimental investigation. The ability of
certain ions to promote nucleation more readily than others is also
reflected in the nature of their bonding at the small cluster sizes.

Kebarle (Kebarle, 1972) has shown, for the case of water clustered about alkali metal ions, that the difference between the total
enthalpy of hydration of a given ion under consideration, compared
to a reference ion, approaches a constant value. Interestingly
enough, in the case of alkali metal ions, these values were also
found to rapidly approach those predicted from a consideration of
the Randle heats of solvation for the individual ions in the bulk
liquid phase. We have employed similar considerations to deduce
the cluster size at which the successive addition of ammonia molecules becomes independent of the nature of the central ion. Referring to Figure 10, this is seen to be the situation above cluster size 5 for Li^+, Na^+, NH_4^+, and K^+.

In the case of data plotted for negative ions in Figure 11,
it can be seen that there is also a tendency for the relative integral heat of formation of the nth cluster of water to level off.
However, it is evident that the degree of clustering necessary before a leveling-off occurs is greater in the case of the negative

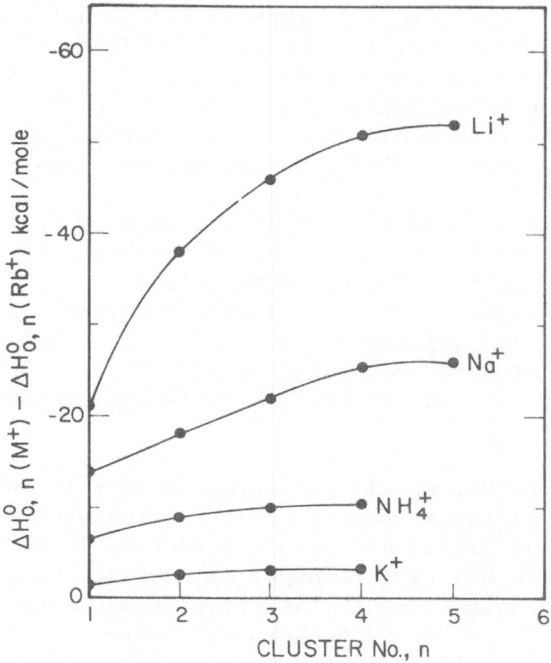

Fig. 10. Integral enthalpies of ammoniation for various ions referenced to those for Rb^+. All data from Castleman, Holland, Lindsay,
Peterson, 1978 except NH_4^+ data taken from Tang and Castleman, 1975.

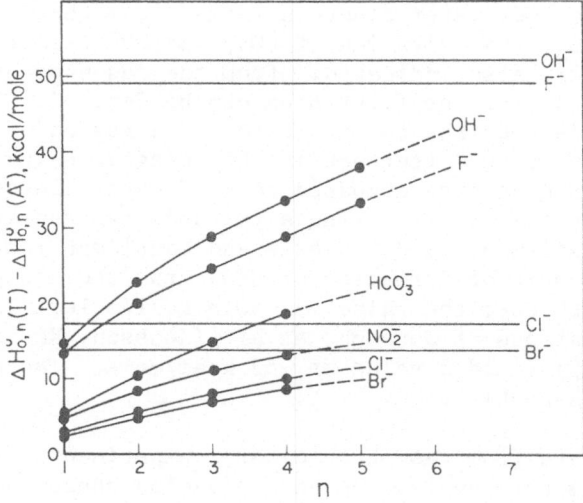

Fig. 11. Integral enthalpies of hydration for certain negative ions
referenced to those for I^-. Adapted from Keesee, Lee, Castleman, 1978.

ions than positive ones. This is interpreted as being due to the difference in the structure, and therefore bonding, of anions compared to cations. In the case of the former, it is apparently the hydrogen atom rather than oxygen which is directed toward the charged site. This presumably enables the ligands to cluster in greater number in close proximity to the negative ion.

It has sometimes been noted that the interaction of a given ligand series with an ion in the gas phase differs from that in solution. Arnett, et al. (Arnett, Jones, Taagepera, Henderson, Beauchamp, Holtz, Taft, 1972) have given a detailed analysis of this problem. With due recognition of the fact that both ligand and ion solvation must be properly accounted for in making a Born-Haber cycle, a direct comparison between gas phase and solvation energies is possible.

Recently, Jönsson, et al. (Jönsson, Karlström, Wennerström, 1978) reported ab initio molecular orbital calculations for the water-carbon dioxide system and have considered, in particular, the reaction of OH^- with CO_2 to form HCO_3^-. In recent gas phase experiments, we have investigated the clustering of water molecules to HCO_3^-. These data, together with those available for OH^-, are plotted in Figure 11. It is interesting to use a Born-Haber cycle to compare the theoretical calculations with the experimental trends in relative solvation energies derived from Figure 11, and data available for the heat of solvation of CO_2 in water.

The requisite Born-Haber cycle is given in Table II, which represents a sequence of reactions between OH^- and CO_2 in the gas phase, leading to the formation of HCO_3^-. Sequential clustering reactions involving HCO_3^- water clusters formed from the reaction of $OH^-(H_2O)_n$ with CO_2, the solvation of CO_2, and OH^- reaction in solution with CO_2, are also indicated. From the sequence of reactions, a value of the heat for the first step can be derived. It is apparent that this requires the existence of a region of cluster size where the relative difference in the integral heat of formation of OH^--water clusters, compated to HCO_3^--water clusters, approaches a constant value. Using a suitable extrapolation (Keesee, Lee, Castleman, 1978), Keesee and Castleman (unpublished) have deduced the heat of formation of HCO_3^- from the gas phase reaction of OH^- with CO_2; the value is -40.5 kcal/mole. The recent ab initio calculations of Jönsson, et al. (Jönsson, Karlström, Wennerström, 1978) yield a value of -53 kcal/mole. The agreement is seen to be reasonably good.

Clearly, data from gas phase cluster experiments are useful for understanding both nucleation and solvation phenomenon. The important next step in elucidating the molecular aspects of solvation is to determine the effects due to ion pair formation. This

Table II. Born-Haber Cycle for the System OH^-, CO_2, and HCO_3^-:
Gas Phase Reaction and Solvation

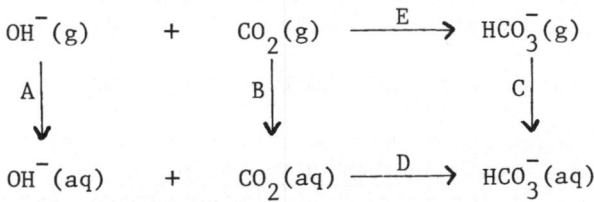

A-C = -24.0 kcal/mole; difference in the total integral heats for the clustering reactions: $OH^-(g)+nH_2O(g)$ $=OH^-(H_2O)_n(g)$ and $HCO_3^-(g)+mH_2O(g)=HCO_3^-(H_2O)_m(g)$. This value was derived from data given in Figure 11 (Keesee, Lee, Castleman, 1978).

B = $\Delta H_f^O \left[CO_2(aq) \right] - \Delta H_f^O \left[CO_2(g) \right] = -4.9$ kcal/mole from: "Selected Values of Chemical Thermodynamic Properties" NBS Technical Note 270-3 (1978).

D = Enthalpy of reaction in the aqueous phase (experimental data reported in Jönsson, Karlström, Wennerström, 1978) = -11.2 kcal/mole.

E = Enthalpy of the gas-phase reaction
  = A-C+B+D

---

requires experiments where ions of both sign are simultaneously present in the gas phase cluster being studied; such experiments are currently under way in our laboratory.

Acknowledgements

Support of the Atmospheric Sciences Section of the National Science Foundation under Grant No. ATM 76-14914, the U. S. Army Research Office under Grant No. DAAG 29-76, the National Aeronautics and Space Administration under Grant No. NSG-2248 and the Department of Energy No. EP-78-S-02-4776 is gratefully acknowledged.

References

Abraham, F. F., "Homogeneous Nucleation Theory, the Pretransition Theory of Vapor Condensation", Supplement I, Advances in Theoretical Chemistry, Academic Press, N.Y., 1974.

Abraham, F. F., Mruzik, M. R. and Pound, G. M., Faraday Disc. of the Chem. Soc., 61, 34 (1976).

Arnett, E. M., Jones III, F. M., Taagepera, M., Henderson, W. G., Beauchamp, J. L., Holtz, D. and Taft, R. W., J. Am. Chem. Soc., 94, 4724 (1972).

Burke, R., remarks made at this NATO Advance Study Institute, 1978.

Burke, R. R., "Shock Layer Measurements of Decomposition Reactions of Water Cluster Ions" in COSPAR Symposium on D- and E-Region Ion Chemistry - An Information Symposium Record, Aeronomy Report, No. 48, Univ. of Ill. (1972), pp. 259-265.

Castleman, A. W., Jr., Advances in Colloid and Interface Science: Nucleation Phenomenon III. (in press) 1978.

Castleman, A. W., Jr., Chem. Phys. Lett., 53, 560 (1978).

Castleman, A. W., Jr., Holland, P. M. and Keesee, R. G., J. Chem. Phys., 68, 1760 (1978).

Castleman, A. W., Jr., Holland P. M., Lindsay, D. M. and Peterson, K. I., J. Am. Chem. Soc., 100, 6039 (1978).

Castleman, A. W. Jr., and Tang, I. N., J. Chem. Phys., 57, 3629 (1972).

Clementi, E. and Barsotti, R., "Study of the Structure of Molecular Complexes. Coordination Numbers for Li^+, Na^+, K^+ and Cl^- in Chem Phys. Lett. (in press) 1978.

Clementi, E., "Lecture Notes in Chemistry", 2, "Determination of Liquid Water Structure, Coordination Numbers of Ions and Solvation for Biological Molecules",Springer-Verlag, 1976.

Clementi, E., (See Kistenmacher, H.)

Coles, D. K., Good, W. E., Bragg, J. K. and Sharbough, A. H., Phys. Rev., 82, 877 (1951).

Eigen, M. and DeMaeyer, L., Proceedings Royal Society A. 247, 505 (1958).

Fenn, J. B. (See Searcy, J. Q.)

Frankel, J., "Kinetic Theory of Liquids", Dover, N.Y., (1946); Khurt, F., Z. Phys., 131, 185 (1952).

Gans., P., and Gill, J. B., J. Chem. Soc.,Dalton Trans.779 (1976).

Haugen, D. E. and Kassner, J. L., Jr., J. Chem. Phys., 61, 4285 (1974).

Heicklen, J., "Colloid Formation and Growth, A Chemical Kinetics Approach", Academic Press, N.Y., (1976).

Hertz, H. C., Z. für Elektrochemie, 60, 1196 (1956).

Holland, P. M. and Castleman, A. W., Jr., (to be published) 1978.

Jönsson, B., Karlström, G. and Wennerström, H., J. of Am. Chem. Soc., 100, 1658 (1978).

Kassner, J. L. and Haugen, D. E., J. Chem. Phys., 64, 1860 (1976).

Kassner, J. L. (also see Haugen, and Plummer and Hale)

Kay, B. D., Hermann, V. and Castleman, A. W., Jr.,(to be published) 1978.

Kebarle, P., "Ion Thermochemistry and Solvation From Gas Phase Ion Equilibria", "Ann. Rev. Phys. Chem.", 28, 445 (1977).

Kebarle, P., "Thermochemical Information from Gas Phase Ion Equilibria", "Interactions Between Ions and Molecules", (NATO Adv. Study Institute Series B: Physics, 6, Plenum Press, N.Y. 1975, 456-487, Ausloos, P. Ed.)

Kebarle, P., "Higher Order Reactions-Ion Clusters and Ion Solvation", Chap. 7, "Ion-Molecule Reactions", 1, Franklin, J. L. Ed., Plenum Press, N.Y. (1972).

Keesee, R., Lee, N. and Castleman, A. W., Jr.,"The Properties of Clusters in the Gas Phase: III. Hydration Complexes of CO_3^- and HCO_3^-" J. Am. Chem. Soc. (in press) 1978.

Kistenmacher, H., Popkie, H. and Clementi, E., J. Chem. Phys., 59, 5842 (1973).

Kistenmacher, H., Popkie, H. and Clementi, E., J. Chem. Phys., 58. 1689 (1973).

Kistenmacher, H., Popkie, H. and Clementi, E., J. Chem. Phys., 61, 799 (1974).

Kortzeborn, N. and Abraham, F. F., J. Chem. Phys., 58, 1529, (1973).

Lee, N., Keesee, R. and Castleman, A. W., Jr., (recent results to be published) 1978.

Lin, S. S., Rev. Sci. Instrum., 44, 516 (1973).

Loeb, L. B., Kip, A. F. and Einarsson, A. W., J. Chem. Phys., 6, 264 (1938).

Lothe, J. and Pound, G. M.,"Nucleation", Chap. 3 (See Zettlemoyer 1969).

Lothe, J. and Pound, G. M., J. Chem. Phys. 45, 630 (1966).

Mason, B. J.,"The Physics of Clouds", Chapter, 1, Oxford Press, London, 2nd Ed. (1971).

Mruzik, M. R., Abraham, F. F. and Schreiber, D. E., J. Chem. Phys., 64, 481 (1976).

Newton, M. D. , J. Chem. Phys., 67, 5535 (1977).

Plummer, P. L. M. and Hale, B. N., J. Chem. Phys., 56, 4329 (1972).

Pound, G. M., J. Physical and Chemical Reference Data, 1, 119 (1972).

Reiss, H., Advances in Colloid and Interface Science, 7, 1 (1977).

Robinson, R. A. and Stokes, R. H., "Electrolyte Solutions",Butterworth Scientific Publications, Ltd., London, 1959, p.1.

Russell, K. C., J. Chem. Phys., 50, 1809 (1969).

Scharrer, I., Ann. Phys., 35, 619 (1939).

Searcy, J. Q. and Fenn, J. B., J. Chem. Phys., 64, 1861 (1976).

Searcy, J. Q., J. Chem. Phys., 63, 4114 (1975).

Searcy, J. Q. and Fenn, J. B., J. Chem. Phys., 61, 5282 (1974).

Siksna, R., "Water Clathrates II", UURIE Report, 53:73, Uppsala Universitet Institutet für Högspänningstorskning (August, 1973).

Tang, I. N. and Castleman, A. W., Jr., J. Chem. Phys., 62, 4576 (1975).

Thomson, J. J., "Applications of Dynamics to Physics and Chemistry" Vol. I, p. 165, McMillan & Co., London 1888; and "Conduction of Electricity Through Gases", 3rd Edition, Vol. 1, Cambridge 1928.

Weast, Ed., "Handbook of Chemistry and Physics", 57th Ed., Chemical Rubber Co., Cleveland, Ohio, 1976.

Zettlemoyer, A. C., Ed., "Nucleation", Marcell Dekker, N. Y.(1969).

CHEMICAL IONIZATION IN FLAMES

D. K. Bohme

Department of Chemistry and C.R.E.S.S., York University,

Downsview, Ontario M3J 1P3, Canada.

The ion chemistry which proceeds in natural hydrocarbon flames, in particular CH_4/O_2 flames, is viewed from the perspective provided by recent room-temperature measurements of individual ion-molecule reactions. Limited reaction schemes are proposed for both positive and negative ions in the preheat region, the reaction zone, and the burnt gas downstream. Attention is drawn to further measurements of ion-molecule reactions which are required to improve our understanding of the ion chemistry proceeding in such extreme and complex environments.

INTRODUCTION

Recent developments in mass spectrometric flame-ion sampling techniques and the design of flame burners have allowed the measurement of complete families of axial profiles for both positive and negative ions all the way from the cooler upstream region through the flame front into the burnt gas downstream. This has been accomplished with a resolution sufficient to separate the sequential features of the ionization into several spatially and therefore chemically distinct regions. In principle such measurements can be exploited in at least two different ways:

1. When the axial evolution of the neutral composition is known from independent measurements or calculations, the ion profiles can provide fundamental information on the kinetics and equilibria of ion-molecule reactions under conditions of pressure and temperature not easily accessible through other techniques.

323

2. Alternatively, given a sufficient understanding of the underlying ion-molecule reactions, these profiles can provide detailed information about the axial evolution of neutral composition in a flame and thereby reveal the stepwise progress of combustion.

In practice, both of these endeavours are limited by the distortion in the relative abundance of ions which may accompany the mass spectrometric sampling of the flame and, of course, also by the sheer complexity of the ion chemistry - the number of different negative and positive ions found naturally in a flame can be very large and we can expect ion-molecule reactions to proceed with virtually every neutral species in the flame. However, sampling effects, particularly those associated with the cooling in the boundary layer around the sampling orifice and in the supersonic expansion inside the sampling cone, are beginning to be understood (Hayhurst and Kittelson, 1977) and there has been a concomitant increase in the authenticity of measured ions and their profiles. Also there continues to be a steady growth in kinetic and thermodynamic information about individual ion-molecule reactions of relevance to flames. Here we attempt to illustrate how these advances have led to a significant increase in our understanding of flame ion chemistry by providing an account of a limited scheme for the ion chemistry proceeding in a CH_4/O_2 flame. This flame was chosen in part because of the availability of extensive profile measurements for positive and negative ions (Sugden et al., 1973; Goodings et al., 1977; Bohme et al., 1977; Goodings et al., 1978a, 1978b) as well as neutral species (Peeters and Mahnen, 1973; Hastie, 1973) for various flame compositions. With these profiles as a guide, the scheme is conceived in terms of our direct laboratory experience with individual ions and ion-molecule reactions proceeding at room temperature, viz. rate constants, branching ratios, heats of formation, proton and electron affinities etc. determined at 300K. No account is taken of excess energy which an ion may acquire in its initial formation and therefore have available in a subsequent reactive collision. Ionization is initiated through the formation of CHO^+ and free electrons by chemi-ionization. The fate of these species is then traced through the evolving neutral bath all the way from their initial formation in the preheat region through the flame front to the burnt gas downstream.

COMPOSITION

In a stationary flame, the fuel, O_2 and diluent (eg. N_2 or Ar) can conveniently be regarded to pass through three regions: the preheat region, the reaction zone and the burnt gas region downstream. The bulk of the oxidation of fuel proceeds by very rapid branched chain reactions in the reaction zone which is often luminous and extremely thin at atmospheric pressure, eg., 0.3 mm for a conical premixed fuel-rich (ϕ = 2.15) CH_4/O_2 flame with a cone base diameter of 2.5 mm and a cone height of 4 mm (Goodings et al, 1978a). The gradients in temperature, concentration, and velocity which exist across the reaction zone are extremely large. The chain reactions are initiated upstream due to the counterflow transport of heat and matter, particularly H atoms and electrons, into the preheat region where they result in the early formation of free radicals, eg.,

$$H + O_2 \rightarrow [HO_2] \rightarrow OH + O$$

$$O\,(OH) + CH_4 \rightarrow CH_3 + OH\,(H_2O)$$

$$CH_3 + O \rightarrow CH + H_2O$$

$$\rightarrow CH_2O + H$$

$$CH_3 + O_2 \rightarrow [CH_3O_2] \rightarrow CH_2O + OH$$

$$CH_2O + H\,(O, OH) \rightarrow CHO + H_2\,(OH, H_2O)$$

as well as the growth of certain stable intermediates such as formaldehyde and methanol. This is evident from the composition profiles shown in Fig.1 for a fuel-lean CH_4/O_2 flame. Other stable intermediates in CH_4/O_2 combustion which have either been established experimentally or involved in models of the neutral chemistry include acetylene, formic acid, acetaldehyde and ketene which has been invoked in the first step in the oxidation of intermediate acetylene (Miller, 1968). Free radicals and stable intermediates may rise to levels in the reaction zone well above their final equilibrium values further downstream. Some radicals, eg., O, OH and H, may persist well downstream, comprising up to a few percent of the gas leaving the reaction zone which consists largely of the final equilibrium products of combustion, eg., H_2O, CO_2 and H_2. The actual composition and temperature profiles will be dependent on the initial composition of the premixed gas and indeed may be altered deliberately by changing the molar ratio of fuel to oxidant.

Hydrocarbon flames are weak plasma containing natural levels of ions up to 1 p.p.m. Fig. 2 shows total positive and negative ion concentration profiles measured recently for a fuel-lean CH_4/O_2 flame (Goodings et al, 1978b). The positive ions peak just downstream of the luminous zone at $\sim 4 \times 10^{10}$ ions cm^{-3}. The electrons apparently have a mobility high enough to diffuse upstream of the positive ions where they may attach to electronegative molecules to form negative ions.

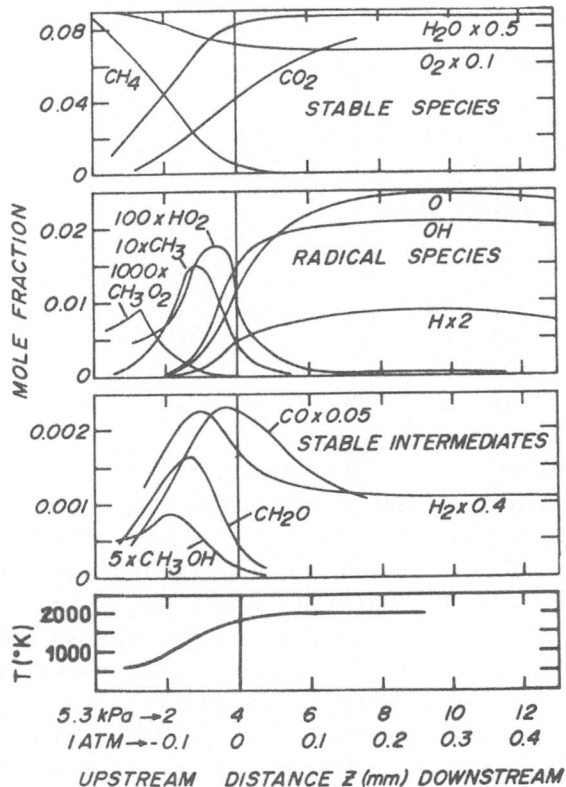

Fig. 1 Neutral concentration profiles obtained from the beam sampling experiments of Peeters and Mahnen (1973) for a fuel-lean CH_4/O_2 flame at 40 torr, together with an approximate temperature profile. The distance has been scaled to atmospheric pressure (Bohme et al, 1977)

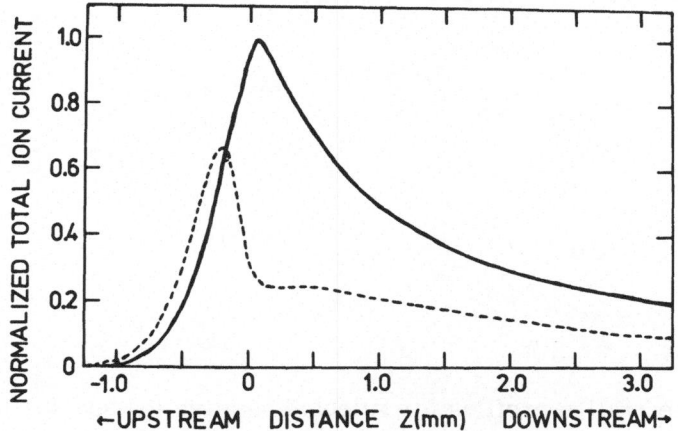

Fig. 2   Total positive (solid curve) and negative (dashed curve) ion concentration profiles observed for a fuel-lean CH_4/O_2 flame at atmospheric pressure (Goodings et al, 1978b).

Under these fuel-lean conditions the negative ions persist well into the burnt gas region downstream.

A wide variety of positive and negative ions have been identified mass-spectrometrically in CH_4/O_2 flames. The ions are not distributed randomly throughout the flame but appear in several spatially distinct groups, for example, as shown in Fig. 3 for the dominant positive ions found in a fuel-rich CH_4/O_2 flame.

POSITIVE ION CHEMISTRY

Ion formation in hydrocarbon flames is generally accepted to proceed initially via the chemi-ionization reaction

$$CH + O \rightarrow HCO^+ + e$$

which is 4 kcal mole^{-1} exothermic. Formation of the COH^+ isomer from ground state CH and O is energetically much less favourable as it is endothermic by $\sim$ 14 kcal mole^{-1} and requires the insertion of O into the C-H bond. The relative contribution of thermal ionization will increase downstream with the build-up of species with low ionization potentials and the rise in temperature. HCO^+ is unreactive towards CH_4 and O_2 at room temperature and so can be expected to seek out the radical species produced during the initial stages of combustion. Laboratory measurements of reactions of HCO^+ with radical species

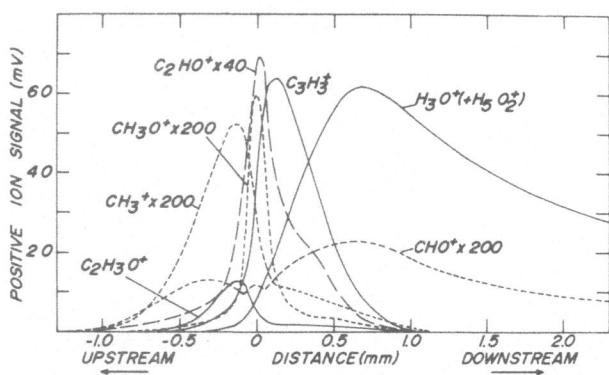

Fig. 3 Concentration profiles for selected positive ion species showing
several spatially distinct regions of flame ionization in a fuel-rich
CH_4/O_2 flame at atmospheric pressure (Goodings et al, 1977).

continue to be elusive but, by analogy with its known reactions with
stable species (Tanner et al, 1978) and given the proton affinity and
stability of CO, we can anticipate the reactions with radicals of higher
proton affinity (see Fig. 4) to proceed predominantly by proton transfer.
The possibility of charge transfer can be ruled out almost completely
given the low ionization potential of HCO. Also there appear to be no
exothermic reaction channels with O and H. With CH_3 we should allow
for the occurrence of the reaction

$$HCO^+ + CH_3 \quad \rightarrow \quad CH_3CO^+ + H$$

which is exothermic by 25 kcal mole^{-1} if the ion product is protonated
ketene.

The reactions of HCO^+ with the early radical species and the
secondary ion chemistry initiated by their ion products in excess O_2
(fuel-lean mixture) and excess CH_4 (fuel-rich mixture) are summarized
in Fig. 5. This scheme makes extensive use of the room-temperature
measurements reported recently by Smith and Adams (1977) and Adams
et al (1978). CH_3^+ which is the terminal ion in the radical proton
affinity ladder is known to associate rapidly with O_2 to yield $CH_3O_2^+$.
The reactions of CH_2^+, CH_2O^+ and CH^+ with O_2 re-establish
CHO^+ and some $H_2O_2^+$. There is some leakage into O^+, CO^+ (from
$CH^+ + O_2$) and O_2^+ (from O^+, CO^+, $H_2O^+ + O_2$) which is known to
react with CH_4 to again produce $CH_3O_2^+$. Reaction with CH_4 will
convert CH_x^+ into $C_2H_x^+$ species by condensation, CH_2O^+ into CH_3O^+
and $C_2H_5O^+$, and H_2O^+ into H_3O^+. $C_2H_2^+$ and $C_2H_3^+$ will continue to

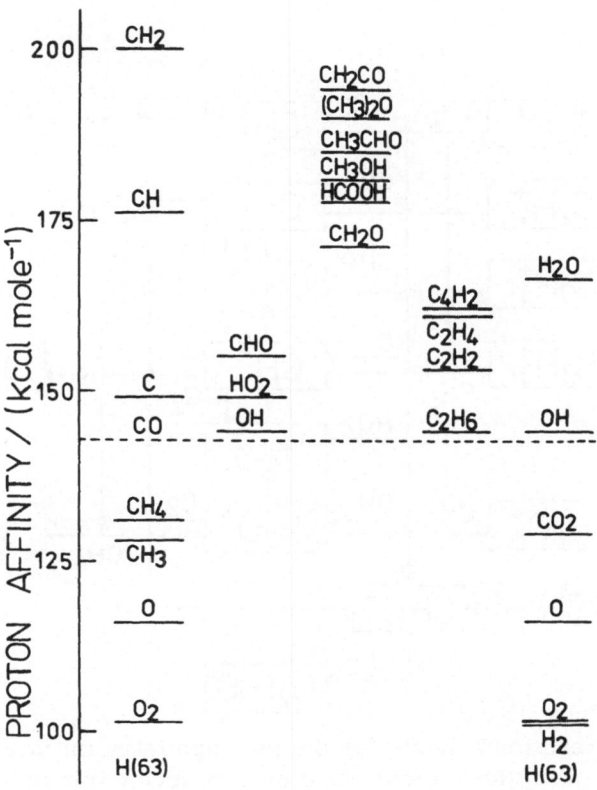

Fig. 4 Proton affinities of selected flame constituents.

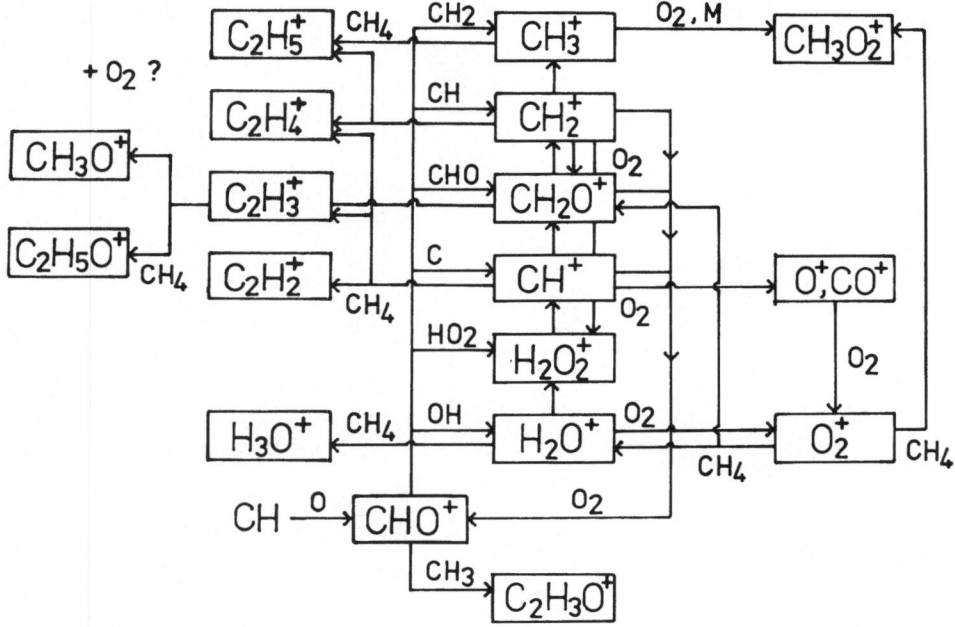

Fig. 5   Limited reaction scheme for the ion chemistry initiated by the reactions of HCO$^+$ with the radical species produced early in a CH$_4$/O$_2$ flame under fuel-rich (left) and fuel-lean (right) conditions.

condense with CH_4 to establish $C_3H_5^+$ and $C_3H_4^+$ but reactions of $C_2H_x^+$ with O_2 are generally unknown.

As the stable intermediate molecules grow in concentration they will compete for the protons but CH_3^+ will remain as the terminal ion because of its high stability. At the same time CH_3^+ may associate with these molecules in fast association reactions of the type recently characterized by Smith and Adams (1978) for CH_2O, CH_3OH, H_2O, CO, CO_2, H_2 and O_2. Simultaneous bimolecular products were observed only with CH_2O and CH_3OH :

$$CH_3^+ + CH_2O \rightarrow CHO^+ + CH_4$$
$$\overset{M}{\rightarrow} C_2H_5O^+$$

$$CH_3^+ + CH_3OH \rightarrow CH_3O^+ + CH_4$$
$$\overset{M}{\rightarrow} C_2H_7O^+$$

The high values of the termolecular coefficients at 300K together with the temperature dependences reported by these authors suggest that reactions of this type may well have effective bimolecular coefficients at atmospheric pressure substantially in excess of 10^{-11} cm^3 molecule^{-1} s^{-1} even at 2000K. The magnitude of these termolecular coefficients is taken as testimony of the high stability of the association products. Indeed, the CH_3^+ adducts have isomers which may be formed by the direct protonation of a stable molecule. For example, isomers of the CH_2O and CH_3OH adducts of CH_3^+ may be formed by the direct protonation of CH_3CHO and C_2H_5OH or $(CH_3)_2O$, respectively. This has obvious implications for the inference of neutral flame composition from observed flame ion profiles. The association reactions of CH_3^+ are included in Fig. 6 except the reactions with CO_2 and the intermediate molecules $HCOOH$, CH_2CO, and CH_3CHO which would generate $C_2H_3O_2^+$, $C_2H_5O_2^+$, $C_3H_5O^+$ and $C_3H_7O^+$, respectively.

Ultimately, with the build-up of O, H and OH and the equilibrium products H_2, CO, CO_2 and H_2O, we can expect the proton to become preferentially associated with H_2O, the species with the highest proton affinity in the burnt gas region. Formation of H_3O^+ may be accomplished in several ways: directly by ion-molecule reactions involving O, H or OH, eg.,

$$CH_3O^+ + O \rightarrow H_3O^+ + CO,$$

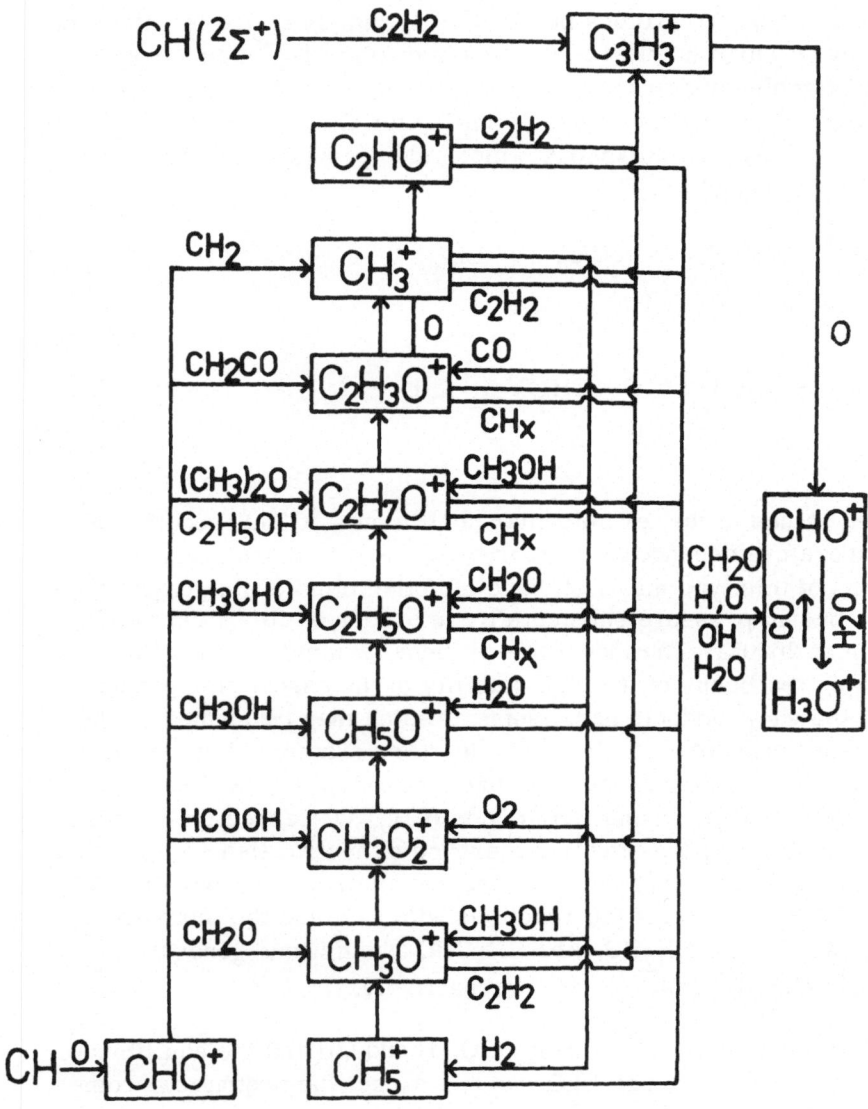

Fig. 6 Limited reaction scheme for the positive ion chemistry involving the stable intermediates and radicals which rise in the reaction zone.

by direct protonation of H_2O through endothermic proton-transfer reactions involving protonated intermediates of slightly higher proton affinity (see Fig. 4), eg.,

$$CH_3O^+ + H_2O \rightarrow H_3O^+ + CH_2O$$

or indirectly through the initial formation of protonated species of lower proton affinity such as HCO^+, H_2O^+, HCO_2^+ etc. (see Fig. 4), eg.,

$$CH_3^+ + O \rightarrow CHO^+ + H_2$$

$$CHO^+ + H_2O \rightarrow H_3O^+ + CO$$

Some of these reactions may be driven towards equilibrium in the burnt gas region where the required neutrals achieve their equilibrium concentrations. The large amounts of equilibrium water vapour can be expected to promote some hydration of ions, particularly H_3O^+, even at the high temperature prevalent in this region:

$$H_3O^+ + H_2O + M \rightarrow H_3O^+. H_2O + M$$

In a fuel-rich CH_4/O_2 flame, C_2H_2 rises in the reaction zone to become the major stable intermediate (Hastie, 1973). Concomitant species with similar profiles but lower concentrations include CH_3 and C_4H_2. The room-temperature reaction of CH_3^+ with C_2H_2 also has a bimolecular reaction channel:

$$CH_3^+ + C_2H_2 \rightarrow C_3H_3^+ + H_2$$

which has been observed to be rapid, $k = 1 \times 10^{-9}$ cm^3 $molecule^{-1}$ s^{-1}, in several laboratories. That $C_3H_3^+$ is an important intermediate ion for such flames is clearly evident from Fig. 3. Other processes which may contribute to its formation are indicated in Fig. 6 and include the chemi-ionization reaction of excited CH with C_2H_2 as well as ion-molecule reactions with radicals of the type CH_x (x = 0-3) which may arise from the decomposition and other reactions of C_2H_2 produced in the reaction zone and from any persistent (unburnt) CH_4. Formation of $C_3H_3^+$ from C_3H_2 by direct proton transfer may be possible but there is only circumstantial evidence for its existence in flames and no indication of the proton affinity of C_3H_2 (Hayhurst and Kittelson, 1978).

The presence of C_2H_2, C_4H_2 and CH_x at the tip of the luminous zone should encourage the formation of large carbonaceous cations in this region. Double carbon ions may be formed by direct ionization or protonation of C_2H_2, or by condensation reactions between single carbon ions and single carbon neutrals as alluded to earlier in connection with the ion chemistry in the preheat region. The $C_2H_x^+$ ions can continue to condense with CH_x neutrals sequentially to build up the hydrocarbon skeleton one carbon at a time, eg.,

$$C_2H_2^+ + CH_4 \rightarrow C_3H_5^+ + H$$
$$\rightarrow C_3H_4^+ + H_2,$$

or by analogous reactions with C_2H_2 two carbon atoms at a time,

$$C_2H_2^+ + C_2H_2 \rightarrow C_4H_3^+ + H$$
$$\rightarrow C_4H_2^+ + H_2,$$

or even with polyacetylenes to build up the skeleton 2n carbon atoms at a time. Alternatively acetylene may simply associate, eg.,

$$C_3H_3^+ + C_2H_2 + M \rightarrow C_5H_5^+ + M$$

Finally another route may be envisaged which will lead to the formation of large carbonaceous ions which are observed to persist particularly in fuel-rich CH_4/O_2 and in C_2H_2/O_2 flames (Goodings et al, 1978a; Hayhurst and Kittelson, 1978). This route involves the build-up of a large uncharged hydrocarbon species by neutral condensation reactions probably involving free radicals, followed by its thermal ionization and/ or chemical ionization by proton or charge transfer. The plausibility of this scheme is suggested by the observation of diacetylene in the CH_4/O_2 flame (Hastie, 1973) and large concentrations of polyacetylenes and other hydrocarbon fragments in C_2H_2/O_2 flames (Bonne et al, 1965).

NEGATIVE ION CHEMISTRY

The free electrons provided by the primary chemi-ionization reaction are produced in a bath rich in electronegative oxygen molecules to which they may attach by three-body association

$$e + O_2 + M \rightarrow O_2^- + M$$

The other electronegative radical species in the preheat region are present in much lower amounts but may, of course, also attach electrons. These include, in order of decreasing electron affinity, HO_2, O_3, OH, O,

C, CH, H, CH_2 and CH_3 (see Fig. 7) which, with the exception of CH_2 and CH_3, may also receive an electron in an exothermic charge transfer from O_2^-. Three-body electron attachment is not restricted to the preheat region but will persist downstream as long as electronegative species are present. The steady state abundance of the negative ions will be determined also be any dissociative attachment which may proceed, as well as by the various loss processes which occur concomitantly such as collisional and associative detachment and ion-ion recombination. Very little experimental information is available for electron attachment even at room temperature and low pressures (Caledonia, 1975). Nevertheless, on theoretical grounds, three-body attachment or three-body attachment followed by charge transfer or ion-molecule reaction is generally to be preferred as a source of negative ions in flames over dissociative attachment. The latter process is often endothermic, eg., the formation of O^- from dissociative attachment to O_2 is almost 50 kcal mole^{-1} endothermic and has been shown to have a rate constant $< 10^{-16}$ cm^3 molecule^{-1} s^{-1} at 2000K (Caledonia, 1975), and may exhibit an activation energy larger than its endothermicity.

Although O_2^- has been used as a chemical ionization reagent at atmospheric pressure (Dzidic et al, 1974) it is expected to have a limited reactivity towards the stable neutral species found in CH_4/O_2 flames in which it should act as a relatively weak base. Proton transfer appears to be exothermic only with H_2O_2 and HCOOH. The latter proton transfer has been observed to be rapid at room temperature in a drift tube (Sugden et al, 1973) and in our flowing afterglow laboratory. O_2^- has been shown in our laboratory to be unreactive towards CH_4 at room temperature. This would imply that the bimolecular chemistry of O_2^- in the preheat region should, as was the case with CHO^+, be dominated by reactions with free radicals: charge transfer as alluded to above, associative detachment, particularly with the atomic species, and possibly exothermic ion-molecule reactions of the type

$$O_2^- + H \rightarrow O^- + OH$$
$$\rightarrow OH^- + O$$

$$O_2^- + CH_3 (CH_2) \rightarrow CHO_2^- + H_2 (H)$$

The situation may be somewhat more complicated if one allows for the internal excitement of O_2^- associated with the first step in its formation by electron attachment (Caledonia, 1975). The vibrational excitation of O_2^-, dominated at room temperature by the resonance with the v = 4 level of O_2^-, may be sufficient to promote reaction with the relatively abundant CH_4 molecules in the preheat region to produce HO_2^- and OH^- (already exothermic for ground state O_2^-), CH_3O^-, O^- and possibly $CH_3O_2^-$ whose energetics are unknown (see Fig. 8).

In any case, it appears that O_2^- may be replaced through chemical reaction by considerably more basic anions such as O^- and OH^- which may abstract protons from most of the intermediate flame species (see Fig. 8). A comprehensive study of the room temperature kinetics of these reactions has recently been completed in our flowing afterglow laboratory. OH^- was observed to react almost exclusively by proton transfer except with CH_2O for which association was observed to be the dominant channel. Although proton transfer was also often a major channel with O^-, this ion also reacted with stable molecules by H atom and H_2^+ transfer, H atom elimination and associative detachment. For

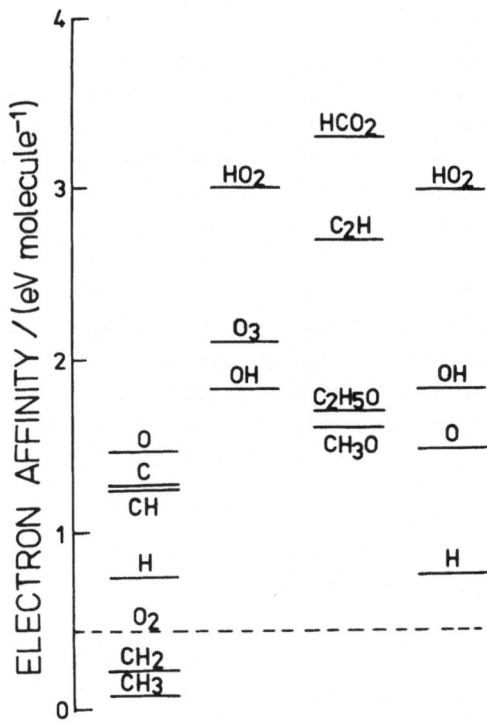

Fig. 7   Electron affinities of selected flame constituents.

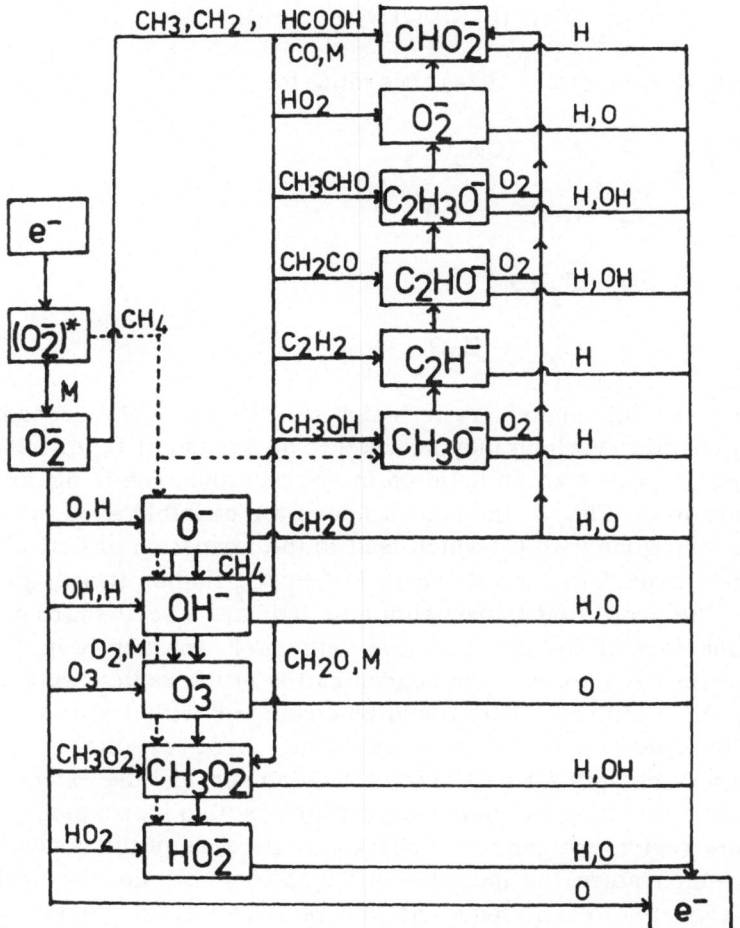

Fig. 8 Limited reaction scheme for the negative ion chemistry involving the stable intermediates and radicals which rise in the reaction zone of a CH_4/O_2 flame.

example, O^- appears to react with CH_2O according to:

$$O^- + CH_2O \;\rightarrow\; CHO_2^- + H$$

$$\rightarrow\; OH^- + CHO$$

$$\rightarrow\; e + (HCOOH)$$

and with C_2H_2 (Bohme et al, 1974) according to:

$$O^- + C_2H_2 \;\rightarrow\; e + (CH_2CO)$$

$$\rightarrow\; C_2H^- + OH$$

$$\rightarrow\; C_2HO^- + H$$

$$\rightarrow\; C_2^- + H_2O$$

Part of the O^- and OH^- chemistry is included in Fig. 8. The secondary negative ion chemistry which may proceed in an excess of O_2 or with the free radical species which build up in the reaction zone is again generally unknown. Fig. 8 includes a few of the possible exothermic channels for reactions with O_2 which lead to the formation of CHO_2^-, one of the major intermediate ions detected in CH_4/O_2 flames (Goodings et al, 1978b). The profile measurements also indicate a very sharp drop in the reaction zone of the total negative ion signal, apparently in coincidence with the growth in the concentration of the radical species, particularly H, O and OH. This would be consistent with a sharp increase in the rate of associative detachment. However, the individual rate constants for associative detachment and the extent to which these are sensitive to temperature which is also increasing rapidly in this region are generally unknown. Because of the **concomitant** increase in temperature, an increase in the rate of collisional detachment can also be expected to contribute and to a relatively larger extent with the species of lower electron affinity.

In those regions of the flame where molecules of low electron affinity and low acidity predominate, such as in the burnt gas downstream, reactions of negative ions are likely to result largely in the termolecular association of the molecule and the ion to form the corresponding 'cluster' ion, eg.,

$$OH^- + CO_2 + M \;\rightarrow\; HCO_3^- + M$$

$$OH^- + H_2O + M \rightarrow OH.^- H_2O + M$$

We can also expect cluster ions to exchange 'adduct' molecules through rapid switching reactions, eg.,

$$OH.^- H_2O + CO_2 \rightarrow HCO_3^- + H_2O$$

Much of this chemistry has been thoroughly characterised in flowing afterglow (Fehsenfeld and Ferguson, 1974) and drift tube (Moruzzi and Phelps, 1966) studies of negative ion reactions with atmospheric gases and trace constituents at room temperature and pressures up to 5 torr. A scheme which is based on these measurements and which should also apply qualitatively under flame conditions is shown in Fig. 9. A series of association reactions based on HO_2^- has been added because of the observation in a fuel-lean CH_4/O_2 flame of ions with m/e corresponding to $HO_2^-.H_2O$ and HCO_4^- (Bohme et al, 1977). These ions have not been included in traditional atmospheric models and appear to have been identified for the first time in flames. HCO_4^- has also recently been observed in an atmospheric pressure $H_2/N_2/$ acetone diffusion flame by McAllister et al (1978) who suggest an alternate route for its formation by association

$$CO_3^- + OH + M \rightarrow HCO_4^- + M$$

Reactions of cluster ions with radical species have not been studied extensively. Flowing afterglow experiments (Ferguson, 1973; Fehsenfeld, 1975) have shown that such reactions can act to establish more stable entities by rearrangement, eg.,

$$CO_4^- + O(H) \rightarrow CO_3^- + O_2 (OH)$$

recycle a parent negative ion, eg.,

$$CO_3^- + O(H) \rightarrow O_2^- (OH^-) + CO_2$$

or recycle electrons by associative detachment, eg.,

$$OH.^- (H_2O)_2 + H \rightarrow 3H_2O + e$$

Just as was the case with positive ions, mass spectrometric studies of fuel-rich CH_4/O_2 and C_2H_2/O_2 flames have indicated that there may also be an extensive negative ion chemistry associated with the build up of unsaturated hydrocarbon species near the tip of the reaction zone. Certainly these studies have indicated the presence of large carbonaceous

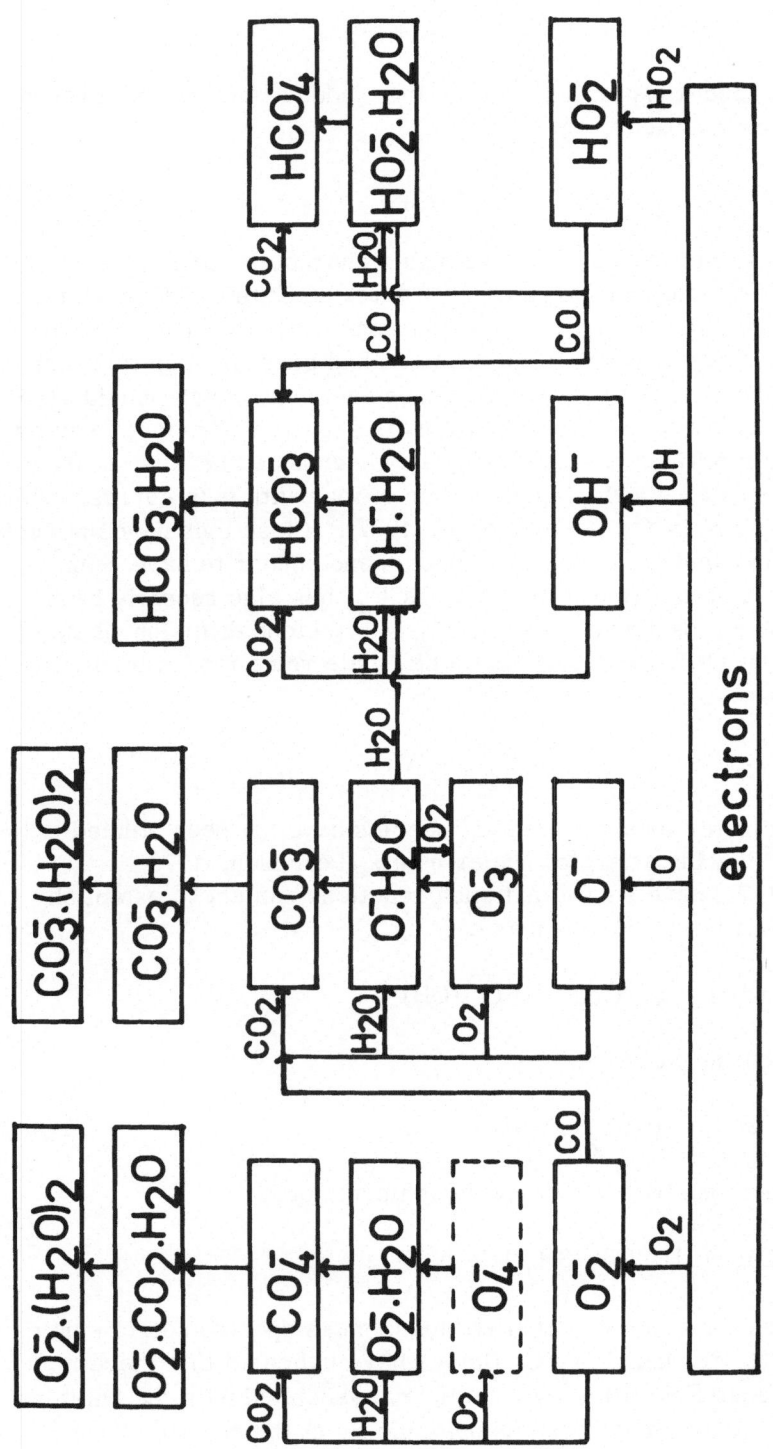

Fig. 9   Limited reaction scheme for the negative ion chemistry of negative ion 'clusters' observed in CH_4/O_2 flames.

anions generally of the type $C_nH_x^-$ (n = 1-8, x = 1-3) with concentration profiles which appear to fall into two general groups depending on the parity of the number of C atoms in the anion (Sugden et al, 1973; Goodings et al, 1978a; Hayhurst and Kittelson, 1978). The formation of these carbanions may well be dominated by proton transfer reactions,eg.,

$$OH^- + C_nH_x \rightarrow C_nH_{x-1}^- + H_2O$$

or charge transfer reactions, eg.,

$$O_2^- + C_nH_x(radical) \rightarrow C_nH_x^- + O_2$$

or even three-body electron attachment reactions, eg.,

$$e + C_nH_x (radical) + M \rightarrow C_nH_x^- + M,$$

with large carbon containing radicals and neutrals built up by the underlying neutral chemistry. However, in analogy with the positive ions, one should allow for the possibility of build up by ion-molecule association, eg.,

$$C_2H^- + C_2H_2 + M \rightarrow C_4H_3^- + M$$

and condensation, eg.,

$$CH_3^- + C_2H_2 \rightarrow C_3H_3^- + H_2$$

Such reactions are virtually unknown at the present time. A few flowing afterglow meaurements of reactions of C^-, C_2^-, and C_2H^- with CH_4 and C_2H_2 have indicated that C-C bond formation in this fashion is kinetically unfavourable at least with stable molecules at room temperature (Schiff and Bohme, 1975; Payzant et al, 1976). Certainly there is a clear need for much more information on the energetics of these carbanions and their kinetics of formation.

SUMMARY

An attempt has been made to identify and pursue the different types of positive and negative ion-molecule reactions which predominate in the various regions of a CH_4/O_2 flame. The limited schemes of ion chemistry which are presented are based largely on room-temperature measurements of individual ion-molecule reactions. Many additional studies of ion-molecule reactions have suggested themselves in their development. We can expect further extensions to these schemes as such studies proceed and as these schemes are adapted to the sequential ionization in other hydrocarbon flames as has been done, for example, in a recent study of CH_3OH/O_2 combustion (Goodings et al, 1978c).

ACKNOWLEDGMENTS

The author wishes to acknowledge the substantial contributions made to this research by his colleagues, Prof. J. M. Goodings, Dr. G. I. Mackay, and Mr. S. D. Tanner. He also wishes to thank Prof. K. R. Jennings for his kind hospitality at the University of Warwick where this manuscript was written while the author was on sabbatical leave.

REFERENCES

Adams, N. G., Smith, D. and Grief, D, (1978) Int. J. Mass Spectrom. Ion Phys. 26, 405.

Bonne, U., Homann, H. and Wagner, H. Gg. (1965) Tenth Symposium (International) on Combustion, The Combustion Institute, Pittsburgh, Penna., 503.

Bohme, D. K., Mackay, G. I., Schiff, H. I. and Hemsworth, R. S. (1974) J. Chem. Phys. 61, 2175.

Bohme, D. K., Goodings, J. M. and Ng, C.-W. (1977) Int. J. Mass Spectrom. Ion Phys. 24, 335.

Caledonia, G. E. (1975) Chem. Rev. 75, 333.

Dzidic, I., Carroll, D. I., Stillwell, R. N. and Horning, E. C. (1974) J. Amer. Chem. Soc. 96, 5258.

Fehsenfeld, F. C. and Ferguson, E. E. (1974) J. Chem. Phys. 61, 3181.

Fehsenfeld, F. C. (1975) J. Chem. Phys. 63, 1686.

Ferguson, E. E. (1973) Atomic Data and Nuclear Data Tables, Academic Press, N.Y. 12, 159.

Goodings, J. M., Bohme, D. K. and Sugden, T. M. (1977) Sixteenth Symposium (International) on Combustion, The Combustion Institute, Pittsburgh, Penna., 891.

Goodings, J. M., Bohme, D. K. and Ng, C.-W. (1978a). Detailed Ion Chemistry in Methane-Oxygen Flames. I. Positive Ions. Combust. Flame. Submitted for publication.

Goodings, J. M., Bohme, D. K. and Ng, C.-W. (1978b). Detailed Ion Chemistry in Methane-Oxygen Flames II. Negative Ions. Combust. Flame. Submitted for publication.

Goodings, J. M., Ng, C.-W. and Bohme, D. K. (1978c) Int. J. Mass Spectrom. Ion Phys. In press.

Hastie, J. W. (1973) Combust. Flame 21, 187.

Hayhurst, A. N. and Kittelson, D. B. (1977) Combust. Flame 28, 137.

Hayhurst, A. N. and Kittelson, D. B. (1978) Combust. Flame 31, 37.

McAllister, T., Nicholson, A. J. C. and Swingler, D. L. (1978) Int. J. Mass Spectrom. Ion Phys. 27, 43.

Miller, W. J. (1968) Oxidation and Combustion Revs. 3, 97.

Moruzzi, J. I. and Phelps, A. V. (1966) J. Chem. Phys. 45, 4617.

Payzant, J. D., Tanaka, K., Betowski, L. D. and Bohme, D. K. (1976) J. Amer. Chem. Soc. 98, 894.

Peeters, J. and Mahnen, G. (1973) Fourteenth Symposium (International) on Combustion, The Combustion Institute, Pittsburgh, Penna., 133.

Schiff, H. I. and Bohme, D. K. (1975) Int. J. Mass Spectrom. Ion Phys. 16, 167.

Smith, D. and Adams, N. G. (1977) Int. J. Mass Spectrom. Ion Phys. 23, 123.

Smith, D. and Adams, N. G. (1978) Chem. Phys. Letters 54, 535.

Sugden, T. M., Goodings, J. M., Jones, J. M. and Parkes, D. A. (1973) Combustion Institute (European) Symposium, Academic Press, London, 250.

Tanner, S. D., Mackay, G. I., Hopkinson, A. C. and Bohme, D. K. (1978) Int. J. Mass Spectrom. Ion Phys. Accepted for publication.

ION MOLECULE REACTIONS IN LOW TEMPERATURE PLASMAS :

FORMATION OF INTERSTELLAR SPECIES

David Smith and Nigel G. Adams

Department of Space Research
University of Birmingham
Birmingham B15 2TT, England

1. INTRODUCTION

The purpose of this paper is to briefly review the role of
ion-molecule reactions in the chemical evolution of low temperature
plasmas. Since this is such a vast subject which cannot be treated
in detail in a paper of this length, a brief over-view of this topic
is presented (Section 2) with reference to the more important
reviews of specific reaction types, and then the lesser-known process
of binary ion-molecule radiative association is discussed in more
detail and the probable importance of this reaction type in molecular
synthesis in the very low temperature ($<$100K) interstellar
molecular clouds is considered. To prepare for this, Section 3
outlines the current ideas concerning the initial steps (through
ion-neutral reactions) in the formation of the observed molecules
in interstellar clouds, referring to the most recent laboratory
data on which many of the ideas are based. In Section 4 radiative
association is discussed and some recent laboratory data on the
ternary collisional association reactions of CH_3^+ ions with several
molecules is presented to show how, through a simple transition
state model, estimates of the binary radiative association rate
coefficients can be obtained. Finally, in Section 5 it is shown
how the deduced rate data for radiative association have been used
to approximately predict the observed relative abundances of
several interstellar molecules.

2. ELEMENTARY INTERACTION PROCESSES IN LOW TEMPERATURE PLASMAS

For the purpose of this paper the term "low temperature plasmas"
refers to those plasmas (and many ionized gases which strictly

speaking would not conform to the rigid definition of a gaseous
plasma) in which the kinetic temperatures of the composite particles
are sufficiently low that significant concentrations of weakly
bound cluster ions and of negative ions of low detachment energy
can exist, and hence that collisions are not so energetic as to
destroy these species. This then encompasses a wide range of
laboratory plasmas, certainly many types of gas discharges and
afterglows including laser plasmas, and some naturally occurring
plasmas such as the Earth's ionosphere and the dark and dense
interstellar molecular clouds. In high temperature plasmas such as
fusion plasmas and stellar atmospheres, few molecules exist and the
chemistry involves only atoms, atomic ions and electrons.

When a neutral gas or gas mixture is exposed to a sufficiently
intense electric field or is irradiated with sufficiently energetic
charged particles or photons, ionization of the neutral atoms or
molecules can occur resulting in the production of a partially or
totally ionized gas, which, if certain criteria are fulfilled
(McDaniel, 1964), can be described as a gaseous plasma. In a gas
mixture this process initiates a complicated chemical scheme
involving many parallel and sequential particle interactions which
can radically change the chemical composition of the medium, i.e.
chemical evolution of the gas mixture occurs. If the source of
excitation of the gas mixture is removed, then the reverse process
of ionization, i.e. recombination, returns the plasma back to the
gaseous state but invariably the composition of the gas will differ
from that which existed before ionization was created within it.
This overall process is summarised in Fig.1. Thus on exposure to
ionizing radiation the neutral gas of initial composition ① is
modified to contain positive ions, electrons, often negative ions
and radicals. The large box in Fig.1 emphasises the central role
of, and the very wide variety of, ion-molecule reactions which may
occur and which can produce a mixture of simple and complex
positive and negative ions in the gas. Also electron-ion and/or
ion-ion recombination will be continuously occurring at a rate
dependent on the ionization density within the gas and on the
nature of the recombining ions (see below). These processes
contribute to the variety of neutral species which are generated
in the gas (often very reactive atoms and radicals) which contribute
to the neutral chemistry which proceeds in parallel with the ion
chemistry. Thus, a neutral gas of composition ② is produced.
It is this gas phase ion and neutral chemistry which is being
successfully invoked to explain the existence of the complex ionic
and neutral species observed in the Earth's ionosphere and in the
interstellar gas clouds.

That a reasonable effort can now be made to explain the complex
chemistry occurring in the above-mentioned natural plasmas is due
to the concerted efforts in several laboratories to accumulate data

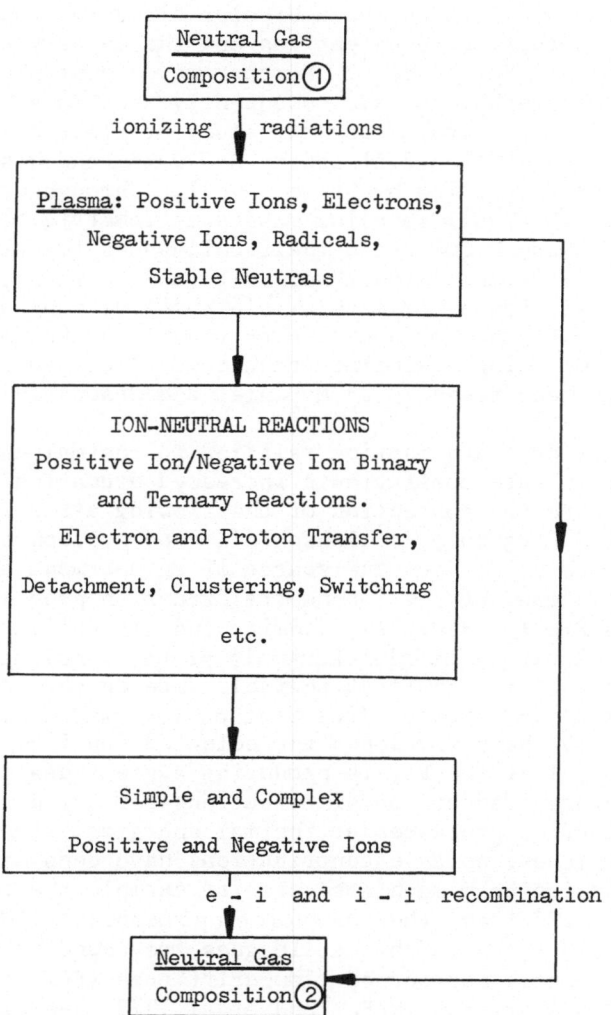

Fig. 1 Chemical evolution in a low temperature plasma.

on each of the many reaction types referred to in Fig.1. Much of
the relevant data has been obtained from studies of afterglow plasmas,
following the pioneering work of S.C.Brown and M.A.Biondi (Biondi
and Brown, 1949; reviewed by Biondi, 1968) in the early 1950's.
This group, using the pulsed afterglow/microwave cavity technique,
initially chose to study the relatively simple afterglow of pure
helium, and obtained data on the ambipolar diffusion rates of the He^+
and He_2^+ ions in their pure parent gas and the ternary conversion
rate coefficient of He^+ to He_2^+. Further major contributions
arising from this technique have subsequently been made by M.A.Biondi
and his collaborators notably for the study of electron-ion
recombination (see Section 2.5) and by H.J.Oskam and his collaborators
(Oskam, 1958 and the review by Oskam, 1969). Throughout the 1960's
the pulsed afterglow/wall sampling mass spectrometric technique was
used by J.Sayers and D.Smith and their colleagues (Dickinson and
Sayers, 1960; Sayers and Smith 1964; Smith and Fouracre, 1968) to
obtain rate coefficients for many simple positive ion-molecule
reactions and by Lineberger and Puckett (1969) and Puckett and
Lineberger (1970) using a similar technique. These earlier
techniques have been reviewed by McDaniel and Mason (1973).

The rate of data acquisition relating to thermal energy ion-
molecule reaction rate coefficients increased dramatically in the
late 1960's due to the conception of the flowing afterglow (FA)
technique by E.E.Ferguson, F.C.Fehsenfeld and A.L.Schmeltekopf
(1969a) which has since been the source of an enormous amount of
data relating to many of the ion-neutral reaction processes
referred to in Fig.1. Recently, D.Smith and his colleagues have
exploited the flowing afterglow/Langmuir probe combination (Smith
et al, 1975) to obtain the first reliable data on the process of
ion-ion mutual neutralization (see Section 2.6) and N.G.Adams and
D.Smith (1976 a,b) have developed the selected ion flow tube (SIFT)
technique which, like the FA, is producing a great deal of data
including accurate product ion distributions for a wide variety of
ion-neutral reaction processes at thermal energies. In addition to
the above techniques, notable contributions have been made using
drift tubes and ion beam machines (see for example the review by
McDaniel et al, 1970) and the ion cyclotron resonance (ICR) technique,
notably by W.T.Huntress and his colleagues (reviewed by Huntress,
1977a). The recently introduced flow-drift tube (FDT) technique
pioneered by D.L.Albritton (McFarland et al 1973, see also the
Chapter by D.L.Albritton in these proceedings) is providing data
over an appreciable energy range.

Brief discussions of a few of the more important reaction
processes which can occur in low temperature plasmas will now be
given, with reference to notable contributions by specific workers
and to the techniques used.

2.1 Charge (Electron) Transfer

This is the most widely studied process at thermal energies; contributions have been made by numerous groups using a variety of techniques. This process almost invariably represents a major component of the chemistry of a plasma. It is exemplified by the much-studied reaction

$$He^+ + N_2 \longrightarrow N^+ + N + He$$

$$\longrightarrow N_2^+ + He$$

(1)

which at 300K exhibits two product channels, the dissociative channel producing N^+ (60%) dominating over the non-dissociative channel (40%). This is one of the few reactions for which the product ion distribution has been obtained using several different experimental techniques and good agreement has been obtained (see Adams and Smith, 1976b). A major contribution has been made to the determination of the rate coefficients by the NOAA group using the FA technique (e.g. Ferguson 1972, 1974), who have also studied several such reactions over the temperature range 80-900K (Ferguson et al, 1969b; Lindinger et al, 1974). An important general principle appears to be that the very fast, gas kinetic reactions show very little variation of the rate coefficient with temperature whereas significant variations occur for slower reactions (Ferguson et al, 1969b). The data of the NOAA group has been summarised in several reviews and data compilations (see, for example, Ferguson, 1973; Albritton, 1979).

The FA has not contributed greatly to the determination of product ion distributions for these reactions, the greatest contributions being made using the ICR technique, notably by Huntress and his co-workers (see the compilation by Huntress, 1977a), and by Adams and Smith using the SIFT technique (see for example Adams and Smith 1976b, 1978; Adams et al, 1978; Smith and Adams, 1977; Smith et al 1978b). The SIFT is particularly useful when multiple ion products result as is illustrated in Table II which shows how complex the chemistry becomes when multiple products result from ostensibly simple interactions. Note how sequential reactions sometimes result in the production of a stable (terminating) ion. This simplifies somewhat the evolved ion composition of a plasma (see Section 3).

Since this class of reactions generally proceed at or near the collision rate, the rate coefficients are not a sensitive indicator of any internal excitation in the reactant ion and under such circumstances, the product ion distribution is a better indicator as has been shown recently using a SIFT to study the reactions of several metastable excited atomic and molecular positive ions

(Glosik et al, 1978; Tichy et al, 1979). Excitation of the product
ions and neutrals is often energetically possible and needs to be
considered in detailed chemical modelling of a plasma. Very clear
evidence has recently been obtained for product ion excitation in a
study of the reactions of the ground and metastable electronic states
of the doubly charged ions Xe^{++} and Ar^{++} (Adams et al 1979a; Smith
et al, 1979b). Two-electron transfer was observed in a single
collision of Ar^{++} with several molecules.

2.2   Proton Transfer

This process involves a heavy charged particle being transferred
between the reactant ion and neutral and as such necessitates a close
encounter between the reactants. It is exemplified by the reactions:

$$NH_3^+ + NH_3 \longrightarrow NH_4^+ + NH_2 \tag{2}$$

in which a proton is transferred from an ion to a molecule, and

$$OH^- + CH_3CHCH_2 \longrightarrow CH_2 = CHCH_2O^- + H_2O \tag{3}$$

in which a proton is transferred from a molecule to a negative ion.
Detailed studies of these phenomena have been carried out by
D.K.Bohme and his associates (reviewed by Bohme, 1975) and more
recently by V.M.Bierbaum et al (1976) using FA's, by P.Kebarle using
his high pressure mass spectrometer (HPMS) technique (Kebarle,1975)
and by J.L.Beauchamp using the ICR technique (Beauchamp, 1975).
Through the efforts of these workers, a comprehensive scale of
relative proton affinities is continuously being constructed which
assists greatly in predicting the important protonated species in a
plasma of known neutral composition. The rate coefficients for this
well-characterised class of reactions have been used by Bowers and
Su (1975) as a check on their ion-molecule reaction rate theories.

2.3   Associative Detachment

This reaction process relates exclusively to negative ions and
is an ion loss mechanism resulting in the production of a free
electron and a molecule, e.g.

$$Cl^- + H \longrightarrow HCl + e \tag{4}$$

$$O^- + CO \longrightarrow CO_2 + e \tag{5}$$

These reactions have been studied almost exclusively using the
FA, notably by F.C.Fehsenfeld and his colleagues (see the review by
Fehsenfeld, 1975). An appreciable number of rate coefficients for
these reactions have been measured. Whilst the rate coefficient
varies greatly from reaction to reaction, it is often very large
and must therefore always be considered as a potentially serious
loss process for negative ions in a plasma and an equivalent
production process for free electrons and new molecules.

2.4 Ternary (Collisional) Association: Clustering

As the gas pressure in a plasma is increased, the probability
of ternary (3-body) collisions increases and hence collision-
stabilised association reactions may proceed at significant rates.
Such reactions can build-up significant concentrations of weakly-
bonded ions in the plasma. e.g.

$$O_2^+ + O_2 + N_2 \rightleftharpoons O_2^+ \cdot O_2 + N_2 \tag{6}$$

This reaction is an important process in the Earth's lower
ionosphere and generates the weakly bonded O_4^+ ion (bond energy
$\sim 0.42eV$). In fact such reactions dominate the chemistry of the
D-region of the ionosphere as has been very well described by
E.E.Ferguson and others (e.g. Thomas, 1974; Ferguson, 1975;
Ferguson et al, 1979; Sechrist, 1975; Reid, 1976) largely based
on FA studies of this class of reactions (e.g. see Ferguson, 1973,
1974).

As might be expected, the forward and reverse rates of these
reactions are usually very temperature dependent and studies of the
equilibrium constants for such reactions as a function of temperature
has provided a great deal of critical thermodynamic data relating
to the nature of the association ions (Kebarle, 1972, 1975, 1977;
Porter, 1975). At low temperatures and high pressures, very large
'cluster' ions are formed and shell structures have been inferred
for such ions (Castleman, 1978). It is thought that in the case of
water molecules clustered around an ion, eventually a critical size
will be reached beyond which rapid nucleation occurs to produce
water droplets (see the Chapter by A.W.Castleman in these
proceedings).

It is well known that (at least for moderately sized clusters)
the cluster ions also undergo switching reactions (Adams et al, 1970)
of the type:

$$O_2^+ \cdot O_2 + N_2O \rightleftharpoons O_2^+ \cdot N_2O + O_2 \tag{7}$$

and so in a relatively high pressure, low temperature plasma created
in a gas mixture, a very large number of different positively and
negatively charged cluster ions may exist, although the lifetimes
of many may be small, the plasma evolving towards one containing
relatively few, more stable terminating ions. A relatively recent
review and data compilation by Good (1975) is available, but a great
deal of effort is being made to accumulate more data on these
reactions in several laboratories.

2.5 Electron-Ion Recombination (Electronic Recombination)

Whilst the reactions discussed above can dramatically change the
ionic composition of a plasma they do not reduce the degree of
ionization directly. Such can only result from the recombination
of positive and negative charges. In a plasma devoid of negative
ions, then the only process of de-ionisation is electron-ion
recombination: e.g.,

$$He^+ + e \rightarrow He + h\nu \qquad : \text{ radiative recombination} \qquad (8)$$

$$NO^+ + e \rightarrow N + O \qquad : \text{ dissociative recombination} \qquad (9)$$

Process (8) is relatively very slow and is usually insignifcant
whereas process (9) is some five orders of magnitude faster
(recombination coefficient typically $\gtrsim 10^{-7}$ cm^{3}s^{-1}) and dominates
the loss of ionization. Thus any reactions which convert atomic
positive ions to molecular ions can enhance greatly the loss of
ionization from a plasma via dissociative recombination. Almost all
of the data which has been obtained for this class of reaction and
which includes data on electron temperature dependences is due to
M.A.Biondi and his colleagues (e.g. Bardsley and Biondi, 1970;
Shiu and Biondi, 1978). These workers have also obtained data
for the electronic recombination of the hydrated hydronium ions
$H_3O^+(H_2O)_{0,1,2,3}$ (Leu et al, 1973; Huang et al, 1978) which shows
an increasing recombination coefficient with increasing degree of
hydration, the largest coefficient measured approaching 10^{-5} cm^{3}s^{-1},
i.e. an increase of about one or two orders of magnitude above the
recombination coefficients for the common diatomic positive ions.
Thus, the formation of significant concentrations of cluster ions
in a plasma can result in a considerable increase in the rate of
loss of ionization. The recombination also adds to the richness
of the neutral content of the plasma although little or no data
are available on the neutral products of such reactions.

2.6 Ion-Ion Mutual Neutralization (Ionic Recombination)

When significant concentrations of negative ions exist in a
plasma then the process of binary or ternary mutual neutralization

(see the review by Moseley et al, 1975) may result in loss of ionization, e.g.

$$NO^+ + NO_2^- \longrightarrow NO + NO_2 \tag{10}$$

$$H_3O^+(H_2O)_3 + NO_3^-(HNO_3) \longrightarrow products \tag{11}$$

Very recently the first reliable data at thermal energies on the binary process exemplified by reactions (10) and (11) have been obtained by D.Smith and his colleagues (Smith and Church, 1976; Smith et al, 1976). The rate coefficients for binary ionic recombination have been shown to be $\sim 5 \times 10^{-8} cm^3 s^{-1}$ at 300K and, somewhat surprisingly, almost independent of the complexity of the positive and/or negative ion, at least for moderately sized ions (Smith et al, 1978a). That the cluster ions have ionic recombination coefficients much smaller than those for electronic recombination indicates that the conversion of electrons to negative ions in a plasma will, at least in the presence of clustered positive ions, reduce the rate of loss of ionization from the plasma. In any event, this process will also add to the complexity of the neutral composition of the medium. Little data is available on the neutral products of such reactions, although Smith et al (1979a) have recently shown that reaction (10) above proceeds via simple electron transfer resulting in the products indicated.

As the gas pressure is increased, collisionally enhanced (ternary) ionic recombination becomes more likely. Ternary recombination coefficients greater than $10^{-26} cm^6 s^{-1}$ have been measured (Mahan, 1973) albeit for ions of unknown masses (but probably clusters), equivalent to effective binary recombination coefficients of about $10^{-6} cm^3 s^{-1}$. In the Earth's atmosphere below about 60km, ionic recombination is the major loss process for ionization, the upper half being dominated by the binary process and the lower half being dominated by the ternary process (Smith and Church, 1977).

3. THE ROLE OF ION-MOLECULE REACTIONS IN MOLECULAR SYNTHESIS IN INTERSTELLAR CLOUDS

The above is a brief outline of the individual reaction processes which can occur in low temperature plasmas. Whilst many experimental problems have had to be overcome to obtain data on these processes, problems of a much greater magnitude are presented when one attempts to construct from laboratory data satisfactory models of the chemistry of a real plasma in which a large number of different reactions are usually occurring in parallel. However, such has been the stimulus for much of the work done over the last two decades and, in fact, conspicuous success has been obtained in describing in particular the chemistry of the Earth's ionosphere. This has involved the comparison of in-situ observations of ionic composition and electron density (and their diurnal variations) in the ionosphere and satisfactorily explaining the observations by reaction rate models based on ion-molecule reaction rate data, electronic recombination data etc. obtained from laboratory experiments. Several reviews have been written on this topic, (e.g. Reid, 1976; Ferguson et al, 1979).

Natural plasmas which are currently attracting considerable interest are those associated with the massive molecular clouds which pervade the galaxy and in which many different molecular species of varying complexity have been detected. A considerable amount of laboratory data on ion-molecule reactions, thought to be relevant to chemical synthesis in these clouds, are currently being obtained, principally due to the application of the ICR and SIFT techniques, and real progress in the understanding of the chemistry of molecular clouds is being made. In view of this, and in the light of some recent SIFT data (Smith and Adams 1977a, b, 1978a; Adams and Smith 1977, 1978; Adams et al, 1978), it seems pertinent here to briefly consider some of the problems involved in explaining how the molecules observed in these gas clouds are synthesised, particularly since it seems that ion-molecule reactions play an important and perhaps a dominant role. Interest in this area of science began as long ago as 1941 with the observation of absorption bands of CH^+ in the diffuse clouds surrounding hot stars (Douglas and Herzberg, 1941), but interest has grown in the last decade with the discovery by millimetre-wave astronomy of much more complex molecules in the massive dense and dark molecular clouds such as the Orion A and Sagitarrius B2 complexes. The list of observed molecules is already long and is continuously being lengthened as the microwave spectra of more molecules are characterised by theory and laboratory spectroscopy and as the observational techniques are improved. Currently, in excess of 50 molecules of varying complexity have been detected, including three molecular positive ions (CH^+, HCO^+, N_2H^+). The list in Table I has been compiled from several reviews (Watson, 1977; Huntress, 1977b; Herbst, 1978) and recent research paper (Ulich et al, 1977; Broten et al, 1978; Guélin et al, 1978; Liszt and Turner, 1978; Fox and Jennings, 1978). Other common molecules not listed in Table I such as N_2, O_2 and CO_2 undoubtedly exist, but they are not readily detected due to the very weak microwave emissions from these symmetrical molecules.

A considerable amount of thought has been given to the mode of synthesis of these molecules in the hostile environments of these clouds starting from the atomic species H,C,N,O...(the end points of nuclear synthesis in stellar atmospheres). Significantly, only diatomic species have been detected in the diffuse clouds in which the gas number density is low (a few tens per cm^3, mostly atomic hydrogen) and the temperature relatively high (up to a few hundred

TABLE I

Observed interstellar molecules and molecular ions listed in
the order of atomicity. Note the appearance of the series of
cyano-acetylenes HC_3N, HC_5N, HC_7N and HC_9N, the last mentioned
being the most complex molecule observed to date.   In many
cases isotopic species are also observed.

H_2, CH, $\underline{CH}^{+*}$, C_2, CN, CO, CS, NO, NS, OH, SiO, SiS, SO

HCN, HNC, HCO, $\underline{HCO}^+$, $HNO^{\not{}}$, H_2O, H_2S, C_2H, COS, $\underline{N_2H}^+$, SO_2

HNCO, H_2CO, H_2CS, NH_3, C_2H_2, C_3N

HCOOH, HC_3N, CH_2NH,NH_2CN, CH_2CO, CH_4, C_4H

CH_3CN, CH_3OH, NH_2CHO

HC_5N, CH_2CHCN, CH_3C_2H, CH_3CHO,CH_3NH_2

$HCOOCH_3$, CH_3C_3N

HC_7N, $(CH_3)_2O$, C_2H_5CN, C_2H_5OH

HC_9N

* indicates a species which has been observed in diffuse
  clouds only

$\not{}$ indicates a tentative identification

^{O}K) through which starlight can penetrate and photodissociate larger
molecules. The larger molecules are only observed in the higher
gas density (a few thousands per cm,^{3} mostly molecular hydrogen),
low temperature (a few tens ^{O}K) dark and dense clouds with which
large accumulations of dust are often associated. Thus the
molecules are protected from dissociating radiations by H_2 molecular
absorption and scattering by the dust grains. (For a recent summary
of the conditions within the various types of clouds see Myers,
1978). It is to be expected that the mechanisms of molecular
synthesis will be dependent on the physical conditions within the
clouds, and ultimately each particular cloud (or region within the
cloud) will have to be considered separately. However the possible
production mechanisms are clear, these being gas phase binary
ion-neutral reactions and neutral-neutral reactions, and reactions
catalysed on the surfaces of dust grains followed by molecular
desorption. Many authoritative reviews have been written on this
topic (e.g. Lang, 1974; Dalgarno and Black, 1976; Watson 1976).
The consensus view is that whilst gas phase neutral-neutral reactions
cannot be ruled out, they are in general likely to be slow at the
low temperatures of the clouds (due to activation energy barriers)
whereas both gas phase ion-neutral reactions and surface grain
catalysis are important processes for molecular synthesis, their
relative importance being dependent on the conditions within the
particular cloud. Much has been written on the importance of grain
surface catalysis (see, for example, Pickles and Williams, 1977) but
it will not be considered further here. Rather the extent to which
gas phase ion-neutral reactions can at least qualitatively explain
the existence of the observed molecules will be considered. Several
workers have addressed themselves to this problem, notably A.Dalgarno,
W. Klemperer, W.D. Watson and E. Herbst and their colleagues (see, for
example, Solomon and Klemperer, 1972; Herbst and Klemperer,1973,
1976; Dalgarno, 1975; Dalgarno and Black, 1976; Watson, 1976,
1977) and the chemical models carefully thoughtout by these workers
form the basis for thinking in this subject. However, although
considerable advances have been made, progress has often been
hampered by the lack of relevant laboratory data, and so some of the
proposed reaction paths are inevitably speculative. Happily, this
unfortunate circumstance is now being gradually corrected by the
involvement of more laboratory workers in these challenging problems.

Although the reactions by which even diatomic molecules are
produced are still not confidently understood, some initial stages
are illustrated in Fig.2. This diagram was constructed from a
consideration of the ideas discussed in the reviews by Herbst and
Klemperer (1973), Dalgarno (1976) and by Huntress (1977b). The
reaction schemes are essentially self-explanatory but brief
comments on some of the particular reaction types are necessary.
No quantitatively satisfactory gas phase mechanism has yet been
invoked to explain the production of H_2 from H atoms, and so dust

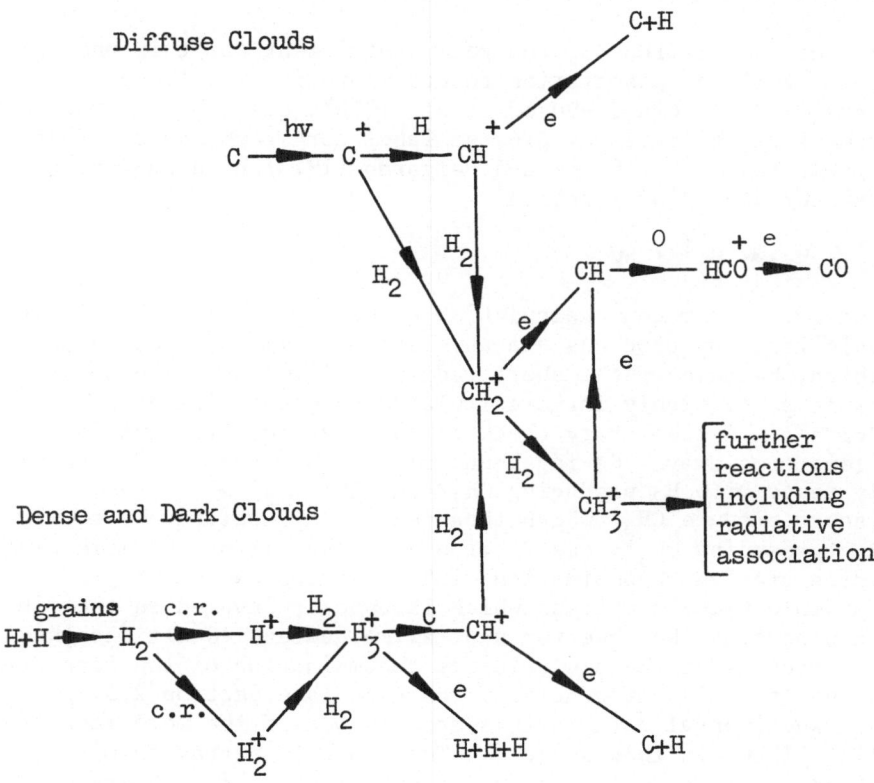

Fig. 2 Some initial reactions in molecular synthesis in interstellar clouds. c.r. indicates cosmic ray ionization.

grain catalysis has to be invoked here. Cosmic ray ionization of H_2 (producing H_2^+) followed by the well-known reaction of H_2^+ with H_2 to produce H_3^+ is acceptable. The reaction of H_3^+ with carbon atoms to produce CH^+ is likely but this has not been observed in the laboratory. This illustrates an important general point that molecular positive ion/atom reactions could be important in these clouds but very little laboratory data is available for such reactions at appropriate energies. Fortunately the FA and very recently the SIFT measurements of Fehsenfeld (Fehsenfeld, 1976, 1978) are now beginning to provide important data on such reactions. The addition reactions producing CH^+ and CH_2^+ from C^+ and H and H_2 respectively in diffuse clouds have not been observed experimentally but it is generally accepted that the process of radiative association occurs, first proposed by Bates and Spitzer (1951) for the following reaction:

$$C^+ + H \longrightarrow CH^+ + h\nu \tag{12}$$

Recent calculations of the rate coefficient for this photon-emission stabilized association reaction predict a value of $\sim 10^{-17} cm^3 s^{-1}$ (e.g. Giusti-Suzor et al, 1976), quite beyond the measurement capabilities of present laboratory techniques. Where appreciable H_2 exists, Black and Dalgarno (1973) have suggested that the corresponding reaction

$$C^+ + H_2 \longrightarrow CH_2^+ + h\nu \tag{13}$$

could occur and a recent calculation of Herbst et al (1977) indicates that this reaction proceeds via an electronically excited state of CH_2^+ which, because of the short radiative lifetime of the state, results in a relatively large calculated rate coefficient of $\sim 10^{-14} cm^3 s^{-1}$. A laboratory check on this rate coefficient is surely not a long time away. Having been formed, the stable CH^+ ions will rapidly react with H_2 producing CH_2^+ and the CH_2^+ will react likewise to produce CH_3^+ which then reacts only relatively slowly with H_2 (producing CH_5^+, see Section 4). For all of the molecular ions, dissociative recombination with electrons can occur producing atoms or molecules, a process which is usually invoked as the final step in producing the observed neutral molecules. Sometimes the neutral products of the dissociative recombination of the ions are obvious but this is not generally the case (see Section 2.5) and lack of experimental data in this area is one of the more important stumbling blocks in this subject. The neutral-neutral chemi-ionization reaction $CH + O$ producing HCO^+ (one of the observed interstellar ions) and an electron (MacGregor and Berry, 1973; Dalgarno et al, 1973b) is included to illustrate the concurrent neutral chemistry.

So, although problems still remain, even in explaining how some of the simplest of the observed molecules are formed, progress has been made and this process will accelerate as more critical data becomes available. The ion-molecule reaction rate and product ion distribution data have largely been obtained from FA (e.g. Burt et al, 1970; Fehsenfeld et al, 1973, 1974; Schiff et al, 1974; Liddy et al, 1977), ICR and, more recently, SIFT experiments. The relevant ICR data have recently been compiled by Huntress (1977a).

The SIFT experiment (Adams and Smith 1976a, b), has much to offer in the detailed study of ion-neutral reactions at thermal energies and is proving to be especially useful in the study of reactions of interstellar significance. In essence, the technique involves the injection of a mass-selected current of ions into a flowing gas at pressures sufficiently high to collisionally relax the kinetic and internal excitation of the ions (except for long lived metastable states, see Section 2.1) before a reactant gas is introduced into the flow tube. The operation of the apparatus is then quite similar to the well-established flowing afterglow. The relative merits of the FA and SIFT and a detailed discussion of the techniques has been given by Smith and Adams (1979). It is sufficient to say here that the SIFT can accurately determine product ion distributions in ion-neutral reactions even when several products result, some of which may be very minor channels ($<1\%$) and that the rate coefficients and product ion distributions can be obtained over an appreciable temperature and energy range. The technique can be used to study reactions of a very wide variety of positive and negative ions including very reactive radical ions (e.g. CH^+, CH_2^+) (Smith and Adams, 1977a,b; Adams and Smith 1977, 1978) and cluster ions (e.g. $H_3O^+(H_2O)_n$) (Smith et al, 1979c). The pressure and temperature ranges accessible are such that ternary association reactions can also be usefully studied (Smith et al, 1977; Smith and Adams, 1978a; Adams et al, 1979b,c; and see Section 4).

An especially valuable feature of the SIFT technique, particularly useful in studies directed towards the chemistry of naturally occurring plasmas like the interstellar clouds, is the relative ease by which the reactions of the ions in a regular series can be studied, for example the methane-derived ions CH_n^+ or the ethylene-derived ions $C_2H_n^+$ (n=0 to 4) (Adams and Smith, 1977). An illustrative sample of the data obtained from such a study is shown in Table II from which for example the evolution of the "terminating ions" CH_5^+, $C_2H_5^+$ and $C_3H_5^+$ can be seen. Such observations are vital to an understanding of the complex chemistry in a plasma containing gas mixtures. Note also the multiple products (including association products) in some of the reactions. From such comprehensive studies one can more confidently construct limited chemical models for interstellar clouds. In Fig.3 a limited qualitative model is presented, based on the data in Table II, which indicates likely channels for the production of some observed interstellar molecules. In this simple model we assume that electronic recombination of the positive ions proceeds by the ejection of an hydrogen atom only. This is clearly an over-simplification and in most cases it is probable that more than one fragmentation channel will appear (H atoms will not necessarily be produced at all). This again emphasises the dire necessity for experimental work in this area. In a recent theoretical paper, Herbst (1978a) has shown for the electronic recombination of several positive ions of astrophysical interest (i.e. H_3O^+, CH_3^+, NH_4^+ and H_2CN^+) that whilst H-atom ejection is not always the major channel, it is always a significant channel. Clearly, each recombination reaction should be dealt with individually with due regard to the exoergicity of the reaction and bonding arrangements in the reactants and products. Some experimental data on the neutral products of electronic recombination of a few specific positive ions has been obtained by Rebbert and Ausloos (1972) and Rebbert et al (1973). Proton transfer will, in effect, produce the same

TABLE II

The rate coefficients and product ion distributions for the
reaction of the series of ions CH_n^+ and $C_2H_n^+$ $(n = 0$ to $4)$
determined in a SIFT at 300K.   Rate coefficients in cm^3s^{-1},
except those asterisked which are in cm^6s^{-1}. † assumed to be
the product of radiative association, reaction not observed
in the laboratory.

Reactant Ion	H_2 Products	H_2 Rate Coefficient	CH_4 Product Distribution (%)	CH_4 Rate Coefficient
C^+	$(CH_2^+)^†$	$\sim 1(-14)$ theory	$C_2H_3^+(67), C_2H_2^+(33)$	$1.2(-9)$
CH^+	CH_2^+	$1.2(-9)$	$C_2H_3^+(84), C_2H_2^+(11)$ $C_2H_4^+(5)$	$1.3(-9)$
CH_2^+	CH_3^+	$1.6(-9)$	$C_2H_4^+(70), C_2H_5^+(30)$	$1.2(-9)$
CH_3^+	CH_5^+	$1.3(-28)^*$	$C_2H_5^+(100)$	$1.2(-9)$
CH_4^+	CH_5^+	$3.3(-11)$	$CH_5^+(100)$	$1.5(-9)$
C_2^+	C_2H^+	$1.4(-9)$	$C_3H_2^+(41), C_2H^+(17)$ $C_3H_3^+(15), C_3H^+(14)$ $C_2H_2^+(13)$	$1.4(-9)$
C_2H^+	$C_2H_2^+$	$1.7(-9)$	$C_2H_2^+(34), C_3H_3^+(34)$ $C_3H_4^+(12), C_3H_5^+(20)$	$1.1(-9)$
$C_2H_2^+$	$C_2H_3^+$ $C_2H_4^+$	$1.0(-11)$ $1.2(-27)^*$	$C_3H_5^+(78), C_3H_4^+(22)$	$1.0(-9)$
$C_2H_3^+$	$C_2H_5^+$	$\sim 2(-29)^*$	$C_3H_5^+(100)$	$1.7(-10)$
$C_2H_4^+$	—	$<1(-13)$	—	$<1(-13)$

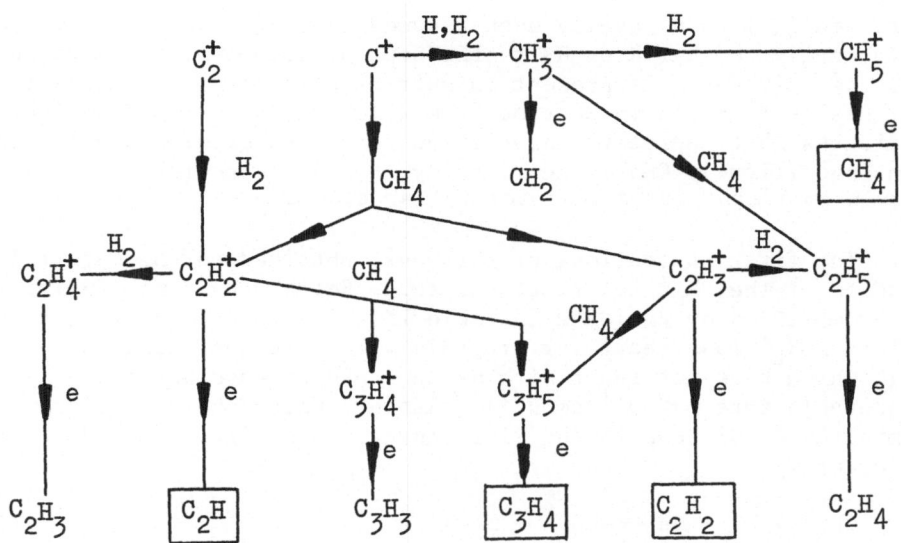

Fig. 3 Limited qualitative reaction scheme deduced from data
in Table II. Square boxes indicate that the enclosed
molecules have been detected in interstellar clouds.

molecular fragment as the process we have called H-atom ejection and
may well be an important process in molecular clouds in which the
electron density is relatively small and in which appreciable
concentrations of molecules of high proton affinity exist.

In summary, a great deal of relevant laboratory data is now
being produced, and detailed modelling of particular gas clouds is
currently being carried out taking into account all that is known of
the local physical conditions, composition, etc. (e.g. Black and
Dalgarno, 1977; Black et al, 1978).

4. LABORATORY STUDIES OF COLLISIONALLY - STABILIZED ASSOCIATION REACTIONS: DEDUCTION OF RADIATIVE ASSOCIATION RATE COEFFICIENTS

At the low particle number densities existing in interstellar
clouds, 3-body collisions do not occur and so all gas-phase theories
of molecular synthesis need only consider binary reactions. However
this need not exclude ion-neutral association reactions if binding
energy can be effectively removed by radiation emission, and indeed,
as referred to in Section 3 this process of radiative association is
invoked in the initial stages of molecular synthesis in gas clouds
(equations (12) and (13)). Since theoretical studies have indicated
that the rate coefficients for radiative association reactions are

very small, it has usually been assumed that such reactions could
only involve either H or H_2 as the neutral component since these
are the only neutrals present in sufficiently large concentrations.
In this Section and in Section 5 we will present evidence which
indicates that radiative association reactions between a polyatomic
ion (specifically CH_3^+) and minority molecular species can occur at
very significant rates in interstellar clouds.

The first indications of this were obtained during our SIFT
studies of the CH_n^+ ion reactions (e.g. Smith and Adams, 1977c) such
as exemplified by the data in Table II. It can be seen that the
CH^+ and CH_2^+ ions react very rapidly with H_2 channelling through to
CH_3^+ which does not react with H_2 in a binary fashion at a
measurable rate ($<10^{-13} cm^3 s^{-1}$). Rather, under the collision-
dominated conditions in the SIFT, ternary association of CH_3^+ with
H_2 occurs:

$$CH_3^+ + H_2 + He \longrightarrow CH_5^+ + He \tag{14}$$

The stable CH_5^+ ion is produced and the ternary association
rate coefficient is readily determined (see Table II). Since the
binary rate coefficient for $CH_3^+ + H_2$ reaction is very small, then
the CH_3^+ is available for reaction with minority species in
interstellar clouds and this becomes an especially interesting ion
whose reactions are worthy of a more detailed study. Such a study
with several molecular species revealed that at thermal energies
CH_3^+ reacted either by: (i) a "pure binary" process, i.e. with
COS, H_2S, C_2H_2 and CH_4, at gas kinetic rates; (ii) a "pure ternary"
association, i.e. with H_2, N_2, O_2, CO, CO_2, and H_2O, some of these
association reactions being very rapid even at 300K, or (iii)
rapid parallel binary and ternary channels, i.e. with NH_3, H_2CO,
CH_3OH and CH_3NH_2 (see Smith and Adams, 1978a). The category (iii)
reactions are exemplified by:

$$CH_3^+ + NH_3 \left\{ \begin{array}{l} \nearrow CH_2NH_2^+ + H_2 \\ \rightarrow NH_4^+ + CH_2 \\ \searrow [CH_3^+ \cdot NH_3]^* \xrightarrow{+ He} CH_3 \cdot NH_3^+ \end{array} \right. \tag{15}$$

That the effective binary rate coefficient for the association
(ternary) channel ($\sim 5 \times 10^{-10} cm^3 s^{-1}$) is comparable with that for
the gas kinetic binary channels ($\sim 1.8 \times 10^{-9} cm^3 s^{-1}$) under the
conditions of the SIFT experiment implies that the excited
intermediate complex ion $[CH_3^+ \cdot NH_3]^*$ has a long lifetime against
unimolecular decomposition, τ_d, and is thus effectively stabilized

by He atom collisions before it can dissociate. The equivalent
ternary association rate coefficient for the association channel is
correspondingly large ($>10^{-26}cm^6s^{-1}$), a determination of which can
be used to estimate the magnitude of τ_d (see below). All of the
association reactions of CH_3^+ studied indicated rate coefficients
which were always appreciable and often very large even at 300K.
Measurements made at 220K indicated that the reactions of CH_3^+
with the molecules listed in (ii) above (excluding the H_2O reaction
which could not be studied readily at this lower temperature), were
much greater than at 300K and a very recent series of measurements
of the rate coefficients for these reactions over the temperature
range 100-300K (Adams et al 1979b) has confirmed this trend. The
data obtained is summarised in Fig.4. It can be seen that the
ternary rate coefficients rapidly increase in a power law fashion
with decreasing temperature (i.e. $k_c = AT^{-n}$) and approach very
large values at the lowest temperatures. The magnitude of the
exponent, n, is typically about 4, considerably smaller than
suggested by the RRK theory which predicts a power law dependence
but with n+1=s=3N-6, where s is the number of degrees of freedom and
N is the number of atoms in the intermediate complex. Adams et al
(1979b) have tentatively suggested that this major discrepancy
indicates that only the inter-molecular vibrational modes associated
with the weak $[CH_3...X]^+$ association bond are involved in dissipating
the energy available and that the relatively high frequency bonds in
the CH_3^+ and in the associating molecule X are essentially decoupled
and do not take part. Theory and experiment are then approximately
reconciled.

In order to obtain an estimate for τ_d from the measured value
of k_c the following model outlined by Ferguson (1972) is adopted.

The overall reaction described by:

$$A^+ + B + M \xrightarrow{\quad k_c \quad} AB^+ + M \qquad (16)$$

is envisaged as proceeding via an excited intermediate complex $(AB^+)^*$
formed at a rate described by the coefficient k_1 and with a lifetime
against unimolecular decomposition τ_d as before:

$$A^+ + B \underset{\tau_d}{\overset{k_1}{\rightleftharpoons}} (AB^+)^* \qquad (17)$$

In collision with a chemically passive atom or molecule M (the
helium carrier gas in these experiments), $(AB^+)^*$ can either be
stabilized or dissociated back to the reactants A^+ and B:

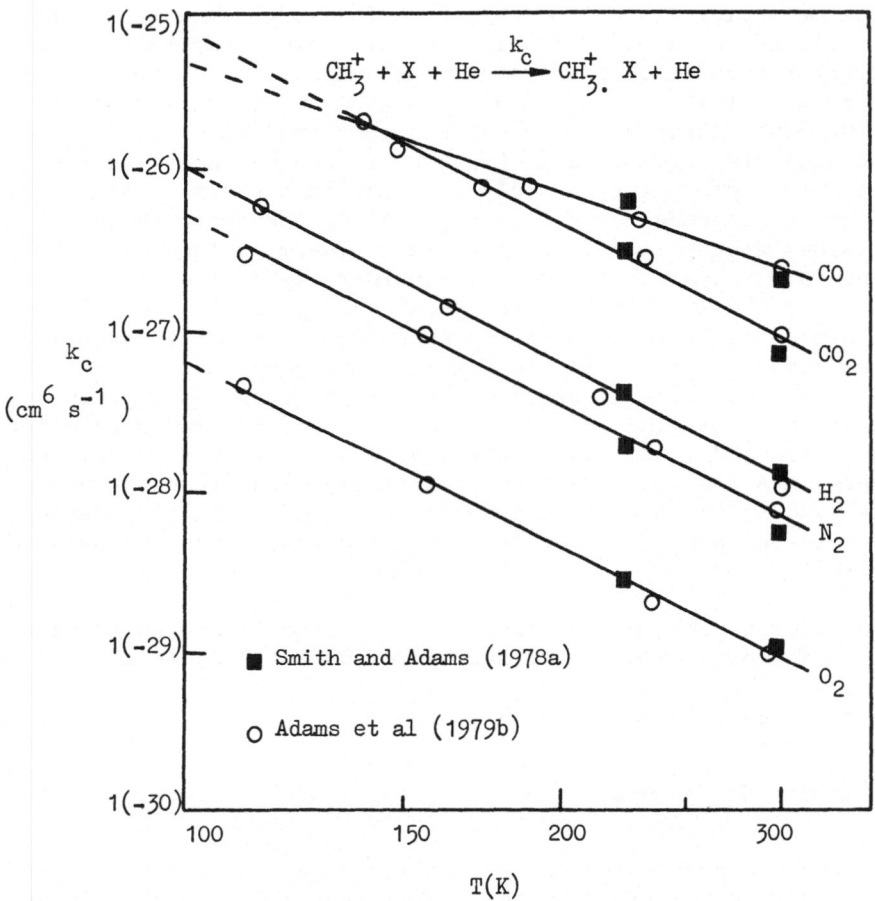

Fig. 4 Ternary association rate coefficients, k_c, as a
function of temperature, T, for the reactions of
CH_3^+ ions with the molecules indicated. In each
case helium atoms are the third body.

$$(AB^+)^* + M \xrightarrow{} \begin{array}{c} \overset{fk_2}{\nearrow} AB^+ + M \\[1em] \underset{(1-f)k_2}{\searrow} A^+ + B + M \end{array} \qquad (18)$$

k_2 is the collisional rate coefficient and f is the fraction of the collisons which result in stabilization. It can be readily shown from equations (17) and (18) that

$$k_c = \frac{k_1 k_2 f \, \tau_d}{1 + k_2 \tau_d [M]} \qquad (19)$$

The apparent or effective binary rate coefficient is given by $k_2^{eff} = k_c[M]$ and it is k_2^{eff} which is actually measured in the experiments. At low pressures (i.e. $k_2 \tau_d [M] \ll 1$), $k_c = k_1 k_2 f \, \tau_d$, i.e. $k_2^{eff} = k_1 k_2 f \, \tau_d [M]$. At high pressures (i.e. $k_2 \tau_d [M] \gg 1$) $k_c = k_1 f/[M]$, i.e. $k_2^{eff} = k_1 f$, that is k_2^{eff} independent of pressure (binary kinetics). So from a study of the pressure dependence of k_2^{eff} at different temperatures, estimates of both τ_d and f as a function of temperature can be obtained if the reasonable assumption is made that both k_1 and k_2 are the collisional temperature independent rate coefficients (Langevin or ADO, Su and Bowers, 1973) (strictly speaking fk_1 and $k_2\tau_d$ are obtained as a function of temperature). For the reactions referred to in Fig.4, the values of f generally vary between 0.1 and $\sim$1 (for details see Adams et al 1979b).

Of greater interest to interstellar chemistry are the estimated magnitudes of τ_d at the temperatures appropriate to interstellar clouds. Thus τ_d has been obtained as a function of temperature and assuming that the temperature dependences observed between 100K and 300K are maintained down to the lower molecular cloud temperatures the data has been extrapolated to 20K. The summarised data for the reactions of CH_3^+ ions with a few interstellar molecules is given in Table III in which it can be seen that the values of τ_d differ by several orders of magnitude for different reactions, are predictably very dependent on temperature and can be as large as a second! The significance of these results is clear. If τ_d is an appreciable fraction of τ_r, the radiative lifetime of the excited molecule $(AB^+)^*$, then radiative association becomes significant. A simple model of such a reaction can be constructed:

$$A^+ + B \xrightarrow{k_r} AB^+ + h\nu \qquad (20)$$

is the overall reaction which again can be described by two component steps

$$A^+ + B \underset{\tau_d}{\overset{k_1}{\rightleftharpoons}} (AB^+)^* \xrightarrow{\quad\tau_r\quad} AB^+ + h\nu \qquad (21)$$

and it is a simple procedure to show that k_r, the binary radiative association rate coefficient, is given by:

$$k_r = k_1 \left(\frac{\tau_d}{\tau_d + \tau_r} \right) . \qquad (22)$$

So if a value for τ_r can be assumed then from the experimentally derived τ_d, values of k_r can be obtained. Since only a low energy ($E \gtrsim k'T$) photon needs to be emitted for the complex to become bound, a single transition between vibrational energy levels in the complex (characteristic time $\sim 10^{-3}$ seconds, Herbst, 1976) is all that is required in principle to produce a stable ion. Thus the values of k_r in Table III have been calculated on the assumption that τ_r is $\sim 10^{-3}$ seconds. The resulting values are appreciable fractions of a typical collisional rate coefficient ($\sim 10^{-9} cm^3 s^{-1}$) particularly at the lower temperature. For the H_2O reaction, the k_r has reached the limiting (ADO) value at 20K. If stabilization of the excited complex can proceed via an electronic transition ($\tau_r \sim 10^{-8}$ seconds) in any of these reactions, then the upper limiting value of k_r will be reached at relatively higher temperatures.

Notwithstanding the assumptions made throughout this analysis, it seems likely that the process of radiative association could be an important reaction mechanism in the low temperature interstellar cloud plasmas. It is important to note that the estimated value of k_r for the $CH_3^+ + H_2$ reaction at 100K is some 50 times greater than the recent upper limit theoretical estimate of Herbst (1976). Obviously, radiative association reactions are not limited to CH_3^+ ions, other ions such as H_3CO^+ are very likely to undergo such reactions in interstellar clouds. Radiative association could well be the dominant process by which very large molecules are synthesised in these regions of the galaxy.

5. THE SIGNIFICANCE OF RADIATIVE ASSOCIATION TO MOLECULAR SYNTHESIS IN INTERSTELLAR CLOUDS

Having presented experimental evidence that the process of radiative association could be an important reaction mechanism in the exceptional environment of interstellar molecular clouds, quantitative estimates are presented here of the probable abundances of the neutral products of several such reactions and these estimates are compared to astronomical observations. The deductions and suggestions presented here are a sample of those

TABLE III

Rate coefficients and product ions for the reactions of CH_3^+ with several molecules, X. k_b (cm^3s^{-1}) and k_c (cm^6s^{-1}) are the measured binary and ternary values respectively at 300K. $\tau_d(s)$ are the derived lifetimes against unimolecular decomposition of the excited intermediate complex, $CH_3^+.X$, and k_r $(cm^3\ s^{-1})$ are the estimated radiative association rate coefficients.   The temperatures, T, were chosen as the approximate upper and lower limits to the temperature in interstellar clouds.

Reactant Molecule	k_b	k_c	T = 100K τ_d	k_r	T = 20K τ_d	k_r
H_2	-	$CH_3^+.H_2$ $1.3(-28)$	$2(-8)$	$3(-14)$	$8(-6)$	$1(-11)$
CO	-	$CH_3^+.CO$ $2.2(-27)$	$3(-7)$	$3(-13)$	$8(-5)$	$8(-11)$
H_2O	-	$CH_3^+.H_2O$ $>3(-26)$	$>3(-5)$	$>8(-11)$	$>5(-1)$	$2(-9)$
NH_3	$H_4CN^+(88\%)$ NH_4^+ (12%) $1.8(-9)$	$CH_3^+.NH_3$ $>7(-26)$	$>8(-5)$	$>4(-11)$	$>1(0)$	$5(-10)$
H_2CO	HCO^+ $1.6(-9)$	$CH_3^+.H_2CO$ $3.5(-26)$	$\sim3(-5)$	$\sim4(-11)$	$\sim5(-1)$	$1(-9)$
CH_3OH	H_3CO^+ $2.3(-9)$	$CH_3^+.CH_3OH$ $>4(-26)$	$>1(-4)$	$>3(-11)$	$>1(-1)$	$2(-10)$

TABLE IV

The calculated relative abundances of some interstellar species,
formed via the reactions of CH_3^+ ions, using the data given in
Table III. The order-of-magnitude relative abundances determined
from astronomical observations are shown in square brackets after
each species.

Reactant Molecule	Possible Recombination Products	Probable Interstellar Analogues	Calculated Relative Abundances, a_c	
			T=100K	T=20K
$H_2[0]^*$	CH_4	CH_4 [-4]	1(-6)	6(-4)
$CO[-4]$	C_2H_2O	CH_2CO [?]	1(-9)	4(-7)
$H_2O(-5)$ to $(-6)^{**}$	CH_4O	CH_3OH [-7]	[-7]	[-7]
$NH_3[-6]$	CH_5N	$CH_3NH_2[-10]$	>2(-9)	2(-8)
	CH_3N	CH_2NH [-10]	7(-8)	7(-8)
$H_2CO[-8]$	C_2H_4O	$CH_3CHO[-10]$	~2(-11)	6(-10)
	CO	CO [-4]	7(-10)	7(-10)
$CH_3OH[-7]$	C_2H_6O	$\begin{bmatrix} C_2H_5OH[-10] \\ (CH_3)_2O[-10] \end{bmatrix}$	>1(-10)	1(-9)
	CH_2O	H_2CO [-8]	[-8]	[-8]

* [x] and (x) represents 10^x. ** calculated assuming that the a_c
values are equal to the observed relative abundance for CH_3OH.

contained in the recent papers of Smith and Adams (1977c, 1978b) but also included are references to very recent observational data which are in general support of the deductions.

Using the calculated values of k_r given in Table III in conjunction with the relative abundances (with respect to that of H_2) of the reactant neutrals $[m]$ (when known) in the dense and dark interstellar clouds, the <u>relative</u> time constants, $\tau_c = (k_r[m])^{-1}$ can be calculated for the <u>radiative</u> association reactions of CH_3^+ with each reactant neutral. The calculated relative abundance of the product neutrals, a_c, formed via electronic recombination of the product ions, is then assumed to be proportional to τ_c^{-1} i.e. $a_c \propto k_r[m]$. This assumes that the electronic recombination rates are greater than the ion production rates and therefore that electronic recombination is not the rate determining step for the production of neutrals. The calculated relative abundances are given in Table IV and have been normalised to the observed relative abundance of H_2CO (10^{-8} relative to H_2) assuming that the H_2CO is formed exclusively via the reaction sequence:

$$CH_3^+ + CH_3OH \nearrow \quad H_3CO^+ \xrightarrow{+e} H_2CO + H \qquad (23a)$$
$$\searrow \quad CH_3^+CH_3OH \xrightarrow{+e} \text{see Fig.5} \qquad (23b)$$

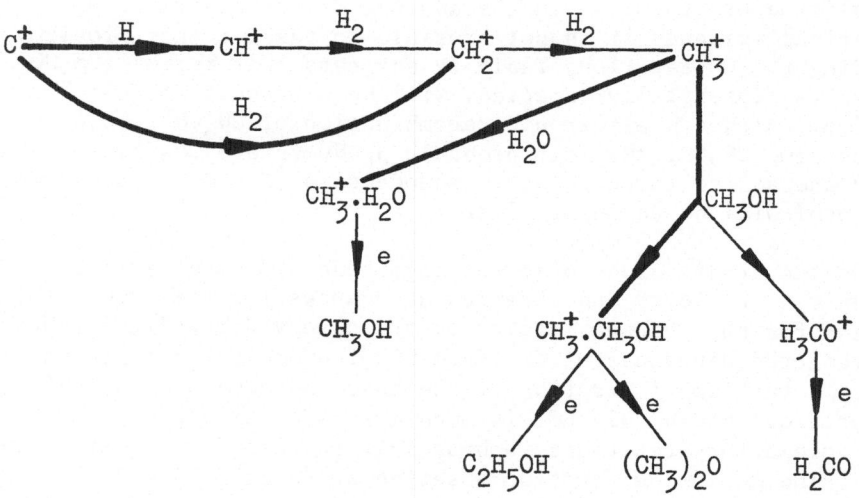

Fig. 5 A suggested mechanism for the production of H_2CO in dark and dense interstellar clouds. The reaction steps involving radiative association are indicated by the thick lines.

The channel producing H_3CO^+ is a gas kinetic binary channel
the rate of which is expected to be insensitive to temperature
and because of this, coupled with the fact that H_2CO is widely
observed throughout the galaxy, normalisation to the H_2CO relative
abundance is the most suitable procedure. The rate coefficient of
the association channel (23b) in the above reaction will, of course,
be temperature dependent (further discussion of this reaction is
given below).

The reactions of CH_3^+ ions listed in Tables III and IV and were
selected from a more extensive list, since these are sufficient to
illustrate some of the successes, anomalies and weaknesses of the
proposed model for molecular synthesis in interstellar clouds.
Firstly, the anomalies. It can be seen from Table IV that the
predicted relative abundances, a_c, of the CH_3NH_2 and CH_2NH –
assumed to be the products of the NH_3 reaction – are much larger than
the observed order-of-magnitude relative abundances. An order-of-
magnitude spread in both the neutral reactant (NH_3) and the neutral
products (CH_3NH_2 and CH_2NH) abundances determined from radio
observations can be expected, which could help to reconcile this
discrepancy, but it is more likely in this case that the simple
assumption that H-atom ejection is the major product channel of
electronic recombination is unjustified. Huntress (1977b) has
suggested that HCN could be an important product, this being a
very abundant species in molecular clouds. However, Ausloos (1978)
has pointed out that since about 78kcal/mole is required to break
the C–N bond in the $CH_3NH_2^*$ formed during electronic recombination
of the $CH_3NH_3^+$, then CH_3NH_2 may indeed be a small fraction of the
neutral products formed. In the absence of any experimental or
theoretical evidence it is not possible to resolve this problem.
Regarding the CO reaction, Table IV suggests that ketene (CH_2CO),
an observed interstellar species, will be a product of this
reaction. Although electronic recombination of CH_3CO^+ could
produce some CH_2CO, the most probable products are CH_3 radicals
and CO and so the large relative abundance of CH_2CO predicted by
this model will be an upper limit.

On the credit side, order-of-magnitude agreement exists
between the predicted and observed abundances for the H_2CO reaction,
at least for the channel leading to the observed species CH_3CHO.
The predicted abundance of CO (from HCO^+) produced in this reaction
is only a very small fraction of the observed abundance but HCO^+,
and therefore presumably CO, is much more effectively produced
in the elementary reactions such as that indicated for example in
Fig.2. The following reaction observed by Fehsenfeld (1976):

$$CH_3^+ + O \longrightarrow HCO^+ + H_2 \tag{24}$$

followed by electronic recombination of the product ion, may also
be an efficient source of CO.

The association channel in the CH_3OH reaction adequately predicts the observed concentrations of C_2H_5OH and CH_3OCH_3. Some structural re-arrangement is clearly necessary during the reaction if dimethyl ether is to be produced.

The most striking prediction in the data of Table IV is that the H_2 reaction should lead to the production of CH_4 in relatively large concentrations second only to H_2 and comparable with that of CO. Subsequent to the publication of the paper which predicted this (Smith and Adams, 1978b), CH_4 was detected in both IRC + $10^{\circ}216$ by infra-red spectroscopy (Hall and Ridgway, 1978) and in the Orion A molecular cloud from observations of weak microwave emissions (Fox and Jennings, 1978). In the latter observation, the CH_4 column density determined is consistent with the high relative density predicted by the radiative association model. From this evidence alone, it would seem that H-atom ejection is at least a significant channel in the electronic recombination of CH_5^+. Rebbert et al (1973) have shown experimentally that, under high pressure conditions, electronic recombination of CH_5^+ produces mainly CH_4 and H. Under the low pressure conditions of interstellar clouds however, further dissociation of the $(CH_4)^*$ could result in a significant fraction of CH_3 radicals being produced.

In the reaction of CH_3^+ with H_2O the only observed product is the association ion $CH_3H_2O^+$ which on recombination is assumed to produce CH_3OH. Since the relative abundance of CH_3OH in interstellar clouds is better known than that of H_2O, we have reversed the procedure discussed above and calculated the relative abundance of the H_2O using the known value for CH_3OH. This procedure indicates a relatively large abundance of H_2O, of the order of 10^{-6} to 10^{-5} of that of H_2, which is consistent with the upper-limit estimate of Snyder et al (1977). More recently, Phillips et al (1978) have estimated the density of H_2O in Orion from observations of the isotopic species $H_2^{18}O$ assuming the terrestrial oxygen isotopic ratio, and have deduced that $[H_2O] \approx 10^{-5}[H_2]$, again in good agreement with the model estimate.

Earlier in this Section we suggested that the CH_3^+ + CH_3OH reaction (23a) resulted in the production of the widely observed H_2CO. A mechanism for production of H_2CO has long been sought and several have been suggested, notably by Herbst and Klemperer (1973), Dalgarno et al (1973b) and by Millar and Williams (1975), although none have been shown to be satisfactory (discussed by Watson, 1977). The data of Table IV would appear to offer a solution to this problem. The first step is the radiative association of CH_3^+ with H_2O leading to CH_3OH. The CH_3OH then reacts with CH_3^+ (reaction (23a)) producing H_3CO^+ and hence H_2CO. That the relative abundances of the neutral species involved are in the order $[H_2O] > [CH_3OH] > [H_2CO]$ lends support to this scheme. The suggested overall reaction scheme starting with the most elementary reactions and eventually producing H_2CO is shown in Fig.5. In a recent detailed astronomical study of CH_3OH and H_2CO column densities in 14 galactic sources, Gottlieb et al (1978) conclude that the reaction of CH_3 with H_2O leading to CH_3OH is consistent with their observations and that the notion that the observed CH_3OH and H_2CO are intimately connected through their chemistry is at least not contradicted by their observations. It is this kind of correlation which will eventually either disprove or support any reaction scheme. In this case it would appear that some progress towards an understanding of the chemistry occurring in these regions is being made and that radiative association is playing an important part in molecular synthesis.

6. CONCLUDING REMARKS

It has been shown that an appreciation of the complex chemistry of a plasma, with the many concurrent and consecutive reactions which occur, can only be obtained from detailed studies of the individual reaction processes. In the large majority of low temperature plasmas, ion-molecule reactions play a central role in the chemical evolution and it was the recognition of this fact, particularly with respect to ionosphere, which provided much of the stimulus for the study of ion-molecule reactions during the last two decades. Additional stimulus is being provided by the desire to appreciate the complex chemistry occurring in interstellar clouds and in gas lasers. However much more experimental data are required particularly relating to reactions occurring under high gas pressure conditions such as in the stratosphere and troposphere and in high pressure gas lasers (association and clustering reactions). Data concerning the neutral products of ion-neutral, ion-electron and ion-ion reactions are also essential if an adequate appreciation of chemical evolution in such plasmas is to be obtained.

The process of radiative association is unimportant in most plasma situations except those in which the particle number densities and temperatures are very low such as the interstellar gas clouds. Nevertheless, the process is of considerable fundamental interest and direct experimental observations of radiative association (which should be possible in ion traps with variable draw-out times coupled with suitable spectroscopic techniques) would contribute greatly to a better understanding of ion-molecule interactions.

REFERENCES

Adams,N.G. and Smith,D.(1976a) Int.J.Mass Spectrom.Ion.Phys. 21, 349
Adams,N.G. and Smith,D.(1976b) J.Phys.B. 9, 1439
Adams,N.G. and Smith,D.(1977) Chem.Phys.Letts. 47, 383
Adams,N.G. and Smith,D.(1978) Chem.Phys.Letts. 54, 530

Adams,N.G., Bohme,D.K., Dunkin,D.B., Fehsenfeld,F.C. and Ferguson,E.E.
     (1970) J.Chem.Phys. $\underline{52}$, 3133
Adams,N.G., Smith,D. and Grief,D. (1978) Int.J.Mass Spectrom.Ion.
     Phys. $\underline{26}$, 405
Adams,N.G., Smith,D. and Grief, D. (1979a) J.Phys.B. (In Press)
Adams,N.G., Lister,D.G., Rakshit,A.B., Smith,D., Tichy,M. and
     Twiddy,N.D. (1979b). Chem.Phys.Letts. (In Press)
Adams,N.G., Lister,D.G., Rakshit,A.B., Smith,D., Tichy,M. and
     Twiddy,N.D. (1979c) Chem.Phys.Letts. (In Press)
Albritton,D.L. (1979) Atom. Data Nucl. Data Tables (In Press)
Ausloos, P. (1978) Private Communication

Bardsley,J.M. and Biondi,M.A. (1970) Adv.Atom.Molec.Phys. $\underline{6}$, 1
Bates,D.R. and Spitzer,L.Jr. (1951) Ap.J. $\underline{113}$, 441
Beauchamp,J.L. (1975) In "Interactions between Ions and Molecules"
     ed. P.Ausloos, Plenum Press, New York, 413
Bierbaum,V.M., Depuy,C.H., Shapiro,R.H., and Stewart,J.H. (1976)
     J.Amer.Chem.Soc. $\underline{98}$, 4229
Biondi,M.A. (1968) In "Methods of Experimental Physics" ed.
     B.Bederson and W.L.Fite, Academic Press, New York, Vol.7B, 78
Biondi,M.A. and Brown S.C. (1949) Phys. Rev. $\underline{75}$, 1700
Black,J.H. and Dalgarno,A. (1973) Astrophys. Letts. $\underline{15}$, 79
Black,J.H. and Dalgarno,A. (1977) Ap.J. Suppl.Series $\overline{34}$, 405
Black,J.H.,Hartquist,T.W. and Dalgarno,A. (1978) Ap.J. $\overline{224}$, 448
Bohme,D.K. (1975) In "Interactions between Ions and Molecules"
     ed.P.Ausloos, Plenum Press, New York, 489
Bowers,M.T. and Su,T. (1975) In "Interactions between Ions and
     Molecules" ed P.Ausloos, Plenum Press, New York, 163
Broten,N.W., Oka,T., Avery,L.W., MacLeod,J.M. and Kroto,H.W. (1978)
     Ap.J. $\underline{223}$, L105
Burt,J.A., Dunn,J.L., McEwan,M.J., Sutton,M.M., Roche,A.E. and
     Schiff,H.I. (1970). J.Chem.Phys. $\underline{52}$, 6062

Castleman,A.W.Jr. (1978) Adv.in Colloid and Interface Sci.
     "Nucleation" ed. A.Zettlemoyer (In Press)

Dalgarno,A. (1975) In "Interactions between Ions and Molecules"
     ed. P.Ausloos, Plenum Press, New York, 341
Dalgarno,A. (1976) In "Atomic Processes and Applications"
     ed. P.G.Burke and B.L.Moiseiwitsch, North Holland Publ.Co.
     Amsterdam, Ch.5, 110
Dalgarno,A. and Black,J.H. (1976) Rept.Prog.Phys. $\underline{39}$, 573
Dalgarno,A., Oppenheimer,M. and Berry,R.S. (1973a) Ap.J. $\underline{183}$, L21
Dalgarno,A., Oppenheimer,M. and Black,J.H. (1973b) Nature $\overline{245}$, 100
Dickinson,P.H.G. and Sayers,J. (1960) Proc.Phys.Soc. $\underline{76}$, 137
Douglas,A.E. and Herzberg,G. (1941) Ap.J. $\underline{94}$, 381

Fehsenfeld,F.C. (1975) In "Interactions between Ions and Molecules"
     ed.P.Ausloos, Plenum Press, New York, 387
Fehsenfeld,F.C. (1976) Ap.J. $\underline{209}$, 638

Fehsenfeld,F.C. (1978) Private Communication
Fehsenfeld,F.C., Dunkin,D.B., Ferguson,E.E. and Albritton,D.L.
     (1973) Ap.J. 183, L25
Fehsenfeld,F.C., Dunkin,D.B. and Ferguson,E.E. (1974) Ap.J. 188, 43
Ferguson,E.E. (1972) In "Ion-Molecule Reactions" ed.J.L.Franklin,
     Plenum Press, New York, Ch.8, 363
Ferguson,E.E. (1973) Atom.Data Nucl.Data Tables 12, 159
Ferguson,E.E. (1974) Rev.Geophys.Space. Sci. 12, 703
Ferguson,E.E. (1975) In "Atmospheres of Earth and the Planets"
     ed.B.M.McCormac, D.Reidel Publ.Co., Dordrecht, Holland, 197
Ferguson,E.E., Fehsenfeld,F.C. and Schmeltekopf,A.L. (1969a) In
     "Advances in Atomic and Molecular Physics" ed.D.R.Bates
     and I.Estermann, Academic Press, New York, Vol 5, 1
Ferguson,E.E., Bohme,D.K., Fehsenfeld,F.C. and Dunkin,D.B. (1969b)
     J.Chem.Phys. 50, 5039
Ferguson,E.E., Fehsenfeld,F.C. and Albritton,D.L. (1979) In
     "Ion-Molecule Reactions" ed. M.T.Bowers, Academic Press,
     New York. (In Press)
Fox,K. and Jennings, D.E. (1978) Ap.J.Letts. (In Press)

Giusti-Suzor,A., Roueff, E. and van Regemorter,H. (1976) J.Phys. B.
     9, 1021
Glosik,J., Rakshit,A.B., Twiddy,N.D., Adams,N.G. and Smith,D.
     (1978) J.Phys.B. (In Press)
Good,A. (1975) Chem. Revs. 75, 561
Gottlieb,C.A., Ball,J.A., Gottlieb,E.W. and Dickinson,D.F. (1978)
     Ap.J. (Submitted)
Guélin,M., Green,S. and Thaddeus,P. (1978) Ap.J. 224, L27

Hall,D.N.B. and Ridgway,S.T. (1978) Private Communication
Herbst,E. (1976) Ap.J. 205, 94
Herbst,E. (1978a) Ap.J. 222, 508
Herbst,E. (1978b) Private Communication
Herbst,E. and Klemperer,W. (1973) Ap.J. 185, 505
Herbst,E. and Klemperer,W. (1976) In "The Physics of Electronic
     and Atomic Collisions" ed.J.S.Risley and R.Geballe, Univ.
     Washington Press, Seattle, 62
Herbst,E., Schubert,J.G. and Certain,P.R. (1977) Ap.J. 213, 696
Huang,C-M., Whitaker,M., Biondi,M.A. and Johnsen,R. (1978)
     Phys.Rev. A18, 64
Huntress,W.T.Jr. (1977a) Ap.J. Suppl.Series 33, 495
Huntress,W.T.Jr. (1977b) Chem.Soc.Revs. 6, 295

Kebarle,P. (1972) In "Ion-Molecule Reactions" ed.J.L.Franklin,
     Plenum Press, New York, Vol.1, 315
Kebarle,P. (1975) In "Interactions between Ions and Molecules"
     ed.P.Ausloos, Plenum Press, New York, 459
Kebarle,P. (1977) Ann.Rev.Phys.Chem. 28, 445

Lang,K.R. (1974) "Astrophysical Formulae" Springer-Verlag, Berlin.

Leu,M.T., Biondi,M.A. and Johnsen,R. (1973) Phys.Rev. A7, 292
Liddy,J.P., Freeman,C.G. and McEwan,M.J. (1977) Mon.Not.R.A.S.
    180, 683
Lindinger,W., Fehsenfeld,F.C., Schmeltekopf,A.L. and Ferguson, E.E.
    (1974) J.Geophys.Res. 79, 4753
Lineberger,W.C. and Puckett,L.J.(1969) Phys.Rev. 186, 116
Liszt,H.S. and Turner,B.E. (1978) Ap.J. 224,L73
MacGregor,M. and Berry,R.S. (1973) J.Phys.B. 6, 181
Mahan,B.H. (1973) In "Advances in Chemical Physics" ed.I.Prigogine
    and S.A.Rice,J.Wiley,New York, Vol.23, 1
McDaniel,E.W. (1964) "Collision Phenomena in Ionized Gases"
    J.Wiley, New York
McDaniel,E.W. and Mason,E.A. (1973) "The Mobility and Diffusion
    of Ions in Gases" J.Wiley, New York
McDaniel,E.W., Čermák,V., Dalgarno,A., Ferguson,E.E. and
    Friedman,L. (1970) "Ion-Molecule Reactions" Wiley-Interscience,
    New York
McFarland,M., Albritton,D.L., Fehsenfeld,F.C., Ferguson,E.E. and
    Schmeltekopf,A.L. (1973) J.Chem.Phys. 59, 6610
Millar,T.J. and Williams,D.A. (1975) Mon.Not.R.A.S. 170,51P
Moseley,J.T., Olsen,R.E. and Peterson,J.R. (1975) In "Case Studies
    in Atomic Physics" ed. M.R.C.McDowell and E.W.McDaniel, North
    Holland Publ.Co., Amsterdam, 5, 1
Myers,P.C. (1978) Ap.J. 225, 380

Oskam,H.J. (1958) Philips Res.Rept. 13, 335
Oskam,H.J. (1969) In "Case Studies in Atomic Collision Physics"
    ed. E.W.McDaniel and M.R.C.McDowell, North Holland Publ.Co.,
    Amsterdam, Vol.1

Phillips,T.G., Scoville,N.Z., Kwan,J., Huggins,P.J. and Wannier,P.G.
    (1978) Ap.J. 222, L59
Pickles,J.B. and Williams,D.A. (1977) Astrophys. and Space Sci.
    52, 443
Porter,R.F. (1975) In "Interactions between Ions and Molecules"
    ed. P.Ausloos, 231
Puckett,L.J. and Lineberger,W.C. (1970) Phys. Rev. 1A, 1635

Rebbert,R.E. and Ausloos,P. (1972) J.Res.Nat.Bur.Stand.(US) 76A, 329
Rebbert,R.E., Lias,S.G. and Ausloos,P. (1973) J.Res.Nat.Bur.Stand
    (US) 77A, 249
Reid,G.C. (1976) In "Advances in Atomic and Molecular Physics"
    ed.D.R.Bates and B.Bederson, Academic Press, New York,Vol.12, 375

Sayers,J. and Smith,D. (1964) Disc.Farad.Soc. 37, 167
Schiff,H.I., Hemsworth,R.S., Payzant,J.D. and Bohme,D.K. (1974)
    Ap.J. 191, L49
Sechrist,C.F.Jr. (1975) Rev.Geophys.Space Sci. 13, 894
Shiu,Y-J. and Biondi,M.A. (1978) Phys.Rev. A17, 868

Smith,D. and Adams,N.G. (1977a) Int.J.Mass Spectrom. Ion. Phys. 23, 123

Smith,D. and Adams,N.G. (1977b) Chem.Phys.Letts. 47, 145

Smith,D. and Adams,N.G. (1977c) Ap.J. 217, 741

Smith,D. and Adams,N.G. (1978a) Chem.Phys.Letts. 54, 535

Smith,D. and Adams,N.G. (1978b) Ap.J. 220, L87

Smith D. and Adams,N.G. (1979) In "Ion-Molecule Reactions"
        ed. M.T.Bowers, Academic Press, New York (In Press)

Smith D. and Church,M.J. (1976) Int.J.Mass Spectrom. Ion. Phys. 19, 185

Smith,D. and Church, M.J. (1977) Planet.Space Sci. 25, 433

Smith,D. and Fouracre,R.A. (1968) Planet. Space Sci. 16, 243

Smith,D., Adams,N.G., Dean,A.G. and Church,M.J. (1975)
        J.Phys.D. 8, 141

Smith,D., Adams,N.G. and Church,M.J. (1976) Planet. Space Sci. 24, 697

Smith,D., Adams,N.G. and Grief,D. (1977) J.Atmos.Terres.Phys. 39,513

Smith,D., Church,M.J. and Miller,T.M. (1978a) J.Chem.Phys. 68, 1224

Smith,D., Adams,N.G. and Miller,T.M. (1978b) J.Chem.Phys. 69, 308

Smith,D., Adams,N.G. and Church,M.J. (1979a) J.Phys.B. (In Press)

Smith,D., Grief,D. and Adams,N.G. (1979b) Int.J.Mass Spectrom
        Ion.Phys. (In Press)

Smith,D., Adams,N.G. and Henchman,M.J. (1979c) (In Preparation)

Snyder,L.E., Watson,W.D. and Hollis,J.M. (1977) Ap.J. 212, 79

Solomon,P.M. and Klemperer,W. (1972) Ap.J. 178, 389

Su,T. and Bowers,M.T. (1973) Int.J.Mass Spectrom. Ion. Phys. 12,347

Thomas,L. (1974) Radio Sci. 9, 121

Tichy,M., Rakshit,A.B., Lister,D.G., Twiddy,N.D.,Adams,N.G. and
        Smith,D. (1979) Int.J.Mass Spectrom Ion.Phys. (In Press)

Ulich,B.L., Hollis,J.M. and Snyder,L.E. (1977) Ap.J. 217, L105

Watson,W.D. (1976) Rev.Mod.Phys. 48, 513

Watson,W.D. (1977) Acc.Chem. Res. 10, 221

ION-MOLECULE REACTIONS IN THE ATMOSPHERE

Eldon E. Ferguson

Aeronomy Laboratory, NOAA

Boulder, Colorado, 80303, U. S. A.

Efforts to understand the natural radiation chemistry of the earth's atmosphere have made a major impact on the development of atomic and molecular physics generally and ion molecule chemistry particularly. This has happened in two ways. First, direct observation of atmospheric properties has led to a great deal of knowledge about ion-molecule reactions. For example, Bates and Nicolet (1960) pointed out, prior to the laboratory studies, that O^+ ions must have reaction rate constants with O_2 and N_2 much lower than the Langevin collision rate constant, otherwise the electrons in the ionosphere below 200 km would rapidly disappear after sunset, contrary to observation. Current analyses of detailed atmospheric measurements obtained with satellites are leading to many reaction rate constants not yet measured in the laboratory, as will be described below. Secondly, a great deal of the modern technological development in laboratory ion-neutral reaction kinetics has been directly motivated by an interest in understanding atmospheric ion chemistry, particularly the flowing afterflow and its various spin-off's which will be referred to elsewhere in these proceedings. This paper will emphasize the present status of understanding of atmospheric ion chemistry, as derived from in-situ ion composition using laboratory reaction rate constants and other data when both laboratory and in-situ measurements are available. In other cases, particularly in the troposphere or lowest altitude region of the atmosphere, present considerations are quite speculative and do not have an adequate basis in either laboratory or in-situ observational data.

It is useful to divide the atmosphere into high altitude (low pressure) and low altitude (high pressure) regimes for purposes of ion chemistry discussion. An altitude of $\sim$ 100 km (p $\sim$ 2 × 10^{-4} torr) represents a generally convenient dividing line. The in-situ observations are simpler to carry out and therefore much more de-tailed as well as more precise in the higher altitude regime, where satellites, as well as rockets, serve as experimental platforms. The complexity of atmospheric measurements below 100 km is very much greater.

High Altitude Ion Chemistry (> 100 km)

The major ionization source in the upper atmosphere above 100 km (the so-called ionosphere) is photoionization by solar ultraviolet light. Since molecular oxygen is largely photodissociated in this altitude region, and since atomic oxygen is lighter than the major constituents N_2 and O_2, atomic oxygen becomes the dominant neutral constituent of the atmosphere above $\sim$ 200 km. The reactions which convert O^+ ions to molecular ions play an exceptionally important role in ionospheric behavior because conversion of atomic ions to molecular ions is rate controlling for electron loss in the region of the ionosphere ($\sim$ 250 km) where the electron density is a maximum ($\sim 10^6$ cc^{-1}). The electron loss processes are dissociative recombina-tion with NO^+ and O_2^+. Because the electrons in the ionosphere re-flect electromagnetic waves, the atmospheric ion chemistry of O^+ has been a matter of some practical importance as well as scientific interest. The final development in acquiring appropriate laboratory measurements has been to bridge the gap between the 300 K temperature characterizing most laboratory studies and the higher temperatures of the upper ionosphere, $\sim$ 600 $\rightarrow$ 1500 K or even higher.

Atomic oxygen ions are destroyed primarily by the reactions

$$O^+ + N_2 \rightarrow NO^+ + N \tag{1}$$

and
$$O^+ + O_2 \rightarrow O_2^+ + O \tag{2}$$

Both k_1 and k_2 have been measured in a heated flowing afterglow from 80 K to 900 K (Lindinger et al, 1974) and found to decrease over this temperature range. Recently Chen et al, (1978) have measured (1) to 900 K and (2) to 700 K in a static drift tube, with increased ac-curacy. Recently, cross-sections for (1) and (2) have been obtained from thermal energy to 3 eV relative kinetic energy, using a Flow-Drift technique (Albritton et al., 1977). In order to interpret these measurements, Lin and Bardsley (1977) deduced the O^+ ion vel-ocity distribution as a function of E/N in the drift tube experiments, in both helium and argon buffer gas, using the experimental mobilities measured by Albritton et al. to determine the intermolecular inter-action potential. The cross-sections were folded into Maxwellian velocity distributions to yield the reaction rate constants as a

function of kinetic temperature. The determination of velocity distributions and hence rate constants for Maxwellian speed distributions in drift tubes represents an important advance which is described in the paper by Albritton in these proceedings.

The several measurements all agree to within their stated experimental uncertainties. In particular, the agreement between the
flow drift tube, in which only kinetic energy is varied, and the
heated flow tube data shows that internal excitation does not play
a significant role in (1) and (2) below 900 K. It is known (Schmeltekopf et al, 1968) that at higher temperatures, vibrational excitation does play a role in (1). From an examination of laboratory
data on (1) and (2), we judge that k_1 and k_2 are known to about the
10% accuracy level from 80 to 900 K.

Current analyses of detailed satellite observations are making
a major impact on our understanding of the upper atmosphere ion
chemistry. The Atmosphere Explorer-C satellite launched late in
1973 carried mass spectrometers to measure both the neutral and ion
composition, optical spectrometers to measure the solar ionizing
flux, photometers to measure various optical emissions from excited
state ions and neutrals and other instrumentation to measure temperature and other parameters. The analyses of these simultaneous
measurements have allowed detailed information about the ion chemistry to be deduced. Figure 1 shows a comparison of calculated and
observed O^+ profiles for four satellite passes (Oppenheimer et al,
1977a). Below the O^+ concentration peaks, $\sim$ 240 km, a photochemical
stationary state exists and agreement is extremely good. Since the
calculated profiles include the laboratory measurement on (1) and
(2), this agreement very effectively supports the laboratory measurements. Above $\sim$ 240 km the O^+ ions are lost by diffusion and agreement with a photochemical model is not expected.

The remaining reactions controlling the major ion species have
all been measured in the laboratory in one or more experiments and
the rate constants have also been checked by analyses of ion profiles
measured with Atmosphere Explorer-C. Figure 2 shows one such comparison.

One of the more difficult F-region reactions to measure is

$$N_2^+ + O \rightarrow NO^+ + N \tag{3}$$

due to the unstable neutral reactant. The laboratory measurement of
McFarland et al. (1974) in a flow-drift tube as a function of energy
has been supported by the analysis of Fig. 2 and similar analyses
for different satellite passes and also by an independent analysis
by Torr et al, (1977b) in which they isolated some 400 satellite

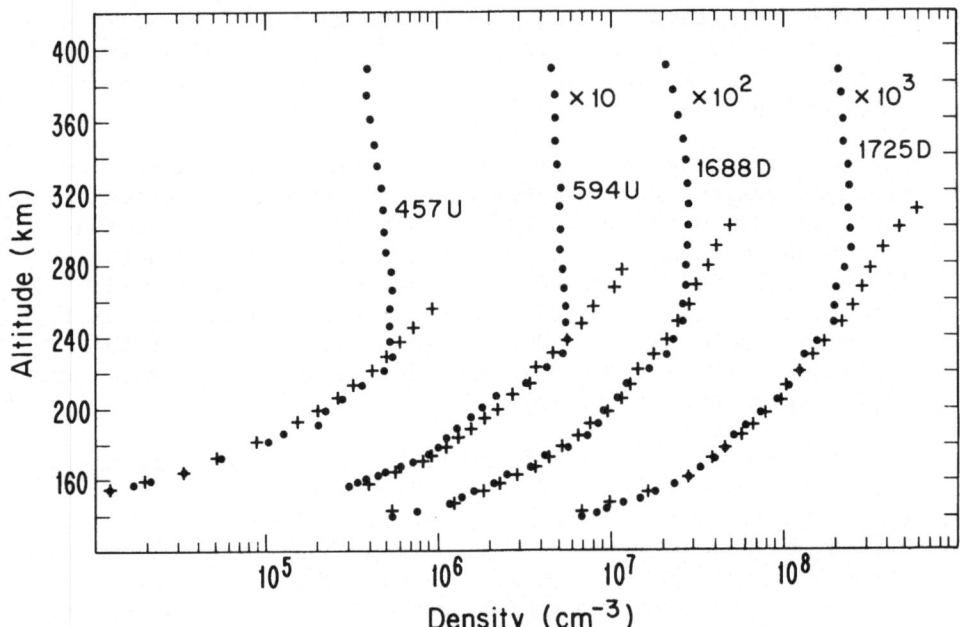

Figure 1. Comparison of calculated (plus) and (circle) O$^+$ ion densities as a function of altitude for four Atmosphere Explorer-C satellite orbital passes (from Openheimer et al., 1977a).

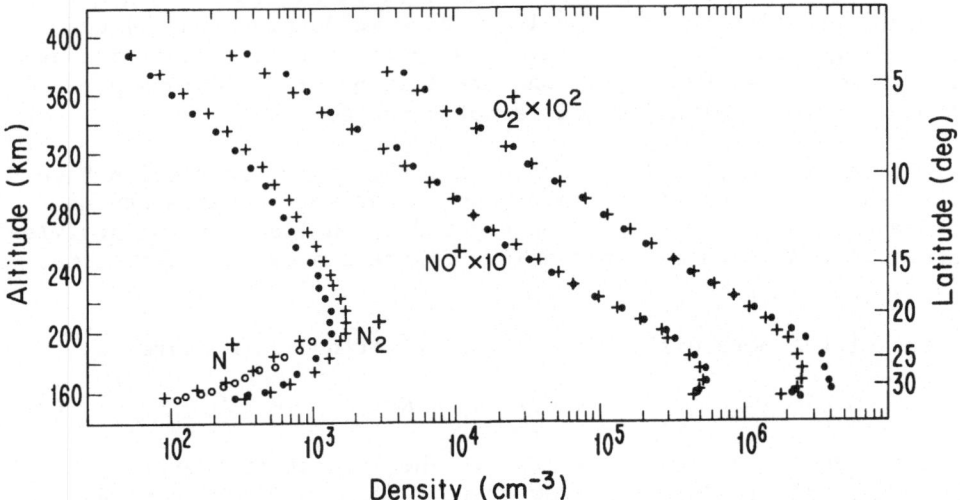

Figure 2. Comparison of calculated (plus) and observed (circle) N$^+$, N$_2^+$, NO$^+$, and O$_2^+$ ion densities as a function of altitude and latitude for an orbit pass of Atmosphere Explorer-C satellite (from Oppenheimer et al., 1977a).

observations of N_2^+ concentration for which (3) was the only significant loss. Using the simultaneous determination of N_2, O and the
solar ionizing flux, they were able to deduce k_3 for the 600-700 K
temperature range involved. The value k_3 = 1.1 ± 15% × 10^{-10} cm^3
s^{-1} agrees very well with the lab value k_3 = 1.0 ± 50% × 10^{-10} cm^3
s^{-1}. In regions of lower electron density where vibrationally excited N_2^+ ions are expected to be present, Orsini et al., (1977)
have found lower N_2^+ densities than expected and have attributed
this to an enhanced dissociative recombination coefficient for
vibrationally excited N_2^+. Biondi, however, (1978) rejects this
interpretation and suggests instead the possibility that the rate
constant for reaction (3) may be enhanced by N_2^+ vibrational excitation. This is an important ionospheric reaction to measure but
obviously a difficult one. The complexities of reaction studies of
vibrationally excited ions and of chemically unstable neutral reactants have not yet been combined in one experiment.

From analyses such as shown in Fig. 1 and 2, as well as the
internal consistency of laboratory measurements themselves, we now
believe the ion chemistry of the upper-atmospheric major ion species
to be quite well understood. This represents a culmination of over
15 years research in the laboratory and in in-situ atmospheric
measurements.

Atmosphere Explorer measurement of the $O^+(^2P) \rightarrow O^+(^2D)$ + 7320 Å
airglow have been used to deduce the $O^+(^2P)$ quenching rates with
atomic oxygen and nitrogen,

$$O^+(^2P) + O \rightarrow O^+(^4S) + O \tag{4}$$

k_4 = 1.8(-10) cm^3 s^{-1} (Oppenheimer, et al. 1977b), k_4 = 5.2±2.5(-10)
(Rusch, et al. 1977b),

$$O^+(^2P) + N_2 \rightarrow O^+(^4S) + N_2 \tag{5a}$$

$$\rightarrow N_2^+ + O \tag{5b}$$

k_5 = 4.8±1.4(-10) (Rusch, et al. 1977), k_5 = 4.5(-10) (Oppenheimer,
et al. 1977b). Oppenheimer et al. have deduced that k_{5b} = 5.0(-11)
from AE-C N_2^+ measurements. The radiative lifetime of $O^+(^2P)$ is
4.7s and it lies 5.0 eV above the $O^+(^4S)$ ground state. Dalgarno and
McElroy (1965) calculated that 20% of the O^+ ions in the F-region
are produced in the ^2P state. No laboratory measurements of O^+ ions
specifically identified as $O^+(^2P)$ have yet been reported.

The $O^+(^2D)$ state lies 3.3 eV above the ground state and has a
radiative lifetime of 3.h. Some 38% of the O^+ ions in the F-region
are produced in the ^2D state. Torr and Orsini (1978) have deduced
the limits on

$$O^+(^2D) + N_2 \rightarrow N_2^+ + O, \quad 0.5 \times 10^{-10} < k_{6a} < 3 \times 10^{-10} \text{ cm}^3 \text{ s}^{-1} \quad (6a)$$

$$O^+(^2D) + N_2 \rightarrow O^+(^4S) + N_2, \quad 0 < k_{6b} < 0.5 \times 10^{-10} \text{ cm}^3 \text{ s}^{-1} \quad (6b)$$

$$O^+(^2D) + O \rightarrow O^+ + O, \quad k_7 << 3(-11) \text{ cm}^3 \text{ s}^{-1} \quad (7)$$

and Torr et al. (1977b)

$$O^+(^2D) + O_2 \rightarrow O_2^+ + O, \quad k_8 = 2(-9) \text{ cm}^3 \text{ s}^{-1} \quad (8)$$

with an uncertainty of a factor of two on k_8.

This is a change from an earlier estimate by Torr and Orsini (1977) for k_6, also from an evaluation of AE data. This points up the need for laboratory verification of these rate constants.

Molecular metastable ion reactions have been measured as a function of kinetic energy in the NOAA flow drift tube, including

$$O_2^+(a^4\Pi_u) + N_2 \rightarrow N_2^+ + O_2, \quad 4.1 \pm 1.6(-10) \text{ cm}^3 \text{ s}^{-1} \quad (9)$$

$$O_2^+(a^4\Pi_u) + O_2 \rightarrow O_2^+(X^2\Pi_g) + O_2, \quad 3.1 \pm 1.5(-10) \text{ cm}^3 \text{ s}^{-1} \quad (10)$$

Lindinger et al. (1975) and

$$NO^+(a^3\Sigma^+) + N_2 \rightarrow N_2^+ + NO, \quad 3.2 \pm 1.3(-10) \text{ cm}^3 \text{ s}^{-1} \quad (11a)$$

$$\rightarrow NO^+(X^1\Sigma^+) + N_2, \quad 3.2 \pm 1.3(-10) \text{ cm}^3 \text{ s}^{-1} \quad (11b)$$

$$NO^+(a^3\Sigma^+) + O_2 \rightarrow NO^+(X^1\Sigma^+) + O_2, \quad 5.0 \pm 2.0(-10) \text{ cm}^3 \text{ s}^{-1} \quad (12)$$

Dotan et al. (1979). Recently Glosik et al. (1978) have measured $O_2^+(a^4\Pi_u)$ and $NO^+(a^3\Sigma^+)$ reactions in good agreement with the above and they have also measured reactions of an undetermined mixture of O^+ excited states. Torr and Torr (1978) have reviewed the rate constate determinations from the Atmosphere Explorer satellite.

Low Altitude Positive Ion Chemistry ($\lesssim 100$ km)

The ion chemistry below ~ 100 km is qualitatively different than the higher altitude ionosphere chemistry because of the relatively high pressure. Three-body reactions become important, reactions of trace species become significant because their absolute concentrations become large, and electron attachment to O_2 initiates a complex negative ion chemistry. The ion chemistry of the D-region ($\sim 100-60$ km), the stratosphere ($\sim 60-10$ km) and the troposphere (below 10-15 km) are qualitatively similar. Ion composition determinations are largely restricted to the D-region with only a few

recent stratospheric positive ion observations and as yet no tropospheric ion composition determinations.

The ionization sources vary with altitude, the major D-region
source is photoionization of the trace constituent NO by solar Ly-α
with a smaller contribution from $O_2(^1\Delta_g)$ photoionization by solar
UV. In the stratosphere and troposphere, the major ionization
source is galactic cosmic rays, principally high energy protons.
This ionization source has a latitudinal dependence because of the
earth's magnetic field and a solar cycle variation.

Figure 3 shows the very familiar positive ion chemistry scheme
of the D-region. Association reactions to major species N_2 and CO_2,
followed by rapid switching reactions (or ligand exchange) reactions
play a key role. Since the cluster ions such as $NO^+ \cdot N_2$ are very
weakly bound (D ~ 0.2 eV) thermal dissociation is very important.
For this reason, the temperature of the D-region, which varies with
altitude and latitude and season (from ~ 120 K to ~ 250 K) is critical and the temperature dependence of the association and dissociation rate constants are required. Since association is often
nearly balanced by dissociation, equilibrium constant data may be
almost as useful as rate constant data.

The reaction

$$NO^+ + 2N_2 \rightarrow NO^+ \cdot N_2 + N_2 \tag{13}$$

is now rather well measured by Johnsen et al. (1975), Turner and
Conway (1976), and Smith et al. (1977) in the laboratory and by
Arnold and Krankowsky (1978) from ionospheric ion composition
measurements. Exothermic switching reactions such as

$$NO^+ \cdot N_2 + CO_2 \rightarrow NO^+ \cdot CO_2 + N_2 \tag{14}$$

and
$$NO^+ \cdot CO_2 + H_2O \rightarrow NO^+ \cdot H_2O + CO_2 \tag{15}$$

are invariably fast, k $\sim 10^{-9}$ cm^3 s^{-1} and the sequence (13, 14, and
15), is a more rapid way to hydrate NO^+ in the atmosphere for low
temperatures (T $\lesssim 215$ K) than direct hydration. Measurements of
reactions like (13) are extremely difficult in the laboratory and
the association reactions of N_2 and CO_2 with $NO^+ \cdot H_2O$ and $NO^+ \cdot 2H_2O$
have not yet been measured. No precise detailed analysis of D-
region ion chemistry has yet been carried out. This would require
quantitative ion composition measurements, simultaneous neutral
composition, ionization rate, and temperature measurements and more
detailed laboratory data. This is a formidable enterprise which will
probably not be forthcoming in the immediate future. Chakrabarty et
al. (1978) have deduced from an analysis of D-region ion composition
measurements made with a rocket-borne mass spectrometer flown from
Wallops Island in June 1973 that reactions of the type

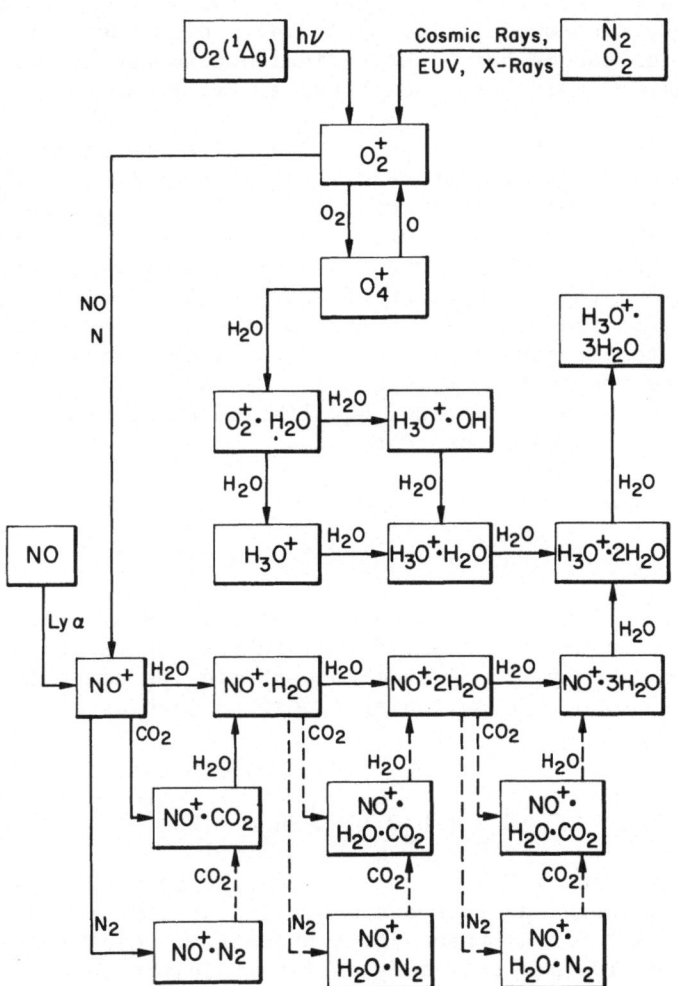

Figure 3.   D-region positive ion reaction scheme.

$$H^+(H_2O)_n + X + M \rightleftarrows H^+(H_2O)_n X + M \qquad (16)$$

followed by

$$H^+(H_2O)_n X + H_2O \rightleftarrows H^+(H_2O)_{n+1} + X \qquad (17)$$

are necessary to rationalize the observed relative abundances of $H^+(H_2O)_n$ species around 85 km.

In the stratosphere and troposphere where cosmic ray ionization of N_2 and O_2 initiates the ion chemistry, the relatively simpler reaction scheme from O_2^+ to $H_3O^+(H_2O)_n$ is dominant. Our approach to the ion chemistry of this region is to start with the D-region O_2^+ scheme that has already been developed and ask two questions. First, are there any neutral constituents of the troposphere or stratosphere which will interupt this flow from O_2^+ to $H_3O^+(H_2O)_n$? It would require a fairly abundant species to do this since the reaction sequence is quite rapid. Roughly speaking, it would require a neutral species comparable in abundance to H_2O, i.e. in the ppm or greater range. The species in this category in the stratosphere or troposphere are CO_2, CH_4 and O_3 and the reactions of O_2^+ and O_4^+ with these species therefore need to be considered.

The other question to be investigated is whether trace atmospheric species exist which might react with the $H_3O^+(H_2O)_n$ ions after they are produced and before they are removed by positive ion-negative ion recombination. This lifetime is of the order of 100-1000 sec so that reactive species in 10^{-9} to 10^{-12} mixing ratio concentrations could control the ion composition.

Pursuing the first question, the reaction scheme does not appear to be vulnerable at the O_2^+ level. There are no exothermic reaction channels with CO_2, H_2O or O_3. The exothermic reaction

$$O_2^+ + CH_4 \rightarrow CH_3O_2^+ + H \qquad (18)$$

has been found to be relatively slow by many workers (Franklin and Munson, 1965; Hollebone and Bohme, 1973; Nestler and Warneck, 1977 and Dotan et al. 1978a), $k_{18} = 6.8 \times 10^{-12}$ cm^3 s^{-1}. While of no particular concern atmospherically, Dotan et al. find k_{18} to increase rapidly with both translational energy and with O_2^+ vibrational energy. At higher translational energy, the products change to $CH_3^+ + HO_2$ which is 0.2 eV endothermic.

The next point of vulnerability is the O_4^+ ion. Any O_4^+ reaction must compete with

$$O_4^+ + H_2O \rightarrow O_2^+ \cdot H_2O + O_2 \qquad (19)$$

where $k_{19} = 1.5 \times 10^{-9}$ cm^3 s^{-1}. It has been found that O_4^+ reacts only slowly with CH_4, the rate constant being $3 \pm 1 \times 10^{-12}$ cm^3 s^{-1}, so that CH_4 does not compete (Dotan et al. 1978b). This $O_4^+ - CH_4$ reaction rate measured at NOAA is six times larger than a recent measurement reported by Nestler and Warneck (1977). It has been found that the reaction

$$O_4^+ + CO_2 \rightarrow CO_4^+ + O_2 \tag{20}$$

is nearly thermoneutral, $\Delta H = 0.3 \pm 1.0$ kcal mol^{-1} and $\Delta S = 4.3 \pm 2.6$ eu so that the greater abundance of O_2 keeps the equilibrium to the left and CO_4^+ will not have a significant concentration. On the other hand, the reaction of O_4^+ with O_3 is significantly exothermic,

$$O_4^+ + O_3 \rightarrow O_5^+ + O_2 \tag{21}$$

$\Delta H = -3.7 \pm 1.0$ kcal mol^{-1} and $\Delta S = 4.5 \pm 2.6$ eu so that a substanial amount of O_5^+ will be produced. The O_5^+/O_4^+ ratio will actually approach unity around 30 km, near the stratospheric ozone peak. However, O_5^+ reacts rapidly with H_2O

$$O_5^+ + H_2O \rightarrow O_2^+ \cdot H_2O + O_3 \tag{22}$$

$k_{22} = 1.2 \times 10^{-9}$ cm^3 s^{-1} and does not react with CH_4, $k < 5 \times 10^{-13}$ cm^3 s^{-1}. Therefore the diversion by way of O_5^+ does not have much significance, since $O_2^+ \cdot H_2O$ quickly results just as it would if there were no O_3 present. This is shown in Fig. 4. We tentatively conclude, therefore, that $H_3O^+(H_2O)_n$ ions will be produced in the stratosphere and troposphere, just as they are in the D-region, and thus move on to our second question; are there trace atmospheric species which will further react with the $H_3O^+(H_2O)_n$ ions? Rather large hydrations are expected, the dominant n being ~ 3 at 40 km and increasing to ~ 6 at 10 km, depending critically on the atmospheric temperature profile, which varies with time and place.

An essential criteria to be satisfied for a reaction to occur is that it be exothermic. This is quite a restrictive criteria since the $H_3O^+(H_2O)_n$ ions are quite stable. Since there are not likely to be a sufficient concentration of species with lower ionization potentials present charge-transfer is unlikely. The most favorable prospect involves species with greater proton affinities than H_2O. Several such species exist in the atmosphere, NH_3, CH_3OH, CH_2O, HNO_3 and N_2O_5. These species have all been investigated with the following results.

Reaction of H_3O^+ with nitric acid has been found to be fast (Fehsenfeld et al. 1975). However, protonated nitric acid is chemically equivalent to hydrated NO_2^+ and it is known that the second hydrate

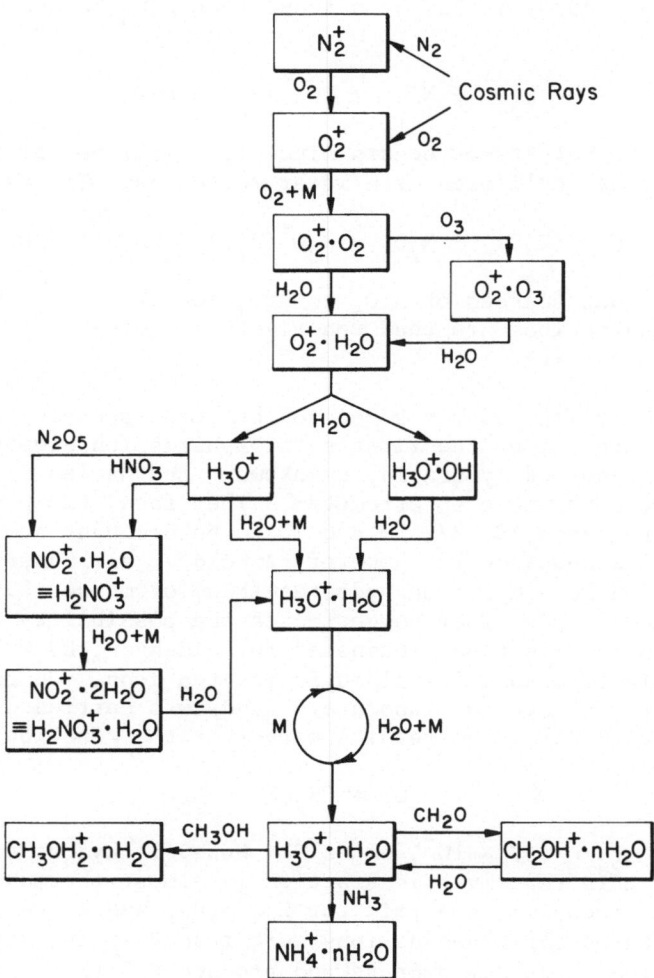

Figure 4.   Stratospheric and tropospheric positive ion reaction scheme.

of NO_2^+ reacts with H_2O,

$$NO_2^+ \cdot 2H_2O + H_2O \rightarrow H_3O^+ \cdot H_2O + HNO_3 \qquad (23)$$

to regenerate HNO_3 and hydrated hydronium ions so that nitric acid has no net effect on atmospheric ion composition. More recently, (Davidson et al. 1978) it has been found that N_2O_5 behaves in the same manner,

$$H_3O^+ + N_2O_5 \rightarrow NO_2^+ \cdot H_2O + HNO_3 \qquad (24)$$

so that again no net effect occurs since (24) will be followed by (23) after further collision with water molecules. The reactions

$$H_3O^+(H_2O)_n + N_2O_5 \rightarrow H_3O^+(H_2O)_{n-1}HNO_3 + HNO_3 \qquad (25)$$

may be significant sources of N_2O_5 to HNO_3 conversion, although preliminary indications are that reactivity is quenched for n greater than about 2.

Considerable impetus was given to the stratospheric positive ion chemistry problem by the first stratospheric ion composition measurements, reported by Arnold, Krankowsky and Marien in 1977, using a rocket borne mass spectrometer. They found the proton hydrates (PH) of masses 19, 37, 55 and 73 to be dominant above $\sim$ 40 km altitude. Below about 40 km, however, Arnold et al. observed a rapid conversion to non-proton hydrates (NPH) of masses 29 ± 2, 42 ± 2, 60 ± 2 and 80 ± 2. They suggested as one possibility that the ion of mass 29 ± 2 might be protonated formaldehyde, CH_2OH^+, mass 31, and the higher masses formaldehyde related ions. On the basis of their report, a detailed laboratory study was carried out (Fehsenfeld et al. 1978a). The results were that the reaction

$$H_3O^+ + H_2CO \rightarrow CH_2OH^+ + H_2O \qquad (26)$$

is very fast, $k_{26} = 2.2 \times 10^{-9}$ cm^3 s^{-1}. However, hydration of the H_3O^+ quenched this reactivity. Reaction is almost thermoneutral for $H_3O^+(H_2O)$ and becomes endothermic for $H_3O^+ \cdot 2H_2O$ and higher hydrates. Even if protonated formaldehyde ions were formed in the stratosphere, they would rapidly hydrate and produce $H_3O^+(H_2O)_n$ ions and neutral formaldehyde.

Very recently Arnold, Bohringer and Henschen (1978), using a balloon borne mass spectrometer floating at 37 km, found a similar conversion of proton hydrates to non-proton hydrates. They found 57% PH's and 43% NPH's at 37 km. They did not find the ions of mass 29, 42, and 60 reported earlier. An ion at mass 77 ± 1 may be the same ion as earlier reported at 80 ± 2 amu. Stratospheric measurements of Arijs et al. (1978) and Olson et al. (1977) also did not show the 29, 42, and 60 peaks. At present, no case exists for a role for formaldehyde in stratospheric positive ion chemistry.

In the same investigation, Fehsenfeld et al. (1978a) also studied the atmospheric ion chemistry of methanol. Methanol, like formaldehyde, is an oxidation product of methane in the atmosphere. While neither species has been measured in the stratosphere, the neutral chemistry leading to methanol is much less certain that that leading to formaldehyde, so that the expected concentration of methanol is far less certain. It was found that the reactions

$$H_3O^+ \cdot nH_2O + CH_3OH \rightarrow products \tag{27}$$

all had approximately 2×10^{-9} cm^3 s^{-1} rate constants for n = 0, 1, 2, 3 and for energies from thermal to well above thermal. The products were protonated methanol ions, but the extent of hydration was undetermined. The failure of any of the stratospheric experiments to observe hydrated protonated methanol ions, having masses 33 + n18, leads to a very low upper limit on the concentration of stratospheric CH_3OH, $< 10^6$ cm^{-3} at 40 km, or a mixing ratio less than 10^{-11}.

Finally, another atmospheric species with large proton affinity is NH_3. It has been found (Fehsenfeld and Ferguson, 1973) that the reactions

$$H_3O^+ \cdot nH_2O + NH_3 \rightarrow NH_4^+ \cdot mH_2O + (n + 1 - m)H_2O \tag{28}$$

are fast for n = 0, 1, and 2. Appreciable concentrations of NH_3 are not expected in the stratosphere because of its high solubility, which would lead to its washout in the troposphere. Arnold et al. (1978) infer a concentration of less than $\sim 10^5$ cm^{-3} or a mixing ratio of less than 10^{-12} from their failure to observe NH_4^+ hydrates at 37 km.

In the troposphere, where NH_3 is present in appreciable concentrations, the ions $NH_4^+ \cdot nH_2O$, possibly with other species clustered as well, are presumably abundant. Figure 4 summarizes our present understanding of the lower atmospheric positive ion chemistry which appears to be validated down to about 40 km by the observations of Arnold and his colleagues.

The nonproton hydrates reported below about 40 km by Arnold et al. have recently been interpreted (Ferguson, 1978) as being due to the presence of protonated sodium hydroxide ions. The interpretation is given in Table I. It has been well known for over seventy years

TABLE I

Mass Spectra of Stratospheric Positive Ions

Observed			Proposed NaOH Ion Clusters	KOH Ion Clusters
(1)	(2)	(3)	(4)	
29±2				
42±2			41 $NaOH_2^+$	
60±2	60±2		59 $NaOH_2^+ \cdot H_2O$	57 KOH_2^+
80±2	78±2	78±2	77 $NaOH_2^+ \cdot 2H_2O$	75 $KOH_2^+ \cdot H_2O$
	82±2		81 $NaOH_2^+ \cdot NaOH$	
	96±2	96±1	95 $NaOH_2^+ \cdot 3H_2O$	93 $KOH_2^+ \cdot 2H_2O$
	99±2	100±1	99 $NaOH_2^+ \cdot NaOH \cdot H_2O$	
		114±2	113 $NaOH_2^+ \cdot 4H_2O$	111 $KOH_2^+ \cdot 3H_2O$
				113 $KOH_2^+ \cdot KOH$
		118±1	117 $NaOH_2^+ \cdot NaOH \cdot 2H_2O$	
				129 $KOH_2^+ \cdot 4H_2O$
		136±1	135 $NaOH_2^+ \cdot NaOH \cdot 3H_2O$	131 $KOH_2 \cdot KOH \cdot H_2O$
		140±1	139 $NaOH_2^+ \cdot 2NaOH \cdot H_2O$	

(1) Arnold, Krankowsky and Marien, Nature, 267, 39 (1977).

(2) Arijs, Ingels and Nevejans, Nature, 271, 642 (1978)
Arijs, private communication

(3) Arnold, Bohringer and Henschen, Geophys. Res. Letters 5, 653 (1978).

(4) Ferguson, Geophys. Res. Letters, in press (1978).

that a layer of atomic sodium exists in the atmosphere from its
resonant scattering of sunlight which has been detected spectro-
scopically. This sodium exists in a layer near 90 km and is due to
the ablation of meteorites in the atmosphere upon entry. It is
argued that the sodium will be in the form of NaOH below about 40 km
in the atmosphere. The proton affinity of NaOH is exceedingly high,
$\sim$ 248 kcal mol^{-1} (Kebarle, 1977). It is presumed therefore that
reactions

$$H_3O^+(H_2O)_n + NaOH \rightarrow NaOH_2^+(H_2O)_m + (n - m + 1)H_2O \quad (29)$$

would occur at the collision rate, $k_{29} \sim 2 \times 10^{-9}$ cm^3 s^{-1}. A concentration of NaOH $\sim 10^5$ cm^{-3} at 40 km is required for (29) to compete with positive ion - negative - ion recombination and this
agrees well with calculations of Reid and Liu (private communication)
on the total expected sodium concentration in this altitude range.
It is also predicted that at somewhat lower altitudes KOH$_2^+$ clusters
will become dominant since the proton affinity of KOH $\sim$ 263 kcal
mol^{-1} (Kebarle, 1977) exceeds that of NaOH. The atmospheric abundance of K is an order of magnitude or more less than that of Na so
that at 40 km the K concentration is too low to be involved in the
ion chemistry. MgO has a large proton affinity, $\sim$ 212 kcal mol^{-1}
and it may well be that MgOH$^+$ clusters are formed by

$$H_3O^+(H_2O)_n + MgO \rightarrow MgOH^+(H_2O)_m + (n - m + 1)H_2O \quad (30)$$

since Mg is an order of magnitude more abundant than Na in meteorites. However the MgOH$^+$ hydrates would presumable proton transfer
to NaOH,

$$MgOH^+ + NaOH \rightarrow NaOH_2^+ + MgO \quad (31)$$

so that Mg, while relatively abundant, probably does not contribute
to the ion composition.

Low Altitude Negative Ion Chemistry (< 100 km)

The ion chemistry of the D-region has been actively studied in
the laboratory prior to the first D-region negative ion composition
measurements, detailed reaction schemes had been developed over a
number of years in our laboratory. This scheme, with only minor
changes over the years, is shown in Fig. 5. The atmospheric negative
ion composition measurements (Narcisi et al., 1971, Arnold et al.,
1971) are rather sparse, somewhat qualitative and the results are
somewhat contradictory. The negative ion measurements are not
accompanied by detailed measurements of the atmospheric trace constituents, temperature, etc. so that detailed comparisons of observation with theoretical prediction have not been carried out and
do not appear to be imminent.

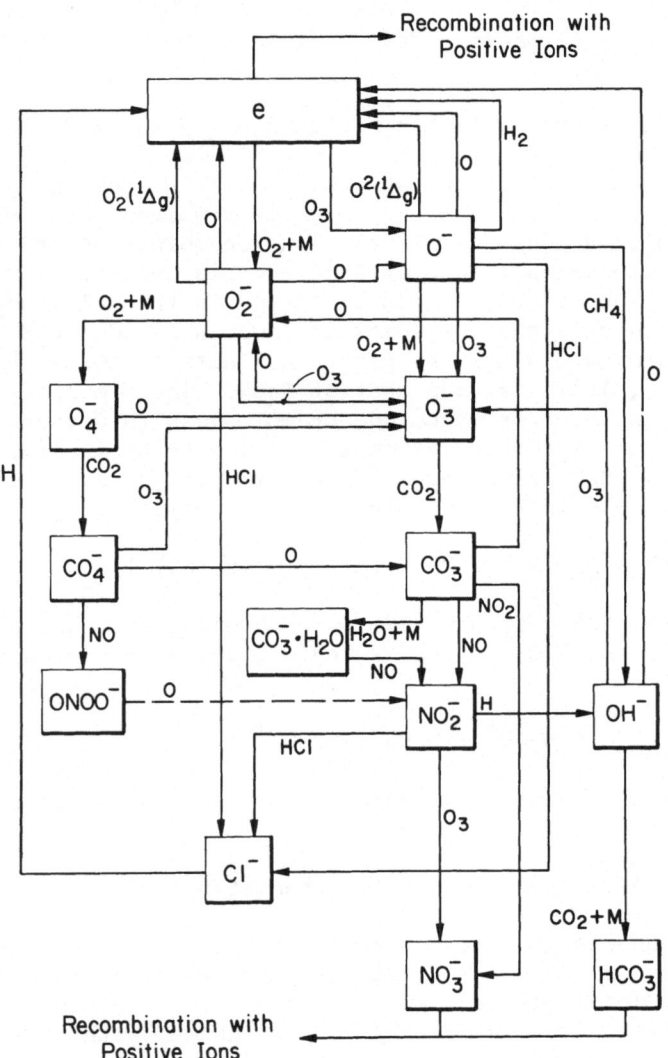

Figure 5.   Reaction scheme for D-region negative ion chemistry.

There have been a number of additional laboratory measurements of the reactions indicated in Fig. 5 and these are listed in Table II. The major unpublished changes in this scheme are the following: In addition to the reaction

$$O_2^- + O_2 + M \rightarrow O_4^- + M, \tag{32}$$

the reaction

$$O_2^- + N_2 + M \rightleftharpoons O_2^- \cdot N_2 + M \tag{33}$$

followed by

$$O_2^- \cdot N_2 + O_2 \rightarrow O_4^- + N_2 \tag{34}$$

is a significant source of atmospheric O_4^- production. Fehsenfeld et al. (1978b) finds for He third body that $k_{32} = 3.1 \times 10^{-31} (\frac{300}{T})^{2.5}$ cm^6 s^{-1} and $k_{33} = 6.3 \times 10^{-32} (\frac{300}{T})^{4.2}$ cm^6 s^{-1}. The factor of 5 favoring k_{32} is almost offset by the N_2/O_2 ratio in the atmosphere at 300 K although $O_2^- \cdot N_2$ thermal breakup will work against (33). However, at lower temperatures, (33) becomes relatively faster than (32) and collisional breakup is diminished so that very likely (33) and (34) will sometimes be the dominant O_4^- production path. This presumes that the enhancement of reactions (32) and (33) when He is replaced by N_2 and O_2 in the atmosphere are similar.

In the scheme of Fig. 5, it has recently been found that the reaction of O_2^- with O branches,

$$O_2^- + O \rightarrow O_3 + e \tag{35a}$$

$$\rightarrow O^- + O \tag{35b}$$

with $k_{35a} \approx k_{35b}$ and the total $k_{35} = 3.0 \times 10^{-10}$ cm^3 s^{-1}. Originally, the charge-transfer channel was not observed. The most serious error in the earlier measurements involved the conversion of NO_2^- to NO_3^-,

$$NO_2^- + O_3 \rightarrow NO_3^- + O_2 \tag{36}$$

$k_{36} = 9.0 \times 10^{-11}$ cm^3 s^{-1}, some 5 times faster than originally reported and

$$NO_2^- + NO_2 \rightarrow NO_3^- + NO \tag{37}$$

$k_{37} < 2 \times 10^{-13}$ cm^3 s^{-1}, in contrast to an originally reported $k_{37} = \sim 4 \times 10^{-12}$ cm^3 s^{-1} which appears to have been an error attributable to HNO_3 impurity in the NO_2.

The reaction

$$CO_4^- + O_3 \rightarrow O_3^- + CO_2 + O_2 \tag{38}$$

Table II

Recently Measured Negative Ion Reaction Rate Constants (300 K)
Aeronomy Laboratory, NOAA.

	Reaction	Rate Constant $(cm^3\ s^{-1})$
1a.	$O_2^- + O \rightarrow O^- + O_2$	1.5(−10)
1b.	$\rightarrow O_3 + e$	1.5(−10)
2.	$O_2^- + O_3 \rightarrow O_3^- + O_2$	6.0(−10)
3.	$O_2^- + HCl \rightarrow Cl^- + HO_2$	1.7(−9)
4.	$O_3^- + O \rightarrow O_2^- + O_2$	2.5(−10)
5.	$O_3^- + CO_2 \rightarrow CO_3^- + O_2$	5.5(−10)
6.	$CO_3^- + O \rightarrow O_2^- + CO_2$	1.1(−10)
7.	$CO_3^- + NO \rightarrow NO_2^- + CO_2$	1.1(−11)
8.	$CO_4^- + O \rightarrow CO_3^- + O_2$	1.4(−10)
9.	$CO_4^- + O_3 \rightarrow O_3^- + CO_2 + O_2$	1.3(−10)
10.	$NO_2^- + O_3 \rightarrow NO_3^- + O_2$	9.0(−11)
11.	$NO_2^- + NO_2 \rightarrow NO_3^- + NO$	< 2 (−13)
12.	$NO_2^- + HCl \rightarrow Cl^- + HNO_2$	1.4(−9)
13.	$Cl^- + H \rightarrow HCl + e$	9.6(−10)
14.	$CO_3^- + HCl \rightarrow products$	< 3 (−11)
15.	$CO_4^- + HCl \rightarrow O_2^-HCl + CO_2$	1.2(−9)
16.	$ClO^- + NO_2 \rightarrow NO_2^- + ClO$	3.2(−10)
17.	$ClO^- + NO \rightarrow NO_2^- + Cl$	2.9(−11)
18a.	$ClO^- + O_3 \rightarrow Cl^- + 2O_2$	6 (−11)
18b.	$\rightarrow O_3^- + ClO$	1 (−11)
19.	$Cl^- + O_3 \rightarrow ClO^- + O_2$	< 5 (−13)

has been added, $k_{38} = 1.3 \times 10^{-10}$ cm^3 s^{-1}. In addition, a number of Cl compound reactions have been added (Dotan et al. 1978c) the potentially most important of these being

$$NO_2^- + HCl \rightarrow Cl^- + HNO_2 \tag{39}$$

with $k_{39} = 1.4 \pm 0.4 \times 10^{-9}$ cm^3 a^{-1}.

Arnold et al. (1971) reported Cl^- ions in the D-region. It is presumed that the chlorine compounds introduced into the stratosphere by both natural and anthropogenic means do lead to chlorine in the D-region, presumably largely as HCl. Turco (1977) has analyzed the D-region chlorine negative ion chemistry but in light of the newly revised rate constants in Table II, this analysis needs to be repeated. Specifically, the larger rate constant for reaction (36) makes HCl much less competitive for NO_2^- and the reaction of CO_3^- with HCl which Turco involved is now known to be endothermic. The relatively slow reaction of CO_3^- with NO,

$$CO_3^- + NO \rightarrow NO_2^- + CO_2 \tag{40}$$

$k_{40} = 1.1 \times 10^{-11}$ cm^3 s^{-1}, when coupled with the low NO concentration, $\sim 10^8$ cc^{-1}, causes reaction (40) to be a bottleneck in the progression to the stable terminal ion NO_3^- (and its hydrates). For this reason, the CO_3^- lifetime may be long enough for hydration to occur at this level. Keesee et al, (1978) have examined this question on the basis of recent studies of CO_3^- and HCO_3^- hydration equilibria in the laboratory. They find, for a typical model atmosphere, that the CO_3^- ion will be largely singly hydrated above ~ 77 km and unhydrated below this down to ~ 36 km where the first hydrate again dominates down to ~ 25 km where the second hydrate dominates. This calculation is critically sensitive to the atmospheric temperature profile, of course, and to a lesser extent, the water profile, both of which are quite variable in time and place. It does not appear that hydration will significantly hinder the reaction sequence of Fig. 5.

Another current input into the laboratory D-region negative in chemistry is the energy dependence studies being carried out at NOAA by Albritton and his colleagues. Figure 6 shows one such example, the reaction of CO_3^- with NO as a function of temperature from about 250 K to 600 K and as function of average kinetic energy from 0.04 eV to 1 eV. It appears that the rate constant as a function of average kinetic energy is the same as the rate constant as a function of temperature and that the rate constant as a function of average kinetic energy is the same in He and Ar buffer gases. This indicates that the reaction is not sensitive to internal energy of the reactants. This is not always found to be the case, at least

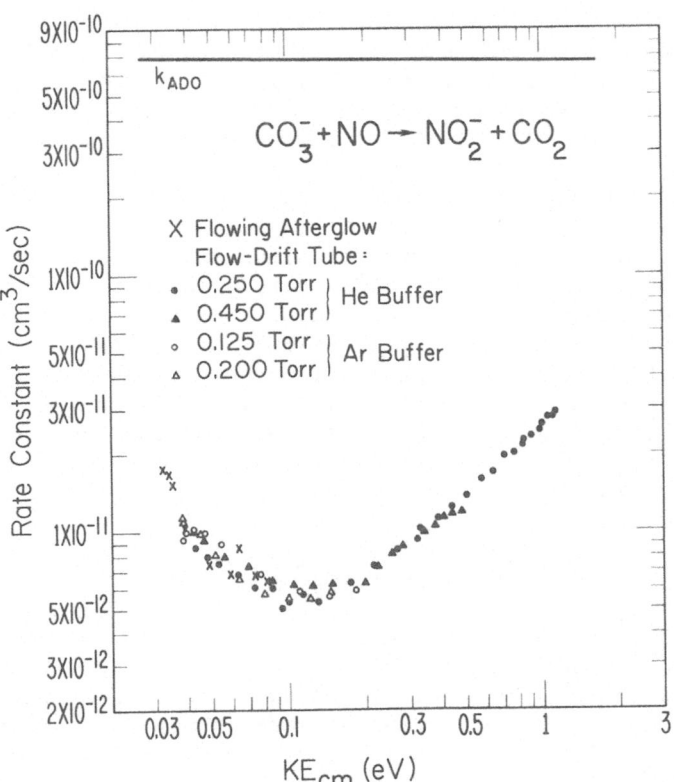

Figure 6.   Rate constants for the reaction $CO_3^- + NO \rightarrow NO_2^- + CO_2$ as a function of mean relative kinetic energy (flow-drift tube) and temperature (flowing afterglow, using $3kT/2 = KE_{cm}$) (D. L. Albritton, NOAA).

for positive ion reactions studied in the flow drift tube. It is, of course, the sub 300 K temperature range that is of concern in atmospheric negative ion chemistry. It might be mentioned in passing that the heat of formation has been determined precisely for NO_3^- (Davidson et al. 1977) from which it is now possible to deduce the electron affinity rather well, $EA(NO_3) = 4.01 \pm 0.02$ eV.

Our strategy in stratospheric-tropospheric negative ion chemistry is the same as it is in positive ion chemistry, namely we start with the D-region ion chemistry and adapt it to the changing pressure, temperature and most importantly, neutral composition. In this endeavor, we are unguided and uninhibited by any direct knowledge of the atmospheric ion composition. Again we ask the same two questions, (1) will the negative ion chemistry initiated by electron attachment to O_2 still proceed to NO_3^-? and; (2) are there trace constituents of the atmosphere which can react with NO_3^- and its stable hydrates?

The answer to the first question seems to be a clear yes. In addition to the chemistry of Fig. 5 for the D-region, there are even more ways to make NO_3^- in the stratosphere. These are indicated in Fig. 7. For one thing, the disruptive effect of atomic oxygen is removed since the atomic oxygen concentration is insignificant below the D-region. In addition, there are additional reaction paths leading to NO_3^-. In the stratosphere and troposphere, the NO_2 concentration becomes comparable to the NO concentration. This removes the CO_3^- bottleneck, since CO_3^- reacts $\sim$ 20 times more rapidly with NO_2,

$$CO_3^- + NO_2 \rightarrow NO_3^- + CO_2 \qquad (41)$$

$k_{41} = 2 \times 10^{-10} \, cm^3 \, s^{-1}$.

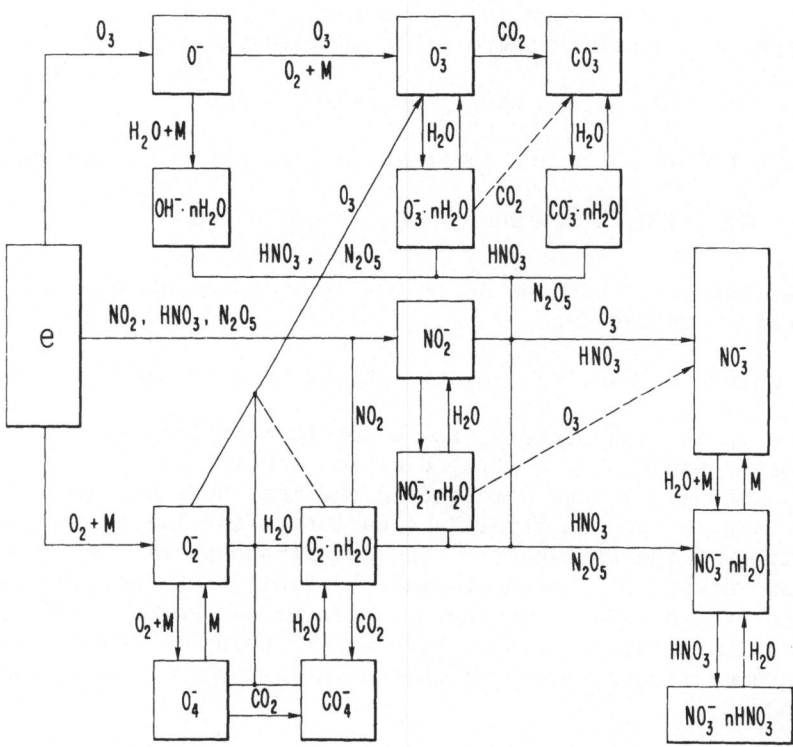

Figure 7. Stratospheric and tropospheric negative ion reaction scheme.

The nitric acid concentration   exceeds the NO and NO_2 concentrations below about 30 km in the atmosphere. Fehsenfeld et al. (1975) found that O^-, O_2^-, CO_3^-, NO_2^- and Cl^- all react rapidly with HNO_3 to produce NO_3^-. Recently, Davidson et al. (1978) have found that Cl^-, CO_3^-, and NO_2^- also react rapidly with N_2O_5 to produce NO_3^-. These reactions all have rate constants in the range of 10^{-9} cm^3 s^{-1}. Thus, the efficiency of NO_3^- production is greatly enhanced in the lower atmosphere.

The second question then is whether there exist species in the lower atmosphere which will react with the NO_3^- produced. Again the approach is to seek out exothermic reaction possibilities for measurement. As with any ion, positive or negative, NO_3^- will readily hydrate in the atmosphere if given time. It is also known that SO_2 will exothermically displace H_2O clustered to negative ions (Fehsenfeld and Ferguson, 1974) and that HNO_3 will displace H_2O (Fehsenfeld et al. 1975). Thus it is not unlikely that the ambient negative ions of the lower atmosphere will be of the form $NO_3^-(H_2O)_n(HNO_3)_m$ where $m > n$ in the stratosphere and $m < n$ in the troposphere. The water concentration decreases abruptly from $\sim$ 1% in the troposphere to a few parts per million in the stratosphere. The HNO_3 concentration is of the order of magnitude of a few parts per billion in both the troposphere and stratosphere. The equilibrium constant for

$$NO_3^- \cdot H_2O + HNO_3 \rightarrow NO_3^- \cdot HNO_3 + H_2O \qquad (42)$$

is $K_{eq} > 2 \times 10^5$ at 300 K and the same is true for the reaction

$$NO_3^- \cdot HNO_3 \cdot H_2O + HNO_3 \rightarrow NO_3^- \cdot 2HNO_3 + H_2O \qquad (43)$$

Other polar molecules such as HCl may also be associated with stratospheric negative ions.

Ion Molecule Reactions as Atmospheric Sources and Sinks

To ionospheric physicists, ion-molecule reactions have a first order interest since they control electron density and therefore radio-propagation. To the atmospheric chemist, however, ion chemistry has largely been an irrelevant curiosity involving a negligible fraction of the atmosphere. The new stratospheric ion composition measurements have begun to make a significant contribution to knowledge of the neutral atmosphere as described above, however, and the potential utility of stratosphere and troposphere ion composition measurements as sensitive probes of atmospheric composition is quite apparent.

One area in which ion chemistry has arisen as a potentially significant atmospheric factor from time to time is as a source or sink of atmospheric neutrals. It was proposed by Ruderman, et al.

(1976) that the reaction

$$NO_3^- + O_3 \rightarrow NO_2^- + 2O_2 + 0.35 \text{ eV} \qquad (44)$$

might destroy ozone in the stratosphere and thereby account for an
observed solar cycle variation in stratospheric ozone at high lati-
tudes. Since the ionization rate varies with the
solar cycle, the negative ion concentration presumably does also.
However, Fehsenfeld et al. (1976a) examined this reaction in the lab
and found the rate constant to be less than 10^{-13} cm^3 s^{-1}, which
makes this reaction too slow to be of concern. In addition, the
NO_3^- ions will be heavily hydrated, which would drive the reaction
endothermic unless the water molecules remained with the NO_2^-.

Another possible sink concerned the tropospheric loss of N_2O by
negative ion reactions. The ions O_2^-, O_3^-, CO_3^- and NO_2^- all have
exothermic reaction channels with N_2O. A few years ago, it was
believed that an unidentified tropospheric sink for N_2O must exist
and these reactions were studied and found to be extremely slow in
each case, an upper limit being 10^{-14} cm^3 s^{-1}, except for NO_2^-
for which the upper limit was 10^{-12} cm^3 s^{-1} (Fehsenfeld and Ferguson, 1976). It now appears, in light of greatly refined tropospheric N_2O measurements, that no tropospheric sink is required.

The most interesting situation arose during the initial stages
of the recent chlorofluoromethane-stratospheric ozone controversy.
The chlorofluoromethanes $CFCl_3$(F-11) and CF_2Cl_2(F-12) are released
at the earth's surface and after tropospheric residence times of the
order of 50-100 years, they are transported into the stratosphere
where they are rapidly photodissociated to produce free chlorine
atoms which catalytically destroy ozone. Therefore, any process
which removes F-11 or F-12 on a time scale less than $\sim$ 100 years
would have a significant impact on the extent of ozone destruction.
The rate of neutral molecule removal by a binary ion-molecule reaction would be given by $\nu(s^{-1}) = k$ [ions] where k is the rate constant for ions of concentration [ions]. Since ion-molecule reaction
rate constants are often as large as 10^{-9} cm^3 s^{-1} and tropospheric
ion concentrations are $10^3 - 10^4$ cm^{-3}, it would be possible to have
neutral lifetimes against ion removal as short as a few days.
Clearly, a small fraction of the ambient ions could play a significant role if they had fast reactions with the chlorofluoromethanes.

In the case of positive ion chemistry, it was found (Fehsenfeld
et al. 1976b) that the ions expected to have significant concentrations in the troposphere, $H_3O^+(H_2O)_n$ and $NH_4^+(H_2O)_n$, did not react
with F-11 or F-12 and indeed the reactions are endothermic. One
exception, the reaction

$$H_3O^+ + CFCl_3 \rightarrow CCl_3^+ + HF + H_2O, \qquad (45)$$

was found to have a large rate constant, $k_{45} = 4 \times 10^{-10} cm^3 s^{-1}$.
Numerous literature values of the heat of formation of CCl_3^+ indicated a substantial endothermicity for (45), obviously in error.
An early measurement of Martin et al. (1966) and a recent measurement of Lias and Ausloos (1978) find a heat of formation of CCl_3^+
low enough to support an exothermic reaction, however.

In the case of negative ions, a similar situation prevailed.
The negative ions which would have any significant abundances were
too stable to have exothermic reaction channels with F-11 and F-12.
It was also found (Fehsenfeld et al. 1976b) that ion-molecule reactions in the stratosphere did not pose a significant sink for the
F-11 and F-12 products, Cl, ClO and HCl. To date, no ion-molecule
reactions have been identified as having significant roles as neutral
molecule sources or sinks in the atmosphere.

REFERENCES

Albritton, D. L., I. Dotan, W. Lindinger, M. McFarland,
J. Tellinghuisen and F. C. Fehsenfeld, J. Chem. Phys. 66, 410
(1977).

Arijs, E., J. Ingels and D. Nevejans, Nature 271, 842 (1978).

Arnold, F., D. Krankowsky and K. M. Marien, Nature 267, 30
(1977).

Arnold, F., H. Bohringer and G. Henschen, Geophys. Res. Lett.
5, 653 (1978).

Arnold, F., J. Kissel, D. Krankowsky, H. Wieder, and J. Zahringer,
J. Atm. Terr. Phys. 33, 1159 (1971).

Arnold, F., and D. Krankowsky, J. Atm. Terr. Phys. 39, 635 (1977).

Arnold, F. and D. Krankowsky, Ion Composition and Electron and
Ion Loss Processes in the Earth's Atmosphere, in Dynamical and
Chemical Coupling of the Neutral and Ionized Atmosphere, B.
Grandal and J. A. Holtet, editors, D. Riedel, 1977.

Bates, D. R. and M. Nicolet, J. Atmos. Terr. Phys. 18, 65 (1960).

Biondi, M. A., Geophys. Res. Letters 5, 661 (1978).

Chakrabarty, D. K., P. Chakrabarty and G. Witt, J. Atm. and
Terr. Phys. 40, 437 (1978).

Chen, A., R. Johnsen and M. A. Biondi, J. Chem. Phys. 69, 2688
(1978).

Dalgarno, A., and M. B. McElroy, Planet. Space Sci. 13, 947 (1965).

Davidson, J. A., F. C. Fehsenfeld, and C. J. Howard, Int. J. Chem. Kinetics 9, 17 (1977).

Davidson, J. A., A. A. Viggiano, C. J. Howard, I. Dotan, F. C. Fehsenfeld, D. L. Albritton and E. E. Ferguson, J. Chem. Phys. 68, 2085 (1978).

Dotan, I., F. C. Fehsenfeld and D. L. Albritton, J. Chem. Phys. 68, 5665 (1978a).

Dotan, I., J. A. Davidson, F. C. Fehsenfeld and D. L. Albritton, J. Geophys. Res. 83, 4036 (1978b).

Dotan, I., F. C. Fehsenfeld, G. E. Streit and E. E. Ferguson, J. Chem. Phys. 68, 5415 (1978c).

Dotan, I., D. L. Albritton, and F. C. Fehsenfeld, J. Chem. Phys. (1979).

Fehsenfeld, F. C. And E. E. Ferguson, J. Chem. Phys. 61, 3181 (1974).

Fehsenfeld, F. C. and E. E. Ferguson, J. Chem. Phys. 59, 6272 (1973).

Fehsenfeld, F. C., C. J. Howard, and A. L. Schmeltekopf, J. Chem. Phys. 63, 2835 (1975).

Fehsenfeld, F. C., E. E. Ferguson, G. E. Streit, and D. L. Albritton, Science 194, 544 (1976a).

Fehsenfeld, F. C. and E. E. Ferguson, J. Chem. Phys. 64, 1853 (1976b).

Fehsenfeld, F. C., I. Dotan, D. L. Albritton, C. J. Howard and E. E. Ferguson, J. Geophys. Res. 83, 1333 (1978a).

Fehsenfeld, F. C., P. J. Crutzen, A. L. Schmeltekopf, C. J. Howard, D. L. Albritton, E. E. Ferguson, J. A. Davidson and H. I. Schiff, J. Geophys. Res. 81, 4454 (1976b).

Fehsenfeld, F. C., T. J. Brown and D. L. Albritton, Proceeding, Gaseous Electronics Conf. Buffalo, NY, Oct. 17-20 (1978b).

Ferguson, E. E., Geophys. Res. Letters, in press (1978).

Franklin, J. L. and M. S. B. Munson, Tenth Int. Symposium on Combustion, Pittsburgh, p. 561 (1965).

Glosik, J. A., A. B. Rakshit, N. D. Twiddy, N. G. Adams and D. Smith, Faraday Society (1978).

Johnsen R., C. M. Huang, and M. A. Biondi, J. Chem. Phys. 63, 3374 (1975).

Hollebone, B. R. and D. K. Bohme, J. Chem. Soc. Faraday Trans. 2, 69, 1469 (1973).

Kebarle, P., Ann. Rev. Phys. Chem. 28, 445 (1977).

Keesee, R. G., N. Lee and A. W. Castleman, J. Geophys. Res., in press (1978).

Lias, S. G. and P. Ausloos, Int. J. Mass Spectrometry and Ion Phys. 28, 273 (1977).

Lin, S. L. and J. N. Bardsley, J. Chem. Phys. 66, 435 (1977).

Lindinger, W., F. C. Fehsenfeld, A. L. Schmeltekopf and E. E. Ferguson, J. Geophys. Res. 79, 4753 (1974).

Lindinger, W., D. L. Albritton, M. McFarland, F. C. Fehsenfeld, A. L. Schmeltekopf and E. E. Ferguson, J. Chem. Phys. 62, 210 (1975).

Martin, R. H., F. W. Lampe and R. W. Taft, J. Amer. Chem. Soc. 88, 1353 (1966).

McFarland, M., D. L. Albritton, F. C. Fehsenfeld, E. E. Ferguson and A. L. Schmeltekopf, J. Geophys. Res. 79, 2925 (1974).

Narcisi, R. S., A. D. Bailey, L. Della Lucca, C. Sherman and D. M. Thomas, J. Atm. Terr. Phys. 33, 1147 (1971).

Nestler, V. and P. Warneck, Chem. Phys. Lett. 45, 96 (1977).

Olson, J. R., R. C. Amme, J. N. Brooks, D. G. Murcray and G. E. Keller, Trans. Am. Geophys. Union 58, 1201 (1977).

Oppenheimer, M., E. R. Constantinidies, K. Kirby-Docken, G. A. Victor, A. Dalgarno and J. H. Hoffman, J. Geophys. Res. 82, 5485 (1977a).

Oppenheimer, M., A. Dalgarno, F. P. Trebino, L. H. Brace, H. C. Brinton and J. H. Hoffman, J. Geophys. Res. 82, 191 (1977b).

Orsini, N., D. G. Torr, H. C. Brinton, L. H. Brace, W. B. Hanson, J. H. Hoffman and A. O. Nier, Geophys. Res. Letters $\underline{4}$, 431 (1977).

Ruderman, M. A., H. M. Foley and J. W. Chamberlain, Science $\underline{192}$, 555 (1976).

Rusch, D. W., D. G. Torr, P. D. Hays and J. C. G. Walker, J. Geophys. Res. $\underline{82}$, 719 (1977).

Schmeltekopf, A. L., E. E. Ferguson and F. C. Fehsenfeld, J. Chem. Phys. $\underline{48}$, 2966 (1968).

Smith, D., N. G. Adams and D. Grief, J. Atm. Terr. Phys. $\underline{39}$, 513 (1977).

Torr, M. R., J. P. St.-Maurice, and D. G. Torr, J. Geophys. Res. $\underline{82}$, 3287 (1977a).

Torr, D. G. and N. Orsini, Planet Space Sci. $\underline{25}$, 1171 (1977).

Torr, D. G., N. Orsini, M. R. Torr, W. B. Hanson, J. H. Hoffman and J. C. G. Walker, J. Geophys. Res. $\underline{82}$, 1631 (1977b).

Torr, D. G. and M. R. Torr, Reviews of Geophys. and Space Phys. $\underline{16}$, 327 (1978).

Torr, D. G. and N. Orsini, Geophys. Res. Letters (1978).

Turco, R. P., J. Geophys. Res. $\underline{82}$, 3585 (1977).

Turner, D. L. and D. C. Conway, J. Chem. Phys. $\underline{65}$, 3944 (1976).

ION-MOLECULE PROCESSES IN LASERS

James B. Laudenslager

Jet Propulsion Laboratory, California
Institute of Technology, Pasadena, CA 91103

1   INTRODUCTION

In the last five years a variety of short wavelength molecular
electronic transition gas lasers have been discovered. These pulsed
lasers all operate at high pressures, above 1 atmosphere, where
collisional processes dominate over radiative processes even fast,
nsec, vacuum ultraviolet resonance radiation from electronically
excited rare gas atoms. At high pressures new classes of reactions
such as recombination, association, energy transfer, charge transfer, and excited state reactions can be used to produce the population inversion for laser oscillation. All of these lasers operate
on similar principles and are examples of a hybrid laser pumping
mechanism. That is, an external source of excitation, such as an
electron-beam, proton beam, or an electric avalance discharge is
used to deposit energy rapidly into a high pressure gas mixture
composed primarily of rare gases. The excitation pulse produces
a highly ionized plasma consisting initially of atomic rare gas
ions, electronically excited rare gas atoms, and a large density
of electrons having a kinetic energy distribution unique to the
method of excitation used. The fast and specific reaction of
these nascent species formed in the plasma, usually in the presence
of a small percentage of molecular additive, gives rise ultimately
to a population inversion between electronic energy levels in a
short-lived molecule.

There are three classes of molecular electronic transition
lasers produced by this hybrid pumping scheme of high pressure rare
gas mixtures. The rare gas dimer lasers, such as Ar_2^*, Kr_2^* and

Xe_2^*, lasing in the vacuum ultraviolet, VUV, were the first of
these lasers to be discovered. The rare gas halide, RGH, or excimer
lasers such as KrF^*, ArF^*, $XeC\ell^*$ etc., which lase in the ultraviolet, UV, are the second class of laser. The third type of laser
is the charge transfer molecular ion laser, e.g., N_2^+, which is a
visible laser.

In order to have a hybrid pumping mechanism yield an efficient
laser, several criteria should be met. First, the kinetic sequence
leading from the primary excited rare gas species to the upper
laser level should be rapid and have a favorable kinetic branching
ratio leading to formation of the upper laser level. Second, the
upper laser level should not be depopulated at a faster rate than
the lower laser level. This condition insures that a population
inversion is maintained and the laser will not terminate itself.
Third, the unexcited gas mixture as well as the transient excited
species formed during the excitation pulse should not absorb
strongly at the lasing wavelength. For VUV and UV lasers, this
condition is very difficult to satisfy. Fourth, stimulated emission and not collisional deactivation should be the dominant loss
process from the upper laser level.

The initial discovery of these high pressure lasers did not
occur until the development of electron-beam (e-beam) and fast discharge technology. Prior to this technology, it was very difficult
to excite lasers uniformly above several 100 torr of pressure. The
efficiency of these new lasers is based on favorable kinetic channelling of excited atoms, ions, and electrons to electronically
excited molecules. In all of these lasers a key kinetic step is a
three body association or recombination reaction which only occurs
efficiently at high pressures. As a consequence, the overall efficiency of these lasers will depend on the electrical efficiency
for plasma formation as well as the overall kinetic scheme. Therefore, in order to understand the ion-chemistry occurring in various
laser systems, one must be familiar with the various excitation
methods. Each excitation method, i.e., fast particle beam or
avalanche discharge, establishes different kinetic pathways even
for the same laser system.

2   LASER EXCITATION METHODS

2.1   E-Beam Excitation

Detailed descriptions of e-beam energy deposition into rare
gases and the properties of e-beam excitation are contained in the
literature (Hurst and Klots, 1976; Daugherty, 1976; Lorents and
Olson, 1972). A schematic diagram of an e-beam excited laser
device is shown in Figure 1.

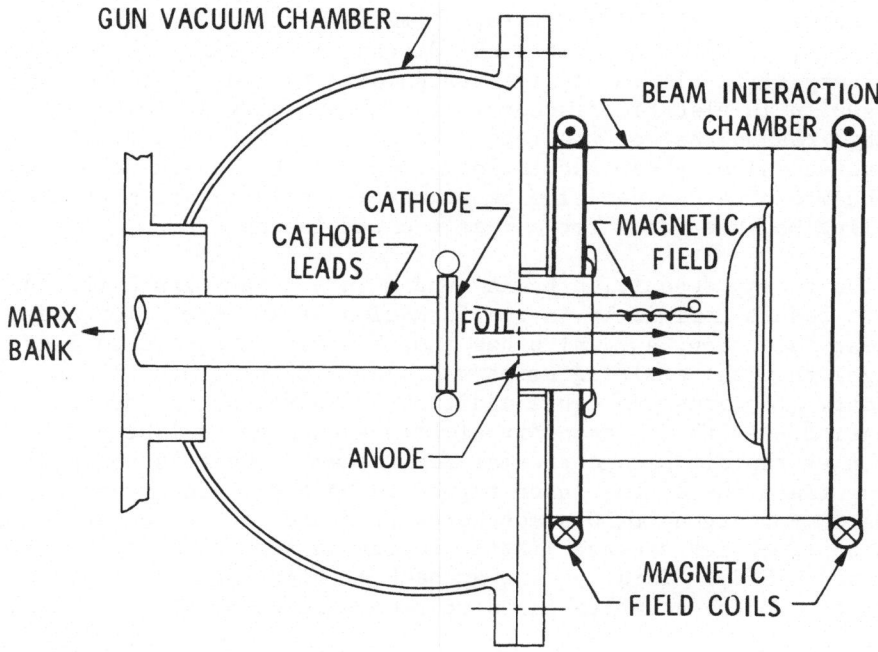

Figure 1   E-Beam Laser Apparatus

        When a high voltage (>300kV) pulse from the Marx generator is
applied to the cathode, an intense electron beam is formed in the
vacuum region between the cathode and a thin foil anode. The
thickness of the anode foil is chosen such that, when supported on
a semitransparent grid structure, it can withstand the pressure
differential between the laser region and the low pressure e-beam
formation region. At the same time the foil should not be too
thick or it will absorb too much of the e-beam energy. For typical
foil thickness which satisfy both conditions of strength and trans-
parency, one needs minimum e-beam energies of around 150 keV to
penetrate the foil. This foil anode also limits large e-beam
machines to very low repetition rates, because of excessive foil
heating which leads to rupture of the foil. The e-beam machines
are typically large expensive devices needing extensive shielding
to protect the operator from X-rays. However, the major advantage
to the e-beam excitation method is the electron beam is generated
independently and separate from the laser excitation region.
Therefore, it is straightforward to deposit known amounts of
electron beam energy into gas mixtures at high pressures for pulse
durations ranging from several nanoseconds, nsec, to several micro-
seconds, μsec. On the otherhand, uniform volumetric excitation of
high pressure gases is extremely difficult to achieve by means of
a self-sustained electric discharge because the exciting electrons

are generated in the laser mixture itself. Control of electron production, proper excitation, and discharge stability is, at present, very difficult to achieve for pulse durations of longer than 100 nsec in an electric discharge of high pressure gas mixtures. For this reason most of the lasers to be discussed were first discovered using e-beam excitation. Because the energy deposition by e-beam excitation can also be accurately calculated, most laser modelling has been done for e-beam excited lasers.

The energy deposition by an e-beam into a rare gas buffered mixture results primarily in the formation of an atomic ion and a secondary electron, e_s, which has a much lower energy (6 to 8 eV average) than the exciting primary e-beam electron, e_p. These secondary electrons can cause additional ionization or neutral excitation. As these secondary electrons undergo inelastic collisions they rapidly become thermalized because there is no accelerating field inside the laser region to heat the electrons further. For argon at 10 atmospheres pressure, the secondary electrons approach 1eV average kinetic energy in several nsec (Lorents and Olson, 1972). This low energy bath of electrons can rapidly attach to electro negative gases or recombine with molecular ions.

The ionization cross sections for rare gases by electron impact is largest for electron energies of several 100 eV and falls to low cross section values above 1 keV. However, the primary e-beam energy must be of the order of >150 keV to penetrate the foil anode and the high pressure gas. Therefore, the energy conversion efficiency of the high energy e-beam into ion-electron pair formation is not as efficient as is ionization produced directly in an electric discharge where the electron energy is much lower. In fact, as is depicted in Table 1, the energy cost per ion-electron pair formation of rare gases is approximately twice the atomic ionization potential.

Table 1   Typical Energy Loss Per Ion Pair
(Hurst and Klots, 1976)

Rare Gas	Electron Volts
He	54.1
Ne	37.3
Ar	28.0
Kr	25.9
Xe	23.2

Consequently, one has to sacrifice overall quantum efficiency for ease in controlling energy deposition in an e-beam excited laser.

2.2   Electric Avalanche Discharge Excitation

Self-sustained avalanche discharges have been commonly used to excite gas lasers at pressures below several hundred torr. A self-sustained discharge occurs when a high voltage is rapidly applied across two transverse electrodes. The high electric field accelerates the small density of free electrons which are always present in a gas and these electrons collide with the gas producing ionization thus creating new electrons which cause additional ionization leading to the so-called Townsend avalanche. Therefore, the electrons which excite the laser gas to initiate the laser pumping reaction sequence must be generated in the laser mixture by the Townsend avalanche process.

For gas pressures above 1 atm it is very difficult to achieve a controlled volumetric discharge using conventional discharge techniques. Since the development of the TEA CO_2 laser in 1970 (Beaulieu, 1970; Dumanchin and Rocca-Serra, 1969), there have been significant advances in discharge technology for high pressure gas mixtures. Stable volumetric self-sustained discharges can now be produced for high pressure lasers when special techniques such as uniform field electrodes, preionization, and low inductance discharge circuitry are used. However, close attention to the laser discharge geometry and electrical circuit design must be maintained because all electric discharges tend to rapidly convert from glow discharges into constricted arc discharges at the slightest provocation. Once arc streamers form, the voltage drops along with the gas resistance and large currents flow through a small constricted volume. The electron energy is too low in the arc discharge at high gas pressures to cause electronic excitation of the gas necessary for laser pumping and volumetric excitation is lost. In order to insure glow discharge operation, the electric discharge pulse width is usually kept shorter than the arc formation time which is usually several avalanche times. For most laser devices to be discussed, this restricts the electric discharge laser pulse width to less than 100 nsec and, typically, most discharge excimer lasers have a 20-30 nsec laser pulse width.

The electrons produced in the avalanche discharge are accelerated in the electric field and ionize and excite the gas mixture. The energy deposition function in the laser gas depends on the kinetic energy distribution of the electrons which changes as a function of time in the discharge. At gas breakdown, the discharge voltage is highest but falls rapidly as current starts to flow.

The actual electron energy distribution obtained in an electric
discharge is strongly coupled to the discharge circuit, the gas
mixture, the gas composition, and the electrode gap separation.
It is, therefore, very difficult to even determine the energy
deposition function let alone to try and vary parameters such as
mean electron energy and density and maintain a controlled volumetric discharge. These are the main limitations for this form
of excitation, however, electric discharge lasers, once designed
properly, are inexpensive, small size, capable of high energy out-
puts and can be operated at high repetition rates.

Figure 2 shows the electric discharge device constructed at
JPL to provide glow discharge excitation at high pressures. Both
the RGH and charge transfer, CT, molecular ion lasers are pumped
through ion kinetic pathways. Therefore, the mean electron energy
in the discharge should be high enough to produce efficient electronic excitation and ionization of the rare gas atoms, but in a
controlled fashion or else the avalanche process will quickly
degenerate into an arc. The breakdown voltage of the gas is
determined by the gas mixture, total pressure, electrode gap, and
to a certain extent by the inductance of the discharge circuit.
A high voltage discharge pulse is required at high pressures to
cause gas breakdown and to establish a high mean electron energy

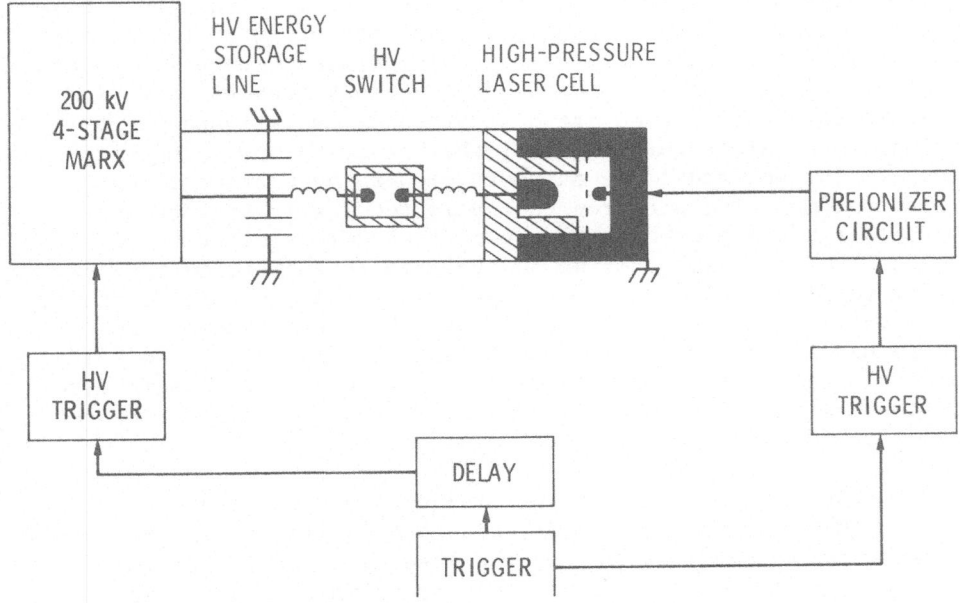

Figure 2   JPL High Pressure Transverse Discharge Laser

for electronic excitation. The necessary high voltage, HV, pulse is produced by a low inductance four stage Marx generator in the JPL laser. The HV output of the Marx generator is used to pulse charge a very low inductance intermediate storage water capacitor network which is isolated from the laser cathode during pulse charging by a self-breakdown HV multichannel surface transfer gap switch (Sarjeant et. al., 1977). This intermediate water capacitor along with all the associated electrical leads are carefully designed to make a low impedance, 0.5 ohm, energy source in order to trans-fer electrical energy efficiently to the low impedance laser plasma. The HV switch has a high resistive phase during pulse charging of the intermediate storage line, but once the switch breaks down it rapidly converts to a low resistance element and this tends to give a rapid rate of rise to the HV pulse appearing on the laser cathode. The rapid rise time of the HV discharge pulse helps to initiate a volumetric glow discharge and the circuit elements are constructed in such a way that the discharge duration is shorter than the time for arc formation.

A preionization pulse consisting of a row of UV arc sources located behind a screen anode is fired before the main discharge pulse. The UV arcs produce a uniform volumetric low density of free charge carriers by photoionization and these electrons start the avalanche discharge process upon application of the HV potential across the cathode and anode. Without some form of preionization, it is extremely difficult to establish glow discharges in a high pressure gas. The geometry of the UV arc sources and the shape and separation distance of the electrodes is also critical with respect to a stable volumetric discharge. Lasing on the XeCℓ* excimer laser at pressures as high as 11 atm total pressure have been obtained in this device.

2.3   E-beam Sustained Discharge Lasers

The major limitation of e-beam excitation is the low quantum efficiency for formation of electronically excited rare gas atoms. These excited atoms react with halogen additives to form the upper excimer state in RGH lasers. With e-beam excitation, most metast-able rare gas atoms are formed through ion-electron recombination reactions. The energy needed to produce an electronically excited rare gas atom is about 4 eV less than the rare gas ionization potential and the energy cost to produce a rare gas ion by e-beam excitation is about twice the atomic ionization potential. In an electric discharge, metastable rare gas atoms are produced directly by electron impact and the energy cost is just the excitation energy of the metastable. Clearly the overall quantum efficiency for metastable pumped RGH lasers is better by a factor of two for a discharge system, but the limited control of self-sustained ava-lanche discharges has favored e-beam excitation for large laser systems.

A combination of e-beam and discharge excitation is possible for RGH lasers which contain small amounts of electron attaching halogen molecules. The e-beam sustained discharge laser uses a long pulse low current density e-beam to produce a volumetric low level of ionization. The secondary electrons produced by the e-beam are accelerated by a DC electric field which is kept below the self-breakdown voltage. The potential of the sustaining field is chosen to provide an electron kinetic energy which will excite the rare gas atoms to their metastable states efficiently, and the e-beam is kept on for the duration of the sustaining field to resupply electrons which are lost due to attachment and recombination.

An analysis of the e-beam sustained discharge excitation method for RGH lasers has shown the discharge to be unstable unless the rate of electron loss is greater than twice the ionization rate (Daugherty et. al., 1976). For RGH lasers, the primary loss of electrons is due to dissociative attachment of lower energy electrons by the additive halogen molecules. The e-beam sustained discharge technique has been used to obtain efficient laser action on RGH laser systems, but long pulse width operation has not as yet been demonstrated. Only when the e-beam does most of the laser pumping has μsec laser operation been obtained. In reality one does not have total control over the discharge field. If the sustaining field, which is lower than the self-breakdown potential, is maximized to produce metastable atoms, eventually a large density of metastables are produced and these low ionization potential species are rapidly ionized by electrons heated in the weak electric field. When this occurs, the ionization rate becomes too large and the discharge collapses into an arc. If weaker sustaining potentials are used, the metastables are produced inefficiently and only weak lasing can be achieved. Another problem associated with long pulse excitation is the conversion of the halogen molecules to halogen atoms by dissociate recombination and excimer formation. The halogen molecules attach electrons which helps prevent ionization run away, but the halogen atoms are weak scavengers of electrons. Therefore, as the halogen molecule is lost to dissociation, the ionization rate starts to exceed the attachment rate and the discharge becomes unstable.

At present the e-beam sustained discharge method has been demonstrated to have only comparable electrical to laser efficiencies as pure e-beam pumped systems and the discharge control has not been as good as anticipated. However, this method may prove superior for high repetition rate lasers since high current density e-beams are not necessary and foil heating is not as severe a problem.

3   RARE GAS DIMER LASERS

The rare gas dimer lasers were the first high pressure short
wavelength lasers discovered. A tabulation of the dimer lasers is
given in Table 2 along with the peak laser wavelength.

The laser transitions for these dimer lasers is from the
bound $A^1\Sigma_u^+$ excited upper level to the repulsive $X^1\Sigma_g^+$ ground state.
The bound-free optical transition results in a broad continuous
emission spectrum and the repulsive lower laser level insures that
a population inversion is always achieved. Table 3 lists the
various kinetic pathways for the Xe_2^* laser and this reaction chain
is similar for the other rare gas dimer lasers

The primary energy deposition in high pressures of Ar, Kr,
or Xe by e-beam excitation is ionization with the simultaneous
production of slower secondary electrons. The secondary electrons,
e_s, cause additional ionization along with neutral excitation, and
at high pressures these secondary electrons are rapidly thermalized
by collisions with the gas and other electrons. The atomic rare
gas ions are rapidly ($\sim$0.1 nsec at 10 atm.) converted into dimer
and trimer ions by fast three body association reactions. Radiative recombination of atomic ions with low energy electrons is
too slow to compete with dimer ion formation at pressures above
1 atmosphere, but once molecular ions are formed dissociative
recombination with the thermalized electrons rapidly removes all
molecular ions and produces electronically excited neutral atoms.
At low pressures (<100 torr) these excited rare gas atoms would
lose their excitation energy by fast radiative transitions, but at
high pressures, these excited atoms are quickly converted into
excited neutral dimer rare gas molecules. Highly excited atoms in
Rydberg states undergo associative ionization reactions rather
than neutral dimerization reactions and lower lying excited atoms
and dimer molecules are dexcited to the lowest lying electronically
excited state by collisions with electrons and gas atoms. Once

Table 2   Rare Gas Dimer Lasers

Molecule	Molecular Transition	Wavelength (nm)
Xe_2	$A^1\Sigma_u^+ \rightarrow X^1\Sigma_g^+$	172.0
Kr_2	$A^1\Sigma_u^+ \rightarrow X^1\Sigma_g^+$	145.7
Ar_2	$A^1\Sigma_u^+ \rightarrow X^1\Sigma_g^+$	126.1

Table 3   Simplified Kinetic Chain for the Xenon Dimer Laser

Reaction Number		Rate Constant (e.g. cm^3/sec, cm^6/sec)
1.	$e_p + Xe \rightarrow Xe^+ + e'_p + e_s$	
2.	$e_s + Xe \rightarrow Xe^* + e_c$	
3.	$Xe^+ + 2Xe \rightarrow Xe_2^+ + Xe$	2.5×10^{-31}
4.	$Xe_2^+ + 2Xe \rightleftarrows Xe_3^+ + Xe$	9.0×10^{-32}
5.	$Xe_2^+ + e_c \rightarrow Xe^{**} + Xe$	2×10^{-7}
6.	$Xe_3^+ + e_c \rightarrow Xe^{**} + 2Xe$	$\sim 9.5 \times 10^{-5}$
7.	$Xe^* + 2Xe \rightarrow Xe_2^* + Xe$	5×10^{-32}
8.	$Xe^{**} + Xe \rightarrow Xe_2^+ + e$	$\sim 10^{-9}$
9.	$Xe^* + Xe^* \rightarrow Xe^+ + Xe + e$	$\sim 5 \times 10^{-10}$
10.	$Xe_2^* + Xe_2^* \rightarrow Xe_2^+ + 2Xe + e$	$\sim 5 \times 10^{-10}$
11.	$Xe_2^{**} + Xe \rightarrow Xe^* + 2Xe$	$\sim 10^{-11}$
12.	$Xe_2^{**} + e \rightarrow Xe_2^* + e$	$\sim 10^{-6}$
13.	$Xe_2^* + e \rightarrow Xe_2^+ + e$	5×10^{-9}
14.	$Xe_2^{**} + e \rightarrow 2Xe + e$	10^{-9}
15.	$Xe_2^* \rightarrow 2Xe + h\nu$	
16.	$h\nu + Xe_2^* \rightarrow 2Xe + 2h\nu$	
17.	$h\nu + Xe_2^* \rightarrow Xe_2^+ + e$	

the lowest excited neutral dimer molecule is formed, it is lost
predominately by radiation because further dexcitation requires
conversion of large amounts of electronic energy to translational
energy and this is generally an inefficient process. Spontaneous
emission leads to stimulated emission as long as the number density
of the excited dimer remains large.

When the excited atomic and dimer molecule densities are
large, collisions between two excited species can occur with large
probability and this results in formation of atomic or dimer ions
by Penning reactions. Collisions between excited neutral species,
therefore, limits the maximum density of upper laser level
molecules.

Because of the broad bandwidth of the dimer fluorescence, the
rare gas dimer lasers have small stimulated emission cross sections
and correspondingly low gain coefficients. Hence, high upper level
densities are necessary for laser oscillation. The VUV laser
radiation at which the dimer lasers operate is absorbed by excited
species as well as the upper laser level and these large absorption
losses further limit the laser efficiencies obtainable from these
systems. In practice the rare gas dimer lasers are not as efficient or scalable as the longer wavelength RGH lasers, but the
VUV fluoresence efficiency per e-beam energy deposited in the gas
is very high for the rare gas dimer systems ($\sim$50%). At present,
the rare gas dimer systems are not being developed as laser sources,
but as very efficient fluorescent sources to be used to photolyti-
cally pump other laser systems (Krupke and George, 1978).

The high VUV fluorescence efficiency for the rare gas dimers
is due to the efficient kinetic sequence which converts atomic
ions, atomic metastables, and electrons produced by the e-beam
into electronically excited rare gas dimer molecules. The poor
laser performance results from the small gain and short inversion
lifetime of the upper level and the absorption losses which occur
for VUV wavelengths.

4   E-BEAM EXCITED RARE GAS HALIDE LASERS

4.1   Background

Rare gas halide molecules are bound ion pair molecules, e.g.
$Kr^+ F^-$, in the upper excited electronic levels and are weakly
bound or repulsive in their ground electronic states. Figure 3
shows the potential energy curves for KrF and the other RGH molecules have similar molecular electronic energy levels. The
repulsive ground state for these molecules gives rise to continuous

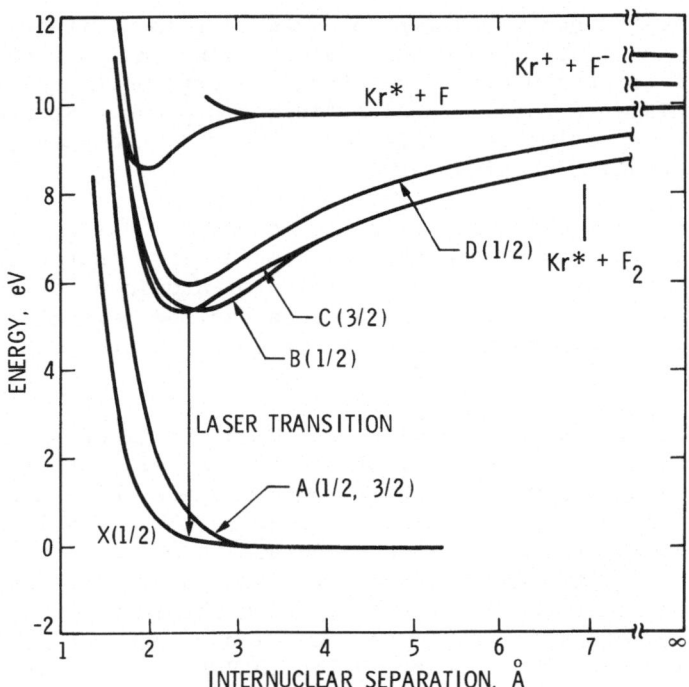

Figure 3   Potential Energy Curves for KrF

emission spectra from the upper bound levels and also insures that
a population inversion between the upper bound laser level and the
lower dissociative ground state can be maintained.

Because the upper laser level is an ion pair, it can be produced by recombination of a rare gas atomic ion with a negative
halide ion in a three body reaction. Another kinetic pathway
which produces the excited RGH molecule is reaction with a metastable rare gas atom and a halogen molecule. This bimolecular
reaction produces excited RGH molecules with large efficiencies
and proceeds through a "harpoon" mechanism which was used to
explain the large reactive cross sections for alkali metal-halogen
reactions (Herschbach, 1966).

$$Kr^* + F_2 \rightarrow Kr^+ - F_2^- \rightarrow Kr^+ F^- + F \tag{18}$$

The electron from the excited rare gas atom jumps to the halogen
molecule at large internuclear separations, and the coulombic field
between the ion pair draws the pair together dissociating the
halide molecule. The "harpoon" electron returns to the rare gas
ion in the form of a negative halide atom to produce an excited
RGH molecule. The similarity between alkali atoms and alkali

halide modecules with rare gas metastable atoms and RGH excited molecules has been used successfully to predict RGH formation kinetics and the spectroscopic properties of excited RGH molecules.

4.2 ArF* Excimer Laser

The simplest RGH laser is the ArF* laser which is produced by e-beam excitation of a 0.2% F_2:99.8% Ar mixture at 1 to 2 atmosphere total pressure. Table 4 lists the kinetic sequence for the e-beam excited ArF* laser at total pressures above and below 1 atm.

Most of the energy deposited into the gas by the e-beam goes into Ar^+ formation with simultaneous production of a secondary electron which produces a small amount of neutral excitation to form Ar^*. The thermalized electrons form F^- rapidly through dissociative attachment reactions.

Therefore, the initial plasma consists of Ar^+, Ar^*, F^-, and electrons. The upper laser level is an excited ArF* molecule and this is formed by two parallel kinetic pathways. The first pathway is by a three body positive-negative recombination reaction 22, and the other kinetic pathway is the neutral bimolecular metastable-halogen "harpoon" reaction 23.

As the total pressure is raised above 1 atmosphere, association of Ar^+ to form Ar_2^+ competes with the recombination reaction 22. However, Ar_2^+ can recombine with F^- to form ArF* (reaction 29). Although the branching ratio for this reaction has not been verified, it is assumed to be unity in order to fit kinetic modelling of the ArF* laser system (Rokni et. al., 1976). Once Ar_2^+ is formed it can be lost through electron-ion recombination, but this produces Ar* which reacts to form ArF*. The fact that the major kinetic pathways for reactions of ions and excited neutrals ends in ultimate formation of excited RGH molecules makes these very efficient lasers. However, as the Ar buffer pressure is increased it quenches the ArF* molecule rapidly, reaction 25. The chemical quenching of ArF* by Ar limits the total pressure at which the ArF* laser can operate even though the kinetic pathways leading to ArF* production are more efficient at higher pressures.

4.3 KrF* Excimer Laser

If 5 to 10 percent of Kr is added to the ArF* laser gas mixture, the ArF* laser emission at 193nm gives way to KrF* emission at 249nm. The kinetic sequence for the KrF* laser, which consists of an Ar, Kr, F_2 mixture, is tabulated in Table 5.

Table 4   Kinetic Sequence for ArF* Laser

Reaction Number		Rate Constant (e.g. cm^3/sec, cm^6/sec)
	<1 atmosphere	2 torr F_2 and balance Ar
19.	e_p + Ar $\rightarrow Ar^+$ + e_s	(55% of e-beam energy deposited)
20.	e_s + Ar $\rightarrow Ar^*$ + e_c	(10% of e-beam energy deposited)
21.	e_c + F_2 $\rightarrow F^-$ + F	5×10^{-9}
22.	Ar^+ + F^- + Ar $\rightarrow$ ArF* + Ar	$\sim 3 \times 10^{-6}$
23.	Ar^* + F_2 $\rightarrow$ ArF* + F	7.5×10^{-10}
24.	ArF* $\rightarrow$ Ar + F + $h\nu$	($\tau = 4 \times 10^{-9}$ sec)
25.	ArF* + 2Ar $\rightarrow Ar_2F^*$	4×10^{-31}
26.	Ar_2F^* $\rightarrow$ 2Ar + F + $h\nu$	($\tau = 185 \times 10^{-9}$ sec)
	>1 atmosphere	
27.	Ar^+ + 2Ar $\rightarrow Ar_2^+$ + Ar	2.5×10^{-31}
28.	Ar_2^+ + e_c $\rightarrow Ar^*$ + Ar	7.5×10^{-7}
29.	Ar_2^+ + F^- + Ar $\rightarrow$ ArF* + 2Ar	1.1×10^{-6}
30.	Ar^* + 2Ar $\rightarrow Ar_2^*$ + Ar	$1\text{-}2 \times 10^{-32}$
31.	Ar_2^* + F_2 $\rightarrow$ ArF* + Ar	5.2×10^{-10}
32.	Ar_2^* $\rightarrow$ 2Ar + $h\nu$	($\tau = 3 \times 10^{-6}$ sec)

Table 5   Kinetic Sequence for KrF* Laser

Reaction Number			Rate Constant (e.g. cm^3/sec, cm^6/sec)
33. $Ar^+ + Kr$	$\rightarrow Kr^+ + Ar$		$\leq 10^{-14}$
34. $Ar^+ + Kr + Ar$	$\rightarrow Kr^+ + 2Ar$		
35. $Ar_2^+ + Kr$	$\rightarrow Kr^+ + 2Ar$		7.5×10^{-10}
36. $Ar_2^+ + Kr + Ar$	$\rightarrow Kr^+ + 3Ar$		$\leq 2 \times 10^{-30}$
37. $Ar^* + Kr$	$\rightarrow Kr^* + Ar$		6.3×10^{-12}
38. $ArF^* + Kr$	$\rightarrow KrF^* + Ar$		1.6×10^{-9} (3×10^{-10})
39. $Kr^+ + F^- + Ar$	$\rightarrow KrF^* + Ar$		$\sim 2.9 \times 10^{-6}$
40. $Kr^+ + Kr + Ar$	$\rightarrow Kr_2^+ + Ar$		$\sim 10^{-31} - 10^{-32}$
41. $Kr^+ + 2Ar$	$\rightarrow ArKr^+ + Ar$		$\sim 10^{-31} - 10^{-32}$
42. $ArKr^+ + F^-$	$\rightarrow$ products		$\sim 10^{-6}$
43. $ArKr^+ + Kr$	$\rightarrow Kr_2^+ + Ar$		3.2×10^{-10}
44. $Kr_2^+ + e$	$\rightarrow Kr^* + Kr$		1.2×10^{-6}
45. $Kr_2^+ + F^- + (Ar)$	$\rightarrow KrF^* + F + (Ar)$		2.5×10^{-6}
46. $Kr^* + Kr + Ar$	$\rightarrow Kr_2^* + Ar$		$\sim 10^{-32}$
47. $Ar_2^* + Kr$	$\rightarrow Kr^* + 2Ar$		$\sim 10^{-10}$
48. $Kr^* + F_2$	$\rightarrow KrF^* + F_2$		7.2×10^{-10}

Table 5   Kinetic Sequence for KrF* Laser (Continued)

Reaction Number			Rate Constant (e.g. cm^3/sec, cm^6/sec)
49.	$KrF* + F_2$	→Products	7.8×10^{-10}
50.	$KrF* + 2Kr$	→$Kr_2F* + Kr$	6.7×10^{-31}
51.	$KrF* + Kr + Ar$	→$Kr_2F* + Ar$	6.5×10^{-31}
52.	$KrF* + 2Ar$	→Products	7×10^{-32}
53.	$KrF*$	→$Kr + F + h\nu$	τ = 5 nsec
54.	$h\nu + KrF*$	→$Kr + F + 2 h\nu$	
55.	$F + F + Ar$	→$F_2 + Ar$	6×10^{-31}

At pressures around 1 atm the initial laser plasma species caused by e-beam excitation into a <90:<10:0.2 percentage mixture of Ar:Kr:F_2 are Ar^+, Ar^*, F^-, and some Kr^+ and Kr^* along with low energy secondary electrons. The same reactions for a Ar/F_2 mixture in Table 4 occur but collisions between Kr and ArF*, which is still formed, produces KrF* by a fast displacement reaction 38. The KrF* production through kinetic pathways starting from Ar^+ has to occur by this displacement reaction 38 because charge transfer of Ar^+ to Kr is too slow to compete with ArF* formation through recombination reaction with F^-. Likewise conversion of Ar^* to Kr^* is too slow to compete with ArF* formation by the harpoon reaction 23. The estimated value of the displacement reaction of ArF* by Kr to form KrF* depends on the choice of kinetic model, but this reaction must occur at a faster rate than radiation from ArF* (5 nsec) or else laser oscillation from KrF* at 249 nm would not be observed. At pressures above 1 atm, Ar_2^+ is formed and this ion is rapidly converted to Kr^+ by a charge transfer reaction. This charge transfer reaction is the major loss of Ar_2^+ and once Kr^+ is formed rapid recombination of Kr^+ with F^- produces KrF*. The dimerization of Kr^+ to Kr_2^+ at higher total pressures is too slow for reaction with Kr and Ar, reaction 40, for percentages of Kr of 10% or less. However, the formation of $ArKr^+$ can be rapid at high pressures reaction 41, and $ArKr^+$, once formed, is very quickly converted to Kr_2^+ through reaction 43 (Bohme et al., 1970). Therefore, efficient production of Kr_2^+ through three body association can only

compete with ion recombination of Kr^+ with F^- if the $ArKr^+$ reaction
intermediate is formed. Once Kr^+, Kr^*, and Kr_2^+ are produced,
efficient parallel kinetic pathways reactions 39, 45, and 48 all
lead to KrF* production.

As in the case of the ArF laser, high Ar pressures can lead
to rapid quenching of KrF* and thus the total pressure at which
this laser can be operated using an Ar buffer gas is limited by
quenching reactions.

Both e-beam excited ArF* and KrF* lasers are pumped predominately through kinetic pathways involving ion reactions. The
association and recombination reactions of these ion species are
three body reactions which favor high pressure operation. A problem
which effects laser efficiency is absorption losses. Even small
non-saturable absorption losses limits the overall energy which
can be extracted from a laser. Ionic species produced in the
initial kinetic sequence in the laser plasma can absorb at the UV
wavelengths that RGH lasers operate. For example, the absorption
of Ar_2^+ is continuous and extends from 225nm to 450nm with the
peak absorption occurring at 325nm. Negative ions such as F^-,
$C\ell^-$, and Br^- absorb throughout the UV. The XeF* and XeCℓ* lasers
which lase at 351nm and 308nm respectively, operate better with a
neon buffer rather than an argon buffer because Ne_2^+ does not
absorb strongly at these laser wavelengths and Ar_2^+ does. At the
same time, the kinetic sequence going from Ne^+ and Ne^* to XeF* and
XeCℓ* is still very efficient and Ne does not quench the RGH molecules as efficiently as does Ar. Consequently, neon buffered
XeF* and XeCℓ* lasers can operate at high total pressures which
should enhance ion recombination reactions.

In the KrF* laser, the neutral F_2 molecule absorbs at the
laser wavelength, 249nm, while NF_3 does not. However, in spite of
absorption losses using F_2, the KrF* laser operates better with
F_2 than with NF_3 as the halogen donor. The reason for the poor
KrF* laser efficiency with NF_3 is because Kr^+ can charge transfer
to NF_3 to form NF_3^+ and, consequently, some Kr^+ is removed
from kinetic pathways leading to KrF* production. On the other
hand, Kr^+ does not charge transfer to F_2 and this accounts for the
better performance of the KrF* with F_2 additive.

5   ELECTRIC DISCHARGE EXCITED LASERS

5.1   The XeCℓ* Excimer Laser

The XeCℓ* laser has been studied in a high pressure electric
discharge device, Figure 2, at JPL. In most electric discharge
excited RGH lasers, helium is used as a buffer gas. Helium is

not normally used as a buffer gas with e-beam excitation since
helium does not absorb much of the e-beam energy unless the pressure is very high. However, helium is a poor quencher of electronically excited RGH molecules and no major absorption losses
due to ionic species of helium have been identified as yet.
Therefore, helium is a very good buffer gas for electric discharge
lasers.

Figure 4 shows the laser output energy for the JPL electric
discharge excited XeCℓ* laser as a function of total and partial
pressures of a gas mixture containing He, Xe, and HCℓ. The output
energy of the laser increases with helium buffer pressure, but at the
same time the partial pressures of Xe and HCℓ must be decreased as
the total pressure is increased. In effect, this means that the

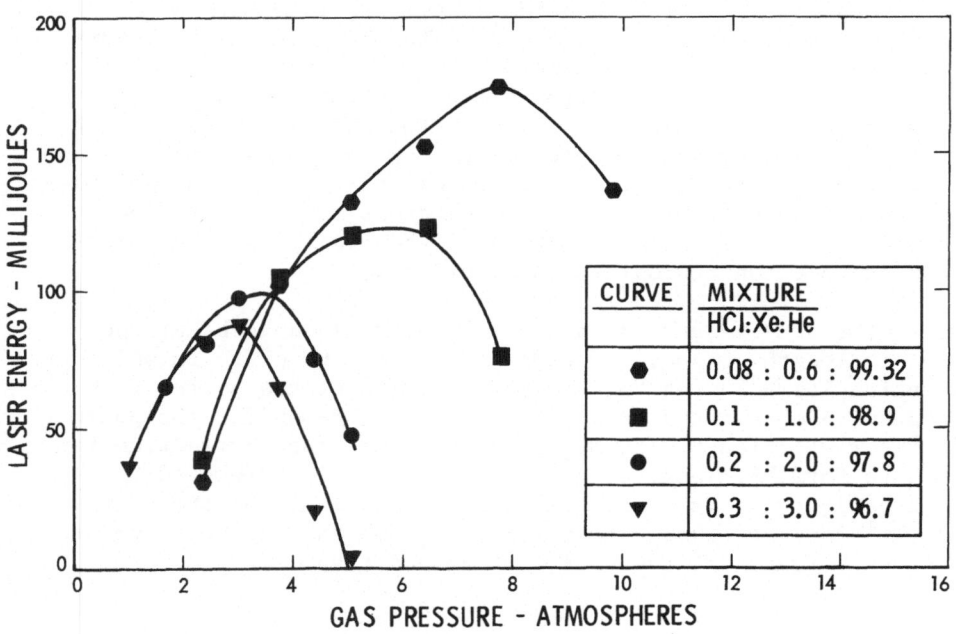

CURVE	MIXTURE HCl:Xe:He
●	0.08 : 0.6 : 99.32
■	0.1 : 1.0 : 98.9
●	0.2 : 2.0 : 97.8
▼	0.3 : 3.0 : 96.7

Figure 4   Pressure Scaling of XeCℓ* Laser

Xe and HCl number density is held fixed at an optimum value and the laser energy increases with helium buffer pressure. Table 6 lists a probable kinetic sequence for the discharge excited XeCl* laser.

Unlike e-beam excitation, the initial plasma produced by electric discharge excitation is not well characterized. The primary excited species produced in discharge excitation of a He:Xe:HCl mixture is probably predominately Xe^+, Xe^*, and Cl^- with some He^+ and He^* produced efficiently only at gas breakdown where the mean electron energy is highest. The cross section for dissociative attachment of electrons to HCl is small for thermal energy electrons, but the cross section is high for electrons having energies of 0.8 eV. In the electric discharge, a distribution of electron energies exist and this energy distribution changes throughout the discharge as the voltage across the electrodes drops. The formation of Cl^- should be rapid due to the high density of electrons having energies of 1 eV. The major kinetic pathways producing XeCl* are ion recombination of Xe^+ with Cl^-, reaction 74, and the metastable reaction channel, reaction 72, of Xe^* with HCl. The ion recombination channel for XeCl* proceeds through a three body reaction and this reaction should be enhanced at high helium pressures as is experimentally observed in Figure 4. The branching ratio for the neutral metastable reaction has not been measured and only the total quenching rate constant for $Xe^*(^3P_2)$ with HCl is known (J.E. Velazco, et. al., 1976). The production of XeCl* by reaction 72 is endothermic by 0.2eV and, consequently, the ion recombination channel for XeCl* formation may be the dominant kinetic pathway. On the other hand, the reaction of Xe^* with Cl_2 has been measured and this reaction rapidly produces XeCl*, but unfortunately Cl_2 absorbs strongly at 308nm where the XeCl* laser oscillates.

The kinetic pathways converting He^+ and He^* to XeCl* do not appear favorable. The only way to convert He^+ to Xe^+ is to have a fast charge transfer reaction of He_2^+ with Xe proceeding by either a bimolecular, reaction 69, or a termolecular, reaction 71, process. Reaction 71, if it had a reasonable rate constant, would proceed rapidly only at high total pressures.

The XeCl* laser was first studied using e-beam excitation with argon as the buffer gas but lasing was weak compared to KrF* and, thus, XeCl* was considered to be poor laser candidate. But in electric discharge excited helium buffered devices, the XeCl* laser is very efficient. In order to explain this fact, it has been argued that absorption by Ar_2^+ and other ion species in an e-beam excited XeCl* laser was responsible for poor laser performance. It has also been postulated that the high laser efficiency for the XeCl* laser using electric discharge excitation is because the reaction occurs through the neutral metastable reaction channel

Table 6   Kinetic Sequence for XeCℓ* Laser

Reaction Number			Rate Constant (e.g. cm^3/sec, cm^6/sec)
56.	He + e	$\rightarrow He^+$ + 2e	
57.	He + e	$\rightarrow$ He* + e	
58.	Xe + e	$\rightarrow Xe^+$ + 2e	
59.	Xe + e	$\rightarrow$ Xe* + e	
60.	HCℓ + e	$\rightarrow C\ell^-$ + H	$\sim 1.1 \times 10^{-9}$ @ 0.8eV
61.	HCℓ + e	$\rightarrow HC\ell^+$ + 2e $\rightarrow C\ell^+$ + H + 2e	
62.	He^+ + Xe	$\rightarrow Xe^+$ + He	$< 10^{-11}$
63.	He^+ + HCℓ	$\rightarrow C\ell^+$ + H	3.3×10^{-9}
64.	He* + Xe	$\rightarrow Xe^+$ + He + e	1.2×10^{-10}
65.	He* + HCℓ	$\rightarrow HC\ell^+$ + e $\rightarrow C\ell^+$ + H + e	8.2×10^{-10}
66.	He^+ + 2He	$\rightarrow He_2^+$ + He	1×10^{-31}
67.	He* + 2 He	$\rightarrow He_2^*$ + He	2.5×10^{-34}
68.	He_2^+ + 2He	$\underset{3}{\overset{2}{\rightleftharpoons}} He_3^+$ + He	$k_{23} = 1 \times 10^{-31}$ $k_{32} = 1.6 \times 10^{-10}$
69.	He_2^+ + Xe	$\rightarrow Xe^+$ + 2He	
70.	He_2^* + Xe	$\rightarrow Xe^+$ + 2He	5.9×10^{-10}

Table 6   Kinetic Sequence for XeCl* Laser (Continued)

Reaction Number		Rate Constant (e.g. cm^3/sec, cm^6/sec)
71.	$He_2^+ + Xe + He \rightarrow Xe^+ + 3He$	
72.	$Xe^* + HCl \rightarrow XeCl^* + H$	5.6×10^{-10}
73.	$Xe^+ + HCl \rightarrow HCl^+ + Xe$	2.9×10^{-11}
74.	$Xe^+ + Cl^- + He \rightarrow XeCl^* + He$	$\sim 10^{-25}$
75.	$Xe^+ + Xe + He \rightarrow Xe_2^+ + He$	1.1×10^{-31}
76.	$Xe^{**} + Xe \rightarrow Xe_2^+ + e$	
77.	$Xe_2^+ + e \rightarrow Xe^* + Xe$	
78.	$He_2^+ + HCl \begin{cases} \rightarrow HCl^+ + 2He \\ \rightarrow Cl^+ + H + 2He \end{cases}$	1.7×10^{-9}
79.	$He_2^+ + HCl + He \rightarrow Products$	30×10^{-30}
80.	$XeCl^* \rightarrow Xe + Cl + h\nu$	
81.	$XeCl^* + h\nu \rightarrow Xe + Cl + 2h\nu$	
82.	$XeCl^* + He \rightarrow Products$	
83.	$XeCl^* + 2He \rightarrow Products$	
84.	$XeCl^* + Xe \rightarrow Products$	
85.	$XeCl^* + Xe + He \rightarrow Products$	

and, therefore, absorbing ion species are not produced. (Ewing, 1978). However, the real difference is the helium buffer does not form ionic absorbing molecules as does argon in e-beam excited systems. For example, replacement of the argon buffer by neon and replacement of Cl_2 by nonabsorbing HCl in e-beam excited $XeCl^*$ lasers increases the overall laser efficiency (Rothe and West, 1978; Champagne, 1978). Ionic pathways leading to RGH formation are also important in electric discharge excited lasers. For example, lasing from KrF^* in an electric discharge excited laser has been seen even when the F_2 halogen donor was replaced by SF_6 or CF_4 (Rothe and Gibson, 1977). The neutral metastable "harpoon" reactions of Kr^* with both SF_6 and CF_4 are endothermic and, therefore, KrF^* production most likely occurs by the ion recombination channel.

For the $XeCl^*$ discharge excited laser, the ion kinetic pathways leading to $XeCl^*$ formation involve termolecular reactions and these reactions should be more favorable at higher helium pressures. The positive-negative recombination process has been theoretically evaluated for RGH laser systems (Flannery and Yang, 1978) in the presence of various buffer gases. In practice the recombination rate is a three body process at pressures of 1 atm, but as the pressure is increased the rate of ion encounter is limited by the mutual diffusion of the positive and negative ions through the dense buffer gas. For example, in heavier rare gases buffers, the Kr^+ and F^- recombination coefficient maximizes at several atmospheres of pressure due to the small ion diffusion coefficients. However, the ion diffusion coefficients in helium buffered mixtures are large and the recombination coefficient for Kr^+ and F^- in helium does not reach a maximum until 20 atm total pressure. Again high pressure operation should enhance RGH formation rates through ionic kinetic channels. The limitation for scaling high pressure RGH lasers depends on the quenching of the upper laser level by helium collisions. These collisional quenching rate coefficients for $XeCl^*$ have not as yet been measured, but preliminary studies at JPL indicate that the efficiency of the $XeCl^*$ laser is improved when operated at 7 to 8 atm.

Finally, although arguments for the increased $XeCl^*$ laser efficiency with pressure can be made in light of the more favorable three body reaction kinetics at high pressure, it must also be realized that the electrical discharge efficiency may also be improved by high pressures. As the pressure of the gas in the discharge excited laser is increased, the voltage required for gas breakdown also increases. At higher breakdown voltages, the electrical energy deposited into the gas is greater because electrical energy goes as the voltage squared. Also the gas impedance may be higher at higher pressures and a closer impedance match between the

electric discharge driving circuit and the laser plasma leads to
more efficient energy deposition. This again points out the major
limitation for electric discharge excited lasers, that is, it is
very hard to parameterize the laser kinetics since the energy
deposition function is unknown and can change drastically with
even small variations in gas composition and pressure.

5.2   Charge Transfer Molecular Ion Lasers

For RGH and rare gas dimer lasers the initial energy deposition is the production of ions, yet some of the important kinetic
pathways involve reactions of excited neutral species which are
produced from ion-electron recombination reactions. Although the
kinetic pathways may be involved, the overall efficiencies leading
to excited molecules for these laser systems are usually very good.
However, it would be desirable to proceed directly from the initial
ions formed in the plasma to an excited molecule without going
through a long kinetic sequence. The charge transfer reaction of
He_2^+ with N_2 is such a reaction.

Chemiluminescent reactions from ion-molecule collisions have
been known for some time, but most chemiluminescent ion-molecule
studies have been done with high energy ion beams. Luminescent ion-
molecule processes at high kinetic energies are not directly applicable to high pressure lasers because the kinetic energy of the
ions is near thermal in the laser plasma. It is extremely difficult
to study ion-molecule luminescent reactions at thermal energies
although some flowing afterflow (Collins and Robertson, 1964) and
ICR experiments have been done at thermal energies (Marx, 1975;
Govers et. al., 1976). Recently, Leventhal and co-workers have
measured charge transfer, CT, luminescent cross sections in an
intermediate ion energy range of 5eV to 11eV (Bearman et. al.,
1976a; Bearman et. al., 1976b). The luminescent cross sections for
reactions of He_2^+ with N_2, CO, NO, O_2 and CO_2 and for Ne_2^+ with N_2
are very large and these higher energy data correlate with the
thermal luminescence data of Collins and Robertson, 1964.

The mechanism for thermal energy CT reactions has been thought
to depend on energy resonance matches between the recombination
energy of the ion and a molecular ion level in the target molecule
for low energy encounters. However, molecules have a large density
of energy states and an energy resonance match should always exist
for an ion molecule CT reaction. However, experimentally it is
observed that a variety of ion-molecule CT reactions are extremely
slow. It has also been argued that CT at thermal energies depends
on both energy resonance and favorable Franck-Condon, F-C, factors
coupling the molecular ground state to the molecular ion state produced in the CT reaction (Bowers and Elleman, 1972; Laudenslager
et. al. 1976). An extensive kinetic study of the thermal charge

transfer reactions of atomic rare gas ions with small molecules is in the literature (Laudenslager et. al., 1976; Anicich et. al., 1977). The results of these studies confirm the dual requirements for energy resonance and favorable F-C factors for efficient charge transfer reactions with molecular targets at thermal energies. On the other hand, if large F-C factors are not present, the rate coefficient for the reaction is usually much smaller, but there are several notable exceptions such as the Ar^+/NO_2 and Ar^+/CO_2 CT reactions. Figures 5 and 6 illustrate the mechanism of energy resonance and F-C factor requirements for thermal energy CT reactions. The peaks in the photoelectron spectrum of oxygen in Figure 5, are the locations of O_2^+ energy levels which are connected to O_2 ground state with large F-C factors for optical transitions. From the recombination energies of the rare gas ions, shown below the energy scale, it is evident that only He^+ and Xe^+ $(^2P_{3/2})$ have coincident match ups with energy resonant levels of O_2^+ having favorable F-C factors. As expected, the He^+/O_2 and the Xe^+/O_2 CT reactions have larger rate constants than do the Ne^+/O_2, Ar^+/O_2, and Kr^+/O_2 reactions. Also the excited $Xe^+(^2P_{1/2})$ state should react inefficiently with O_2 since it's recombination energy does not correspond to an energy resonant level having a large F-C factor. A photoionization experiment, Figure 6, clearly indicates that the relative cross section for O_2^+ production by Xe^+ drops when $Xe^+(^2P_{1/2})$ state is produced.

An experimental determination of the product ion channels produced by highly exothermic CT reactions of He^+ and Ne^+ ions with small molecules has shown that dissociative CT processes are favored (Anicich et. al., 1977). Clearly then, atomic CT reactions would be ineffective in forming excited molecular ions by CT reactions. However, high pressure operation of actual laser devices would favor the conversion of atomic rare gas ions into dimer rare gas ions. The reactions of these dimer rare gas ions are not only very efficient, but they also produce excited molecular ions very efficiently.

The recombination energies of He_2^+ and Ne_2^+ are several electron volts wide because recombination forms the neutral ground state dimer molecule which has a repulsive potential energy curve. The value of the recombination energy, therefore, depends on the steepness of the repulsive lower curve and the internuclear distance of the dimer ion when electron capture occurs. This range of recombination energies relaxes both the energy resonance requirement as well as the F-C restrictions.

The first CT molecular ion laser was excited by an e-beam (Collins et. al., 1974). But laser action on the same CT N_2^+ laser has also been demonstrated in electric discharge devices (Laudenslager et. al. 1976; Rothe and Tan, 1976). Table 7 lists the dominant kinetic reactions for the He_2^+/N_2 CT laser.

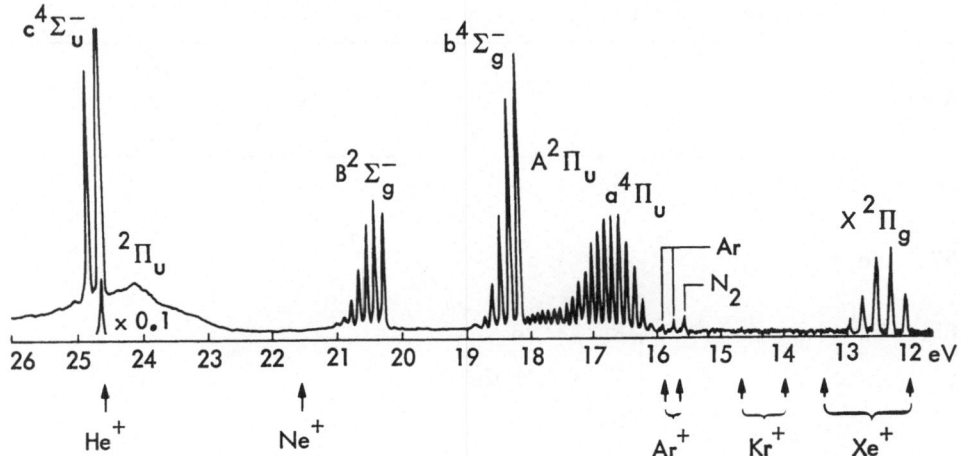

Figure 5   Photoelectron Spectrum of Oxygen
            The Recombination Energies of the Rare Gas Ions
            are Located on the Energy Scale by The
            Arrows

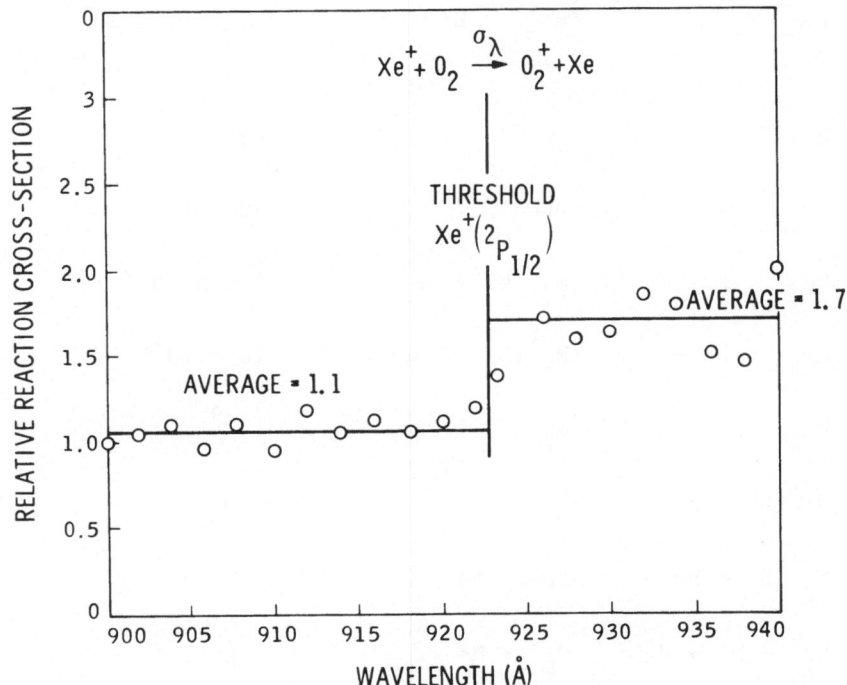

Figure 6   Relative CT Formation of O_2^+ by Xe^+ as a Function
            Photoionization Wavelength (Ajello and Laudenslager,
            1976).

Table 7   Kinetic Sequence for N_2^+ Laser

Reaction Number		Rate Constant (e.g. cm^3/sec, cm^6/sec)
86.	$He^+ + e \rightarrow He^+ + 2e$	
87.	$He + e \rightarrow He^* + e$	
88.	$N_2 + e \begin{cases} \rightarrow N_2^+ + 2e \\ \rightarrow N^+ + N + 2e \\ \rightarrow N_2^* + e \end{cases}$	
89.	$He^+ + N_2 \rightarrow N_2^+ + He$	3.7×10^{-10}
	$\rightarrow N^+ + N + He$	8.3×10^{-10}
90.	$He^+ + 2He \rightarrow He_2^+ + He$	1×10^{-31}
91.	$He_2^+ + 2He \underset{3}{\overset{2}{\rightleftharpoons}} He_3^+ + He$	$k_{23} = 1 \times 10^{-31}$
		$k_{32} = 1.6 \times 10^{-10}$
92.	$He_2^+ + e \rightarrow He_2^*$	$<10^{-8}$
93.	$He_3^+ + e \rightarrow 3He$	10^{-7}
94.	$He_2^+ + N_2 \rightarrow N_2^+(B) + 2He$	1.3×10^{-9}
95.	$He_2^+ + N_2 + He \rightarrow N_2^+(B) + 3He$	16×10^{-31}
96.	$He_3^+ + N_2 \rightarrow N_2^+(B) + 3He$	10^{-9}
97.	$N_2^+(B) + e \rightarrow 2N$	$\sim 10^{-7}$
98.	$N_2^+(B) + He \rightarrow Products$	3×10^{-11}
99.	$N_2^+ + N_2 + He \rightarrow N_4^+ + He$	5×10^{-30}

Table 7   Kinetic Sequence for N_2^+ Laser (Continued)

Reaction Number		Rate Constant (e.g. cm^3/sec, cm^6/sec)
100. $N_2^+(B)$	$\rightarrow N_2^+(X) + h\nu$	τ = 66nsec
101. $h\nu + N_2^+(B)$	$\rightarrow N_2^+(X) + 2\ h\nu$	
102. $N_2^+(X)$, v''=1 + e	$\rightarrow N_2^+(X)$, v''=0 + e	10^{-5}
103. He* + N_2	$\rightarrow N_2^+(X)$	$k_T = 6.9 \times 10^{-11}$
	$\rightarrow N_2^+(B)$	

In an e-beam excited laser, the initial energy deposition produces He^+ and secondary electrons. However, the stopping power of helium for high energy electron beams is low and high pressures (>10 atm) of helium are necessary for efficient production of He^+. An alternate source of excitation might be the use of proton beams for helium excitations, because the stopping power of helium is better for high energy protons.

In an electric discharge, N_2^+ laser, however, the helium-nitrogen mixture consists almost entirely of helium and this mixture tends to break down at low voltages. Therefore, it is very difficult to establish a high electron energy in order to ionize the helium. The efficiency of the CT N_2^+ laser is not as good as the discharge pumped RGH lasers and this is not due to a poor kinetic sequence, but is a direct consequence of not being able to establish a high degree of ionization in high pressure helium by conventional electric discharge excitation. Once He^+ is formed, it rapidly dimerizes to He_2^+ and further associates to He_3^+. The actual percentage of He_3^+ to He_2^+ is important because He_3^+ can be lost by dissociative electron recombination while He_2^+ recombines with electrons by radiative recombination and this is a slow process. The partial pressure of N_2 must be kept low in order that reaction 89, He^+/N_2 CT, does not compete with the dimerization, reaction 90, of He^+. In electric discharge excitation, the optimum percentage of N_2 in the helium mixture has to be kept below 1%, because high partial pressures of N_2 lower the mean electron energy in the discharge and lowers the efficiency of He^+ production. Figures 7 and 8 depict some of the operating parameters of a 3 atm N_2^+ CT laser developed at JPL. Figure 7 shows the N_2^+ discharge laser maximizes at about 0.1% N_2 additive and the laser power decreases with increasing N_2 partial pressures. In fact, at partial

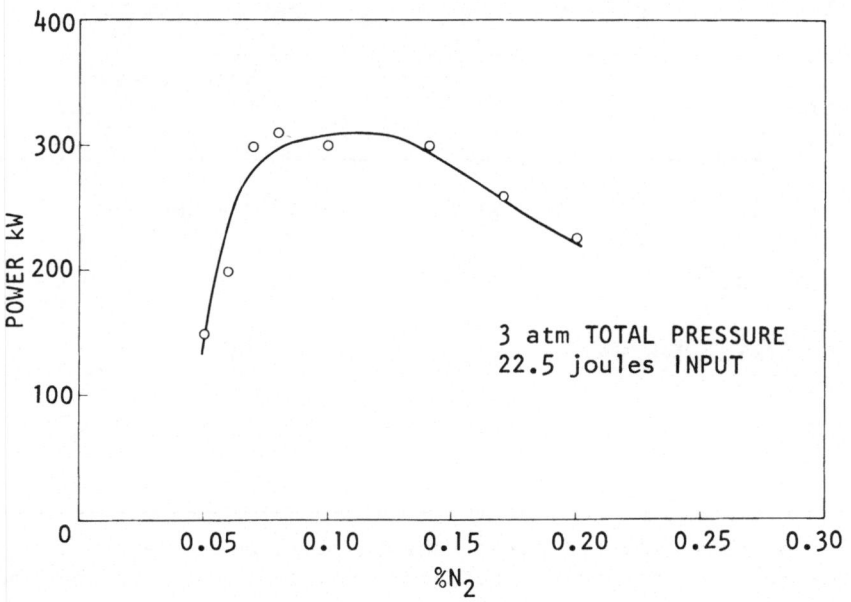

Figure 7   Output Power of N_2^+ Laser as a Function of
             N_2 Partial Pressure in a Helium Buffer

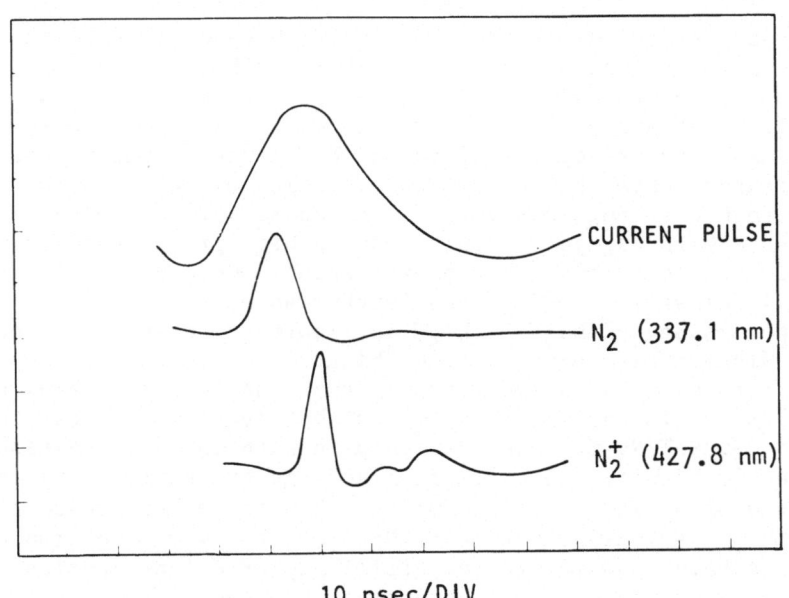

10 nsec/DIV

Figure 8 N_2^+ Laser System Pulse Shapes

pressures of N_2 greater than 1% N_2, the violet lasing from N_2^+ at
427.8nm gives way to lasing in the UV from neutral nitrogen at
337.1nm. This is a clear indication that at these pressures of
N_2 the discharge electrons are exciting N_2 and not the helium.
Without efficient ionization of helium the kinetic pathways leading to $N_2^+(B)$ state can not occur.

Figure 8 shows the discharge current pulse temporal shape,
and the shape of the N_2 laser UV pulse when the pressure of N_2
is above 1%. The N_2 laser is produced by direct electron impact
excitation of N_2 and closely follows the current pulse before
lasing action self-terminates. The N_2^+ laser pulse temporal shape
at low N_2 pressures is delayed from the onset of the discharge
current pulse. This pulse delay is due to the finite reaction time
needed to produce N_2^+ from the kinetic sequence outlined in Table 7
and indicates that $N_2^+(B)$ state is formed through a kinetic chain
and not direct electron impact ionization of N_2.

An interesting feature of the N_2^+ laser kinetics is the
dominance of termolecular reaction pathways over very fast bimolecular channels as the total pressure is raised. Collins and
co-works have studied termolecular CT reactions of rare gas dimer
ions with a variety of small molecules, and for many reaction systems these termolecular rate constants are large enough such that
these reactions will effectively dominate over bimolecular reactions at pressures above several atmospheres (Collins and Lee,
1978; Lee et. al. 1976).

High helium pressures are needed to effectively convert He^+
ions into He_2^+ ions to initiate the CT reaction. This means a
high voltage discharge device such as shown in Figure 4 has to be
used for N_2^+ lasers. The high helium pressures needed for the
formation of dimer ions can also lead to rapid deactivation of
excited $N_2^+(B)$ ions, reactions 98 and 99. There is very little
data on electronic quenching of excited molecular ion states by
atoms, but a rate constant of 3×10^{-11} cm^3/sec for $N_2^+(B)$ state
deactivation by helium has been measured by Tellinghuisen et. al.,
1972. If this is a good deactivation rate coefficient, then collisional deactivation of $N_2^+(B)$ state would be the major loss of the
upper laser level. The radiative lifetime of $N_2^+(B)$ state is
66 nsec and it is this spontaneous radiation which initiates laser
action. At 3 atm helium pressure $N_2^+(B)$ state is deactivated by
reaction 98 in about 1 nsec. Therefore the CT systems should be
poor laser oscillators because the long spontaneous emission lifetimes delays the onset of laser oscillation and collisional deactivation is the major loss from the upper laser level. These CT
lasers should operate more efficiently as laser amplifiers for
injected laser pulses, because stimulated emission would occur on
the injected signal and stimulated dexcitation of $N_2^+(B)$ state
could then be made to compete with collisional deactivation.

6   SUMMARY

The rare gas dimer and rare gas halide excimer lasers and the charge transfer molecular N_2^+ ion laser all are produced by kinetic pathways involving ion reactions. Although rate constants are listed in the various tables, the accuracy of these constants is not always well known. There is a need for careful experimental investigation into specific ionic reactions occurring at high pressures. Unfortunately, this may represent a step backward from the microscopic kinetic experiments which have been developed over the last decade. A great many of the high pressure kinetic data in laser plasmas are being obtained from stationary afterglow experiments. It is well known that stationary afterglow experiments can be extremely difficult to interpret because of the variety of reactions which occur simultaneously. The insight into the mechanisms of ion-molecule bimolecular reactions has been largely derived from the microscopic kinetic studies using experimental instruments which can isolate a particular reaction. Such isolated experiments may prove difficult at high pressures.

Formation of mixed rare gas ions such as $ArKr^+$ and the reactions of such cluster ions have been ignored in most kinetic models of RGH lasers. Likewise, termolecular reactions have also been neglected in kinetic models, most likely due to the presence of fast parallel bimolecular reaction channels for RGH formation. In the past, argon buffered e-beam excited lasers have been operated below 2 atm and for long excitation pulse durations (several μsecs). Under these conditions fast neutral bimolecular harpoon reactions occur with high probability. In electric discharge lasers the excitation pulse width is much shorter and pressure operation above 2 atm is preferred for higher laser output. Under these conditions, termolecular ion reactions will be the dominant reaction pumping steps, because the bimolecular channels occur too slowly. Most kinetic laser models are based on lower pressure e-beam excited lasers and these models have been applied directly to discharge laser devices without taking into account the different operating parameters for each excitation system. Even for the same laser transition, the two excitation methods, e-beam and electric discharge, produce different initial plasma species and therefore, the formation of the upper laser level can occur by different kinetic pathways.

In the future, it is hoped that accurate kinetic models will be based on a more sophisticated knowledge of both plasma formation and ion chemistry. There is a great amount of work which still needs to be done in understanding high pressure ion-molecule interactions.

REFERENCES

Ajello, J. M. and Laudenslager, J. B., 1976, Chem. Phys. Lett., 44, 344.

Anicich, V. G., Laundenslager, J. B., Huntress, W. T., and Futrell, J. H., 1977, J. Chem. Phys., 67, 4340.

Bearman, G. H., Earl, J. D., Pieper, R. J., Harris, H. H., and Leventhal, J. J., 1976a, Phys. Rev. A, 13, 1734.

Bearman, G. H., Earl, J. D., Hams, H. H., and Leventhal, J. J., 1976b, Appl. Phys. Lett. 29, 108.

Beaulieu, A. J., 1970, Appl. Phys. Lett. 16, 504.

Bowers, M. T. and Elleman, D. E., 1972, Chem. Phys. Lett., 16, 482.

Bohme, D. K., Adams, N. G., Mosesman, M., Dunkin, D. B., and Ferguson, E. E., 1970, J. Chem. Phys. 52, 5094.

Champagne, L. F., 1978, Appl. Phys. Lett. 33, 523.

Collins, C. B. and Robertson, W. W., 1964, J. Chem. Phys. 40, 701.

Collins, C. B., Cunningham, A. J., and Stockton, M., 1974, Appl. Phys. Lett., 25, 344.

Collins, C. B. and Lee, F. W., 1978, J. Chem. Phys., 68, 1978.

Daugherty, J. D., 1976, in Principles of Laser Plasmas, G. Bekefi Ed., John Wiley and Sons, New York, p. 369.

Daugherty, J. D., Mangano, J. A., and Jacob, J. H., 1976, Appl. Phys. Lett., 28, 581.

Dumanchin, R. and Rocca-Serra, J., 1969, C. R. Acad. Sci., 269, 916.

Ewing, J. J., 1978, "Physics Today" May, 32.

Flannery, M. R. and Yang, T. P., 1978, Appl. Phys. Lett., 33, 574.

Govers, T.R., Gerard, M., Marx, R., 1976, Chem. Phys., 15, 185.

Herschbach, D. R., 1966, in Advances In Chemical Physics Vol X, J. Ross, Ed., Interscience, New York, p. 319.

Hurst, G. S. and Klots, C. E., 1976, in Advances in Radiation Chemistry, M. Burton and J. L. Magee, Eds., Vol. 5, John Wiley & Sons, New York, p. 1.

Krupke, W. F. and George, E. V., 1978, Optical Engineering, 17, 238.

Laundenslager, J. B., Huntress, W. T., and Bowers, M. T., 1974, J. Chem. Phys. 61, 4600.

Laudenslager, J. B., Pacala, T. J., and Wittig, C., 1976, Appl. Phys. Lett., 29, 580.

Lee, F. W., Collins, C. B., and Waller, R. A., 1976, J. Chem. Phys., 65, 1605.

Lorents, D. C. and Olson, R. E., 1972, SRI Semiannual Tech. Rep. No. 1, Stanford Research Institute.

Marx, R., 1975, in Interactions Between Ions and Molecules, P. Ausloss, Ed., Penum Press, New York, p. 563.

Rokni, M., Jacob, J. H., and Mangano, J. A., 1976, Phys. Rev. A 16, 2216.

Rothe, D. E. and Tan, K. O., 1977, Appl. Phys. Lett., 30, 152.

Rothe, D.E. and Gibson, R.A., 1977, Optics Commun., 22, 265.

Rothe, D. E. and West, J. B., 1978, "Efficient Excitation of XeCℓ laser by E-Beam and Discharge Pumping", presented at the 31st Annual Gaseous Electron. Conf., Buffalo, N. Y.

Sarjeant, W. J., Alcock, A. J., and Leopold, K. E., 1977, Appl. Phys. Lett., 30, 635.

Velazco, J. E., Klots, J. H., and Setser, D. W., 1976, J. Chem. Phys. 65, 3468.

DECAY PROCESSES OF THE LOWEST EXCITED ELECTRONIC STATES OF POLYATOMIC RADICAL CATIONS

J. P. Maier

Physikalisch-Chemisches Institut der
Universität Basel, Klingelbergstr. 80
Ch-4056 Basel/Switzerland

The purpose of this review is to describe the decay behaviour of the lowest excited electronic states of radical cations of polyatomic molecules in the gas phase and to outline the techniques developed for such studies. The emphasis will be on investigations of excited organic cations using their optical emission spectra, although coincidence techniques relying either on fragment or photon detection will also be considered. The latter methods have been introduced within the past few years. The relation to the contents of the other chapters of this book concerned with ion-molecule reactions is twofold. Firstly, it is desirable to know as much as possible about the energetics and structure of species encountered in ion-molecule reactions, and about the decay rates and pathways for their excited electronic states. Secondly, it is the aim of this chapter to indicate the possibilities which now exist to use the characteristics of the radiative channel of polyatomic cations as a probe in ion-molecule reactions in a fashion similar to that exercised hitherto for diatomic species (1) and recently for some triatomic radical cations (2).

The majority of cations encountered in ion-molecule reactions are closed shell species, e.g. CH_5^+, $C_2H_5^+$ and our knowledge of their structures in the gas phase, geometric and electronic, is scarce, forthcoming mainly from theoretical calculations. This review is restricted to a discussion of open shell cations for the reason that

there are techniques available to study their electronic
states. Photoelectron spectroscopy is the principal method
(3) and from such data optical emission spectra of polyatomic radical cations can be assigned. The coincidence
techniques to be described also rely on the principles of
photoelectron spectroscopy.

The use of photoelectron spectroscopy to infer ionisation energies and to assign the symmetries of the cationic states produced is amply discussed elsewhere (3,4).
It suffices here to summarize broadly the energy distribution of the electronic doublet states of organic cations. For readers interested in the ionisation energies
of hydrocarbons in the valence region (i.e. up to ≈ 26
eV), recent compilations (5,6) and surveys (7) should be
consulted. In larger saturated hydrocarbon radical cations there is a quasi-continuum of states from ionisation
threshold to about 25 eV. This is also the case with unsaturated radical cations except that the first excited
state lies typically 1-3 eV above the ground cationic
state and the "continuum" of states sets in at around 10
eV. In polyunsaturated species a few other excited electronic states precede the apparent continuum, as revealed
by well separated bands in their photoelectron spectra
(7). It is for such radical cations, in their first excited electronic states, that emission spectra have been
observed.

For closed shell molecular species there is always
a triplet state lying at lower energy to the lowest excited singlet state and is thus an energetically accessible relaxation channel. In contrast, in radical cations
the lowest excited state, which lies 1-3 eV above the
ground state, is a state of doublet multiplicity. This
state is reached in the photoionisation process. The
states of quartet multiplicity lie to higher energy as
these configurations correspond in the orbital picture to
electron excitation to antibonding orbitals. The complementary doublet state, ionising transitions to which are
dipole forbidden, often lies even higher in energy (8).
The latter doublet states may, however, be reached (dipole selection rules being favourable) from the cationic
ground states using techniques such as photodissociation
spectroscopy in the gas phase (9) or absorption measurements of radical cations generated in low temperature
matrices (10). These considerations are discussed at
length elsewhere (8,9). Thus, not all the doublet states

of the radical cations are located. It is also worth re-
marking that due to the development of theoretical calculations, which incorporate the effects of electron correlation and reorganisation (11), the energy data obtainable from photoelectron spectroscopy can, nowadays,
be calculated with almost the same accuracy (12). The
calculated spectra inherently yield the assignments,
whereas the photoelectron spectra can also provide information on vibrational excitation in the cationic states.

For radical cation internal energies greater than,
say 3-5 eV, fragmentation decays are dominant. This is
the realm of mass-spectroscopic techniques which will not
be considered here (13). In such measurements one may
well know the initial internal energy and the final fragment ions with some indication of the time scale of the
processes. However, it is questionable whether it is
meaningful to discuss such radical cations in terms of
individual electronic states and structures as so many
are accessible on a very rapid timescale ($\geqslant 10^{12}$ s^{-1})
following ionisation. Furthermore, according to the recent theoretical investigations on unsaturated hydrocarbons, in the ionisation energy region $\approx 15-25$ eV, the
states initially formed can not be represented by single
configurations and thus the oscillator strengths are
shared by many ionising transitions (14). Therefore, in
the following sections studies of only the lowest excited electronic states shall be considered.

FATE OF ISOLATED RADICAL CATIONS

Fragmentation decay. - The techniques developed for
the study of the decay processes of radical cations may
be divided into those relying either on fragment ion or
photon detection. The former category encompasses the
long established mass-spectroscopic appearance potential
measurements, preferably using photoionisation (13)

$$M + h\nu \rightarrow M^+(\tilde{J}) \rightarrow A^+ + B.$$

and the more recently developed approach using photodissociation spectroscopy with ions in their ground electronic states ($\tilde{X}$)

$$M^+(\tilde{X}) + h\nu \rightarrow M^+(\tilde{J}) \rightarrow A^+ + B.$$

The latter technique has also been extended to include
multiphoton processes (9). Of particular interest to the
present topic are the results of photoelectron-photoion
coincidence methods (15) where the fragmentation products
of state ($\tilde{J}$) selected radical cations are studied:

$$M + h\nu \rightarrow M^+(\tilde{J}) + \boxed{e_{K.E}}$$
$$\downarrow$$
$$\boxed{M^+ \text{ or } A^+ + B.}$$

The electronic and vibrational state of the cation, populated in the photoionisation process is defined by the
kinetic energy of the electrons ($e_{K.E}$) and these are detected in delayed coincidence with resultant ions, e.g.
M^+ or A^+. These data allow one to obtain ion breakdown
diagrams and, in addition, from the ion time-of-flight
peaks some indications regarding the dissociation rate
and excess kinetic energy (16). Here it suffices to list
the polyatomic radical cations which have been investigated by these techniques over the past ten years and
have been reported in the literature. In table 1 are
presented the cations studied.

Essentially two approaches have been used. The molecules are photoionised either using He(Iα) resonance
line radiation, 21.22 eV, and the photoelectron kinetic
energy is measured (16); or nominally zero energy electrons are collected as the wavelength of a dispersed
continuum photon source (e.g. helium continuum or synchrotron radiation), is scanned (17). The ions are usually
dispersed by a time-of-flight tube or by a quadrupole
spectrometer. It is also worth noting that corresponding
coincidence measurements can also be carried out using
fast electrons (E_0) to simulate the process of photoionisation (18)

$$e(E_0) + M \rightarrow M^+ + e_{ejected} + e_{scattered} \quad (E_0-E)$$

This process is equivalent to ionisation with photons of
energy E (19). These measurements are included in table
1. Examples, where the information forthcoming from these
coincidence techniques is used, are given in subsequent
sections.

Radiative Decay. - If fragmentation pathways are not
detectable, then one has to rely on the detection, or
lack of it, of the radiative channel or relaxation of
excited electronic states. In table 2 are collected the

TABLE 1: Electron-Ion Coincidence Studies of Polyatomic
Radical Cations

3-ATOMIC	REFERENCE	8-ATOMIC	REFERENCE
CO_2^+	(20)(21)	$C_2H_6^+, C_2D_6^+$	(30)(33)
CS_2^+	(22)	$CF_3CF_3^+$	(45)
COS^+	(23)(24)	$C_2H_5Cl^+$	(44)
H_2O^+	(25)(30)	$CH_2ClCH_2Cl^+$	(44)
D_2O^+	(26)	**9-ATOMIC**	
N_2O^+	(23)(24)(27)	$C_2H_5OH^+$	(40)
SO_2^+	(28)(29)	$(CH_3)_2Hg^+$	(46)
4-ATOMIC		**10-ATOMIC**	
$C_2H_2^+$	(30)	$CH_2CHCHCH_2^+$	(47)(48)
NH_3^+	(30)(31)(32)	$C_2H_5CCH^+$	(47)
5-ATOMIC		$CH_3CCCH_3^+$	(47)
CH_4^+, CD_4^+	(18)(30)(31) (33)(34)	$CH_2CCHCH_3^+$	(47)
		Cyclobutene$^+$	(47)
CH_2CO^+	(34)	$CH_3COCH_3^+$	(45)(49)
$CHOOH^+$	(30)	**11-ATOMIC**	
CH_3X^+ X=F,Cl,Br,I	(35)(36) (37)	$C_3H_8^+, CH_3CD_2CH_3^+$	(50)(51)
CD_3I^+	(37)	$CD_3CH_2CD_3^+, C_3D_8^+$	(51)
$CH_2X_2^+$ X=Cl,Br,I	(36)(38)	**12-ATOMIC**	
CHX_3^+ X=Br,I	(36)	Benzene$^+$	(52)(53)
CF_4^+	(24)	$CH_3CCCCCH_3^+$	(54)
6-ATOMIC		$C_6H_5X^+$ X=Cl,Br,I	(55)
$C_2H_4^+, C_2D_4^+$	(30)(39)	**13-ATOMIC**	
CH_3OH^+	(30)(40)	$C_6H_5CN^+$	(53)
CH_2CHF^+	(41)	**17-ATOMIC**	
7-ATOMIC		$C(CH_3)_4^+$	(49)
$CH_2CCH_2^+$	(42)(43)		
$CHCCH_2X^+$, X=Cl,Br	(44)		

triatomic radical cations for which emission spectra have
been observed from the given excited electronic state to
the ground state (56). The most recent additions to this
set are the halocyanide (57) and sulphur dioxide radical
cations (58). In the case of polyatomic radical cations,
with more than three atoms, the emission spectrum of diace-
tylene cation, $\tilde{A}^2\Pi_u \rightarrow \tilde{X}^2\Pi_g$, was the only one known for a
long time (59) until the recent studies following the de-
velopment of photoelectron spectroscopy (3) when an emis-
sion band was suggested to be the $\tilde{A}^2A_{2u} \rightarrow \tilde{X}^2E_{1g}$ transition of hexafluorobenzene cation (60). Since then emission
spectra from $\tilde{A}$ (or $\tilde{B}$) excited states of over fifty organic radical cations have been obtained in our laboratory
using a crossed electron-sample beam apparatus. Many of
these are listed in table 3. The experimental details and
procedures have been given previously (61). The electronic transitions have been identified on the basis of the
photoelectron spectroscopic data. The radical cation
emission spectra range from the tetraatomic systems,
e.g. of haloacetylenes, to 18-atomic, e.g.

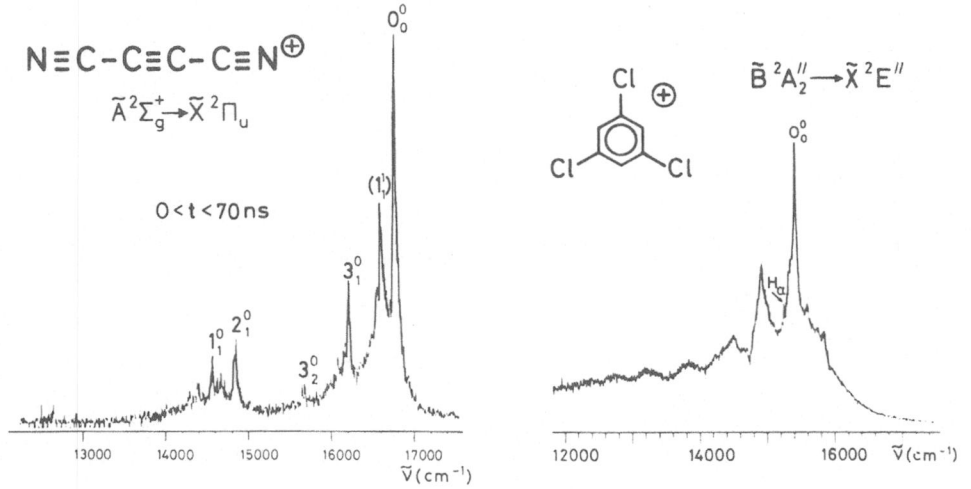

Figure 1. Emission spectra of dicyanoacetylene (62)(left)
          and 1,3,5-trichlorobenzene (63)(right) radical
          cations recorded with optical resolutions of
          0.4 nm and 0.16 nm (fwhm) respectively. The
          spectrum of dicyanoacetylene cation is, in
          addition, time resolved in the indicated in-
          terval in order to eliminate the overlapping
          $A^2\Pi \rightarrow X^2\Sigma^+$ bands of the cyanogen radical.

TABLE 2: Lifetimes of Triatomic Radical Cations in Excited Electronic States Measured via Photon Emission

CATION/STATE[a]	LIFETIME (ns)/TECHNIQUE			
	COINCIDENCE		ELECTRON EXCITATION	
	e-hν	M$^+$-hν (64)	Other Laboratories	This Laboratory[b) c)]
CO_2^+ $\tilde{B}^2\Sigma_u^+$	117±2 (65) 138±12 (68)	120±10	118±12 (66) 118±4 (67)	139±14
$\tilde{A}^2\Pi_u$	108±1 (65) 102±8 (68) 120 (69)	120±10	113±12 (66) 116±1 (67)	123±12
N_2O^+ $\tilde{A}^2\Sigma^+$ 0^0	240 (69) 220±15 (71)	224±10	223±20 (70) 231±2 (72)	245±12
1^1	166±15 (71)		187±2 (72) 165±16 (70)	187±19
COS^+ $\tilde{A}^2\Pi$		<30	175±50 (70)	100±15 (s) ≈500 (ℓ)
CS_2^+ $\tilde{B}^2\Sigma_u^+$		360±40	294±30 (70)	300±30 (s) ≈1.4μs (ℓ)
$\tilde{A}^2\Pi_u$		2.4-15μs		≈2.5μs
H_2O^+ $\tilde{A}^2A_1$ 2^{13}			3.0±.3μs (73) 10.5± 1μs (74)	
H_2S^+ $\tilde{A}^2A_1$ 2^2			4.3±.6μs (75)	(57)
$ClCN^+$ $\tilde{B}^2\Pi$		≈150		205±21 (s)
$\tilde{A}^2\Sigma^+$				≈4.4μs
$BrCN^+$ $\tilde{B}^2\Pi$		300±30		270±27 (s)
$\tilde{A}^2\Sigma^+$				≈3.0μs
ICN^+ $\tilde{B}^2\Pi$				300±30 (s)
$\tilde{A}^2\Sigma^+$				≈1.2μs
SO_2^+ $\tilde{B}^2A_1$[d]				

a) When the lifetime is dependent on the vibrational level of the excited state, the level is given. b) (s) and (ℓ) refer to short and long component lifetimes respectively when a non-exponential decay curve is observed. c) Lifetimes of >1μs given are the lower limits due to escape of ions from the observation regions. d) Lifetime not yet measured.

TABLE 3: Lifetimes of Polyatomic Organic Radical Cations in Excited Electronic States (zeroth vibrational level) Measured via Photon Emission

CATION $\tilde{A}^2\Pi\ 0^0$	LIFETIME[a] (ns)	REF.	CATION $\tilde{A}^2\Pi\ 0^0$	LIFETIME (ns)	REF.
$H\text{+}C\equiv C\text{+}_2H^+$	71 ± 3	(76)	$CF_3\text{+}C\equiv C\text{+}_2F^+$	30 ± 3	(79)
$H\text{+}C\equiv C\text{+}_3H^+$	17 ± 2	(76)	$CF_3\text{+}C\equiv C\text{+}_2CF_3^+$	48 ± 2	(79)
$H\text{+}C\equiv C\text{+}_4H^+$	<6	(76)	$CH_3\text{+}C\equiv C\text{+}_2CH_3^+$	24 ± 2	(81)
$Cl\text{-}C\equiv C\text{-}H^+$	$17\pm3(s)$[b]	(61)	$C_2H_5\text{+}C\equiv C\text{+}_2H^+$	<6	(81)
$Br\text{-}C\equiv C\text{-}H^+$	$12\pm3(s)$[b]	(61)	$\tilde{A}^2\Sigma_g^+\ 0^0$		
$I\text{-}C\equiv C\text{-}H^+$	$15\pm3(s)$[b]	(61)	$NC\text{-}C\equiv C\text{-}CN^+$	13 ± 2	(62)
$Cl\text{-}C\equiv C\text{-}Cl^+$	$13\pm2(s)$	(77)	$\tilde{A}(\pi^{-1})0^0$		
$Br\text{-}C\equiv C\text{-}Br^+$	$28\pm3(s)$	(77)	$\underline{t}\text{-}1,3,5\text{-hexatriene}^+$	17 ± 3	(82)
$I\text{-}C\equiv C\text{-}I^+$	$51\pm5(s)$	(77)	$\underline{c}\text{-}1,3,5\text{-hexatriene}^+$	<6	(82)
$Cl\text{+}C\equiv C\text{+}_2H^+$	43 ± 4	(78)	$\tilde{B}(\pi^{-1})0^0$		
$Br\text{+}C\equiv C\text{+}_2H^+$	26 ± 5	(78)	Hexafluorobenzene$^+$	48 ± 2	(83)
$I\text{+}C\equiv C\text{+}_2H^+$	<6	(78)	Pentafluorobenzene$^+$	47 ± 2	(83)
$F\text{+}C\equiv C\text{+}_2F^+$	28 ± 2	(79)	1,2,3,4-tetrafluoro-benzene$^+$	50 ± 2	(83)
$Cl\text{+}C\equiv C\text{+}_2Cl^+$	$21\pm3(s)$	(80)	1,2,3,5- "	50 ± 2	(83)
$Br\text{+}C\equiv C\text{+}_2Br^+$	$12\pm2(s)$	(80)	1,2,4,5- "	30 ± 2	(83)
$I\text{+}C\equiv C\text{+}_2I^+$	$<6(s)$	(80)	1,3,5-trifluoro-benzene$^+$	58 ± 2	(83)
$CH_3\text{-}C\equiv C\text{-}Cl^+$	$16\pm3(s)$	(78)	1,2,4- "	10 ± 2	(83)
$CH_3\text{-}C\equiv C\text{-}Br^+$	$14\pm3(s)$	(78)	1,3-difluorobenzene$^+$	<6	(83)
$CH_3\text{+}C\equiv C\text{+}_2Cl^+$	21 ± 3	(78)	1,3,5-trichloro-benzene$^+$	22 ± 2	(63)
$CH_3\text{+}C\equiv C)_2Br^+$	12 ± 3	(78)	1,3-dichlorobenzene$^+$	<6	(63)
			1,4-dichlorobenzene$^+$	<6	(63)

[a] (s) refers to short component lifetime when non-exponential decay curve is observed. [b] Long component lifetimes $\approx 300\text{--}500$ ns

of all-trans-octatetraene. These spectra have until now
been recorded with optical resolutions down to ≈ 0.1 nm.
On the other hand the emission spectra of all but the
last four triatomic cations given in table 2 have been,
to a certain degree, rotationally analysed (56). A typical emission spectrum of a small and of a large polyatomic radical cation is shown in figure 1. Most of the
emission spectra of the open shell species given in table
3 exhibit a wealth of vibrational fine structure (c.f.
fig. 1) and in the linear, or pseudo linear, cations the
assignments of the spectra often yield most of the totally
symmetric fundamental frequencies for the cationic ground
states. Hitherto, these have been inferred to an accuracy
of about 10 cm^{-1}.

LIFETIME MEASUREMENTS

The detection of the radiative channel enables one
to measure the lifetime (τ) of the excited state directly.
This yields the sum of the rate constants (radiative (k_r)
and non-radiative (k_{nr})) of all the decay channels de-
pleting the state

$$\tau^{-1} = k_T = k_r + k_{nr}$$

In this formulation the non-radiative rate itself is composed of the various processes accessible, i.e. fragmentation, k_f, internal conversion, k_{ic}, and isomerisation k_{is}.

Essentially, three approaches can be considered for
the lifetime measurements of radical cations, relying
either on pulsed or modulated electron excitation, or
pulsed photon excitation, or on coincidence-techniques.
The most limited as yet, but becoming more relevant involves pulsed photon beams, either the synchrotron radiation from storage rings or from pulsed dye lasers, with
which a time resolution of a few nanoseconds can be
achieved. Such lifetime measurements for polyatomic radical cations have not yet appeared in the literature.
However, the advantages of the time structure of the synchrotron radiation have been illustrated for molecular
species (84), and laser induced fluorescence studies of
polyatomic radical cations have also, recently, been
successfully carried out (85). Electron excitation methods have by far been the most widely used for lifetime

measurements, e.g., by phase shift or by gating of a
high or low energy beams (86). The third approach is
based on coincidence techniques. In one method undispersed photons from the electronically excited ion $M^+(\tilde{J})$
are detected in coincidence with energy selected electrons $(e_{K.E.})$. The latter are produced either using the
He(Iα) resonance line (69) or alternatively, at threshold
$(e_{K.E.} = 0)$ by means of dispersed synchrotron radiation
(68).

$$M + h\nu \rightarrow M^+(\tilde{J}) + \boxed{e_{K.E.}}$$
$$\downarrow \rightsquigarrow \boxed{h\nu}$$
$$M^+(\tilde{X})$$

These experiments identify unambiguously the internal
energy of the cation which results in the fluorescence.
By this means, lifetimes of selected vibrational levels
(excited in the photoionisation process) of the excited
states of the cations have been measured (69,87) as well
as the quantum yields of emission ϕ_v of vibrational levels of CO_2^+ in its $\tilde{A}$ and $\tilde{B}$ electronic states (68) and
of N_2O^+ in the $\tilde{A}^2\Sigma^+$ state (71). The quantum yield of
fluorescence from a vibrational level v in the excited
electronic state is defined as

$$\phi_v = \frac{k_r^v}{k_r^v + k_{nr}^v}$$

where the rate constants refer to the vibrational level v.

On the other hand, the coincidence measurements between mass selected ions (M^+) and undispersed photons $(h\nu)$
from the excited ion in a state $\tilde{J}$

$$M + h\nu \rightarrow M^+(\tilde{J}) + e_{K.E.}$$
$$\downarrow \rightsquigarrow \boxed{h\nu}$$
$$\boxed{M^+(\tilde{X})}$$

identify the cation giving rise to the fluorescence (64),
although the internal energy of the ion is not known.
These coincidence curves yield an averaged value for the
lifetimes, $\tau_{\tilde{J}}$, and the quantum yield $\phi_{\tilde{J}}$. These are composed of the individual values, τ_v and ϕ_v, of the radiative vibrational levels of the excited electronic state
populated in the photoionisation process. Analogous experiments have been carried out using electron impact to
measure the lifetimes of the $\tilde{A}^2\Pi_u$ and $\tilde{B}^2\Sigma_u^+$ states of

CO_2^+ from the inelastic electron-photon coincidences (65).

In table 2 are collected the measured lifetimes of triatomic radical cations in their excited states and the values from the above mentioned coincidence techniques are included. However, not all the lifetimes measured using electron impact have been included. The selected values given were judged to be the most reasonable in view of the other results. Included are the values of the measurements carried out in our laboratory. In these, emission is excited by a gated, 20-30 eV, electron beam and the decay curves of the transitions selected by the monochromator are accumulated by means of the single photon delayed coincidence technique (61). The lifetimes are extracted from the decay curves (e.g. c.f. fig. 2) by a least-squares linear fit of the semilogarithmic plot after first subtracting the random events. For the non-exponential decays the curves were fitted to two exponentials by first stripping off the long-lifetime component (61).

In table 3 are collected the lifetimes of the zeroth vibrational level of the $\tilde{A}$, or $\tilde{B}$, states for many of the radical cations, larger than triatomic, whose emission spectra have been found in the author's laboratory.

NON-EXPONENTIAL DECAY

The decay curves which have been measured for the smaller linear radical cations, e.g. X-CN$^+$ $\tilde{B}^2\Pi$ (57,88), X$(C\equiv C)_n$H$^+$ (61), X$(C\equiv C)_n$X$^+$ n= 1,2 $\tilde{A}^2\Pi$ (77,80), with X=Cl, Br, I, show a non-exponential behaviour. This is especially striking in case of the $\tilde{A}$ states of the halocyanide and haloacetylene radical cations, where the ratio of the amplitudes of the short to long component lifetime is found to be merely about ten (c.f. fig. 2). In the decay curves of the other radical cations, this ratio is two orders of magnitude. The manifestation of this effect in the latter situation requires accumulation over a large time interval (abscissa) as well as sufficient statistics.

It has been shown, by recording time resolved emission spectra (fig. 2 upper) as well as by ion ejection experiments (fig. 2 lower), that both components are due to the radiative decay of the cations in their $\tilde{A}^2\Pi$ states.

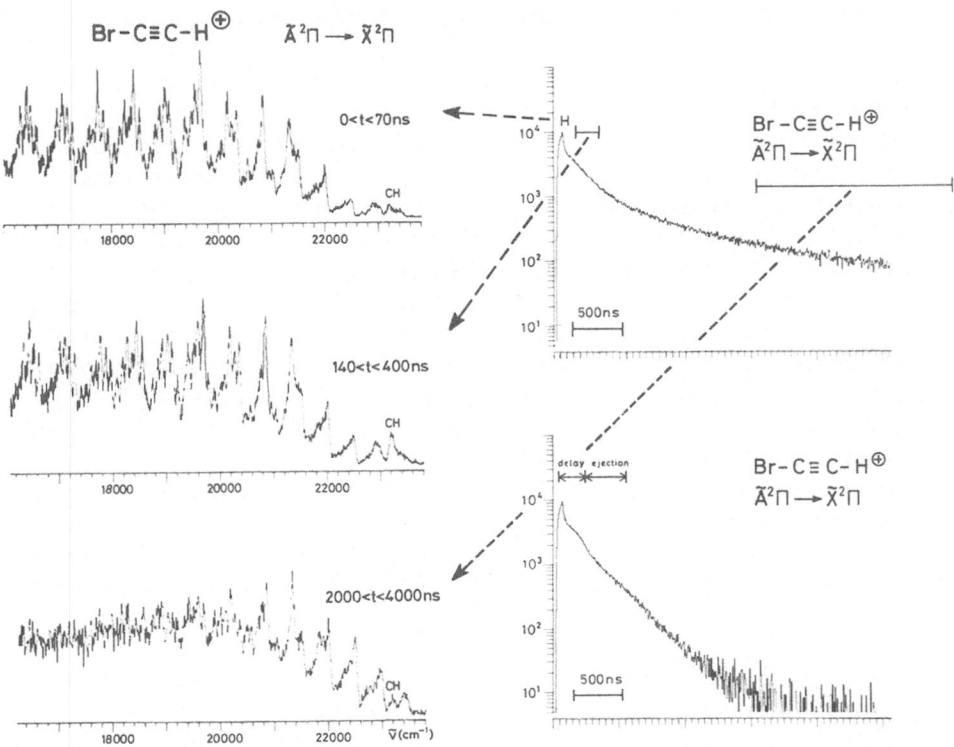

Figure 2: Time resolved emission spectra of bromoacety-
lene radical cation (left) recorded with 0.8
nm resolution in the indicated time intervals
and their correspondence to the decay curve
(right). The decay curves are plotted on a
semilogarithmic scale (with background subtracted) and the bottom curve (right) shows
the effect when ions are ejected from the
observation region under the indicated conditions.

In addition, the non-exponential decay curve of chloro-
acetylene cation in its $\tilde{A}^2\Pi$ state has been reproduced
using He(Iα) and Ne(I) photon excitation in conjunction
with the photon-photoion coincidence approach (64) as
indicated in a scheme above. From the photoionisation
fragment appearance potentials, it can be concluded that
the $\tilde{A}^2\Pi$ states of the haloacetylene cations lie below
fragmentation thresholds whereas the $\tilde{B}$ states are predissociated (61). Thus, cascading from these states can
be excluded. These are the only doublet states accessible
in the photoionisation process as indicated by the photo-

electron spectra (89). The non-exponential behaviour has, therefore, been·ascribed to a strong coupling of the radiative state, $\tilde{A}^2\Pi$, with the manifold of the non-radiative vibrational levels of the ground state, $\tilde{X}^2\Pi$. Theoretical treatments of such phenomena, referred to as "intermediate case coupling" (90), have been successful in explaining the non-exponential decay of molecular species (91), where the density of the non-radiative levels is relatively small. This condition also prevails for the $\tilde{A}^2\Pi$ states of the radical cations as the number of vibrational degrees of freedom is small and these states lie merely 2-3 eV above the ground state. However, in order to interpret the data in terms of the models proposed, vibrational analyses of the emission spectra are a prerequisite.

ABSENCE OF DETECTABLE EMISSION

In all the cases of radical cations where the radiative channel of relaxation has been evident (cf. tables 2,3), the electronic transition has been dipole allowed. There are some cations, however, where the first excited states (or at least their lowest vibrational levels) lie below fragmentation thresholds; the optical transition to the ground state is allowed but the radiative channel is not detected. As the experimental limit for the detection of emission has been estimated to be about 10^{-5} in quantum yield, and taking 10^7-10^8 s^{-1} for the typical radiative rates, the "negative" observation suggests that the non-radiative decays proceed at rates $>10^{11}$ s^{-1}. Additionally, the observation of vibrational fine structure on the band in the photoelectron spectrum associated with the electronic state of concern allows one to conclude that the lifetime of this state is $\geqslant 10^{-12}$ s. Thus, the non-radiative decay rate is bracketed in such cases. Some examples are the first excited states of the cations of acetylene, ammonia, cyanoacetylene and cyclopentadiene. Other cations in this category are given in table 4. In the case of acetylene cation, the Franck-Condon factors are probably unfavourable for the $\tilde{A}^2\Sigma_g^+ \to \tilde{X}^2\Pi_u$ transition as the acetylene cation in the $\tilde{A}$ state has a bent configuration, whereas in the $\tilde{X}$ state it is linear (92). Furthermore, in a bent geometry a quartet state lies below the $\tilde{A}$ state (93) (the energy gap between the $\tilde{X}$ and $\tilde{A}$ states is large ≈ 5 eV), and may therefore provide the non-radiative pathway. The situation with ammonia and hydrogen cyanide radical cations may well

TABLE 4: Some Examples of Polyatomic Radical Cations in
their $\tilde{A}$, or $\tilde{B}$, excited Electronic States for
which the Radiative Relaxation is not Detected.
In categories a) and b) the (lowest vibrational
levels of the) excited states lie below fragmentation thresholds; the transition to the
ground cationic state is dipole allowed in cases
a) and c) but forbidden in case b). In category
c) fragmentation decay is detected.

a)		b)		c)	
$HCCH^+$	$\tilde{A}^2\Sigma_g^+$	$NCCN^+$	$\tilde{A}^2\Sigma_g^+$	$CH_2CCH_2^+$	$\tilde{A}^2E$
NH_3^+	$\tilde{A}^2E$	Benzene$^+$	$\tilde{A}^2E_{2g}$	CH_2CO^+	$\tilde{A}^2B_2$
HCN^+	$\tilde{A}^2\Sigma^+$	$C_6H_5X^+$ X=F,Cl	$\tilde{B}^2B_2$	$CH_3(C{\equiv}C)_nI^+$ n=1,2	$\tilde{A}^2E$
$XCCCN^+$ X=F,H	$\tilde{A}^2\Sigma^+$	$C_6H_4F_2^+$(p)	$\tilde{B}^2B_{1g}$	Styrene$^+$	$\tilde{B}^2B_1$
Cyclopenta-diene$^+$	$\tilde{A}^2A_2$	$C_6H_5CCH^+$	$\tilde{B}^2B_2$	$CH_2CHCHCH_2^+$	$\tilde{A}^2A_u$
Benzene$^+$	$\tilde{B}^2A_{2u}$				

be similar. On the other hand, for the cations in this
category where the $\tilde{A}$ states lie merely 2-3 eV above the
ground state the dominant non-radiative channel is probably to the ground state vibrational manifold (internal
conversion).

Another category included in table 4 is for radical
cations for which the symmetry of the $\tilde{A}$ states is such
that the transition to the ground state is dipole forbidden. Cyanogen and benzene radical cations belong to
this group as well as the fluoro-, chloro- and difluoro-,
dichloro-benzene cations where the radiative channel is
not apparent (63,83). The third category defined in table
4 encompasses the cases where the cations in their $\tilde{A}$
states relax by fragmentation. There belong the radical
cations of allene, 1,3 butadiene or sulphur dioxide in
their $\tilde{A}$ states. That these radical cations in the $\tilde{A}$
states are completely predissociated is shown by the
photoelectron-photoion coincidence breakdown diagrams
(cf. table 1). Whereas the photoelectron band associated
with the $\tilde{A}$ state of allene cation shows well resolved
vibrational fine structure (3), ($\tau \geqslant 10^{-12}$ s), it is ab-

sent for butadiene cation (fig. 3). However, the rate of
$C_3H_3^+$ formation following photoionisation of butadiene
to its $\tilde{A}^2A_u$ state has been inferred to be $\approx 10^6$ s^{-1} (47).
Thus, the initial fast step ($\approx 10^{12}$ s^{-1}) involves internal
conversion (i.e. conversion of electronic energy to
vibrational energy of the ground state). Isomerisation
can also take place, and then the fragmentation to $C_3H_3^+$
takes place with a rate constant many orders of magnitude
slower than the initial step (cf. fig. 3).

COMPETITION OF FRAGMENTATION AND EMISSION

During our studies of the decay processes of organic radical cations in the gas phase, it has been found
that the radiative and fragmentation channels can be in
competition and that both pathways can be detected (81,
82). This unique situation allows one to apply the techniques outlined above relying both on fragment and on
photon generation to probe the decay mechanisms.

Such a situation has been known for some time for
the triatomic radical cations N_2O^+ and COS^+ in their $\tilde{A}$
states (23) and has also recently been established for
the $\tilde{B}$ state of SO_2^+ (29). The observed fragmentation and
radiative processes are summarised in table 5, together
with the data for some of the organic cations which were
found to exhibit such a behaviour. These species consist
of two structural groups, of polyenes and of alkyl substituted diacetylene radical cations. Among the latter
one finds two benzene cation isomers (81).

In case of trans and cis-1,3,5 hexatriene radical
cations, the fragmentation channel of the $\tilde{A}$ states has
been explored by photodissociation spectroscopy (94). The
emission spectra, $\tilde{A} \rightarrow \tilde{X}$ electronic transition, for the
two isomers are distinct; the 0_0^0 band of the cis species
lies 0.1 nm to the red of the 0_0^0 band of the trans (82).
In addition, the absorption spectra of these two cations
in a matrix have been obtained (95), which yield some
vibrational frequencies for the $\tilde{A}$ state as do the photodissociation spectra; the emission spectra provide frequencies of some of the ground cationic state fundamen-
tals.

The quantum yields of emission can be estimated in
this case from either the absorption, or the photodisso-

TABLE 5: Some examples of Polyatomic Radical Cations in their Ã Electronic States for which Emission and Fragmentation are both detected.

Fragmentation	M$^{\oplus}$	Emission
N + NO$^{\oplus}$ ←	N$_2$O$^{\oplus}$	Ã$^2\Sigma^+$ → X̃$^2\Pi$
O + CS$^{\oplus}$ ←	COS$^{\oplus}$	Ã$^2\Pi$ → X̃$^2\Pi$
O + SO$^{\oplus}$ ←	SO$_2^{\oplus}$	B̃^{2}A$_1$ → X̃^{2}A$_1$

H + C$_6$H$_7^{\oplus}$ ← (hexatriene / methylenecyclopentene cation structures)

H + C$_7$H$_9^{\oplus}$ ←
CH$_3$ + C$_6$H$_7^{\oplus}$ ← (methyl-heptatriene cation, CH$_3$)

H + C$_8$H$_9^{\oplus}$ ←
CH$_3$ + C$_7$H$_7^{\oplus}$ ← (octatetraene cation)

} Ã(π^{-1}) → X̃(π^{-1})

H + C$_6$H$_5^{\oplus}$ ← { CH$_3$−≡−≡−CH$_3^{\oplus}$
C$_6$H$_4^{\oplus}$, C$_4$H$_4^{\oplus}$ ← { C$_2$H$_5$−≡−≡−H$^{\oplus}$

H + C$_5$H$_3^{\oplus}$ ← CH$_3$−≡−≡−H$^{\oplus}$

CH$_3$ + C$_7$H$_7^{\oplus}$ ← C$_2$H$_5$−≡−≡−C$_2$H$_5^{\oplus}$

} Ã(π^{-1}) → X̃(π^{-1})

ciation data, and the optically measured lifetimes of the zeroth vibrational levels of the Ã state (17±2 ns for the trans species). The results, which are depicted in figure 3 (where butadiene and trans-1,3,5 hexatriene radical cations are compared), are that for the Ã 0^0 state of trans-1,3,5 hexatriene cation the quantum yield of emission is ≈ 6x10^{-2} whereas for the cis isomer it is about a factor of twenty lower, as estimated from the relative emission intensities (82). A possible reason for the increase in the non-radiative rate in the cis, compared to the trans, cation may be the different symmetry thus relaxing the selection rules for internal conversion for example. One of the measurements as yet outstanding are the breakdown diagrams for the Ã states of these radical cations which would indicate if there are differences in the quantum yields to fragmentation for the two isomers.

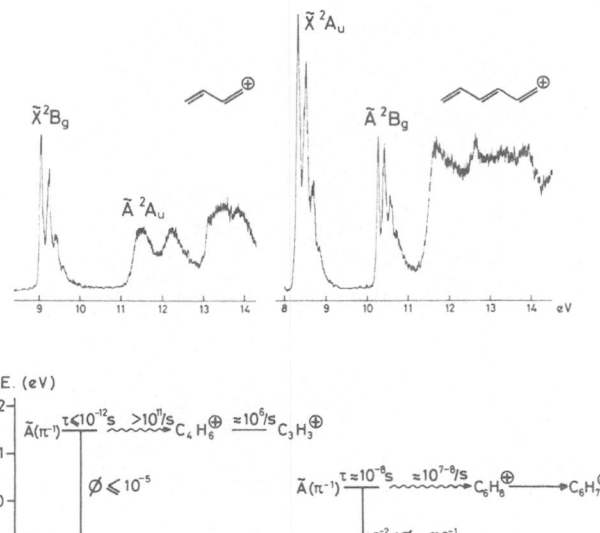

Figure 3. Summary of the energy and decay rate data for the radical cations of butadiene and trans 1,3,5-hexatriene deduced from their photoelectron spectra (top) (96) and from the radiative relaxation measurements (82) (see text).

As a second example the decay processes of 2,4-hexadiyne radical cation, a $C_6H_6^+$ isomer, prepared in its $\tilde{A}(\pi^{-1})$ state are considered. In figure 4 is shown a high resolution photoelectron spectrum of 2,4-hexadiyne which provides the energy data (and vibrational frequencies) for the processes leading to the ground and first excited states of its cation, designated as ^2E_g and ^2E_u respectively (with assumed D_{3d} symmetry). The photoionisation appearance potentials of $C_6H_5^+$ and $C_6H_4^+$ fragment ions (97) (marked in fig. 4) show that the internal energy of 2,4-hexadiyne radical cation in its $\tilde{A}^2E_u$ state is sufficient for fragmentation pathways to be accessible. The radiative pathway is manifested by the $\tilde{A} \rightarrow \tilde{X}$ emission band system, shown in figure 4, which has been recorded with an optical resolution of 0.16 nm (fwhm) (81). Thus, the lifetime of these cations in the $\tilde{A}$ state can be directly measured and is found to be 24 ± 2 ns, which corresponds to the total rate of 4.2×10^7 s^{-1} for the de-

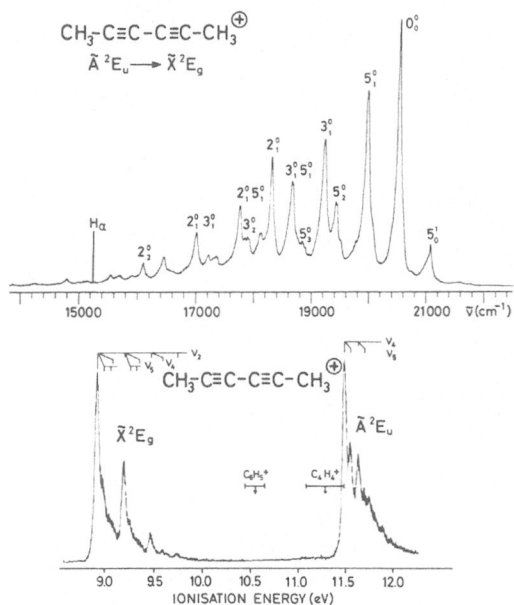

Figure 4. Emission spectrum of 2,4-hexadiyne radical cation (top) recorded with 0.16 nm resolution and
the He(Iα) excited photoelectron spectrum (bottom) recorded with a constant half-width band
pass of 25 meV. The suggested vibrational assignments of the bands to totally symmetric
modes are indicated.

pletion of this state. As fragmentation is a decay channel, the photoelectron-photoion coincidence technique
can be used as a further tool, and this yields the breakdown diagram (fig. 5) of the cationic states populated
by He(Iα) photoionisation (54). It can be seen that the
branching ratio of $C_6H_6^+$ ions, in the energy interval
where the Ã state is populated, has a constant value of
0.60 with a standard deviation of 0.05. This yields, however, only an upper limit for the quantum yield of emission as the structure of the $C_6H_6^+$ ions detected in coincidence is not specified and these can thus be isomers or
vibrationally excited 2,4-hexadiyne cations in their
ground state. Related measurements with 2,4-hexadiyne-d_6
have also been carried out (98). In the following summary, the rates of the radiative and non-radiative processes have been evaluated by assuming that the quantum
yield of emission is given by the branching ratio (0.60)
for the $C_6H_6^+$ ions for the Ã state. A consistent

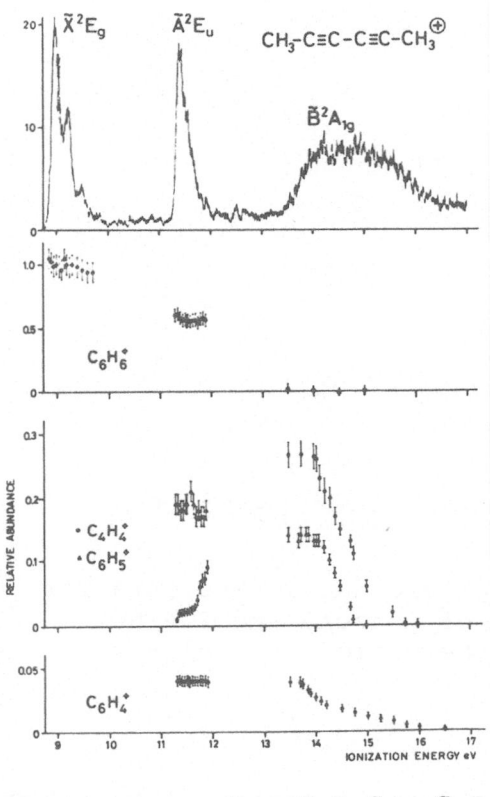

Figure 5. Breakdown diagram for 2,4-hexadiyne radical cation and the resulting fragment ions corresponding to the states generated on photoionisation (cf. top). Redrawn from ref. (54).

$$R - C \equiv C - C \equiv C - R^+ \quad \tilde{A}^2E_u$$

	B.R	τ (ns)	$k_r(s^{-1})$	$k_{nr}(s^{-1})$
R = CH_3	0.60	24±2	2.5×10^7	1.7×10^7
R = CD_3	0.81	32±3	2.5×10^7	0.6×10^7

picture is obtained with the data on the perdeutero derivative. The rate of formation of the $C_6H_5^+$ fragments has also been estimated from the time-of-flight peak shape (54), indicating that the slowest step proceeds at a rate $>10^7$ s^{-1}. Taking in consideration the measured lifetime of the $\tilde{A}$ state and the branching ratio for fragmentation (0.4), the formation rate of $C_6H_5^+$ must be between 1.0×10^7 s^{-1} and 1.7×10^7 s^{-1}. The overall picture is presented schematically in figure 6 where the energy data for benzene radical cation in its lowest electronic states are also summarized. For the latter cation the radiative channel could not be detected (83,99) which, as well as the two photon laser photodissociation experiments (100) indicate, that the ions formed initially in the $\tilde{B}^2A_{2u}$ state decay non-radiatively ($\geqslant 10^{12}$ s^{-1}) to the $\tilde{X}^2E_{1g}$ or to/via the $\tilde{A}^2E_{2g}$ state.

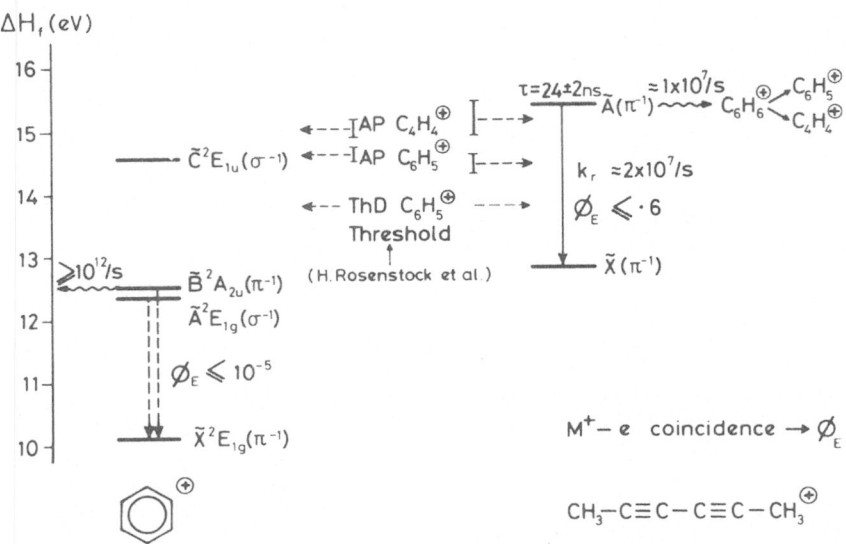

Figure 6. Summary of the energy and decay rate data for the radical cations of benzene and 2,4-hexadiyne deduced from the various techniques applied (see text).

The case of 2,4-hexadiyne radical cation illustrates the sort of detailed studies that can be undertaken when the ancillary data forthcoming from different techniques are combined to provide a quantitative insight into the decay mechanisms. It should be noted, however, that many aspects remain unanswered. These concern, for example, the structures of the outgoing fragments as well as of the precursor ion which yields them. In the example given above, it has been inferred from photoionisation (97) and mass-spectrometric studies (101) of benzene and 2,4-hexadiyne (and some other $C_6H_6^+$ isomers) that the $C_6H_5^+$ and $C_4H_4^+$ fragments, respectively, have the same structure in each case. For the former a phenyl structure is favoured. These observations can be taken to indicate that on formation of 2,4-hexadiyne cation in its $\tilde{A}$ state, the non-radiative process competing with the radiative channel is the internal conversion, which leaves the cation with enough internal vibrational energy ($\approx 2\frac{1}{2}$ eV) that many isomeric forms are accessible. When the internal energy is pooled into specific vibrational modes, fragmentation can take place from different isomeric structures, e.g. elimination of a hydrogen atom to yield the phenyl cation.

CONCLUSIONS

This chapter has tried to provide an overview of the methods now available, and the sort of information one can hope to obtain, in the study of the decay processes of electronically excited radical cations of polyatomics, with emphasis on their lowest states. Many of the techniques, especially those relying on photon detection, have been developed recently. With the discovery of the emission spectra of a variety of organic radical cations, many of which having been listed in this survey, the importance of these methods will grow.

As far as ion molecule reactions are concerned, the importance of excited states (electronic and vibrational) of doublet states of polyatomic radical cations can now be more closely considered. Such data are forthcoming from photoelectron spectroscopy, and in cases where emission spectra have been observed, vibrational frequencies of the totally symmetric modes in the ground state of these cations are also becoming available. In ion-molecule luminescence studies the radiative relaxation of the lowest excited electronic states of polyatomic organic radical cations may now also be used as a diagnostic tool of internal energy partitioning.

ACKNOWLEDGEMENTS

The studies described which have been carried out in the author's laboratory have been financed by the Schweizerischer Nationalfonds zur Förderung der wissenschaftlichen Forschung (Project No. 2.759-0.77, E13).

The author would like to take this opportunity of expressing his appreciation to Prof. E. Heilbronner for all his constructive comments and critical reading of our manuscripts.

REFERENCES

(1) D. Brandt, Ch. Ottinger and J. Simonis, Ber. Bunsen Gesellschaft, 77, 648 (1973),

(2) G.H. Bearman, H.H. Harris and J.J. Leventhal, Appl. Physics Letters, 28, 345 (1976); M. Gérard, T.R. Govers and R. Marx, Chem. Phys., in press.

(3) D.W. Turner, C. Baker, A.D. Baker and C.R. Brundle, "Molecular Photoelectron Spectroscopy" (Wiley-Inter-science, London, 1970).

(4) See for example "Electron Spectroscopy: Theory, Techniques and Applications" eds. C.R. Brundle and A.D. Baker (Academic Press 1977) and references therein.

(5) G. Bieri, F. Burger, E. Heilbronner and J.P. Maier, Helv. Chim. Acta, 60, 2213 (1977).

(6) H.M. Rosenstock, K. Draxl, B.W. Steiner and J.T. Herron, J. of Phys. and Chem. Ref. Data, 6, Suppl. 1 (1977).

(7) E. Heilbronner and J.P. Maier, Chpt. 5 of ref. 4.

(8) P. Carsky and R. Zahradnik, Topics in Current Chemistry, 43, 1 (1973).

(9) R.C. Dunbar, Chpt. in this book.

(10) T. Shida and S. Iwata, J. Amer. Chem. Soc., 95, 3473 (1973).

(11) L.S. Cederbaum and W. von Niessen, J. Chem. Phys., 62, 3824 (1975); W. Meyer in "Modern theoretical chemistry", ed. F. Schaefer III, to be published, and references therein.

(12) W. von Niessen, G.H.F. Diercksen and L.S. Cederbaum, J. Chem. Phys., 67, 4124 (1977).

(13) See for example, W.A. Chupka in "Chemical spectros-copy and photochemistry in the vacuum ultraviolet", eds. C. Sandorfy, P. Ausloos and M.R. Robin (Reidel, Dordrecht 1974) p. 433.

(14) L.S. Cederbaum, W. Domcke, J. Schirmer, W. von Niessen, G.H.F. Diercksen and W.P. Kraemer, J. Chem. Phys., 1978 in press.

(15) B. Brehm and E.v. Puttkamer, Z. Naturforschg. 22a, 8 (1967).

(16) J.H.D. Eland "Photoelectron Spectroscopy" (Butter-worths, London 1974) and references therein.

(17) T. Baer, W.B. Peatman and E.W. Schag, Chem. Phys. Letters, 4, 243 (1969).

(18) C. Backx and M.J. Van der Wiel, J. Phys. B: Atom. Molec. Phys., 8, 3020 (1975).

(19) C.E. Brion, Radiation Research, 64, 37 (1975).
(20) J.H.D. Eland, Int. J. Mass Spectrom. Ion Phys., 8, 143 (1972).
(21) R. Frey, B. Gotchev, O.F. Kalman, W.B. Peatman, H. Pollak and E.W. Schlag, Chem. Phys. Letters, 51, 406 (1977).
(22) B. Brehm, J.H.D. Eland, R. Frey and A. Küstler, Int. J. Mass Spectrom. Ion Phys., 12, 213 (1973).
(23) J.H.D. Eland, Int. J. Mass Spectrom. Ion Phys., 12, 389 (1973).
(24) B. Brehm, R. Frey, A. Küstler and J.H.D. Eland, Int. J. Mass Spectrom. Ion Phys., 13, 251 (1974).
(25) K.H. Tan, C.E. Brion, Ph.E. van der Leeuw and M.J. van der Wiel, Chem. Phys., 29, 299 (1978).
(26) J.H.D. Eland, Chem. Phys., 11, 41 (1975).
(27) P.M. Guyon, I. Nenner, T. Baer, A. Tabché-Fouhailé, T. Govers, L.F.A. Ferreira and R. Botter, to be published.
(28) B. Brehm, J.H.D. Eland, R. Frey and A. Küstler, Int. J. Mass Spectrom. Ion Phys., 12, 197 (1973).
(29) M.J. Weiss, T.-C. Hsieh and G.G. Meisels, J. Chem. Phys., 1978/79 in press.
(30) E.v. Puttkamer, Z. Naturforsch., 25a, 1062 (1970).
(31) B. Brehm and E.v. Puttkamer, "Advances in Mass Spectrometry" Vol. IV, ed. E. Kendrick (Institute of Petroleum 1968) p. 591.
(32) G.R. Wight, M.J. van der Wiel and C.E. Brion, J. Phys. B: Atom. Mol. Phys., 10, 1863 (1967).
(33) R. Stockbauer, J. Chem. Phys., 58, 3800 (1973).
(34) R. Stockbauer, Int. J. Mass Spectrom. Ion Phys., 25, 401 (1977).
(35) J.H.D. Eland, R. Frey, A. Küstler, H. Schulte and B. Brehm, Int. J. Mass Spectrom. Ion Phys., 22, 155 (1976).
(36) B.P. Tsai, T. Baer, A.S. Werner and S.F. Lin, J. Phys. Chem., 79, 570 (1975).
(37) D.M. Mintz and T. Baer, J. Chem. Phys., 65, 2407 (1976).
(38) C.J. Danby and J.H.D. Eland, Int. J. Mass Spectrom. Ion Phys., 8, 153 (1972).
(39) R. Stockbauer and M.G. Inghram, J. Chem. Phys., 62, 4862 (1975).
(40) B. Brehm, V. Fuchs and P. Kebarle, Int. J. Mass Spectrom. Ion Phys., 6, 279 (1971).
(41) J. Dannacher, A. Schmelzer, J.-P. Stadelmann and J. Vogt, Int. J. Mass Spectrom. Ion Phys., 1979 in press.

(42) A.C. Parr, A.J. Jason and R. Stockbauer, Int. J. Mass Spectrom. Ion Phys., 26, 23 (1978).

(43) J. Dannacher and J. Vogt, Helv. Chim. Acta, 61, 361 (1978).

(44) B.P. Tsai, A.S. Werner and T. Baer, J. Chem. Phys., 63, 4384 (1975).

(45) I.G. Simm, C.J. Danby and J.H.D. Eland, Int. J. Mass Spectrom. Ion Phys., 14, 285 (1974); I.G. Simm and C.J. Danby, J.C.S. Faraday II, 72, 860 (1976).

(46) C.S.T. Cant, C.J. Danby and J.H.D. Eland, J.C.S. Faraday II, 71, 1015 (1975).

(47) A.S. Werner and T. Baer, J. Chem. Phys., 62, 2900 (1975).

(48) C.E. Klots, D. Mintz and T. Baer, J. Chem. Phys., 66, 5100 (1977).

(49) R. Stockbauer, Int. J. Mass Spectrom. Ion Phys., 25, 89 (1977).

(50) B. Brehm, J.H.D. Eland, R. Frey and H. Schulte, Int. J. Mass Spectrom. Ion Phys., 21, 373 (1976).

(51) R. Stockbauer and M.G. Inghram, J. Chem. Phys., 65, 4081 (1976).

(52) J.H.D. Eland and H. Schulte; J. Chem. Phys., 62, 3835 (1975).

(53) J.H.D. Eland, R. Frey, H. Schulte and B. Brehm, Int. J. Mass Spectrom. Ion Phys., 21, 209 (1976).

(54) J. Dannacher, Chem. Phys., 29, 339 (1978).

(55) T. Baer, B.P. Tsai, D. Smith and P.T. Murray, J. Chem. Phys., 64, 2460 (1976).

(56) G. Herzberg, Quart. Rev. Chem. Soc., 25, 201 (1971); S. Leach in "The Spectroscopy of the Excited State", NATO Advanced Study Institute, (Plenum Press, 1976) for reviews of the subject and individual references.

(57) M. Allan and J.P. Maier, Chem. Phys. Letters, 41, 231 (1976).

(58) K.T. Wu and A.J. Yencha, Can. J. Phys., 55, 767 (1977).

(59) J.H. Callomon, Can. J. Phys., 34, 1046 (1956).

(60) J. Daintith, R. Dinsdale, J.P. Maier, D.A. Sweigart and D.W. Turner, "Molecular Spectroscopy" 1971, (Institute of Petroleum, 1972) p. 16.

(61) M. Allan, E. Kloster-Jensen and J.P. Maier, J.C.S., Faraday II, 73, 1406 (1977).

(62) J.P. Maier, O. Marthaler and F. Thommen, Chem. Phys. Letters, 1978 in press.

(63) J.P. Maier and O. Marthaler, Chem. Phys., 32, 419 (1978).

(64) J.H.D. Eland, M. Devoret and S. Leach, Chem. Phys. Letters, 43, 97 (1976).
(65) A.J. Smith, F.H. Read and R.E. Imhof, J. Phys. B: Atom. Molec. Phys., 8, 2869 (1975).
(66) J.E. Hesser, J. Chem. Phys., 48, 2518 (1968).
(67) P. Erman, J. Brzozowski and B. Sigfridson, Nucl. Instr. and Meth., 110, 471 (1973).
(68) E.W. Schlag, R. Frey, B. Gotchev, W.B. Peatman and H. Pollak, Chem. Phys. Letters, 51, 406 (1977).
(69) M. Bloch and D.W. Turner, Chem. Phys. Letters, 30, 344 (1975).
(70) W.H. Smith, J. Chem. Phys., 51, 3410 (1969).
(71) R. Frey, B. Gotchev, W.B. Peatman, H. Pollak and E. Schlag, Chem. Phys. Letters, 54, 411 (1978).
(72) E.H. Fink and K.H. Welge, Z. Naturforsch., A23, 358 (1968).
(73) L.S. Curtis and P. Erman, J. Opt. Soc. Am., 67, 1218 (1977).
(74) G.R. Möhlmann, K.K. Bhutani, F.J. de Heer and S. Tsurubuchi, Chem. Phys., 31, 273 (1978).
(75) G.R. Möhlmann and F.J. de Heer, Chem. Phys. Letters, 36, 353 (1975).
(76) M. Allan, E. Kloster-Jensen and J.P. Maier, Chem. Phys., 17, 11 (1976).
(77) M. Allan, E. Kloster-Jensen and J.P. Maier, J.C.S. Faraday II, 73, 1417 (1977).
(78) J.P. Maier, O. Marthaler and E. Kloster-Jensen, to be published.
(79) M. Allan, J.P. Maier, O. Marthaler and J.-P. Stadelmann, J. Chem. Phys., 1979 in press.
(80) M. Allan, E. Kloster-Jensen, J.P. Maier and O. Marthaler, J. Electron Spectr., 14, (1978) in press.
(81) M. Allan, J.P. Maier, O. Marthaler and E. Kloster-Jensen, Chem. Phys., 29, 331 (1978).
(82) M. Allan and J.P. Maier, Chem. Phys. Letters, 43, 94 (1976).
(83) M. Allan, J.P. Maier and O. Marthaler, Chem. Phys., 26, 131 (1977).
(84) R. Lopez-Delgado, A. Tramer and I.H. Munro, Chem. Phys., 5, 72 (1974).
(85) V.E. Bondybey and T.A. Miller, J. Chem. Phys., 67, 1790 (1977); T.E. Miller and V.E. Bondybey, to be published, personal communication.
(86) A. Corney, Adv. Electronics and Electron Phys., 29, 115 (1970).

(87) D.L. Ames, M. Bloch, H.Q. Porter and D.W. Turner,
      "Molecular Spectroscopy" (Institute of Petroleum,
      London 1976) p. 399.
(88) M. Allan, PhD Thesis, University of Basel 1976.
(89) H.J. Haink, E. Heilbronner, V. Hornung and E.
      Kloster-Jensen, Helv. Chim. Acta, 53, 1073 (1970).
(90) J. Jortner, J. Pure and Appl. Chem., 24, 165 (1970).
(91) F. Lahmani, A. Tramer and C. Tric, J. Chem. Phys.,
      60, 4431 (1974).
(92) R.J. Buenker and S.D. Peyerimhoff, Chem. Rev., 74,
      127 (1974)
(93) P. Rosmus, personal communication.
(94) R.C. Dunbar, J. Amer. Chem. Soc., 98, 4671 (1976);
      E.W. Fu and R.C. Dunbar, J. Amer. Chem. Soc., 100,
      2283 (1978).
(95) T. Shida, T. Kato and Y. Nosaka, J. Phys. Chem., 81,
      1095 (1977).
(96) M. Beez, G. Bieri, H. Bock and E. Heilbronner,
      Helv. Chim. Acta, 56, 1028 (1973).
(97) H.M. Rosenstock, K.E. McCulloh and F.P. Lossing,
      Int. J. Mass Spectrom. Ion Phys., 25, 327 (1977).
(98) J. Dannacher, E. Heilbronner, J.-P. Stadelmann and
      J. Vogt, unpublished results.
(99) M. Allan and J.P. Maier, Chem. Phys. Letters, 34,
      442 (1975).
(100) B.S. Freiser and J.L. Beauchamp, Chem. Phys. Letters,
      35, 35 (1975); T.E. Orlowski, B.S. Freiser and
      J.L. Beauchamp, Chem. Phys., 16, 439 (1976).
(101) F. Borchers and K. Levsen, Org. Mass Spectrom. 10,
      584 (1975); R.G. Cooks, J.H. Beynon and J.F. Litton,
      Org. Mass Spectrom., 10, 503 (1975).

ION PHOTODISSOCIATION

Robert C. Dunbar

Chemistry Department
Case Western Reserve University
Cleveland, Ohio 44106

INTRODUCTION

Photodissociation of ions is a new field of interest whose rising popularity perhaps reflects the exceptional satisfaction it has brought to its practitioners. The experimental techniques have moved in a very brief period from being few and heroic to being both diverse and straightforward. As this has happened it has been found that the data available are relevant to progress in a surprisingly wide variety of active areas of interest in ions and their reactions. Rather than duplicating a very recent review of work in the area [Dunbar, 1979], it will be more interesting here (following the presentation at La Baule) to organize some chosen examples of recent work into a framework intended to make clear the different points of view that have been taken in planning and interpreting photodissociation experiments. For a more complete bibliography (and for a less onesided stress on results from our own laboratory) the review cited above can complement the present presentation.

METHODS

Serious ion photodissociation experiments date from the work of Dunn [1964] on H_2^+, but at that time neither the technique nor the physical and chemical background understanding of gas-phase ions were sufficiently developed to make the further pursuit of these experiments very attractive. A fresh start came with the application of ICR techniques [Dunbar, 1971], and the number of active laboratories and different experimental approaches has grown rapidly.

PULSED ICR SEQUENCE

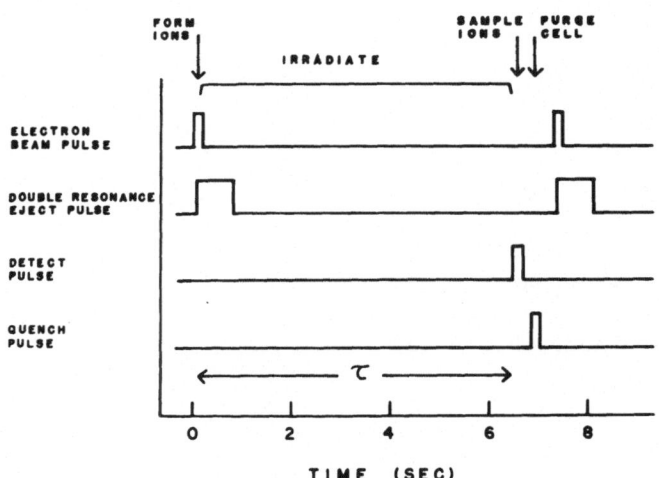

Fig. 1. Typical pulse sequence for ion photodissociation in a
trapped-ion ICR spectrometer.

ICR methods, in which the ICR cell is used as an ion trap to
confine the ions during photodissociation, and ICR detection methods
are used to monitor the occurrence of photodissociation and photochemical reactions, are in widespread use. Groups at CWRU
(ourselves), Stanford (Brauman et al.), Cal Tech (Beauchamp et al.),
Florida (Eyler et al.), Purdue (Freiser et al.), and Leiden (Van
der Hart et al.) are all active. Most work now is done in a pulsed
mode of operation, and Fig. 1 shows a typical sequence of pulse
events. The trap is initially filled with ions by a pulse of the
electron beam. A double resonance pulse may be applied briefly (or
continuously) to remove an unwanted species from the trap, or to
identify a reaction sequence. The cell is illuminated by a suitable
light source during a period which is typically several seconds,
following which the contents of the cell are analyzed by pulsed ICR
detection. Finally a pulse sweeps the cell clean ready for a new
cycle.

On our instrument at CWRU, the data on abundance of a particular
ion with light on and light off are acquired, averaged, and compared
on a microprocessor, normalization is applied for variation of light
intensity with wavelength, and a photodissociation (or photoproduction)
spectrum as a function of wavelength is automatically produced and

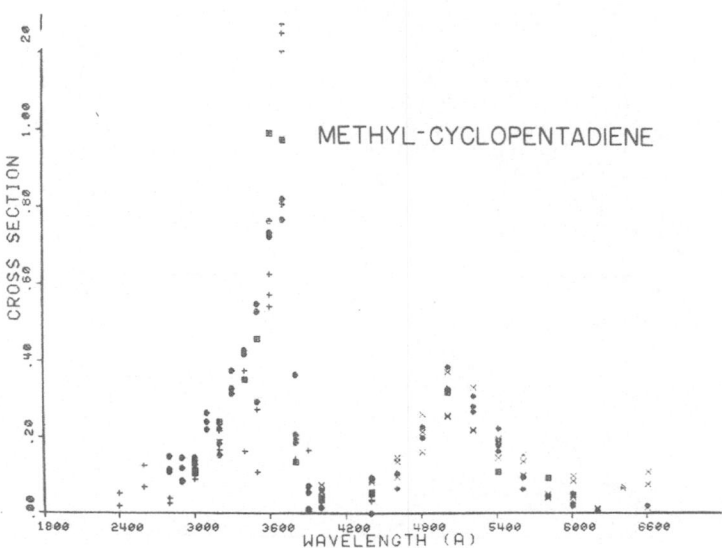

Fig. 2.   Computer-generated photodissociation spectrum of methyl-
cyclopentadiene parent ion. . (Points from three different
experimental runs.)

plotted. Such a spectrum is displayed in Fig. 2. As will be
clear below, both arc-lamp and laser light sources have been used,
each having its own particular advantages.

A small amount of work using rf quadrupole ion traps (largely
by Jefferts at Bell Laboratories) has been reported, but the other
principal approach to observing dissociation has been the intersection of a light beam with a collimated ion beam. Usually the
convenient beam characteristics and easy focusing of a laser source
are used to give high flux in the interaction region. Instruments
at the University of Houston (Vestal et al.) and at La Trobe
University (Morrison et al.) use quadrupole mass filters for ion
beam containment, while those at Amsterdam (Los et al.) Hanscomb
Air Force Base (Paulson et al.), Edgewood Arsenal (Vanderhoff et
al.), Southampton University (Carrington et al.) and SRI (Moseley
et al.) use various combinations of magnetic and electric sectors
along with quadrupole and/or time-of-flight mass analysis. An
example of some of the finest ion beam results is shown in the
methyl iodide results of McGilvery et al. [1977] in Fig. 3.

An exciting recent development is the Doppler-tuned
spectroscopy technique developed by Carrington et al. [1976]. A
highly monochromatic, fixed wavelength laser beam is directed

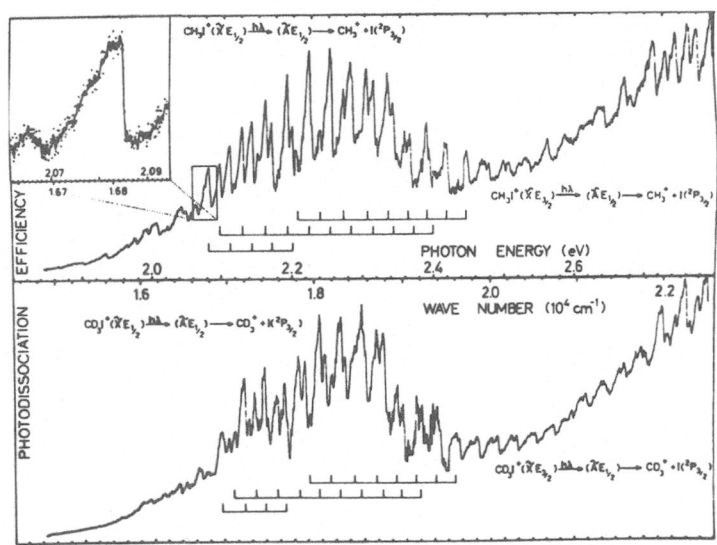

Fig. 3. High resolution photodissociation spectra of methyl
iodide ions. [McGilvery and Morrison, 1977]. (Reproduced, by
permission, from J. Chem. Phys. 67, 368 (1977)).

collinear with a fast ion beam along the ion flight direction, and
the light is brought into resonance with various ion spectroscopic
transitions by varying the ion velocity. Doppler shifting of the
wavelength in the moving ion reference frame gives tuning across
several cm^{-1}. Optical resolutions of the order of 50 MHz have
been reported, and this technique may rival the highest resolution
optical spectra available for any molecules by any method.

HIGH RESOLUTION SPECTRA

Photodissociation spectroscopy relies on the truism that
before an ion can dissociate, it must absorb a photon. Thus, the
photodisappearance spectrum of a given ion can be expected to show
peaks at wavelengths corresponding to optical absorption peaks.
(Obviously, not every photon absorption need lead to dissociation,
so the photodissociation spectrum need not duplicate the optical
absorption spectrum, but it provides a lower limit to the optical
absorption at a given wavelength.) Since the photodissociation
spectrum is usually the only approach one has to obtaining optical
spectroscopic information about gas-phase ions, these spectra have
been extensively measured and interpreted in terms of optical
transitions in the ions.

"High resolution" means various things to various investigators,
but may usefully be taken here as defining work in which vibrational

structure is (or could have been) resolved. Laser sources are
nearly essential to obtain both sufficient power and monochromaticity.
Probably the first reported high resolution photodissociation
spectrum reported was that of hexatriene ion, shown in Fig. 4. Two
vibrational progressions can be distinguished, corresponding to
normal modes at 350 cm^{-1} and 1200 cm^{-1}, which can be accounted for
as, respectively, a symmetric skeletal bend, and a symmetric
skeletal stretch. No difference was seen between spectra of <u>cis</u>
and <u>trans</u> isomers, suggesting either very similar spectra, or
interconversion of the two isomeric ions. (Although the emission
spectra of the two ions are distinguishable [Maier, 1978], they are
not very different.)

The laser spectrum of N$_2$O$^+$ in the 300 nm region obtained by
Thomas, Dale and Paulson [1977] was the first reported high
resolution spectrum in the ultraviolet. Many vibrational peaks
appear, and the spectrum is in good agreement with a very careful
theoretical analysis. The group of Moseley at SRI have obtained
laser spectra of a large number of other ions of atmospheric
interest, using a drift-tube instrument [Smith <u>et al</u>., 1978, for
example]. One surprising result of this work has been the
discovery that many weakly-bound cluster ions have strong photodissociation peaks in the visible. Also interesting is the
discovery of many negative ions (of which CO$_3^-$ was an early example
having a beautifully structured high resolution spectrum [Moseley
<u>et al</u>., 1976]) which dissociate on photon absorption, either in
addition to, or instead of, the expected electron photodetachment
process.

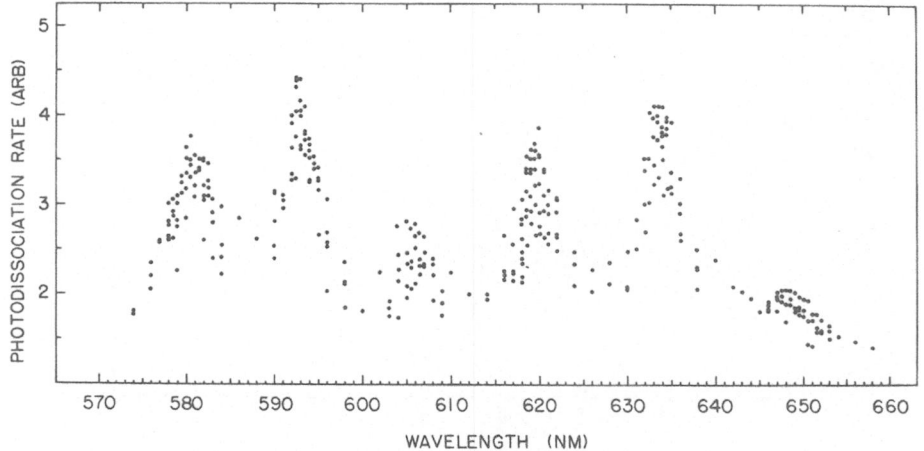

Fig. 4. Photodissociation spectrum of hexatriene ions at 10 Å
resolution showing resolved vibrational progressions [Dunbar and
Teng, 1978] (Reproduced by permission, from J.Am.Chem.Soc. <u>100</u>,
2279 (1978)).

LOW-RESOLUTION SPECTRA AND ION CHROMOPHORES

A great deal of photodissociation spectroscopy performed at a resolution of the order of 100 Å has been guided by two motivations: one is the opportunity to locate, and begin to account for in a systematic way, the many excited ion states newly accessible with this technique. The other is the expectation that the spectra will be related in a regular and useful way to various identifiable chromophoric groups in the ions, so that, just as with neutrals, the spectroscopy of ions will be systematized by grouping various ions and their spectra according to the chromophores involved.

The exploration of ion excited states has been helped immensely by the results of photoelectron spectroscopy. Photoelectron spectra give peaks corresponding to "hole-promotion" excited states of ions, and it has been fruitfully recognized [Dymerski et al., 1974] that when these states are also optically allowed from the ion ground state, direct comparison of the two types of spectra can be made.

An illustration of such a comparison is shown in Fig. 5 for iodobenzene ion, [Dunbar, et al., 1979], where the photodissociation

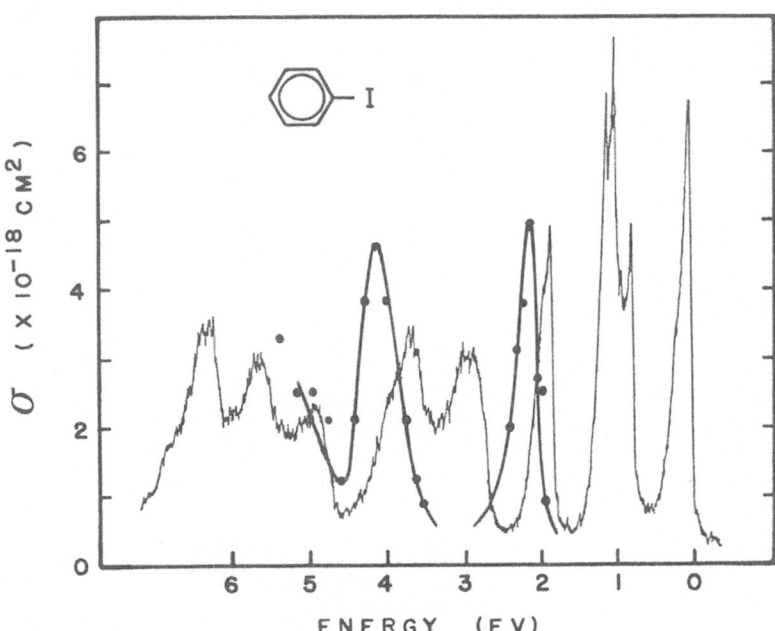

Fig. 5. Photodissociation spectrum of iodobenzene ion, superimposed on the photoelectron spectrum of iodobenzene.

and PES spectra are superimposed with appropriate adjustment of
the PES energy scale. This particular spectrum is interesting in
several ways. The photodissociation peak near 2.2 eV is in the
two-photon region (see below), and it is notable that so a good
spectrum could be obtained with an arc lamp source. This peak
corresponds to an allowed hole promotion to the out-of-plane
iodine lone-pair orbital, and is directly comparable to the sharp
peak at 2 eV in the PES spectrum corresponding to the same hole-
promotion excited state. Obviously the photodissociation peak is
broadened and shifted to higher energy, indicating that the
Franck-Condon overlap is much less favourable in the optical
transition, so that substantial vibrational excitation occurs.
The photodissociation peak near 4 eV is almost certainly a $\pi \rightarrow \pi$
ring type of hole promotion excited state, which is rather
difficult to pick out from the (optically unobservable) sigma
orbitals seen in this region of the PES spectrum. The rise in
the photodissociation spectrum above 5 eV is probably the first
$\pi \rightarrow \pi^*$ electron promotion transition, not expected to be observed
in the PES spectrum. This example may suggest the insights, only
just beginning to be explored, which result from comparing these
two types of spectra for a given radical cation.

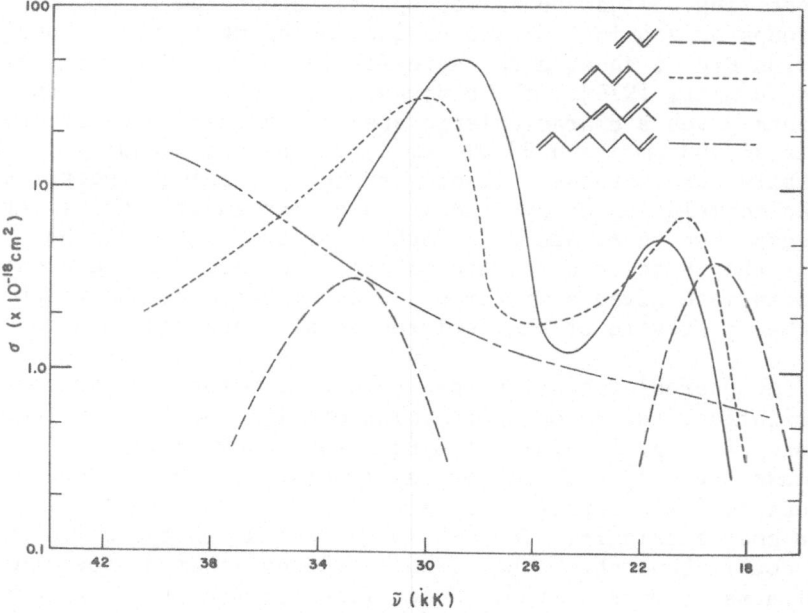

Fig. 6. Photodissociation spectra of three conjugated and one unconjugated diene cations. (Reproduced, by permission, from Anal.
Chem. 48, 723 (1976)).

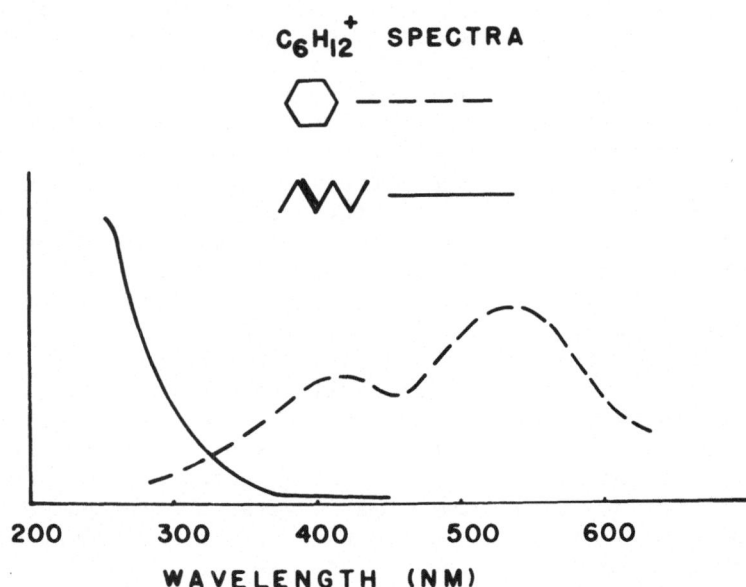

Fig. 7. Photodissociation spectra of two isomeric $C_6H_{12}^+$ ions.

Organizing photodissociation spectra according to chromophoric groups is a point of view of increasing utility. As one
illustration Fig. 6 shows a characterization of diene-containing
molecules [Dunbar, 1976]. The conjugated diene chromophore in
radical ions gives a characteristic spectral pattern with strong
peaks near 20,000 cm^{-1} and 30,000 cm^{-1}, the UV peak usually being
substantially more intense. (There are not yet enough spectra to
permit the formulation of systematic rules for substituent shifts
in the peaks, but there seems no doubt that such rules can be
derived as they have been for neutrals.) The unconjugated diene
ion 1,7-octadiene gives a different spectral pattern, one which is
in fact characteristic of the isolated double-bond chromophore.

As ion photodissociation spectra are used for ion structure
characterization, the chromophore point of view should be valuable,
as suggested in Fig. 7. The two $C_6H_{12}^+$ ions shown give very
similar mass spectra, and are not easily distinguished. However,
the photodissociation spectra are at the same time totally distinct
and quite characteristic. The 2-hexene spectrum is typical of the
isolated double-bond chromophore, while the cyclohexane spectrum is
a typical alkane ion spectrum. Note, finally, the striking fact
[Benz et al. 1979] that alkane radical ions are strongly coloured.

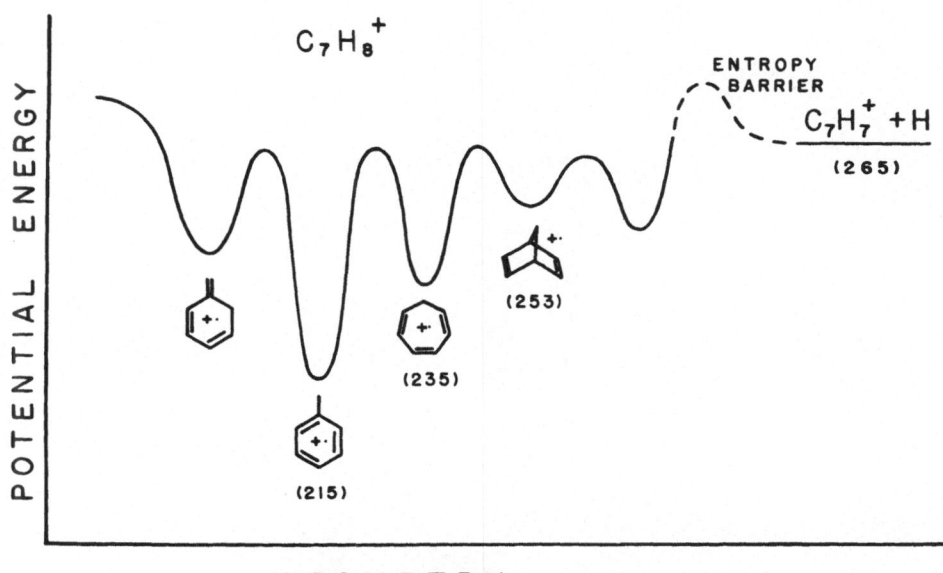

Fig. 8. Schematic $C_7H_8^{+\cdot}$ potential surface. The horizontal axis represents collectively the internal coordinates of the molecule.

ION STRUCTURES

Photodissociation spectra provide an excellent ion structural tool: the ion species to be examined can be prepared in any desired manner – by electron impact or photoionization, by fragmentation, by ion-molecule reaction – and thermalized by collisions or by radiation, following which a spectrum is obtained which is entirely characteristic of the ion structure. The greatest success of this approach has been in comparing isomeric ions, for while controversy may arise about the structural significance of a particular photo-dissociation spectrum, the observation of dissimilar spectra for ions of the same mass having different provenance gives truly unequivocal evidence for non-identical structures.

$C_7H_8^+$ ions provided an early demonstration of these ideas [Dunbar and Fu, 1973], looking at isomers obtained from three different parent neutrals (to which a fourth isomer, resulting from a McLafferty rearrangement fragmentation, was later added [Dunbar Klein, 1976]). Contrary to some prior notions, the non-dissociating $C_7H_8^{+\cdot}$ ions are not structurally labile, and do not interconvert. These conclusions are summarized in Fig. 8, showing a very schematic $C_7H_8^{+\cdot}$ potential surface, along with known energy values. The various isomers have stable potential wells; however, when any of

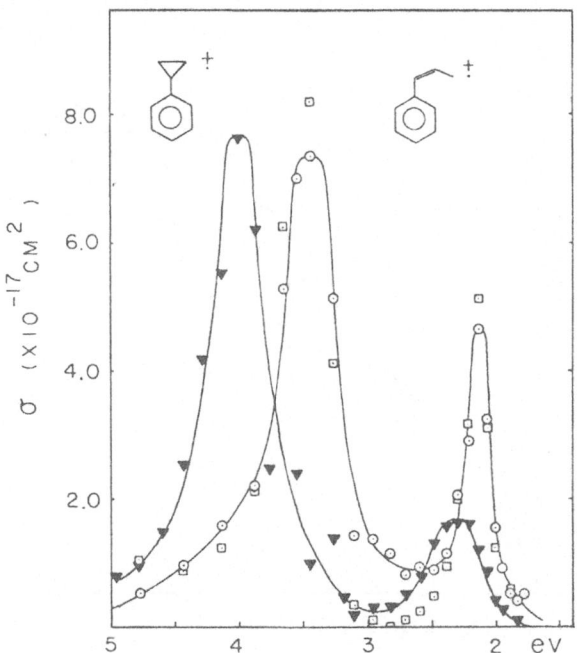

Fig. 9. Spectra of $C_6H_5C_3H_5^{+\cdot}$ ions from cyclopropylbenzene ( ),
1-phenylpropene ( ) and 3-phenylpropene ( ) (the latter not to
scale). (Reproduced, by permission, from J.Am.Chem. Soc. 100,
5949 (1978)).

the ions is raised (as in a mass spectral cracking pattern study)
above the dissociation threshold at 265 kcal, then extensive
traversal of the potential surface may occur before the entropy
barrier (dashed part of curve) leading to $C_7H_7^+$ + H· is crossed.
This illustrates the important point that observation of identical
mass spectra for the various isomeric neutrals does not in any
way preclude the existence of stable isomeric ion structures.

An interesting recent set of isomeric ions is the $C_6H_5C_3H_5^+$
ions [Fu and Dunbar, 1978], shown in Fig. 9. The double bond in
3-phenylpropene obviously migrates into conjugation upon
ionization, so that the spectrum obtained is identical with that of
1-phenylpropene ion. Curiously enough, however, the cyclopropylbenzene structure does not rearrange, and gives a quite distinct
spectrum.

Styrene, cyclooctatetraene and barrelene $C_8H_8^{+\cdot}$ parent ions all
give distinct spectra [Dunbar, Kim and Olah, 1979], but the
protonated molecules, $C_8H_9^+$, give identical spectra, as shown in
Fig. 10. The spectrum is almost certainly that of styryl cation,

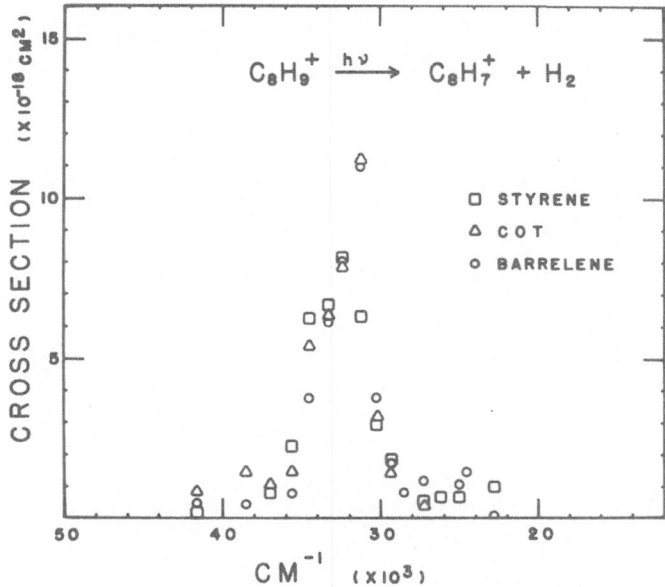

Fig. 10. Spectra of $C_8H_9^+$ ions formed by H_3O^+ protonation of isomeric neutrals. (Reproduced, by permission, from J.Am.Chem. Soc., in press).

so that on protonation these molecules rearrange to the stable styryl structure.

The added information afforded by resolved vibrational structure in laser photodissociation spectra promises to be an important aid in ion structure characterization. Illustrating this, spectra of three isomeric $C_3H_5Cl^{+\cdot}$ ions are shown in Fig.11, all having clear and characteristic vibrational detail [Orth and Dunbar, 1979]. With these spectra of authentic ions in hand, structure identification was undertaken on the products of the ion–molecule reactions

$$n - C_3H_7Cl + C_3H_6^{+\cdot} \rightarrow C_3H_5Cl^{+\cdot} + C_3H_8 \tag{1}$$

$$i - C_3H_7Cl + C_3H_6^{+\cdot} \rightarrow C_3H_5Cl^{+\cdot} + C_3H_8 . \tag{2}$$

The product ion of Reaction (1) gave a spectrum matching that of 1-chloropropene ion (cis-trans mixture), while the product ion of Reaction (2) matched the 2-chloropropene ion spectrum. Thus the skeleton of the neutral chloropropane is retained in the reaction,

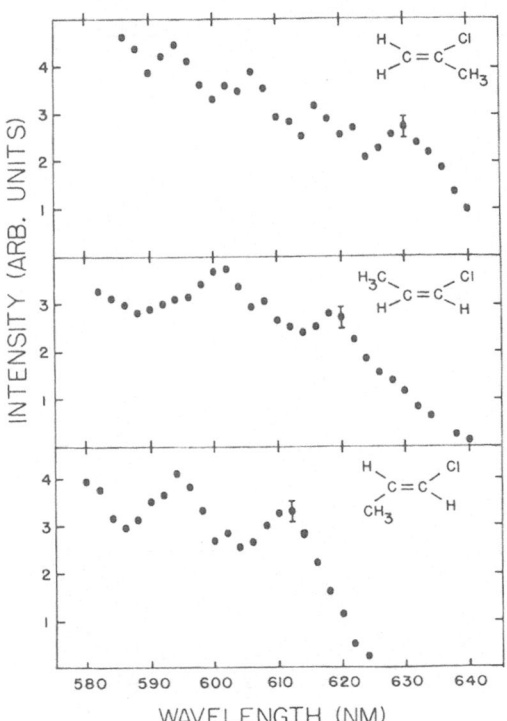

Fig. 11. Laser photodissociation spectra of three isomeric $C_3H_5Cl^{+\cdot}$ ions. (Reproduced by permission, from J.Chem.Soc. <u>100</u>, 2283 (1978)).

which probably proceeds by formal H_2^- abstraction. This subtle structural distinction of reaction products was unambiguous using the laser spectra, but would have been impossible based on low-resolution spectra.

COMPETITIVE DISSOCIATION

Just as with ion fragmentation induced by any other source of excitation energy, competitive photodissociation into more than one product species can be expected, and the very precise control which photon absorption gives over both the amount of energy and the location of its deposition makes this a particularly attractive method for studying such competitions. In fact, however, instances of formation of more than one photoproduct have been few - this reflects the fact that photoexcitation excites the ion only a few eV above the ground state, so that dissociation into any but the most stable product species is, if not energetically impossible, at least statistically unfavourable. One instance in which a

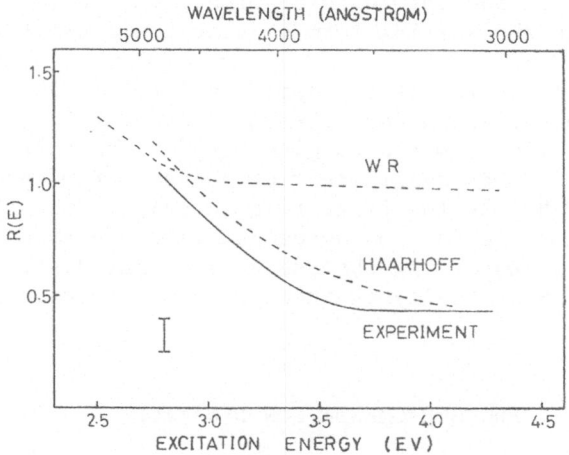

Fig. 12. Comparison of observed ratio R(E) of $C_4H_7^+$ to $C_3H_5^+$ from $C_4H_8^+$ photodissociation with RRKM theory in two approximations. (Reproduced, by permission, from J. Chem. Phys. <u>65</u>, 3365 (1976)).

competition has been studied is the butene ion [Riggin <u>et al.</u>, 1976], which dissociates by the routes

$$C_4H_8^{+\cdot} \xrightarrow{h\nu} \begin{array}{l} C_4H_7^+ + H\cdot \quad (2.17 \text{ eV}) \quad (3) \\ C_3H_5^+ + CH_3\cdot \quad (2.22 \text{ eV}) \quad (4) \end{array}$$

The close similarity in activation energy for the two dissociation channels is favourable to competitive dissociation, and both channels are observed, with comparable abundance at visible and UV wavelengths. If the competition is governed by statistical factors, and if the excitation energy degrades to vibrational excitation, then the assumptions and calculations of RRKM theory may be applied to predicting the ratio of the two products as a function of photon energy. The results of such calculations are shown in Fig. 12. (The two curves marked WR and HAARHOFF use two different approximations to the calculation of state densities, and their difference reflects the fact that such calculations are rather uncertain.) It is clear that the experimental curve is in satisfactory agreement with the predictions, and in particular shows the expected trend at low energy. This is not taken as proving that the particular calculations or parameters used are correct: rather, the fact that good agreement can be obtained with RRKM calculations is a good indication that the competition is indeed statistically governed, and that the photon energy does

become available in the form of vibrational excitation, rather
than remaining in the initial form of electronic excitation.

A further indication of the applicability of RRKM assumptions
comes from the observation that the $C_3H_5^+$ product ion from
Reaction (4) itself fragments to give $C_3H_3^+$. The extent of this
further process depends sensitively on the amount of energy
partitioned to $C_3H_5^+$ in the first fragmentation; the observed
formation of $C_3H_3^+$ is in good agreement with the RRKM
prediction, again supporting the assumption that dissociation
occurs through vibrationally excited ground electronic states of
the ions.

PHOTOFRAGMENTATION DYNAMICS

One would like to characterize photodissociation processes at
a level of detail which would specify, for each initially populated
excited state of the ion, the identities of the fragments, their
angular and energy distributions, and the time dependence of the
fragmentation process. Such a level of information has yet to be

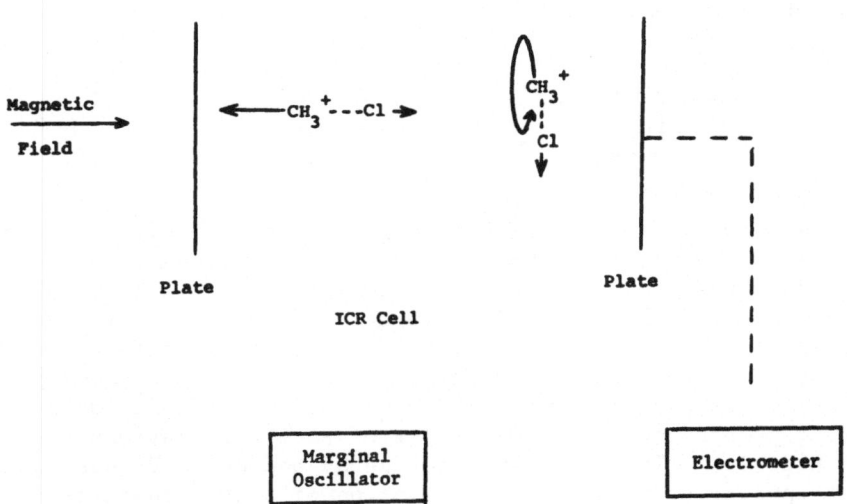

Fig. 13. Determination of the escape of fragment ions (CH_3^+) from
the cell as a function of well depth and light polarization, using
either marginal oscillator or electrometer detection. The two
illustrated CH_3Cl^+ molecules suggest the cell anisotropy, which
traps ions moving perpendicular to the magnetic field more
efficiently than those moving along field lines.

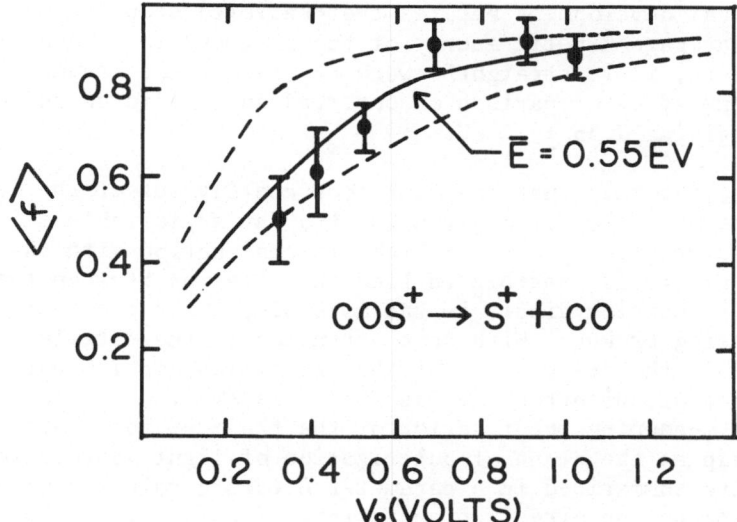

Fig. 14. Fraction <f> of trapped S$^+$ as a function of trapping
well depth. The three curves are theoretical predictions for
three assumed kinetic energies, bracketing the S$^+$ kinetic energy
to .55 $\pm$ .2 eV.

reached for any ion, but techniques are now being used which come
a long way toward providing such a complete picture. One promising
approach uses the anisotropic nature and variable depth of the
ICR ion trap to give access to angular distributions and fragment
energies [Orth et al., 1977]. Fragment ions can escape from the
trap along the magnetic field direction by surmounting the electrostatic trapping potential, which can be varied from zero to
several volts. As suggested, for the example

$$CH_3Cl^{+\cdot} \xrightarrow{h\nu} CH_3^+ + Cl\cdot \tag{5}$$

by Fig. 13, the escape of fragment ions (CH$_3^+$) from the trap can
be measured either by collecting them at the trapping plates and
using electrometer detection, or by following the abundance of
non-escaping ions in the trap with the marginal oscillator detector.
Using either approach, it is possible to determine as a function of
trapping well depth, the fraction of fragment ions escaping from
the trap at a given photodissociation wavelength. Such a plot is
shown as Fig. 14 for the S$^+$ product in the reaction

$$COS^+ \xrightarrow{h\nu} CO + S^+ . \tag{6}$$

Some theoretical development relates the predicted shape of these plots to the average kinetic energy of the fragment ions, and as Fig. 14 suggests, it is straightforward to bracket the fragment ion kinetic energy by comparison of observed and predicted curves, in this case giving 0.55 $\pm$ .2 eV.

Moreover, the fact that fragment ions can fly out of the cell only along the magnetic field gives the trap an exploitable anisotropy; using plane-polarized light in conjunction with the cell anisotropy, it is possible to find the relation between the angle of the transition dipole in the molecule, and the orientation of the bond being broken. With some attention to the detailed theory, not only the orientation of the transition dipole, but also the degree of anisotropy of the photodissociation can be determined by measuring the fraction of the fragment ions which escape the trap as the plane of polarization of light is rotated: the results are summarized in a parameter β taking values from -1 to $+2$ depending on the direction and extent of anisotropy. For Reaction (6), β is found to be near 0, indicating that the COS^+ ion undergoes extensive rotational angle averaging between the time of photon absorption and the time of fragmentation [Orth and Dunbar, 1979]. This is consistent with the hypothesis, suggested on the basis of photoionization coincidence results [Eland, 1973], that the excited state of COS^+ is predissociated by a slow crossing to the $^4\Sigma^-$ state.

On the other hand, methyl chloride ion, Reaction (5), shows a strong angular dependence [Orth and Dunbar, 1978b]. β is near $+2$, indicating a transition dipole along the C–Cl bond, and very little rotational averaging. The dissociation is thus fast compared with a rotational period, suggesting optical excitation to a dissociative excited state, which is probably $\tilde{B}\ ^2E$, based on the transition dipole orientation. The large kinetic energy release (.58 eV) is also in line with this direct-dissociation hypothesis.

The ICR approach can thus yield mean kinetic energies and angular distribution of photodissociation products at selected wavelengths, and these can be used to infer details about the rate and mechanism of the dissociation process. Ion beam experiments have similar capabilities, as has been nicely shown in the study of Ar_2^+ dissociation, for instance, by Moseley et al., [1977].

MULTIPHOTON DISSOCIATION

The discovery of two-photon ion dissociation was made by Freiser and Beauchamp [1975], who observed benzene ion dissociation at   visible wavelengths. From the fact that the dissociation was observed well below the one-photon threshold, and from the I^2

dependence on light intensity, a two photon process was clearly a reasonable assumption, and Freiser and Beauchamp proposed a mechanism which has been widely accepted : Using bromobenzene ion as an example, the Freiser-Beauchamp mechanism can be formulated

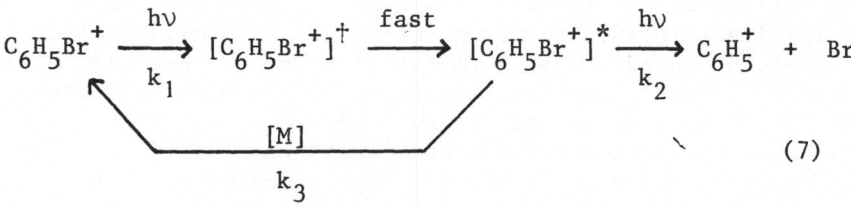

In the first step, absorption of one photon gives the electronically excited species $[C_6H_5Br^+]^\dagger$, which undergoes rapid internal conversion to $[C_6H_5Br^+]^*$, a long-lived vibrationally excited ion. This latter species can be collisionally de-excited back to ground state by collisions with neutral [M], or it can absorb a second photon, carrying it above the dissociation threshold. It is the long-lived $[C_6H_5Br^+]^*$ species which is the unique characteristic of this mechanism: since it can be stable and trapped for times of the order of a second, the absorption of the second photon can be a leisurely process. Accordingly, the mechanism does not require high-intensity light as with most multiphoton chemistry, and in fact two-photon ion dissociations are observed at light fluxes of less than 1 W cm^{-2}

The Freiser-Beauchamp mechanism has been verified by quantitative studies of pressure dependence and light-intensity dependence for bromobenzene [Dunbar and Fu, 1977] and cyanobenzene [Orlowski et al., 1976] as well as benzene. Excellent quantitative agreement is found with the kinetic solution of the scheme (7), which may be approximated at low intensity and high pressure as

$$K = \frac{I^2 \, k_1 \, k_2}{Ik_1 + Ik_2 + [M]k_3} \qquad (8)$$

where K is the unimolecular pseudo-first-order dissociation rate constant. As predicted, the rate is quadratic in I and inverse in [M] at low intensities, falling off toward linear I dependence at high intensities. For bromobenzene, k_1 and k_2 are about 5×10^{-18} cm^2, and k_3 is 1.1×10^{-9} cm^3 molecule^{-1} sec^{-1}, which are very reasonable values.

Naphthalene ion presents an interesting case [Kim and Dunbar, 1979]; it has optical absorptions at 600 nm and at 300 nm. The 300 nm absorption is expected to be a two-photon dissociation, and

does indeed show the expected I^2 intensity dependence. The 600 nm
absorption, however, is expected from known thermochemistry to be
a four-photon transition, and a study of this process has enlarged
our understanding of many-photon dissociations. One might
initially expect to find I^4 intensity dependence, but the observed
dependence is between I^2 and I^3, and the power of I, (designated
LID and defined by the relation $K \propto I^{(LID)}$), decreases with
increasing light intensity. From more detailed consideration of
the four-photon mechansim analogous to the Freiser-Beauchamp
mechanism (7), (with numerical solutions of the most realistic
cases), this behaviour is seen to be reasonable. At light
intensities sufficient for the overall dissociation to be
observable, the ground state becomes highly depleted, giving a
lowering of the observed light intensity dependence. Fig. 15
shows this effect in the calculated curves of LID versus light
intensity, and shows also that the observed naphthalene ion results
correspond to a region of LID substantially less than 4.

The other interesting feature of naphthalene ion photodissociation is that in neither the two-photon nor the four-
photon region is there a strong dependence on pressure of neutral
gas, in marked contrast to previously studied two-photon
dissociations. Apparently relaxation of vibrationally excited ions
by infrared radiation is faster than collisional relaxation, so
that radiative (and thus pressure-independent) relaxation governs

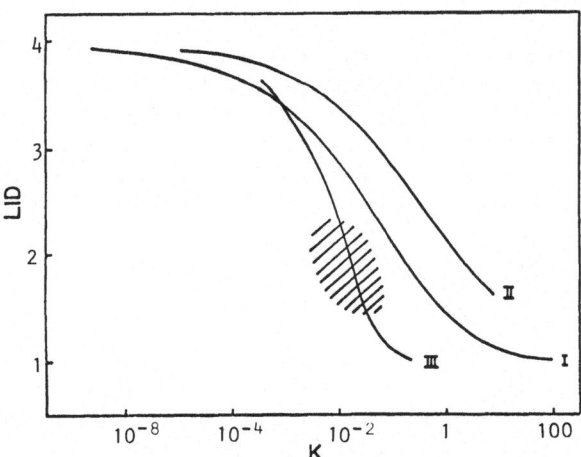

Fig. 15. Light intensity dependence of four-photon dissociation,
showing the strong dependence of the intensity power law exponent
on the rate of photodissociation K (expressed in sec^{-1}). The
three curves are model calculations with increasingly complex and
realistic assumptions, and the cross-hatched area is the range of
experimentally observed points.

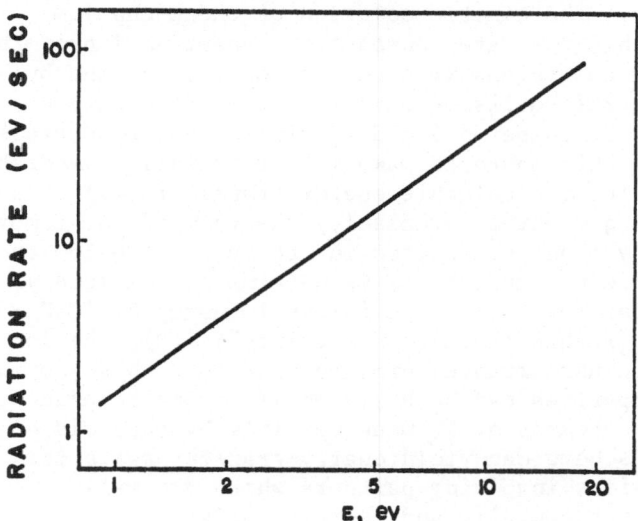

Fig. 16.   Rate of radiative energy dissipation in eV/sec
calculated for naphthalene, as a function of the amount of
vibrational excitation in the molecule.

the k_3 step in Eq. (7).   The rate of radiative cooling of an ion
can be calculated with considerable confidence from known infrared transition probabilities [Dunbar, 1975].   When this is done
for naphthalene, giving the curve shown in Fig. 16, it is found
that, indeed, the rate of radiative cooling of naphthalene
$[C_{10}H_8^+]^*$ ions is rapid (about 6 sec^{-1}) compared with collision
rates in the 10^{-8} torr pressure region.   This appears to be the
first reasonably definite case where infrared radiative cooling
of the ion is implied by experiment and supported by theory.

The Beauchamp group's recent observation of infrared photodissociation by a CO_2 laser [Woodin et al., 1978] completes the
progression from one-photon to many-photon dissociation, since 30
or more photons are required in this case.   Their observations
bear out the trends observed in the naphthalene ion four-photon
case in that the dependence on light intensity is not strong, in
fact approximating first order.

COLLISIONAL QUENCHING OF EXCITED IONS

From the outset [Freiser and Beauchamp, 1975] it was realized
that the two-photon mechanism of Eq.(7) offers a unique opportunity
for studying collisional transfer of internal energy in ion-

molecule collisions. The kinetic solution of the mechanism
yields a value for k_3, the rate constant for deactivation of the
$[C_6H_5Br^{+\cdot}]^*$ ion _via_ collisions with neutral molecules, and by
using different neutral collision partners, the efficiency of
energy transfer can be compared for differing acceptor neutrals.
Several features of this approach make it particularly favourable
as a means of studying ion-molecule energy transfer: (1) The
rate of collisional quenching (precisely, the rate of the removal
of sufficient energy from the excited ion to bring it below the
two-photon thermochemical threshold) is determined absolutely, not
relative to any other rate; (2) The internal energy of $[C_6H_5Br^{+\cdot}]^*$
is known precisely (within thermal uncertainty); (3) The ion is,
almost certainly, an unrearranged bromobenzene ion, whose normal
modes and other properties can be assigned with some assurance.
Thus the direct measurement of k_3 made possible by analysis of the
two-photon kinetic scheme can yield energy-transfer rates for
ion-molecule collisions involving partners which are well
characterized both structurally and energetically.

The bromobenzene ion case was studied with a variety of
neutral collision partners. In the presence of an added gas N,
the kinetic expression acquires an additional term:

$$K = \frac{I^2\sigma^2}{2I\sigma + k_3[M] + k_3'[N]} \qquad (9)$$

where I and σ are the light intensity and absorption cross section,
[M] is the pressure of bromobenzene, [N] is the pressure of added
gas, and k_3' is the quenching rate constant for collisions of
$[C_6H_5Br^{+\cdot}]^*$ with molecules of N. Since k_1, k_2 and k_3 are all
known quantities, k_3' is readily determined, most conveniently by
plotting $1/K$ versus [N]. For comparison, it is useful to
normalize the k_3' values to the orbiting collision rate constant,
that is to calculate the number of orbiting collisions (Langevin
rate corrected for dipole effects) required to quench the ion.
(In this case, the excited ion has 2.4 eV of excess energy, of
which 1.3 eV must be removed before the ion is considered as
quenched.) Data for several neutral partners are presented in
this way in Table I. It is clear that for small neutral collision
partners quenching is rather inefficient, with the efficiency
rising with increasing size and complexity of the neutral. In
contrast to some past suggestions, the strong nature of ion-
molecule collisions does not guarantee wholescale partitioning of
vibrational energy, and for an inefficient acceptor like methane,
a large number of collisions are obviously necessary to thermalize
an energetic ion.

Table I: Number of orbiting collisions required to quench
vibrationally excited bromobenzene ions.

CH_4	>50	SF_6	8
$C_6H_5NO_2$	>30	C_6D_6	5
C_3H_8	>20	C_6H_6	4
CO_2	>10	C_6H_5F	3
		C_6H_5Br	1

Acknowledgements: During the performance of much of the research
described and the preparation of the manuscript, the support of
the National Science Foundation, the US Air Force Geophysical
Laboratory, the donors of the Petroleum Research Fund,
administered by the American Chemical Society, and a John Simon
Guggenheim Memorial Fellowship are gratefully acknowledged.
The author thanks the Physical Chemistry Laboratory of Oxford
University for their hospitality during preparation of this
manuscript.

REFERENCES

R. Benz and R.C. Dunbar, to be published (1979).

A. Carrington, D.R.J. Milverton and P.J. Sarre (1976), Mol.
Phys. 32, 297.

R.C. Dunbar (1971), J. Amer. Chem. Soc. 93, 4354.

R.C. Dunbar and E. Fu (1973), J. Am. Chem. Soc. 95, 2716.

R.C. Dunbar (1975), Spectrochim. Acta 31A, 797.

R.C. Dunbar and R. Klein (1976), J. Am. Chem. Soc. 98, 7994.

R.C. Dunbar (1976), Anal. Chem. 48, 723

R.C. Dunbar and E.W. Fu (1977), J. Phys. Chem. 81, 1531.

R.C. Dunbar and H.H. Teng (1978) , J. Am. Chem. Soc., 100, 2279.

R.C. Dunbar, M.S. Kim and G.A. Olah, (1979), J. Am. Chem. Soc., In Press.

R.C. Dunbar, "Ion Photodissociation", (1979), to appear in "Gas Phase Ion Chemistry", M.T. Bowers, Ed., Academic Press.

G.H. Dunn (1964), "Studies of Photodissociation of Molecular Ions", in "Atomic Collision Processes", M.R.C. McDowell, ed., North Holland, Amsterdam, p.997.

P.P. Dymerski, E. Fu and R.C. Dunbar (1974), J. Am. Chem. Soc. 96, 4109.

J.H.D. Eland (1973), Int. J. Mass Spectrom. Ion Phys. 12, 389.

B.S. Freiser and J.L. Beauchamp (1975), Chem. Phys. Lett. 35, 35.

E.W. Fu and R.C. Dunbar (1978), J. Am. Chem. Soc. 100, 2283.

E.W. Fu, H.H. Teng and R.C. Dunbar (1979), to be published.

M.S. Kim and R.C. Dunbar (1979), to be published.

J.P. Maier (1978), private communication.

D.C. McGilvery and J.D. Morrison (1977), J. Chem. Phys., 67, 368.

T.M. Miller, J.H. Ling, R.P. Saxon and J.T. Moseley (1976). Phys. Rev. A, 13, 2171.

J.T. Moseley, P.C. Cosby and J.R. Peterson (1976) , J. Chem. Phys., 65, 2512.

T.E. Orlowski, B.S. Freiser and J.L. Beauchamp (1976), Chem. Phys., 18, 439.

R.G. Orth and R.C. Dunbar (1978a), J. Am. Chem. Soc. 100, 5949.

R.G. Orth and R.C. Dunbar (1978b), J. Chem. Phys., 68, 3254.

R.G. Orth, R.C. Dunbar and M. Riggin (1977), Chem. Phys., 19, 279.

R.G. Orth and R.C. Dunbar (1979), to be published.

M. Riggin, R. Orth and R.C. Dunbar (1976), J. Chem. Phys., 65, 3365.

G.P. Smith, L.C. Lee, P.C. Cosby, J.R. Peterson and J.T. Moseley, (1978), J. Chem. Phys. 68, 3818.

T.F. Thomas, F. Dale and J.F. Paulson (1977), J. Chem. Phys.,
67, 793.

R.L. Woodin, D.S. Bomse, and J.L. Beauchamp (1978), J. Am. Chem.
Soc. 100, 3248.

AUTHOR INDEX

NEW INSTRUMENTAL METHODS:

SUMMARY OF THE PANEL DISCUSSION

John R. Eyler

Department of Chemistry, University of Florida

Gainesville, Florida   32611

PREPARED BY:

Ben S. Freiser
Department of Chemistry, Purdue University
West Lafayette, Indiana   47907

The panel discussion on new instrumental methods added an important appendix to the material presented in the other sessions of the Institute. Highlighted in the discussion were three topics experiencing particularly rapid growth including ion cyclotron resonance (ICR) spectroscopy, the selected ion flow tube (SIFT), and beam methods. I am indebted to Drs. Melvin Comisarow, John Eyler, Gerard Mauclaire, Nigel Adams, Norman Twiddy, Martin Jarrold, David Hirst, and Ron Gentry for providing abstracts used in preparing this summary.

ION CYCLOTRON RESONANCE SPECTROSCOPY

Fourier Transform Ion Cyclotron Resonance

Professor M. Comisarow of the University of British Columbia (Canada) described new developments and applications using the FT-ICR technique which he pioneered. The title instrument is a device which produces ion cyclotron resonance spectra in a very short period of time by excitation of the cyclotron motion of all of the ions in the ICR cell, detection of the transient time domain signal which results from that excited ICR motion, and finally discrete Fourier transformation to produce the ICR frequency domain spectrum (Comisarow, 1978a). The FT-ICR spectrometer can either produce the whole ICR spectrum in the amount

of time which a conventional scanning ICR spectrometer would
require to observe just a single line in the ICR spectrum or,
by accumulation of many time domain signals prior to Fourier
transformation, produce a spectrum of higher signal to noise
ratio than that given by a scanning ICR spectrometer. Because
of the fixed magnetic field operation and the temporal separation
of the ion excitation and ion detection functions, the FT-ICR
spectrometer is capable of producing very high resolution mass
spectra. These features can also be used to extend the useful
mass range of the ICR spectrometer to m/e=1500 or more (Comisarow,
1978a). In addition fixed magnetic field operation permits double
resonance experiments to be performed such that all product ions
coming from a single reactant ion are determined in a single
experiment rather than in several double resonance experiments
(Comisarow, Grassi, and Parisod, 1978). The FT-ICR retains the capability of conventional pulsed ion cyclotron resonance spectrometers.
Combination of the high speed, high sensitivity, high resolution
and extended mass range resulting from the Fourier method with
the well-known capability of conventional pulsed ICR for studying
ion-molecule reactions opens up an enormous range of compounds of
higher molecular weight whose ion-molecule chemistry can now be
investigated. An example of this is shown in Figure 1 which
shows the FT-ICR mass spectrum of $CpCr(CO)_2NS$ (MW = 219) under
conditions where most of the fragment ions formed by electron impact
have reacted to form ion molecule reaction products. Under the
indicated reaction conditions, sixth order reaction products are
observed at masses as high as m/e = 662.

The desirability of the instrumentation, but sometimes pro-
hibitive cost of FT-ICR led Professor J. Eyler of the University
of Florida (U.S.A.) to test the feasibility of developing a micro-
computer-based instrument (Comisarow, 1978b) utilizing an existing
pulsed ICR mass spectrometer. A KIM-1 microcomputer generates
timing and gating pulses necessary to excite ions in the ICR cell
with pulsed, frequency-swept irradiation, and mixes the resultant
transient response signal with a reference oscillator. All work
to date in Eyler's laboratory has utilized spectral segment extrac-
tion FT-ICR (Comisarow and Marshall, 1974), in which a limited mass
range is observed. In early work the microcomputer has been
used to store the output from an 8-bit A/D computer, with data from
successive transients added with double precision to that already
present. Thus, the microcomputer controls the FT-ICR experiment
and acts as a signal averager with 16-bit word length. In Eyler's
experiments to date, the transient response signals are trans-
mitted via phone line to a campus computer, where both the Fast
Fourier Transform (FFT) and plotting of both the transient response
signal and the resultant transform are carried out.

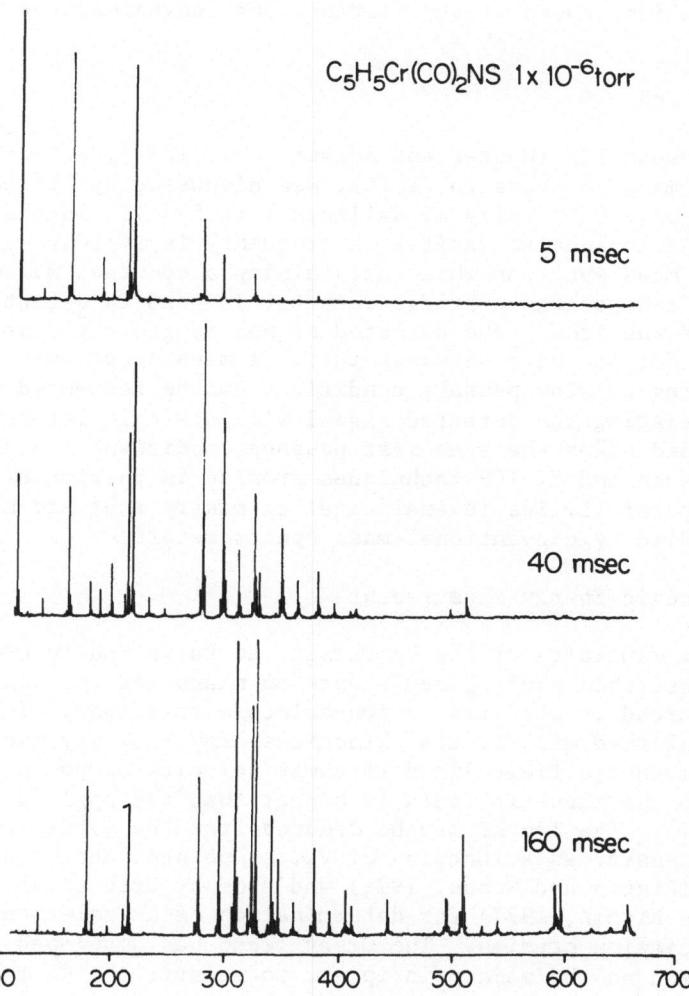

Figure 1.   FT-ICR spectra of CpCr(CO)$_2$NS as a function of reaction
            time.

Eyler's group is actively investigating the feasibility and
the constraints of also performing the FFT with a micro computer.
The use of a microcomputer in the FT-ICR experiment could result
in a substantial cost savings over the commercial FT-ICR system
utilizing a minicomputer, with only moderate sacrifice of the
many advantages of the FT-ICR technique such as wide mass range,
high resolution, speed of acquisition, and convenience of system
operation.

Rapid Scan ICR

Rapid scan ICR (Hunter and McIver, Jr., 1977), a technique
similar in many respects to FT-ICR, was discussed by Professor
R. McIver, Jr., University of California (U.S.A.). In the
rapid scan technique an excitation frequency is rapidly scanned
across the mass spectrum while maintaining a constant magnetic
field, and a capacitance bridge detector is used to detect transient
response of the ions. The detected signal is greatly distorted
by ringing due to the fast sweep rate. A mass spectrum
corresponding to slow passage conditions can be recovered by
cross correlating the detected signal with a single reference
line recorded under the same fast passage conditions. Both
the rapid scan and FT-ICR techniques promise in particular to
open many possibilities in analytical chemistry that are not
easily handled by conventional mass spectrometers.

ICR for Kinetic Energy Measurements

Dr. G. Mauclaire of the Universite de Paris Sud in Orsay
(France) described their group's work on measuring the kinetic
energy imparted to products in ion-molecule reactions. In an ICR
cell, ions formed with initial kinetic energy (KE) may escape
along the magnetic field lines if their velocity component
parallel to the magnetic field is higher than the applied trapping
voltage (V_T). The ion KE may be deduced from the variation of
the ion intensity as a function of V_T, as pointed out by Riggin
and Woods (Riggin and Woods, 1974) and used by Orth et.al. (Orth,
Dunbar, and Riggin, 1977) for determination of KE released in
photodissociation studies. The Orsay group has developed a
method, based on the same principles, to measure the KE released
in thermal charge transfer reactions from rare gas ions. In such
reactions the available energy may be quite high and ions may
be formed with several eV of KE (mainly H^+); therefore, the cell
may have to be operated with trapping voltages as high as 10 V.
For KE measurements it is necessary to vary V_T from 0.1 to 10V,
keeping primary ions very near thermal energies and without
perturbing either the reaction time, or the detection sensitivity.
To achieve this, a five section cell is used in a drift-storage
mode. An ion packet is formed in a separate source section where

V_T is kept at a low value, then the ions are drifted through the second section of the cell. During this drift, all unwanted ions are removed from the ion packet using cyclotron ejection so that only the primary ions of interest are admitted into the storage section. When the packet of ions reaches the center of this third section, drift voltages are switched to a storage configuration similar to that described by McMahon and Beauchamp (McMahon and Beauchamp, 1972). At the end of the reaction time (usually 15 msec) the ions are drifted to the ion collector. Ion detection is performed at constant magnetic field (1.6 Tesla) using a total ion current monitor, (see ref. 15 in R. Marx paper). There is no mass discrimination, even for H^+ ions with several eV of KE. Another advantage using these experimental conditions is that the ion KE determination is simplified since the ions are formed very near the longitudinal axis of the cell with an isotropic velocity distribution. It is easily shown that the fraction of trapped ions vs. $(V_T)^{1/2}$ exhibits a linear increase when V_T < KE and a plateau for V_T > KE. The mean value for KE is given by the position of the break on the experimental curve. A broad KE distribution is expected go give a curvature instead of a clean break. However, due to the square root dependence on V_T, such a curvature will be more or less observable depending on the energy range.

Electron Impact Excitation of Ions from Organics (EIEIO)

A new technique has been developed at Purdue University (U.S.A), electron impact excitation of ions from organics (EIEIO), utilizing trapped ion cyclotron resonance techniques (Cody and Freiser, 1979) in which ions are generated and subsequently excited in a continuous electron beam while being trapped on the order of 150 ms. The phenomenon is particularly evident in appearance potential measurements at high emission currents as illustrated by Figure 2, which shows the intensity of $C_6H_4^+$ generated from cyanobenzene as a function of electron energy. At low emission currents the fragment ion $C_6H_4^+$, arising from loss of HCN from the parent ion, appears at about 3.5 eV above the ionization potential of the parent neutral. This appearance potential measurement is consistent with the estimated thermodynamic threshold. At higher emission currents, however, $C_6H_4^+$ begins to appear with the parent ion at the ionization potential. This striking behavior is attributed to the EIEIO reaction

$$C_6H_5CN^+ \xrightarrow{\quad e^- \quad} C_6H_4^+ + HCN$$

which, as expected, becomes even more apparent as the emission current is increased (Figure 2).

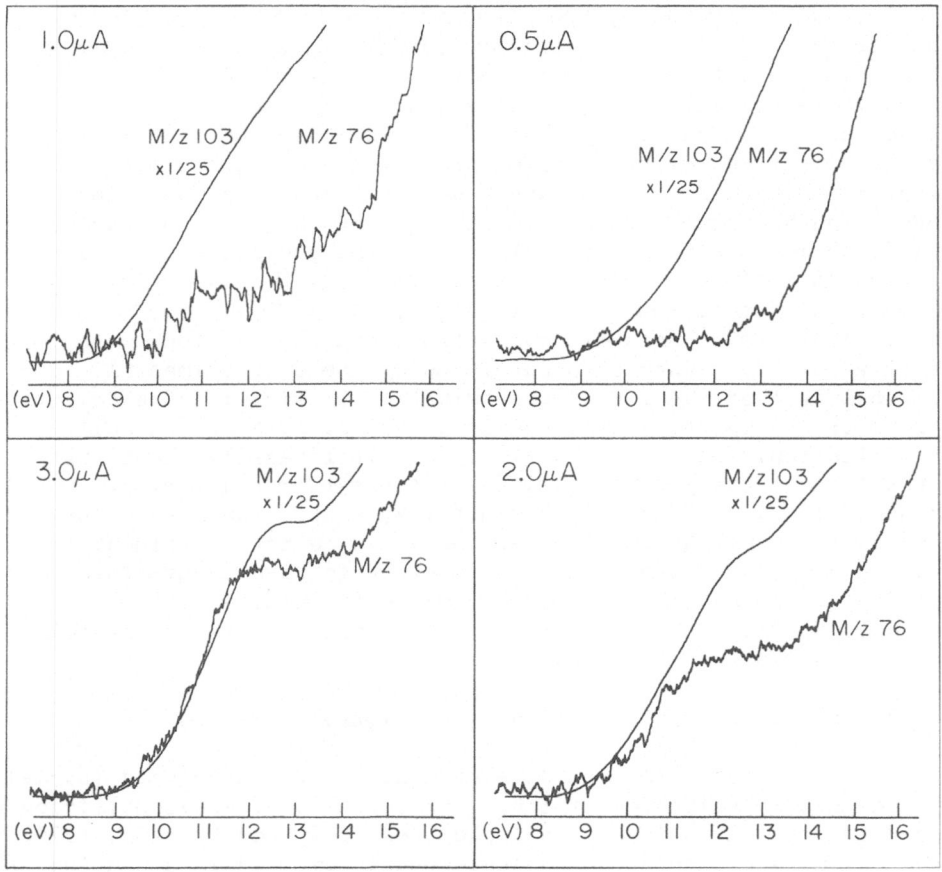

Figure 2.   Appearance potential measurements on cyanobenzene
             at different electron emission currents. As emission
             current is increased, the fragment ion $C_6H_4^+$ (M/z 76)
             begins to appear with the parent ion (M/z 103) at the
             ionization potential. This behavior is attributed to
             EIEIO on $C_6H_5CN^{+\cdot}$

        It is evident from the above results that mass spectra at
a particular electron energy will also be affected by emission
current and trapping conditions. The energy of the exciting
electrons also, as expected, has a profound effect on both the
product ion yields and on the product distribution in cases where
more than one product ion is observed. The EIEIO mass spectrum
is found to be analogous to that obtained by collision induced
dissociation and yields characteristic structural information.

SELECTED ION FLOW TUBE

SIFT Experimental Considerations

   Drs. N. Adams and D. Smith at the University of Birmingham (England) have developed a promising technique in which a mass-selected, low energy, positive or negative ion beam, derived from an ion source in conjunction with a quadrupole filter, is injected into a flowing carrier gas (usually helium) (Adams and Smith, 1976a). The subsequent reaction of these ions with neutral molecules introduced into the carrier gas downstream of the ion injection point is studied using downstream mass spectrometer sampling and ion counting in a manner identical to the well established flowing afterglow technique (Ferguson, Fehsenfeld and Schmeltekopf, 1969). Smith and Adams have termed this apparatus SIFT (Selected Ion Flow Tube) and have discussed the details of the technique and its particular points of merit in a recent review (Smith and Adams, 1979).

   That this technique has not previously been adopted is probably due to the inhibiting experimental factor that the injection of ions from a low pressure quadrupole filter into a high pressure gas requires the ions to move against a back-flow of neutral carrier gas with the attendant loss of ion beam intensity. This major problem has been overcome by introducing the carrier gas into the flow tube through a series of small apertures located around the ion injection point, directing the carrier gas at high velocity parallel to the incoming ion beam and along the flow tube. This "aspirator effect" produces a much lower pressure on the flow tube side of the injection orifice and hence a large reduction in the back-flow of carrier gas below that otherwise expected. Unlike the flowing afterglow which is a plasma experiment, there is only a single species present in the flow tube together with the neutral carrier gas with no source gas, excited neutral species, electrons or photons from the ion source. Thus it is readily possible to calibrate the downstream detection system for mass discrimination and to determine accurately the product ion distributions for ion-neutral reactions (Adams and Smith, 1976b).

   To date, many types of ion species have been injected from microwave, electron impact and flowing afterglow sources including ground state, metastable excited state and doubly charged positive ions, cluster ions and negative ions, and reaction rate coefficients and accurate product ion distributions have been determined (see references in the review by D. Smith and N. G. Adams in these proceedings). Some of these applications were particularly well exemplified by the discussion of Professor N. Twiddy, University College of Wales (Wales) summarized below.

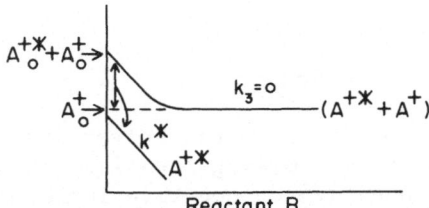

Figure 3.   Exponential decay expected for a composite mixture of
            ground and excited ions assuming no ground state reac-
            tion and no collisional quenching. See text.

Metastable Excited Ion Measurements in a SIFT

A selected ion flow tube apparatus (Adams and Smith, 1976a)
has been employed by Twiddy and Coworkers at U.C.W. Aberystwyth
to measure thermal energy rate coefficients and product ion
distributions for both ground state and metastable electronic
states of O^+_2, NO^+, O^+, C^+, N^+, N^+_2 and S^+ with several neutral
species. In the case of NO^+, N^+_2, and S^+ collisional quenching of
the metastable to the ground state by neutral molecules has been
observed.

The most general reaction scheme can be represented by

$$A^{+*} + B \longrightarrow \left[\begin{array}{l} \xrightarrow{k_1} C^+ + D \\ \xrightarrow{k_2} A^+ + B \end{array} \right.$$

$$A^+ + B \xrightarrow{k_3} E^+ + F$$

where k_1 is the rate constant for reaction of the excited ion
excluding collisional quenching, k_2 is the rate constant for
collisional quenching of the excited ion to the ground state, and
k_3 is the ground state reaction rate constant. The total rate
coefficient for the excited ion-neutral reaction, k^*, is defined
as equaling $k_1 + k_2$. The case in which there is no ground state
reaction ($k_3 = 0$) is illustrated in Figure 3. The mass spectrometer ion signal ($=A^{+*} + A^+$) shows the exponential decay of the
excited species with reactant B. Extrapolation of the constant
ground state ion signal A^+ to the ordinate gives the initial
ground state population A^+_0 and therefore the fraction of excited
species present in the ion beam entering the flow tube. If the
ground state population A^+_0 is subtracted from the composite signal
($A^{+*} + A^+$) this difference can be plotted to yield a linear decay,
the slope of which gives the excited ion rate coefficient k^* as in
a conventional flow tube method.

If collisional quenching of the excited species by the neutral reactant occurs ($k_2 \neq o$) the data will be modified as illustrated in Figure 4a. The quenching by B will result in a conversion of some of the excited species to the ground state resulting in a production of ground state ions represented by the rising dotted curve in Figure 4a. However, the extrapolation and subtraction procedure again yields a linear decay curve which gives the total rate coefficient $k^* = k_1 + k_2$.

The use of a monitor ion method (Bolden et.al., 1970) enables the populations of the ground state and excited ions to be observed separately. In this method sufficient monitor gas M is introduced just before the mass spectrometer sampling orifice so that all the excited ions are converted locally to M^+. The criteria for a suitable monitor gas are that it should (i) react rapidly with the excited species, (ii) not collisionally quench the excited species and (iii) not react with the ground state ion. The mass spectrometer M^+ and A^+ signals are then a measure of the A^{+*} and A^+ populations respectively. The total rate coefficient k^* can be determined from the slope of the M^+ decay and the rise (if any) of the A^+ signal as direct evidence of quenching (Figure 4b). The ratio of k_1 to k_2 can be determined if $(A_o^{+*} + A_o^+)$ is known and this can be obtained from the A^+ signal when both monitor argon and reactant gas flows are zero.

Full details of the experimental procedure and analysis are given in Glosik et. al., J. Phys. B., 1978, 11, 3365, and Tichy et. al., Int. Jnl. Mass Spect. and Ion Physics, 1978/9.

BEAMS

Pulsed Molecular Beams

It is usual for those researchers who do ion-molecule crossed-beam scattering experiments to devote most of their design effort to the "hard" part of the problem, namely, the generation of the ion beam. However, the scattered signal is proportional to the product of the ion and neutral beam intensities, so there is actually just as much to be gained from increasing the neutral beam intensity as from increasing the ion beam intensity. With conventional sources, the neutral beam intensity is limited by the vacuum pumping speed required to keep the background pressure acceptably low. Professor W. R. Gentry at the University of Minnesota (U.S.A.) and his coworkers have recently developed a pulsed molecular beam source which virtually eliminates this constraint, permitting instantaneous neutral beam intensities several orders of magnitude larger than could be sustained continuously (Gentry and Giese, 1978, 1977a, and 1977b). The beam pulse is produced by a small electromechanical valve which seals vacuum-tight and then pulses open to emit supersonic gas bursts having durations, for example, of about 10 μsec for H_2, and 30 μsec for Ar.

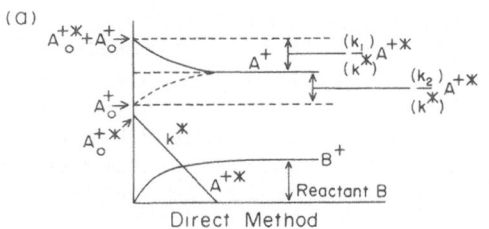

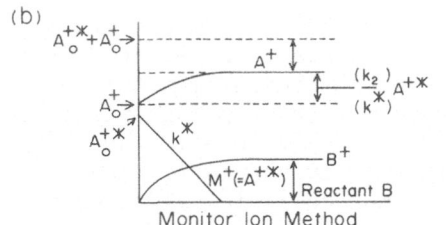

Figure 4 (a).   Exponential decay expected for a composite mixture
                 of ground and excited state ions under conditions
                 where collisional quenching of the excited state
                 species can occur.   (b) Monitor ion method in
                 which a monitor gas M is introduced into the system.

        For ion-molecule scattering studies, the most obvious applica-
tion of this technolgoy is to experiments in which some other part
of the experiment is also pulsed, for example, the ion source, or
a laser used to prepare excited states in one of the beams. A particularly attractive possibility is to use a pulsed uv laser for
2-photon ionization of a molecule to a specific vibrational-rotational state of the molecular ion, then to cross the pulsed state-
selected ion beam thus produced with the pulsed neutral beam,
measuring the product ion speed distribution by time-of-flight.
It is also possible to use a pulsed beam source to produce intense
beams of exotic neutral species.   For example, in recent experiments
by Gentry and coworkers a pulsed beam of CF_3I molecules was
irradiated with the unfocused output beam of a pulsed CO_2 laser
dissociating 80% of all the CF_3I molecules in the beam to CF_3
radicals, plus I atoms (Hoffbauer, Gentry, and Giese).

New Cross-Beam with Quadrupole Source

        Dr. M. Jarrold at the University of Warwick (England) reviewed
the crossed-beam apparatus in operation in his laboratory.   Ions
(eg CO^+, N^+_2, Ar^+) are produced by electron impact in an EAI Quad
150 ioniser, mass selected by a quadrupole mass filter and
focused using an electrostatic octopole lens (Birkinshaw, Hirst,
and Jarrold) to produce a narrow (<3° FWHM), low energy spread
(<0.3 eV) beam of space charge limited intensity.   The performance

of this source is equal to and, in some respects, better than that of the magnetic sector/retarding lens. The simplicity of this system and the consequent time saving in obtaining a good quality ion beam make the method attractive for beam formation.

The neutral beam is produced from a differentially pumped multicapillary array which, for noncondensable gases (eg O_2), gives intensities of approximately 10^{16} molecules s^{-1} at the crossing region with a beam diameter of 3 mm. and an angular divergence of 8°.

The detector consists of a two plate retarding potential difference analyser followed by a Brubaker lens and a quadrupole mass filter. The detector can be rotated both horizontally and vertically about the crossing point. The angular resolution is 3.5° and the energy resolution is less than 10%. Energy spectra of product ions are obtained by time averaged pulse counting. Preliminary investigations using noncondensable neutral beam gases indicated that phase sensitive detection was required and the appropriate modifications are in progress.

REFERENCES

Adams, N. G. and Smith D. (1976a). Int. J. Mass Spectrom. Ion Phys., 21, 349.

Adams, N. G. and Smith D. (1976b). J. Phys. B., 9, 1439.

Birkinshaw, K., Hirst, D. M., and Jarrold, M. F., in press.

Bolden, R. C., Hemsworth, R. S., Shaw, M. J. and Twiddy, N. D. (1970). J. Phys. B., 3, 60.

Cody, R. B. and Freiser, B. S. (1979). Anal. Chem., in press.

Comisarow, M. (1978a). Adv. in Mass Spec., 7, 1042.

Comisarow, M. (1978b). "Transform Teachniques in Chemistry" (Ed. D. R. Griffiths) Plenum Press, New York, p. 257.

Comisarow, M., Grassi, V., and Parisod, G. (1978). Chem. Phys. Lett., 57, 413.

Comisarow, M. and Marshall, A. (1974). Chem. Phys. Lett., 25, 288.

Ferguson, E. E., Fehsenfeld, F. C., and Schmeltekopf, A. L. (1969). Adv. At. Mol. Phys., 5, 1.

Gentry, W. R. and Giese, C. F. (1977a). Phys. Rev. Lett., 39, 1259.

Gentry, W. R. and Giese, C. F. (1977b). J. Chem. Phys., 67, 5389.

Gentry, W. R. and Giese, C. F. (1978). Rev. Sci. Instrum.,49, 595.

Hoffbauer, M. A., Gentry, W. R., and Giese, C. F., unpublished
    results.

Hunter, R. L. and McIver, Jr., R. T. (1977). American Laboratory,
    November.

McMahon, T. B. and Beauchamp, J. L. (1972). Rev. Sci. Instrum.,
    43, 509.

Orth, R., Dunbar, R., and Riggin, M. (1977). Chem. Phys., 19, 279.

Riggin, M. and Woods, I. (1974). Can. J. Chem., 52, 456.

Smith, D. and Adams, N. G. (1979). "Ion-Molecule Reactions" (Ed.
    M. Bowers) Academic Press, New York, in press.

INDEX